Current Topics in Behavioral Neurosciences

Volume 21

About this Series

Current Topics in Behavioral Neurosciences provides critical and comprehensive discussions of the most significant areas of behavioral neuroscience research, written by leading international authorities. Each volume offers an informative and contemporary account of its subject, making it an unrivalled reference source. Titles in this series are available in both print and electronic formats.

With the development of new methodologies for brain imaging, genetic and genomic analyses, molecular engineering of mutant animals, novel routes for drug delivery, and sophisticated cross-species behavioral assessments, it is now possible to study behavior relevant to psychiatric and neurological diseases and disorders on the physiological level. The *Behavioral Neurosciences* series focuses on "translational medicine" and cutting-edge technologies. Preclinical and clinical trials for the development of new diagnostics and therapeutics as well as prevention efforts are covered whenever possible.

More information about this series at http://www.springer.com/series/7854

Veena Kumari · Petr Bob
Nash N. Boutros
Editors

Electrophysiology and Psychophysiology in Psychiatry and Psychopharmacology

Springer

Editors
Veena Kumari
Department of Psychology
Institute of Psychiatry, Psychology and Neuroscience
King's College London
London
UK

Nash N. Boutros
Department of Psychiatry and Neurosciences
University of Missouri Kansas City (UMKC)
Kansas City, MO
USA

Petr Bob
Department of Psychiatry
Center for Neuropsych Research of Traumatic Stress
Charles University
Prague
Czech Republic

ISSN 1866-3370 ISSN 1866-3389 (electronic)
ISBN 978-3-319-12768-2 ISBN 978-3-319-12769-9 (eBook)
DOI 10.1007/978-3-319-12769-9

Library of Congress Control Number: 2014952784

Springer Cham Heidelberg New York Dordrecht London

Printed on acid-free paper

Springer is part of Springer Science+Business Media (www.springer.com)

Preface

Psychiatric disorders are highly complex and multifactorial. Moreover, the disorders are characterized by being heterogeneous with a significant degree of both overlap and co-morbidity. All these characteristics render research attempting to unravel the underlying psychophysiological processes rather complex and difficult. The only way forward is the incremental elucidation of physiological aberrations and attempting to identify the clinical correlates of the identified deviations.

Psychophysiological/electrophysiological methodologies have proven very useful in probing the physiological aberrations in psychiatric disorders and guiding towards effective and/or novel interventions. Given the extremely wide scope of psychiatric aberrations from personality deviations and substance dependence to the frank psychoses with both cognitive disintegration and affective dysregulations a rather large volume of research currently exists.

The first 12 chapters in this volume (Part I) provide updates regarding current understanding of the psychophysiological processes seen to be deviant in a particular disorder or in association with a particular set of symptoms within a disorder spectrum. The last five chapters (Part II) focus on techniques and methodologies that are highly promising as tools to further strengthen the impact of psychophysiological investigations on bringing the field closer to a full understanding of the pathophysiologies of the various neuropsychiatric disorders.

Part I of the volume starts with a contribution from Petr Bob focused on disturbances of neural mechanisms of consciousness which, through attentional mechanisms and memory processes, are linked to specific changes that occur in psychiatric disorders. The disturbances of consciousness and mental disintegration are closely connected with influence of stressful experiences and enable us to understand certain psychopathological mechanisms manifesting in a number of disorders. The contribution by Michael Stone provides a neurophysiological view of the spectrum of Borderline Personality Disorder (BPD). In the last two decades neurophysiological data, including MRI and fMRI, have established correlates in various brain regions, particularly those involving the frontal lobes and various limbic structures, that show promise of providing a more substantial basis for diagnosis relying primarily on identified brain changes. This chapter addresses the possible interrelationships between BPD and Bipolar Disorder. In the next chapter, Tim Outreth, Andrew Kemp and Gin Malhi examine electroencephalogram (EEG) and event-related potential (ERP) measures along with neuroimaging and peripheral physiological measures that both characterize and differentiate Bipolar Spectrum Disorders and their response to treatment. They provide a thoughtful framework for understanding these findings and stress their importance in improving assessment and therapeutic decision making in this population.

The two subsequent contributions are provided by the Galderisi/Vignapiano/Mucci/Boutros group and address, in Chap. 1, the physiological correlates of positive symptoms in schizophrenia. This chapter highlights the findings of electrophysiological studies in schizophrenia dealing with early sensory perception and attention, automatic sensory detection of stimuli changes and cognitive evaluation and integration of information, relevant to the pathophysiological mechanisms underpinning hallucinations and delusions. Results of electrophysiological studies investigating the neural correlates of positive symptoms suggest aberrant intrinsic organization of functional brain networks. The following contribution by the same group highlights the electrophysiological aberrations associated with negative symptoms in schizophrenia. While a number of studies have appeared over the years examining the electrophysiological correlates of the cluster of negative symptoms in schizophrenia, only a few studies have actually focused on the Deficit Syndrome (DS). In this chapter, electrophysiological investigations utilizing EEG, Evoked Potentials (EPs), polysomnography (PSG), or magnetoencephalography (MEG) to probe "negative symptoms", or "Deficit Syndrome" are reviewed.

Dean Acheson, Mark Geyer and Victoria Risbrough then offer a comprehensive review of the current state of knowledge on psychophysiological outcomes in Posttraumatic Stress Disorder (PTSD), with particular attention to their use as markers of current symptoms as well as markers of PTSD-related processes (e.g. fear extinction), and their sensitivity and selectivity for PTSD symptoms versus other anxiety and mood disorders and co-morbid disorders. They highlight potential future avenues for integrating psychophysiology into emerging areas of PTSD research and discuss the use of new wearable physiological monitoring technologies in treatment outcome studies. Wenzel Schicho and Oliver Pogarell provide a review and commentary on the physiological aberrations in Panic Disorder (PD) with a focus on the less frequently explored contribution of isolated epileptic discharges (IEDs) to symptomatology in the absence of epilepsy. It is not known exactly which role IEDs play in the genesis of behavioural aberrations. In this chapter, attention is directed towards this issue and its relevance to managing psychiatric patients suffering from PD, as well as understanding the complex relationship between IEDs and the pathophysiology of PD. The chapter by Christopher Patrick discusses the constructs of psychopathy and antisocial personality disorder (ASPD), their relations with one another and with violent behaviour, and provides an in-depth review of physiological correlates of psychopathy and ASPD with a focus on the features that these conditions share and those that distinguish them.

The two subsequent chapters address a topic that is not commonly included in standard psychophysiology texts, namely premenstrual and postmenopausal physiological and psychophysiological changes. The first of these, by Inger Sundström Poromaa, focuses on premenstrual dysphoric disorder (PMDD). PMDD is common with onset of symptoms in the late luteal phase of the menstrual cycle and provides an important model for our understanding of the influence of ovarian steroids on mood and anxiety in women. She discusses physiological findings in PMDD women (e.g. altered cardiovascular responses to stress) that appear to represent vulnerability traits for PMDD (i.e. also present in asymptomatic menstrual phases), or alternatively, vulnerability traits for the depressive and anxiety disorders that are commonly associated with PMDD. She also presents a number of state-related findings (e.g. lower luteal phase prepulse inhibition) in PMDD. The next contribution by Robert Freedman addresses postmenopausal physiological changes. The hallmark of menopause is the marked reduction of estradiol levels due to ovarian failure. This, among other factors results in hot flashes, the most common menopausal symptom. This chapter reviews the pathophysiology of hot flashes and highlights the contribution of brain structures like the brainstem, the insula and the prefrontal cortex.

The next chapter, by Ian Kodish, Carol Rockhill and Sara Webb, reviews psychophysiological and neuroimaging findings in Autism Spectrum Disorder (ASD), describing alterations in local brain regions as well as coordination of brain activity during both rest and activation paradigms in ASD. They propose that new drug therapies for ASD should aim to realign 'trajectories of network specialization across development' by acting together with behavioural therapies to enhance social and emotional learning by potentiating the effect of experience-induced plasticity on neuronal network connectivity. The last contribution to this part of the volume comes from Timothy Rhoers and colleagues who provided an analysis of the physiological correlates of insomnia. This chapter describes the physiological correlates of insomnia expressed during sleep and during the daytime. Together, the data from nighttime and daytime electrophysiology, event-related brain potential recording, neuroimaging studies, sympathetic nervous system and HPA axis monitoring all suggest insomnia is a 24-h disorder of hyperarousal.

Part II of the volume starts with a contribution from Gregory Light and Neal Swerdlow. They propose a remarkable parading shift (focus more on what is "right" and less on what is "wrong" with the patients) and convincingly argue, with clear examples of more normal-like performance in specific neurophysiological and psychophysiological measures predicting a positive response to specific therapeutic interventions, for an alternative strategy of using psychophysiological measures to identify 'spared neural and cognitive function' and then using this information to optimize clinical outcomes in schizophrenia patients. The next contribution to this section comes from Susan Bowyer and addresses connectivity measurements for network imaging. As is well known, communication across the brain networks is dependent on neuronal oscillations. Detection of the synchronous activation of neurons can be used to determine the well being of the connectivity in the human brain networks. Well connected highly synchronous activity can be measured by

MEG, EEG, fMRI and PET and then analysed with several types of mathematical algorithms. A further contribution by Petr Bob provides a review of topics related to nonlinear measures and dynamics in psychophysiology of consciousness that represent important tools to understand certain specific changes in neural systems implicated in psychiatric disorders. These methods enable us to describe various levels of complex interactions that may influence patterns of temporal and spatial disorganization with decreased or increased functional connectivity and complexity that underlie specific perceptual and cognitive changes in psychopathological states. Martijn Arns and Sebastian Olbrich explore the role of pharmaco-EEG in personalized medicine for Attention Deficit Hyperactivity Disorder (ADHD) and Depression. This chapter summarizes recent developments on personalized medicine in psychiatry with a focus on ADHD and depression and their associated biomarkers and phenotypes. Several neuro-physiological subtypes in ADHD and depression and their relation to treatment outcome are reviewed. The final contribution to this volume comes from Nikolaj Bak and Bob Oranje. They describe the benefits of psychophysiology-informed imaging, an approach also advocated by many others in this volume, in particular how a combination of EEG and fMRI complements each other, allowing both high temporal (EEG) and spatial (fMRI) resolution to be achieved. They also discuss various approaches to combine psychophysiology (EEG, EMG) with fMRI and the issues that need to be dealt with when combining the two methodologies.

We hope this volume, with chapters from leaders in the field, will make a valuable contribution to the literature on proven utility as well as future applications of psychophysiological measures, combined with other methodologies, in the context of improved understanding, prevention and effective treatment of neuropsychiatric disorders.

Veena Kumari
Petr Bob
Nash N. Boutros

Contents

Part I
Psychophysiology in Neuropsychiatric Disorders and Their Treatment: Past, Present and Future

Psychophysiology of Dissociated Consciousness

Petr Bob

Abstract Recent study of consciousness provides an evidence that there is a limit of consciousness, which presents a barrier between conscious and unconscious processes. This barrier likely is specifically manifested as a disturbance of neural mechanisms of consciousness that through distributed brain processing, attentional mechanisms and memory processes enable to constitute integrative conscious experience. According to recent findings a level of conscious integration may change during certain conditions related to experimental cognitive manipulations, hypnosis, or stressful experiences that can lead to dissociation of consciousness. In psychopathological research the term dissociation was proposed by Pierre Janet for explanation of processes related to splitting of consciousness due to traumatic events or during hypnosis. According to several recent findings dissociation of consciousness likely is related to deficits in global distribution of information and may lead to heightened levels of "neural complexity" that reflects brain integration or differentiation based on numbers of independent neural processes in the brain that may be specifically related to various mental disorders.

Keywords Consciousness · Dissociation · Hypnosis · Subliminal perception · Stress · Traumatic memory

P. Bob (✉)
Center for Neuropsychiatric Research of Traumatic Stress,
Department of Psychiatry and UHSL, 1st Faculty of Medicine, Charles University,
Ke Karlovu 11, 128 00 Prague, Czech Republic
e-mail: petrbob@netscape.net

P. Bob
CEITEC, Masaryk University, Brno, Czech Republic

Curr Topics Behav Neurosci (2014) 21: 3–21
DOI: 10.1007/7854_2014_320

Published Online: 22 May 2014

Contents

1 Introduction

According to current findings in psychophysiology and cognitive neuroscience there is growing evidence that neural correlate of consciousness likely represents coherent neural process that connects distributed brain activities into a coherent whole (Bob 2009; Crick and Koch 1992; John 2002). In agreement with this concept, consciousness could be understood as an integrative experience connecting various mental events that enable self-recognition, interpretative activity and related adaptive mental and behavioral responses. Essential characteristics of this self-representational dimension of consciousness is interpretation of certain inner states of own body as mental and somatic identity, while other bodily signals are interpreted as perceptions of the external world. What typically enable to uncover these interpretative processes that provide an understanding of the world on a level of individual experience is a process of "misinterpretation" that disturbs relationship between self and nonself and leads to experience of an inner conflict. For example, typical process of misinterpretation occurs during projection (or transference) when inner psychic states are interpreted as external parts of other persons, or during hallucinations when certain internally generated voices or images may be interpreted as sensory signals from the external world (Feinberg 1978; Feinberg and Guazzelli 1999; Ford et al. 2001). These findings suggest that the self-recognition is a specific cognitive process typically involving conscious activity and experience which is closely related to mechanisms of selective attention and reflects intentionality and interpretation of bodily signals reflecting inner and outside stimuli.

According to the modern definition selective attention can be defined as a selection among potential conscious contents and specific function of attentional mechanisms is to bring different events to consciousness (Baars 1999, 2002). This process enables global distribution of information that is located in brain regions underlying conscious processing and separated from those involved in the selection of visual objects and events (Baars 1997, 1999, 2002). Recent behavioral evidence indicates that perceptual awareness involves not only activation of the relevant perceptual properties but also further constructions of an organized representation in which these visual properties are attributed to their sources in external objects and events that represent basic mechanisms for interpretative activity (Baars 1997; Kanwisher 2001; Hudson 2009).

2 Neural Correlate of Consciousness

More than three centuries ago, Rene Descartes looked for "the seat of the soul" within the brain that could integrate "res cogitans" representing inner world with the outside world– "res extensa." Descartes thought that this special place is involved in sensation, imagination, memory, and the causation of bodily movements, and described the mind as an extracorporeal entity. In his "Passions of the Soul" he intuitively postulated that information from various sensory sources is fused and thought that "...although the soul is joined to the whole body there is a certain part where it exercises its functions more than all the others" (Barrera-Mera and Barrera-Calva 1999; Smith 1998). For example, he thought that when we sense only one image with two eyes, only one sound with two ears or only one object by two hands, the sensations from two sources must be fused somewhere. In this theoretical concept Descartes intuitively anticipated the so-called "binding problem" of consciousness. This problem means that there is a neural correlate of consciousness as a part of the nervous system that transforms neural activity into reportable subjective experiences. A major and still actual hypothesis is that this neural correlate of consciousness can compare and bind activity patterns only if they arrive simultaneously at the neural correlate of a conscious experience (Van De Grind 2002). Based on these processes consciousness may combine the present multimodal sensory information with relevant elements of the past and creates spatio-temporal memory.

The hypothetical center for information convergence was later termed "Cartesian theatre" (Crick and Koch 1992; Dennett 1991; Bob 2009). Recent neuroscience, however, has not located a distinct place in which distributed information in the brain comes together. Additionally, there is evidence that neocortical processing is distributed during all sensory and motor functions (Singer 1993, 2001). Following these findings Dennett (1991) proposed "a multiple drafts" theory of consciousness that does not define consciousness as a unitary process but rather a distributed one. Instead of a single central place called "Cartesian theatre," there are various events of content fixation that occur in various places at various times in the brain [see (Dennett 1991), p. 365]. This new conceptual framework mainly issued from experimental studies that examined relationship between visual consciousness and synchronization among large groups of neurons. Crick and Koch (Crick and Koch 1992) expressed the view that the main problem of visual consciousness in principle may not be resolved without explanation of the so-called binding problem, which means that a seen object in the brain is represented by groups of synchronized excited neurons that are located at different parts of the brain without unifying spatial convergence. This binding problem of consciousness has emerged in connection to findings that features of an object such as color, shape, texture, size, brightness, etc. produce activity in separate areas of the visual cortex (Crick and Koch 1992; Felleman and Van Essen 1991; Singer 1993, 2001). For example, there are only a few neural connections between specific visual areas that correlate with color and motion (Bartels and Zeki 2006; Larock 2006; Zeki 2003). The evidence for this view of consciousness represents a whole series of experimental results in cognitive neuroscience and psychology (Larock 2006; van der Velde and de Kamps 2006; Varela and Thompson 2003; Bob 2009).

Following these findings recent evidence indicates that integration of multiple and disparate neural activities underlying cognitive brain functions requires mechanisms of multiregional functional interaction that enables binding of distributed information (Singer 2001; Varela et al. 2001; John 2002; Bob 2009). Accumulating evidence from experimental studies (Leisman and Koch 2009; Bob 2009; Bob and Mashour 2011; Mashour 2013) also shows that this process of dynamic binding is related to transient and precise synchronization of neuronal activities that is significantly disturbed in certain pathological conditions, such as in schizophrenia (Lee et al. 2003; Tononi and Edelman 2000; Bob et al. 2010; Bob and Mashour 2011).

3 Divided Consciousness

3.1 Threshold of Consciousness and Subliminal Perception

Recent findings in neuroscience of consciousness strongly suggest that a level of synchronization and binding between various parts of the brain to some extent reflects accessibility of various mental contents into the consciousness in accordance with an understanding of consciousness as a gateway to brain integration (Baars 2002). These findings may implicate existence of a threshold of consciousness that reflects levels of information transmission and integration between distributed brain areas that act as a barrier between actual consciousness and unconscious processes.

In this context, modern study of cognitive unconscious phenomena defined explicit and implicit perception, i.e., explicit perception means perception immediately presented to consciousness of a subject while implicit perception is not accessible for the awareness of the subject and cannot be verified immediately in the response of the subject but only indirectly by observation or measurement of subject's behavior or physiological response (Bob 2003a; Kihlstrom 1987, 2004).

Great interest in implicit perception and subliminal phenomena arose from an experiment performed in 1957. During a movie presentation two verbal messages were projected: "Drink Coke" and "Eat popcorn" which led to increased popcorn consumption about 58 % and Coca-Cola consumption about 18 % (Wortman et al. 1992). The experiment caused controversial discussions but subsequent findings confirmed the existence of subliminal perception and information processing (Bob 2003a; Bunge et al. 2001; Crick and Koch 1995; Gawronski et al. 2006; Kanwisher 2001; Kihlstrom 1987, 2004; Marcel 1983).

Although in principle we may consider a threshold of consciousness presenting absolute subliminality at which all discriminative responses disappear (Wortman et al. 1992; Erdelyi 2004a, b; Reingold 2004), it seems to be that various sensory stimuli that have an importance for cognitive processes and adaptive behavioral responses may be influenced by various mechanisms of cognitive modulation and that subliminality may present relative phenomenon characterized by a sensitivity of discriminative responses (Kihlstrom 2004).

3.2 Divided Consciousness and Hypnosis

Typical experimental conditions that enable experimenters to assess modulatory influences on discriminative processes or attentional filtering represent various methods of hypnotic suggestion, and several data indicate that threshold of consciousness may change with respect to experimental conditions during hypnosis. For example, Stross and Shevrin (1962, 1968, 1969) have shown alterations of thought contents under hypnosis in investigation of "freely evoked images" after the subliminal presentation. These and other findings suggest that hypnosis leads to heightened access to subliminal stimuli and that thought organization during hypnosis shares some common elements with thought organization during dreaming (Bob 2004). This finding corresponds to similar reported data when subliminally presented images were found in dreams (Fisher 1954; Poetzl 1960). Another example represents also various methods of manipulation with intensity of subjectively experienced pain in hypnosis (Bob 2008a; Villemure and Bushnell 2002).

These findings indicate that highly hypnotizable persons possess stronger attentional filtering abilities than low-hypnotizable persons and that these differences are reflected in underlying brain dynamics such as an interplay between cortical and subcortical structures (Bob 2008a; Crawford 1994; Eccleston and Crombez 1999; Feldman 2004). Highly hypnotizable persons can better focus and sustain their attention as well as better ignore irrelevant stimuli from the environment (Crawford 1994). This clinical experience corresponds to findings that descend inhibitory pathways, parallel to ascending sensory systems, can modulate quite early responses related to sensory information. Together these findings suggest that high hypnotizables can better inhibit incoming sensory stimuli. This inhibition likely emerges due to influence of the frontal cortex that regulates limbic system structures in processes of active gating of incoming sensory stimuli (Bob 2008a; Crawford 1994).

Modulation of attention in hypnotic states is coupled to global changes in subjective experience and markedly influences regulation and monitoring body and mental state, experiencing of the self and underlying process of self-representation. In this context, self-representation as a mental structure creating identity and awareness can be defined as a result of interpretation of certain inner states of own body as parts of mental and somatic identity, while other bodily signals are interpreted as perceptions of the external world. These alterations in "self-representation" that underlie the changes in subjective experience are likely linked to great and abrupt changes in patterns of neural activity (Bob 2008a). This supports a concept that hypnosis, because of significant attentional shift, leads to a distinct "state" of consciousness with specific neural correlate (Rainville et al. 1997, 2002; Villemure and Bushnell 2009) and the hypnotic lack of the self-representation can be observed as dissociated or divided consciousness (Bob 2007, 2008a; Crawford 1994; Vermetten and Douglas 2004).

3.3 Dissociated Consciousness and Traumatic Stress

Similar conditions that lead to modulation of discriminative processes in hypnosis occur also in cases of traumatic dissociation. Dissociation is defined as a disturbance or alteration of normal integrated functions of consciousness, memory or identity that leads also to characteristic somatoform changes (Bob 2003b, 2008a, b; Hall and Powell 2000; Nijenhuis et al. 1998; Nijenhuis 2000; Putnam 1997; Van Der Hart et al. 1985, 2005). Heightened levels of dissociation in most cases occur due to a traumatic event and typical symptoms include memory losses, fragmentation of knowledge of the self and experience, splitting of emotional and/or cognitive aspects of experiences, numbing of affect, psychological escape from unpleasant stimuli, trance-like states, increased suggestibility and greater hypnotizability (Hall and Powell 2000; Putnam 1997; Spiegel 1997; Van Der Hart and Friedman 1989).

Dysfunctions in accessibility of memory traces that represent traumatic and other negative past experiences as well as intrusive autobiographical memories of childhood abuse are linked to an effort to eliminate these negative memories and intrusive thoughts connected to inner conflict due to contradictory tendencies when unacceptable or traumatic memories are released into consciousness (Bob 2007; Brewin 2007; Vermetten and Douglas 2004). According to recent evidence, stress related conditions frequently affect both episodic and autobiographical memories. For example, Kenardy et al. (2007) reported clinical results in a group of eighty-seven children aged 7–15 years exposed to a traumatic event requiring hospitalization which indicate that specifically, children who showed temporal disorganization, but not absence of emotion or dissociative amnesia, in narrative themes were more likely to report concurrent subsyndromal PTSD symptoms at 4–7 weeks post-trauma (Kenardy et al. 2007). Other similar data also demonstrate that exposure to a significant psychological stressor preserves or even enhances memory for emotional aspects of an event, and simultaneously disrupts memory for nonemotional aspects of the same event (Payne et al. 2006). Further evidence indicates that individuals who are victims of a trauma are unable to register pain during painful affects or self-injury, which is in agreement with clinical evidence that patients with dissociative disorders frequently report amnesia for self-injury (Bob 2008a; Butler et al. 1996; Ebrinc 2002; Orbach et al. 1997; Saxe et al. 2002). These findings suggest that the amnesic barrier likely is due to profound changes in affect state, memory, and sense of identity in response to environmental stress injury (Bob 2008a; Saxe et al. 2002).

According to recent evidence dissociation as a response to psychological stressors and traumas has various neurobiological consequences (Bob 2003b, 2008b; Bremner 1999; Spiegel 1991; Teicher et al. 2003, 2006). Repeated stressors and reexperiencing of the traumatic event in childhood often cause delayed effects of severe psychological trauma that may lead to long lasting enhancement of self-preservative catecholamine states related to anger, fear, meaninglessness and a blunting of emotional responses associated with dysfunction of the locus coeruleus, amygdala, and hippocampal systems (Henry 1997; Schore 1997, 2002). Additionally, there is evidence that stress may significantly influence reparative processes and neurogenesis of hippocampal neurons and cause decreased volume of the hippocampus, corpus callosum, and other brain structures (Bremner 1999, 2006; Bremner et al. 2008; Teicher et al. 2003, 2006). Although in most cases of pathological dissociation, the loss of episodic and/or emotional memories is related to traumatic stress (Bob 2008a; Brewin 2007; Butler et al. 1996; Frankel 1996; Sar 2006; Sar and Ross 2006; Spiegel 1997; Van Der Hart et al. 2005), brain insult, injury, or other organic brain disease may also play a role in this process (Kihlstrom 2005; Spiegel 1991).

4 Dissociation and Traumatic Memory

4.1 History of the Term Dissociation

Historical development of the term dissociation began with work of Theódule Ribot, who in his clinical studies investigated patients with diseases of memory, will, and personality. He adopted basic neurological principle of evolution and dissolution proposed by Hughlings Jackson (Ellenberger 1970). This principle states that functions which appeared last in evolution and that emerge later in human development, are much more fragile with respect to various types of injuries, which cause that due to damages later developed function are lost earlier than developmentally older functions. He called this process "dissolution," which represents a reverse of evolution. Behavior of an individual due to dissolution is more automatic with less voluntary control and performed in a manner that is less complex than in normal state (Ellenberger 1970; Meares 1999). Ribot applied this principle to psychopathology of memory and will, and his reformulated principle states that more recent memories disappear before earlier ones (Ribot's Law). This principle also became a main resource of dissociation theory proposed by Pierre Janet (Ellenberger 1970).

Janet initially elaborated the concept of dissociation in his work Psychological Automatism (Ellenberger 1970; Haule 1983; Havens 1966; Van Der Hart and Friedman 1989), where he outlined his notion of psychic functions and structures and studied psychological phenomena observable in hysteria, hypnosis, and states of suggestion or possession. During complete psychological automatism (Van Der Hart and Friedman 1989), consciousness is totally dominated by repeating past experiences, such as in somnambulism or hysterical crises. In the case of partial automatism, only a part of the consciousness is dominated. In the case of complete or partial automatism systems of unconscious fixed ideas play an important role and repress conscious control and perception. According to Janet's findings the fixed ideas may emerge in many forms of psychopathological or somatoform symptoms, for example paroxysm, which may be understood as a representation of psychological trauma when a fixed idea is transformed into hallucinations and body movements (Van Der Hart and Friedman 1989). Fixed ideas also may be represented in dreams and dissociative episodes (e.g., hysterical attacks) or during hypnosis as a secondary consciousness. A characteristic feature of these states is a lowering of the mental level (abaissement du niveau mental), which is manifested by increased dissociation and mental depression connected to a reduction of psychological tension.

4.2 Dissociation and Abreactive Experience

Recent evidence indicates that pathogenic traumatic memories are at the roots of dissociative disorders. Therapy of these memory disorders is historically linked to the term abreaction (van der and Brown 1992). According to the definition of the American Psychiatric Association abreaction is defined as: "…an emotional release or discharge after recalling a painful experience that has been repressed because it was consciously intolerable. A therapeutic effect sometimes occurs through partial discharge or desensitization of the painful emotions and increased insight (American Psychiatric Association 1980). This definition embodies historical controversy between French school of dissociation and later studies by Joseph Breuer and Sigmund Freud, who for an understanding of abreactive process utilized the concept of repression in their "Studies of Hysteria" (Breuer and Freud 1895). First literary documented utilization of this method is described by Janet in his patient Lucie in 1886 (Breuer and Freud 1895; Ellenberger 1970; Janet 1886) and by Joseph Breuer in his famous patient Anna (Breuer and Freud 1895). Although Freud initially also used the concept of dissociation (i.e., splitting of consciousness) for understanding and treatment of traumatic memories, he later defined principle of constancy, as a consequence of his neurological hypothesis, which states that repressed neural excitation connected to a trauma must be expressed by emotional discharging coupled to verbal and motor activities, i.e., quantity of excitation must be kept constant (Van Der Hart and Brown 1992). Definition of a nature of abreactive experience has key consequences for understanding of therapeutic process as an integration of dissociative state ("double conscience") or on the contrary as discharging of repressed energy linked to traumatic memory. Later Jung, following Janet, suggested that integration of traumatic memory represents a key process in the treatment of traumatized patients. From his point of view therapeutically effective reliving the traumatic memory and emotional discharge need to lead to reintegration of the traumatic complex (Jung 1907). Further research and clinical practice suggest that repeated abreaction without reintegration of dissociated states of mind often has malignant effect without any improvement of the patient state and leads to strengthening intrusive symptoms (Van Der Hart and Brown 1992). In this context, Putnam described an integrative view of abreactive process (Van Der Hart and Brown 1992; Putnam 1992) that includes therapeutically controlled abreactions and also spontaneous abreactions or abreaction-like phenomena such as flashbacks, vivid dreams, and other recalls of traumatic experiences. For the purpose of therapeutically controlled abreactions, Putnam emphasized importance of reliving of traumatic experiences as well as discharge of related affects and integration of traumatic material into conscious awareness during and after the abreactive experience. Similarly Brown and Fromm (1986) critically reviewed the role of abreactive techniques based on Freud's "hydraulic model", in which posttraumatic symptoms are understood as a consequence of repressed emotions which lead to "overpressure". As primary treatment focus of therapeutically controlled abreactions they, similarly as Horowitz (1986), emphasized progressive uncovering and integration which enables to control emotional expressions linked to intrusive experiences. Similarly, Braun (1986) warned against activation of traumatic memories during abreactive experiences without an appropriate cognitive framework which is necessary for adequate defenses or coping skills and recommended to express trauma related emotions in a planned, safer, and controlled manner. According to Ross (1989), a meaningful framework is necessary for successful abreactive treatment and is malignant when it consists of chaotic and uncontrolled screaming linked to self-abusive or regressed behavior.

According to recent clinical evidence abreaction may represent useful clinical instrument for treatment of some dissociative disorders. This evidence also suggests the expression and discharge of affect during the revivification of traumatic memories (often under hypnosis) must enable reintegration of dissociated parts of the traumatic memory and assimilate the traumatic event into the whole of the personality (Van Der Hart and Brown 1992; Bob 2007).

4.3 Dissociation and Memory Consolidation

Recent findings suggest that dissociative states may be understood as consequences of disturbances in memory consolidation and could be explained in the framework of this process. According to these findings stress may influence atypical consolidation of short-term memory into long-term memory (McGaugh 2000; Nadel and Jacobs 1998; Debiec and Altemus 2006) and cause dissociation of memory systems concerned with encoding emotion and context at psychological, physiological, and anatomical levels (LeDoux 1992, 1993, 1994; Nadel and Jacobs 1998; Phillips and LeDoux 1992; Bechara et al. 1995).

Generally, this process on a molecular level is linked to protein synthesis that requires involvement of brain derived neurotrophic factor (BDNF), transcription factor CREB (cAMP response element-binding protein) and other molecular processes that participate in global processes of network consolidation mainly in the hippocampus but also in other brain structures (Debiec and Altemus 2006; Nadel and Jacobs 1998; Debiec et al. 2006; Lee et al. 2004). BDNF plays a crucial role in this process. BDNF is a polypeptide growth factor that influences differentiation and survival of neurons in the nervous system and also plays an important role in regulating synaptic plasticity and connectivity in the CNS and mechanisms of memory storage and mood control (Bath and Lee, 2006; Bramham and Messaoudi 2005). BDNF is an activity dependent modulator of excitatory transmission and synaptic plasticity with predominant effective localization of BDNF and its receptor tyrosine kinase TrkB (tropomyosin receptor kinase B) on glutamate synapses (Bramham and Messaoudi 2005; Soule et al. 2006). Endogenous BDNF-TrkB signaling in synaptic consolidation by long-term potentiation (LTP) needs new gene expression and protein synthesis that enable immediate early gene Arc (activity-regulated cytoskeleton associated protein) (Bath and Lee 2006; Bramham and Messaoudi 2005; Soule et al. 2006). Important factor in this new gene expression is the transcription factor CREB, which is required for hippocampus-dependent long term memory formation (Mizuno and Giese 2005). The CREB is activated by signaling pathways that include Ca(2+)/calmodulin kinases (CaMKs), protein kinase A (PKA) and the mitogen activated protein/extracellular signal-regulated kinases (MAPK or ERKs) (Mizuno and Giese 2005; Rattiner et al. 2005).

Recent molecular genetic and behavioral studies demonstrate that spatial and contextual types of hippocampus-dependent formation of long-term memory require different signaling molecules implicating distinct types of hippocampus-dependent long term memory that differ in their underlying molecular mechanisms (Mizuno and Giese 2005). As a part of these signaling pathways a basic mechanism of BDNF is that it may modulate both excitatory and inhibitory neurotransmitter systems (Savitz et al. 2006), and also influences functions of serotonergic and dopaminergic systems (Savitz et al. 2006; Narita et al. 2003; Mossner et al. 2000).

Recent findings also indicate that this relationship between BDNF and cognitive processes is significantly influenced by stress (Savitz et al. 2006). Stress, and especially chronic stress, influences excessive release of glucocorticoids from the adrenal gland that cause cell death or atrophy of vulnerable neurons through the cortisol action and inhibitory influence on BDNF synthesis and influences modification of synaptic plasticity, transmission and memory formation especially in the hippocampus and neocortex (Thomas and Davies 2005; Savitz et al. 2006; Binder and Scharfman 2004). These processes likely play a significant role in specific formation of dissociative states and according to recent findings an important structure in this process is amygdala that also participates in modulation of memory consolidation and has a specific role in consolidation of the traumatic memory (Bob 2007, 2008b; Payne et al. 2006; Cahill 1997; Cahill and McGaugh 1998).

Typical feature of traumatic memories is that they are not acceptable for conscious awareness because of coupled strong negative emotions (Bob 2007; Nadel and Jacobs 1998; Payne et al. 2006). According to current findings extremely negative emotional experience during traumatic events or inescapable stress likely may block induction of long-term potentiation in medial prefrontal cortex (PFC) and hippocampus, and influences atypical memory consolidation that is characterized by consolidation process predominantly on implicit (subliminal) level in the amagdala. This blocking of higher-order behavior mediated by hippocampus and PFC allows more automatic responses dependent on subcortical structures, mainly the amygdala (Bob 2007; Debiec and Altemus 2006; Nadel and Jacobs 1998; Payne et al. 2006; Vermetten and Douglas 2004; Maroun and Richter-Levin 2003; Bob 2008b). These findings are also in accordance with neuroimaging data which suggest that characteristic changes in the perfusion of limbic brain structures, such as the amygdala and the hippocampus, coincide with the high arousal and/or anxiety during traumatic recall (Vermetten and Douglas 2004) that likely extremely focus attention and this attentional shift may produce fragmented memories (Bob 2007; Vermetten and Douglas 2004), psychological automatisms and lowering of mental level (Bob 2003a, 2008a; Ellenberger 1970; Frankel 1996; Havens 1966; Putnam 1997; Van Der Hart et al. 2005).

These data have important consequences for psychotherapy that in principle influences memory reconsolidation in safe and nondangerous conditions, which lead to neurobiological reprocessing of traumatic memory traces. This reconditioning and reconsolidation is therefore possible only by reexperiencing of the traumatic memory in a new and safe situation that enables integration of the dissociative state (Bob 2007). During this process implicitly consolidated traumatic memory in subcortical structures, mainly in the amygdala, is probably transformed from automatic into higher level of conscious experience by long-term potentiation in higher-level structures of CNS such as medial PFC and hippocampus.

Without memory reconsolidation traumatic memories cannot be reprocessed in an integrated mode of consciousness. In this context, neuroscience research of memory and emotional processes during traumatic recall induced by abreactive process strongly suggests that successful therapeutic work with a dissociative state helps the individual both psychologically and physiologically and that measurable physiology (Kandel 1998, 1999; Gabbard 2000, 2007; Bob 2007, Bob and Mashour 2011; Abbas et al. 2014; Beauregard 2014) is related to these changes induced by psychotherapeutic process. Neural process of reconsolidation in principle may represent potential existence of a new integrated and adaptive level in neurophysiological process, which is actualized, for example, during successful therapy. Memory reconsolidation therefore likely represents a process that enables successful transformation from dissociated, automatic, and implicitly consolidated traumatic memory mainly in the amygdala, to higher level of conscious experience in higher-level structures such as medial PFC and hippocampus. This view corresponds to Janet's definition of dissociative state as an automatic process which does not fit into current cognitive scheme and without successful reprocessing (or reconsolidation) remains dissociated also during recall of dissociative state because of the specific neural substrate of dissociated memories (Bob 2007).

The process of memory reconsolidation is again related to BDNF synthesis and other molecular processes that influence modification of synaptic plasticity, transmission and new memory formation especially in the hippocampus and PFC, which include also new gene expression that is required for hippocampus-dependent long term memory formation (Mizuno and Giese 2005; Morris et al. 2003; Savitz et al. 2006). These findings indicate that learning and memory processes including a wide variety of environmental factors may also influence development of synaptic connections through new gene expression and that psychotherapy represents a special learning process that may specifically influence and modify brain functions, metabolism in specific brain structures and also genetic processes (Kandel 1998, 1999; Gabbard 2000, 2007). From this point of view in the future a new era of psychotherapy research and practice may develop specific modes of psychotherapy that can be designed to target specific sites of brain functioning (Gabbard 2000).

5 Conclusion

According to historical and recent clinical evidence repeated stress and especially traumatic stress experiences may disturb mental integrity and lead to dissociation of memory and mental experience (Bob 2008a; Brewin 2007; Brown et al. 1996; Ellenberger 1970; Frankel 1996; Haule 1983; Kenardy et al. 2007; Sar 2006; Sar and Ross 2006; Spiegel 1997; Van Der Hart et al. 2005). This clinical evidence is consistent with experimental findings of differential effects of stress on brain systems responsible for encoding and retrieving emotional memories in the amygdala and nonemotional memories in the hippocampal formation. Together these findings indicate that memories formed under high-levels of stress are not qualitatively the same as those formed under ordinary emotional circumstances but display typical forms of disorganization, fragmentation, and incompleteness

(Bechara et al. 1995; Brewin 2007; Debiec and Altemus 2006; Kenardy et al. 2007; LeDoux 1992, 1993, 1994; Nadel and Jacobs 1998; Payne et al. 2006).

These findings also suggest a hypothesis that neurophysiological processes related to dissociation of conscious experience decrease levels of synchronization and integration between brain areas involved in memory consolidation, and that active neural assemblies are dynamically segregated, which means that small subsets of brain system tend to behave independently and display decrease in synchronous activity (Seth, Izhikevich, Reeke, and Edelman, Seth et al. 2006; Sporns et al. 2000a, b; Tononi et al. 1994). These changes in neural synchronization related to dissociation may also be reflected using concept of the dynamical complexity that in neural systems could explain and characterize a dynamic equilibrium between differentiation and integration in the complex neuronal dynamics responsible for cognitive processes (Seth et al. 2006; Tononi and Edelman 1998, 2000). This hypothesis is in agreement with findings which indicate that the spatio-temporal structure in certain pathological brain states may be more regular, with excessive order and lower complexity than normal, or more irregular, as uncorrelated randomness with higher complexity (Dawson 2004; Tononi and Edelman 2000). According to these findings increased neural complexity observed using EEG and other psychophysiological measures reflects processes during activity of independent areas that enables fast parallel information processing which runs in a distributed mode (Klonowski et al. 1999; Sammer 1996; Elbert et al. 1992; Svetlak et al. 2009; Bob et al. 2009, 2010) and this desynchronized neural state likely is related to active information processing in the cortex (Tirsch et al. 2004).

An increase in complexity is often associated with symmetry breaking and the ability of a system to have different states, which is also associated with decrease in coherence in space over the long range (Weng et al. 1999). This suggests that more irregular neural states of higher complexity could negatively affect synchronization phenomena in the brain that are closely linked to integration of different neural events into a coherent whole and integrated experience of consciousness. These findings likely could explain why an increase in the number of simultaneously active mental states may be correlated with higher neural fragmentation.

From this point of view dissociation reflects loss of effective connectivity and functional neural interconnections essential for conscious processing (Bob et al. 2009, 2010). Analogically, the mechanisms of "cognitive unbinding" has been proposed to explain loss of consciousness in anesthesia (Mashour 2004, 2005, 2006, 2008), which suggests that a common neurophysiology may underlie dissociative states, loss of consciousness and anesthetic-induced unconsciousness. This connection between neural synchronization and conscious integration is in agreement with data suggesting that consciousness may enable brain integration and access between otherwise separated neuronal functions (Baars 2002). In this context, there are few reported studies that show changes in synchronization and complexity in relationship to certain psychopathological states of conscious disintegration (Bob et al. 2009, 2010; Fingelkurts et al. 2007; Lee et al. 2003;

Tononi and Edelman 2000; Bob 2008c). Together these findings suggest a possibility to integrate neuroscience of consciousness with subjective experiences and clinical evidence that consciousness may be disintegrated due to various conditions related to experimental cognitive manipulations, hypnosis, influences of psychosocial stressors and other conditions related to brain insult or disease.

References

Abbass AA, Nowoweiski SJ, Bernier D, Tarzwell R, Beutel ME (2014) Review of psychodynamic psychotherapy neuroimaging studies. Psychother Psychosom 83:142–147

American Psychiatric Association (1980) A psychiatric glossary, 5th edn. American Psychiatric Association, Washington

Baars BJ (1997) Some essential differences between consciousness and attention, perception, and working memory. Conscious Cogn 6:363–371

Baars BJ (1999) Attention versus consciousness in the visual brain: differences in conception, phenomenology, behavior, neuroanatomy, and physiology. J Gen Psychol 126:224–233

Baars BJ (2002) The conscious access hypothesis: origins and recent evidence. Trends Cogn Sci 6:47–52

Barrera-Mera B, Barrera-Calva E (1999) The Cartesian clock metaphor for pineal gland operation pervades the origin of modem chronobiology. Neurosci Biobehav Rev 23:1–4

Bartels A, Zeki S (2006) The temporal order of binding visual attributes. Vis Res 46:2280–2286

Bath KG, Lee FS (2006) Variant BDNF (Val66Met) impact on brain structure and function. Cogn Affect Behav Neurosci 6:79–85

Beauregard M (2014) Functional neuroimaging studies of the effects of psychotherapy. Dialogues Clin Neurosci 16:75–81

Bechara A, Tranel D, Damasio H, Adolphs R, Rockland C, Damasio AR (1995) Double dissociation of conditioning and declarative knowledge relative to the amygdala and hippocampus in humans. Science 269:1115–1118

Binder DK, Scharfman HE (2004) Brain-derived neurotrophic factor. Growth Factors 22:123–131

Bob P (2003a) Subliminal processes, dissociation and the 'I'. J Anal Psychol 48:307–316

Bob P (2003b) Dissociation and neuroscience: history and new perspectives. Int J Neurosci 113:903–914

Bob P (2004) Dissociative processes, multiple personality, and dream functions. Am J Psychother 58:139–149

Bob P (2007) Hypnotic abreaction releases chaotic patterns of electrodermal activity during dissociation. Int J Clin Exp Hypn 55:435–456

Bob P (2008a) Brain and dissociated mind. Nova Science Publishers, New York

Bob P (2008b) Chaos, cognition and disordered brain. Act Nerv Super (Praha) 50:114–117

Bob P (2008c) Pain, dissociation and subliminal self-representations. Conscious Cogn 17:355–369

Bob P (2009) Quantum science and the nature of mind. J Mind Behav 30:1–14

Bob P, Mashour GA (2011) Schizophrenia, dissociation, and consciousness. Conscious Cogn 20:1042–1049. doi:10.1016/j.concog.2011.04.013

Bob P, Susta M, Chladek J, Glaslova K, Palus M (2009) Chaos in schizophrenia associations, reality or metaphor? Int J Psychophysiol 73:179–185

Bob P, Susta M, Glaslova K, Boutros NN (2010) Dissociative symptoms and interregional EEG cross-correlations in paranoid schizophrenia. Psychiatry Res. doi:10.1016/j.psychres.2009.08.015

Bramham CR, Messaoudi E (2005) BDNF function in adult synaptic plasticity: the synaptic consolidation hypothesis. Prog Neurobiol 76:99–125
Braun BG (1986) Issues in the psychotherapy of multiple personality disorder. In: Braun BG (ed) The treatment of multiple personality disorder. American Psychiatric Press, Washington DC, pp 1–28
Bremner JD (2006) Stress and brain atrophy. CNS Neurol Disord Drug Targets 5:503–512
Bremner JD (1999) Does stress damage the brain? Biol Psychiatry 45:797–805
Bremner JD, Elzinga B, Schmahl C, Vermetten E (2008) Structural and functional plasticity of the human brain in posttraumatic stress disorder. Prog Brain Res 167:171–186
Breuer J, Freud S (1895) Studies on hysteria. Basic, New York
Brewin CR (2007) Autobiographical memory for trauma: update on four controversies. Memory 15:227–248
Brown DP, Fromm E (1986) Hypnotherapy and hypnoanalysis. Lawrence Erlbaum, Hillsdale, NJ
Brown P, Macmillan MB, Meares R, van der HO (1996) Janet and Freud: revealing the roots of dynamic psychiatry. Aust NZ J Psychiatry 30:480–489
Bunge SA, Ochsner KN, Desmond JE, Glover GH, Gabrieli JDE (2001) Prefrontal regions involved in keeping information in and out of mind. Brain 124:2074–2086
Butler LD, Duran REF, Jasiukaitis P, Koopman C, Spiegel D (1996) Hypnotizability and traumatic experience: a diathesis-stress model of dissociative symptomatology. Am J Psychiatry 153:42–63
Cahill L (1997) The neurobiology of emotionally influenced memory. Implications for understanding traumatic memory. Ann NY Acad Sci 821:238–246
Cahill L, McGaugh JL (1998) Mechanisms of emotional arousal and lasting declarative memory. Trends Neurosci 21:294–299
Crawford HJ (1994) Brain dynamics and hypnosis: attentional and disattentional processes. Int J Clin Exp Hypn 42:204–232
Crick F, Koch C (1995) Are we aware of neural activity in primary visual cortex? Nature 375:121–123
Crick F, Koch C (1992) The problem of consciousness. Sci Am 267:153–159
Dawson KA (2004) Temporal organization of the brain: neurocognitive mechanisms and clinical implications. Brain Cogn 54:75–94
Debiec J, Altemus M (2006) Toward a new treatment for traumatic memories. Cerebrum 2–11
Debiec J, Doyere V, Nader K, LeDoux JE (2006) Directly reactivated, but not indirectly reactivated, memories undergo reconsolidation in the amygdala. Proc Natl Acad Sci USA 103:3428–3433
Dennett DC (1991) Consciousness explained. Little, Brown and Company, Boston
Ebrinc S (2002) Hypnotizability, pain threshold and dissociative experiences. Eur Neuropsychopharmacol 12:427–428
Eccleston C, Crombez G (1999) Pain demands attention: a cognitive-affective model of the interruptive function of pain. Psychol Bull 125:356–366
Elbert T, Lutzenberger W, Rockstroh B, Berg P, Cohen R (1992) Physical aspects of the EEG in schizophrenics. Biol Psychiatry 32:595–606
Ellenberger HF (1970) The discovery of the unconscious: the history and evolution of dynamic psychiatry. Basic, New York
Erdelyi MH (2004a) Comments on commentaries: Kihlstrom, Bachmann, Reingold, and Snodgrass. Conscious Cogn 13:430–433
Erdelyi MH (2004b) Subliminal perception and its cognates: theory, indeterminacy, and time. Conscious Cogn 13:73–91
Feinberg I (1978) Efference copy and corollary discharge: implications for thinking and its disorders. Schizophr Bull 4:636–640

Feinberg I, Guazzelli M (1999) Schizophrenia: a disorder of the corollary discharge systems that integrate the motor systems of thought with the sensory systems of consciousness. Br J Psychiatry 17:4196–4204
Feldman JB (2004) The neurobiology of pain, affect and hypnosis. Am J Clin Hypn 46:187–200
Felleman DJ, Van Essen DC (1991) Distributed hierarchical processing in the primate cerebral cortex. Cereb Cortex 1:1–47
Fingelkurts AA, Fingelkurts AA, Rytsala H, Suominen K, Isometsa E, Kahkonen S (2007) Impaired functional connectivity at EEG alpha and theta frequency bands in major depression. Hum Brain Mapp 28:247–261
Fisher C (1954) Dreams and perception: The role of preconscious and primary modes of perception and dream formation. J Am Psychoanal Assoc 2:389–445
Ford JM, Mathalon DH, Heinks T, Kalba S, Faustman WO, Roth WT (2001) Neurophysiological evidence of corollary discharge dysfunction in schizophrenia. Am J Psychiatry 158:2069–2071
Frankel FH (1996) Dissociation: the clinical realities. Am J Psychiatry 153:64–70
Gabbard GO (2000) A neurobiologically informed perspective on psychotherapy. Br J Psychiatry 177:117–122
Gabbard GO (2007) Psychotherapy in psychiatry. Int Rev Psychiatry 19:5–12
Gawronski B, Hofmann W, Wilbur CJ (2006) Are "implicit" attitudes unconscious? Conscious Cogn 15:485–499
Hall JM, Powell J (2000) Dissociative experiences described by women survivors of childhood abuse. J Interpers Violence 15:184–204
Haule J (1983) Archetype and integration: exploring the Janetian roots of analytical psychology. J Anal Psychol 28:253–267
Havens LL (1966) Pierre Janet. J Nerv Ment Dis 143:383–398
Henry JP (1997) Psychological and physiological responses to stress: the right hemisphere and the hypothalamo-pituitary-adrenal axis, an inquiry into problems of human bonding. Acta Physiol Scand Suppl 640:10–25
Horowitz MJ (1986) Stress response syndromes, 2nd edn. Northvale, NJ
Hudson AJ (2009) Consciousness: physiological dependence on rapid memory access. Front Biosci 14:2779–2800
Janet P (1886) Les actes inconscients et le dédoublement de la personnalité pendant le somnambulisme provoqué (Unconscious acts and personality split during induced somnambulism). Revue Philosophique 22:577–592
John ER (2002) The neurophysics of consciousness. Brain Res Brain Res Rev 39:1–28
Jung CG (1907) On psychophysical relations of the associative experiment. J Abnorm Psychol 1:247–255
Kandel ER (1998) A new intellectual framework for psychiatry. Am J Psychiatry 155:457–469
Kandel ER (1999) Biology and the future of psychoanalysis: a new intellectual framework for psychiatry revisited. Am J Psychiatry 156:505–524
Kanwisher N (2001) Neural events and perceptual awareness. Cognition 79:89–113
Kenardy J, Smith A, Spence SH, Lilley PR, Newcombe P, Dob R et al (2007) Dissociation in children's trauma narratives: an exploratory investigation. J Anxiety Disord 21:456–466
Kihlstrom JF (1987) The cognitive unconscious. Science 237:1445–1452
Kihlstrom JF (2004) Availability, accessibility, and subliminal perception. Conscious Cogn 13:92–100
Kihlstrom JF (2005) Dissociative disorders. Annu Rev Clin Psychol 1:227–253
Klonowski W, Jernajczyk W, Niedzielska K, Rydz A, Stepien R (1999) Quantitative measure of complexity of EEG signal dynamics. Acta Neurobiol Exp (Wars.) 59:315–321
Larock E (2006) Why neural synchrony fails to explain the unity of visual consciousness. Behav Philos 34:39–58
LeDoux JE (1993) Emotional memory systems in the brain. Behav Brain Res 58:69–79
LeDoux JE (1994) Emotion, memory and the brain. Sci Am 270:50–57

LeDoux JE (1992) Brain mechanisms of emotion and emotional learning. Curr Opin Neurobiol 2:191–197
Lee JL, Everitt BJ, Thomas KL (2004) Independent cellular processes for hippocampal memory consolidation and reconsolidation. Science 304:839–843
Lee KH, Williams LM, Breakspear M, Gordon E (2003) Synchronous Gamma activity: a review and contribution to an integrative neuroscience model of schizophrenia. Brain Res Brain Res Rev 41:57–78
Leisman G, Koch P (2009) Networks of conscious experience: computational neuroscience in understanding life, death, and consciousness. Rev Neurosci 20:151–176
Marcel AJ (1983) Conscious and unconscious perception: an approach to the relations between phenomenal experience and perceptual processes. Cogn Psychol 15:238–300
Maroun M, Richter-Levin G (2003) Exposure to acute stress blocks the induction of long-term potentiation of the amygdala-prefrontal cortex pathway in vivo. J Neurosci 23:4406–4409
Mashour GA (2006) Integrating the science of consciousness and anesthesia. Anesth Analg 103:975–982
Mashour GA (2004) Consciousness unbound: toward a paradigm of general anesthesia. Anesthesiology 100:428–433
Mashour GA (2005) Cognitive unbinding in sleep and anesthesia. Science 310:1768–1769
Mashour GA (2008) Unconscious processes in psychoanalysis and anesthesiology. Int Anesthesiol Clin 46:195–202
Mashour GA (2013) Cognitive unbinding: a neuroscientific paradigm of general anesthesia and related states of unconsciousness. Neurosci Biobehav Rev 37:2751–2759. doi:10.1016/j.neubiorev.2013.09.009
McGaugh JL (2000) Memory: a century of consolidation. Science 287:248–251
Meares R (1999) The contribution of Hughlings Jackson to an understanding of dissociation. Am J Psychiatry 156:1850–1855
Mizuno K, Giese KP (2005) Hippocampus-dependent memory formation: do memory type-specific mechanisms exist? J Pharmacol Sci 98:191–197
Morris RG, Moser EI, Riedel G, Martin SJ, Sandin J, Day M et al (2003) Elements of a neurobiological theory of the hippocampus: the role of activity-dependent synaptic plasticity in memory. Philos Trans R Soc Lond B Biol Sci 358:773–786
Mossner R, Daniel S, Albert D, Heils A, Okladnova O, Schmitt A et al (2000) Serotonin transporter function is modulated by brain-derived neurotrophic factor (BDNF) but not nerve growth factor (NGF). Neurochem Int 36:197–202
Nadel L, Jacobs WJ (1998) Traumatic memory is special. Curr Dir Psychol Sci 7:154–157
Narita M, Aoki K, Takagi M, Yajima Y, Suzuki T (2003) Implication of brain-derived neurotrophic factor in the release of dopamine and dopamine-related behaviors induced by methamphetamine. Neuroscience 119:767–775
Nijenhuis ERS (2000) Somatoform dissociation: major symptoms of dissociative disorders. J Trauma Dissociation 1:7–29
Nijenhuis ERS, Vanderlinden J, Van Der Hart O, Spinhoven P, Van Dyck R (1998) Somatoform dissociative symptoms as related to animal defensive reactions to predatory imminence and injury. J Abnorm Psychol 107:63–73
Orbach I, Mikulincer M, King R, Cohen D, Stein D (1997) Thresholds and tolerance of physical pain in suicidal and nonsuicidal adolescents. J Consult Clin Psychol 65:646–652
Payne JD, Jackson ED, Ryan L, Hoscheidt S, Jacobs JW, Nadel L (2006) The impact of stress on neutral and emotional aspects of episodic memory. Memory 14:1–16
Phillips RG, LeDoux JE (1992) Differential contribution of amygdala and hippocampus to cued and contextual fear conditioning. Behav Neurosci 106:274–285
Poetzl O (1960) The relationship between experimentally induced dream images and indirect vision. Psychol Issues 2:46–106
Putnam FW (1997) Dissociation in children and adolescents. The Guilford Press, New York
Putnam FW (1992) Using hypnosis for therapeutic abreactions. Psychiatr Med 10:51–65

Rainville P, Duncan GH, Price DD, Carrier B, Bushnell MC (1997) Pain affect encoded in human anterior cingulate but not somatosensory cortex. Science 277:968–971
Rainville P, Hofbauer RK, Bushnell MC, Duncan GH, Price DD (2002) Hypnosis modulates activity in brain structures involved in the regulation of consciousness. J Cogn Neurosci 14:887–901
Rattiner LM, Davis M, Ressler KJ (2005) Brain-derived neurotrophic factor in amygdala-dependent learning. Neuroscientist 11:323–333
Reingold EM (2004) Unconscious perception: assumptions and interpretive difficulties. Conscious Cogn 13:117–122
Ross CA (1989) Multiple personality disorder: diagnosis, clinical features, and treatment. Wiley, New York
Sammer G (1996) Working-memory load and dimensional complexity of the EEG. Int J Psychophysiol 24:173–182
Sar V (2006) The scope of dissociative disorders: an international perspective. Psychiatr.Clin North Am 29:227–244, xi
Sar V, Ross C (2006) Dissociative disorders as a confounding factor in psychiatric research. Psychiatr Clin North Am 29:129–144, ix
Savitz J, Solms M, Ramesar R (2006) The molecular genetics of cognition: dopamine COMT and BDNF. Genes Brain Behav 5:311–328
Saxe GN, Chawla N, Van Der Kolk B (2002) Self-destructive behavior in patients with dissociative disorders. Suicide Life Threat Behav 32:313–320
Schore AN (1997) Early organization of the nonlinear right brain and development of a predisposition to psychiatric disorders. Dev Psychopathol 9:595–631
Schore AN (2002) Dysregulation of the right brain: a fundamental mechanism of traumatic attachment and the psychopathogenesis of posttraumatic stress disorder. Aust N Z J Psychiatry 36:9–30
Seth AK, Izhikevich E, Reeke GN, Edelman GM (2006) Theories and measures of consciousness: an extended framework. Proc Natl Acad Sci USA 103:10799–10804
Singer W (1993) Synchronization of cortical activity and its putative role in information processing and learning. Annu Rev Physiol 55:349–374
Singer W (2001) Consciousness and the binding problem. Ann N Y Acad Sci 929:123–146
Smith CUM (1998) Descartes' pineal neuropsychology. Brain Cogn 36:57–72
Soule J, Messaoudi E, Bramham CR (2006) Brain-derived neurotrophic factor and control of synaptic consolidation in the adult brain. Biochem Soc Trans 34:600–604
Spiegel D (1991) Neurophysiological correlates of hypnosis and dissociation. J Neuropsychiatry Clin Neurosci 3:440–445
Spiegel D (1997) Trauma, dissociation, and memory. Ann N Y Acad Sci 821:225–237
Sporns O, Tononi G, Edelman GM (2000a) Theoretical neuroanatomy: relating anatomical and functional connectivity in graphs and cortical connection matrices. Cereb Cortex 10:127–141
Sporns O, Tononi G, Edelman GM (2000b) Connectivity and complexity: the relationship between neuroanatomy and brain dynamics. Neural Netw 13:909–922
Stross L, Shevrin H (1969) Hypnosis as a method for investigating unconscious thought processes. A review of research. J Am Psychoanal Assoc 17:100–135
Stross L, Shevrin H (1968) Thought organization in hypnosis and the waking state. The effects of subliminal stimulation in different states of consciousness. J Nerv Ment Dis 147:272–288
Stross L, Shevrin H (1962) Differences in thought association between hypnosis and the waking state: An experimental approach. Bull Menninger Clin 26:237–247
Svetlak M, Bob P, Cernik M, Kukleta M (2009) Electrodermal complexity during the Stroop colour word test. Auton Neurosci 152(1–2):101–117
Teicher MH, Andersen SL, Polcari A, Anderson CM, Navalta CP, Kim DM (2003) The neurobiological consequences of early stress and childhood maltreatment. Neurosci Biobehav Rev 27:33–44

Teicher MH, Tomoda A, Andersen SL (2006) Neurobiological consequences of early stress and childhood maltreatment: are results from human and animal studies comparable? Ann N Y Acad Sci 1071:313–323
Thomas K, Davies A (2005) Neurotrophins: a ticket to ride for BDNF. Curr Biol 15:R262–R264
Tirsch WS, Stude P, Scherb H, Keidel M (2004) Temporal order of nonlinear dynamics in human brain. Brain Res Brain Res Rev 45:79–95
Tononi G, Edelman GM (2000) Schizophrenia and the mechanisms of conscious integration. Brain Res Brain Res Rev 31:391–400
Tononi G, Edelman GM (1998) Consciousness and complexity. Science 282:1846–1851
Tononi G, Sporns O, Edelman GM (1994) A measure for brain complexity: relating functional segregation and integration in the nervous system. Proc Natl Acad Sci USA 91:5033–5037
Van De Grind W (2002) Physical, neural, and mental timing. Conscious Cogn 11:241–264
Van Der Hart O, Brown P (1992) Abreaction re-evaluated. Dissociation 5:127–140
Van Der Hart O, Friedman B (1989) A reader's guide to Pierre Janet on dissociation: a neglected intellectual heritage. Dissociation 2:3–16
Van Der Hart O, Nijenhuis ERS, Steele K (2005) Dissociation: an insufficiently recognized major feature of complex posttraumatic stress disorder. J Trauma Stress 18:413–423
Van Der Kolk B, Greenberg M, Boyd H, Krystal J (1985) Inescapable shock, neurotransmitters, and addiction to trauma: toward a psychobiology of post traumatic stress. Biol Psychiatry 20:314–325
van der Velde F, de Kamps M (2006) Neural blackboard architectures of combinatorial structures in cognition. Behav Brain Sci 29:37–70
van der HO, Brown P (1992) Abreaction re-evaluated. Dissociation 5:127–140
Varela F, Lachaux JP, Rodriguez E, Martinerie J (2001) The brainweb: phase synchronization and large-scale integration. Nat Rev Neurosci 2:229–239
Varela FJ, Thompson E (2003) Neural synchrony and the unity of mind: A neurophenomenological perspective. Binding, Integration, and Dissociation, The Unity of Consciousness, pp 266–287
Vermetten E, Douglas BJ (2004) Functional brain imaging and the induction of traumatic recall: a cross-correlational review between neuroimaging and hypnosis. Int J Clin Exp Hypn 52:280–312
Villemure C, Bushnell MC (2002) Cognitive modulation of pain: how do attention and emotion influence pain processing? Pain 95:195–199
Villemure C, Bushnell MC (2009) Mood influences supraspinal pain processing separately from attention. J Neurosci 29:705–715
Weng G, Bhalla US, Iyengar R (1999) Complexity in biological signaling systems. Science 284:92–96
Wortman CB, Loftus EF, Marshall ME (1992) Sensation and perception. In: Wortman CB, Loftus EF, Marshall ME (eds) Psychology. Mc Graw-Hill, New York
Zeki S (2003) The disunity of consciousness. Trends Cogn Sci 7:214–218

The Spectrum of Borderline Personality Disorder: A Neurophysiological View

Michael H. Stone

Abstract Borderline Personality Disorder (BPD) has been defined as a personality disorder in all editions of DSM since 1980; namely, DSM III through V. The criteria are a mixture of symptoms and traits; the etiology, a heterogeneous array of genetic, constitutional, and environmental factors. Until recently the diagnosis relied on clinical descriptions. In the last two decades, neurophysiological data, including MRI and fMRI, have established correlates in various brain regions, particularly those involving the frontal lobes and various limbic structures, that show promise of providing a more substantial basis for diagnosis—relying primarily on (internal) brain changes, rather than on (external) clinical observation. Some of the changes in BPD consist of decreased volume in the orbitofrontal and dorsolateral prefrontal cortices and smaller volume in both the amygdala and hippocampus, though with heightened reactivity in the amygdala. Similar abnormalities have been noted in bipolar disorders (BDs) and in ADHD, both of which often accompany BPD and share certain clinical features. Persons with strong genetic predisposition to BDs can develop BPD even in the absence of adverse environmental factors; those with extreme adverse environmental factors (chiefly, early sexual molestation) can develop BPD in the absence of bipolar vulnerability. In some BPD patients, both sets of factors are present. As ideal treatment depends on careful analysis of these factors, neurophysiological testing may permit both more rational, brain-based diagnostic decisions and more appropriate therapeutic strategies.

Keywords Borderline personality disorder · Bipolar disorder · Attention-deficit disorder · Neurophysiology · Frontolimbic circuitry · MRI

M. H. Stone (✉)
Professor of Clinical Psychiatry, Columbia College of Physicians and Surgeons, 225 Central Park West, New York, NY 10024, USA
e-mail: michaelhstonemd@gmail.com

Curr Topics Behav Neurosci (2014) 21: 23–46
DOI: 10.1007/7854_2014_308

Published Online: 22 May 2014

Contents

1 A Historical Note

Borderline Personality Disorder (BPD) as a clinical diagnosis has its origins a little over a century ago, though it was not originally called a "personality disorder"; rather, a condition intermediate between the then widely used concepts of neurosis and psychosis. The term evolved out of a kind of triage, where it was recognized that in between better-functioning patients who had a good grip on reality and the seriously disturbed (i.e., psychotic) patients whose hold on reality was critically weakened, there was a third group whose symptoms and everyday function fell somewhere in the middle of neurosis and psychosis. Kraepelin, for example, wrote of a *Zwischengebiet*—and in-between territory—where he situated the temperaments noted in some of the relatives in the families of manic-depressive persons (Kraepelin 1905, 1921). Persons exhibiting one of the temperaments: depressive, manic, irritable, and cyclothymic—often showed a clinical picture reminiscent of our contemporary BPD patients. *Irritable temperament* was associated, for example, with irascibility, lability of mood, impulsivity, and mild paranoid tendencies—which map onto four of the criteria for the current definition of BPD (Stone 1980, pp. 326–327). Some patients in this intermediate realm were understood as inhabiting the borderland touching clear-cut manic-depression (such as those with the Kraeplinian temperaments); others, whose symptoms were more cognitive than behavioral in nature, were seen as occupying the borderland of schizophrenia. There was a strong conviction within psychiatry that there was a hereditary component to both the obvious psychoses and to their less serious close-cousins—the "borderline" conditions. Whatever it was that was going on in the brain in these conditions remained, however, elusive. Freud, who started out as a neurologist, had speculated along such lines in his 1895 Project for a Scientific Psychology. Kraepelin was similarly convinced of a hereditary predisposition for both major psychoses [his *dementia praecox*, superseded terminologically by

Bleuler's *schizophrenia*; and *manic-depression*, now more often subsumed under the heading of *Bipolar Disorder* (*BD*)]. But their hunches remained at the speculative level: the neurology of their times could not as yet pinpoint areas of the brain peculiarities of which might underlie the illnesses on which they focused. Not for want of trying: Kraepelin hoped that his neurologist-associate, Alzheimer, might discover some neuroanatomical correlates of the major psychoses, but his success lay only in the area of the eponymous dementia.

Absent superior ways of finding brain correlates to the major psychiatric disorders during the first three-quarters of the last century, diagnostic distinctions tended to remain at the descriptive/phenomenological level. Definitions offered by various groups within psychiatry differed in many instances in accordance with the primary interest and primary treatment methods used by one or another group. The definitions of "borderline" formulated by psychoanalysts relied upon certain qualities that affected one's amenability to psychoanalytic treatment. Stern's definition (Stern 1938) relied on such criteria as *psychic bleeding* (paralysis in the face of crisis), *organic insecurity* (constitutional incapacity to tolerate much stress), and *difficulties in reality testing* (but short of gross psychosis). The term "borderline" also signified that the patient in question was not capable of tolerating conventional psychoanalysis, with its multiple weekly sessions on the couch. Until the 1950s psychiatrists, whether psychoanalytic in their orientation or otherwise, tended to assume that schizophrenia was the psychosis upon whose border their "borderline" patients were situated. Zilboorg (1941) spoke of a borderline variant of schizophrenia that he called "ambulatory schizophrenia"—where patients were able to preserve an adequate social façade and did not require hospitalization. To have any *trait* associated with schizophrenia was more important in Zilboorg's nosology, than was embodying the schizophrenic *state*. This approach dominated psychiatry in the US at mid-century, and led to what we would now see as a widespread overuse of the term "schizophrenia", and to the unrealistic assumption that the so-called borderline cases likewise belonged to the domain of schizophrenia.

The analyst Edith Jacobson realized that this conception was too narrow, and in the 1953 paper expressed the view that certain "borderline" patients were within the penumbra of manic-depression. Some milder cases, that is, of depression and hypomania could best be understood as within the (hereditarily predisposed) province of the manic-depressive disorders. Even so, she understood the psychoses as representative of various stages of psychosexual development, oriented linearly, such that schizophrenia harkened back to a more primitive stage of development, whereas the manic-depressive subtypes answered to a more advanced (and thus less primitive, less "ill") stage—for which reason they seemed more amenable to therapy.

The competing definitions of "borderline" that were ultimately to converge toward our current conception of BPD varied, during the 1960s and 1970s, in their emphasis on hereditary factors, primary observable traits, descriptions of typical signs and symptoms, or on treatment response. Kernberg (1967) deemphasized heredity, placing reliance more on a constellation of signs and symptoms: an enfeebled sense of identity though with adequate preservation of reality testing, along with impulsivity and weakened ability to handle anxiety-laden situations. Gunderson also focused on signs and symptoms: impulsivity, manipulative suicidal threats, mild/brief psychotic episodes, and disturbances in close relationships (Gunderson and Singer 1975). The inclusion of suicidal threats could be understood as an acknowledgement of the importance of an affective factor (in line with Jacobson's view), since patients with depression or other mood disorders are more prone to suicidal gestures than are most other types of patients. Kohut (1971) used the term *borderline* as a label for patients who proved unable to withstand, or to improve via, conventional psychoanalytic treatment. Despite these competing and often parochial views as to what constituted the essence of "borderline" conditions, the red thread that ran through all the definitions was *impulsivity*.

The viewpoints emerging in the 70s culminated in the inclusion of "borderline"—now as BPD—in the next edition of our Diagnostic and Statistical Manual of Psychiatric Disorders [DSM-III] (1980). Those DSM definitions were to be considered *atheoretic*, since there was not enough information in the field to speak authoritatively about underlying causes. There was, nevertheless, a kind of unspoken acceptance that the new BPD was built more along the lines of the mood disorders; viz., manic-depression, than along the lines of schizophrenia. Indeed, aberrations of personality that were more reminiscent of schizophrenia were now divided off into the new definition of Schizotypal Personality Disorder.

As it is not in the nature of scientific enterprises to avoid the search for underlying causes, controversy about the etiology of BPD became animated during the last quarter of the last century, particularly in regard to a possible allegiance to the more severe mood disorders of manic-depressive psychosis—a group of disorders now more often referred to as the BDs. Some investigators espoused the idea that a fair percentage, though by no means all, patients with BPD could be viewed as *formes frustres* of BD (Akiskal 1981; Stone 1981). This led to the emergence of two camps: one group taking a more extreme view and seeing BPD as the other side of the same coin as BD; the other group downplaying the connection to an almost opposite extreme. A more sober note was sounded by (Gunderson 2001; Gunderson et al. 2006) who, though tending to deemphasize the connection, acknowledged that, given an initial diagnosis of BPD, about 10 % of the patients could also (or eventually) be diagnosed with Bipolar-II disorder and another 5 % with Bipolar-I. Conversely, given an initial diagnosis of a BD, 20 % of the Bipolar-II patients were "co-morbid" for BPD and another 15 % of Bipolar-I patients could likewise be diagnosed with BPD (2001, p. 39). As for patients with Major Depression, about 15 % also met criteria for BPD, whereas half the patients in whom BPD was diagnosed first, half could also be considered to suffer from Major Depression. My own impressions (Stone 1990a, p. 74) regarding the overlap between BPD and BDs is similar to the observations of Gunderson just cited, though I have always emphasized the connection more vigorously. That Gunderson cites the percentages that he does reflects the tighter definition he has formulated for BPD—quite similar to that of DSM, both these definitions mapping out a smaller territory on the psychpathological map than is occupied by

Kernberg's "Borderline Personality Organization" (BPO). BPO, with its broad criteria of weakened identity sense, adequate reality-testing, impulsivity and poor ability to handle stress—reaches out to some 10 or 11% of the general population. BPO will include antisocial persons (including psychopaths), narcissistic persons who tend not to indulge in self-harm or suicidal behaviors, and other persons with marked aberrations of personality. The percentage of patients meeting these broad criteria—who in addition show the signs of BD—is of course lower than will be found among the more narrowly defined group with (DSM's or Gunderson's) BPD. These distinctions are important, for if we are to pan for the "gold" of bipolarity in the waters of the "borderline", our yield will be considerably higher (perhaps by a factor of four) if we begin with a sample of BPD patients, rather than of one with BPO. This has relevance when, in the next section, we review the findings of studies on the neurophysiology of borderline conditions (which indeed do rely on samples of BPD patients). There is an added factor pertaining to variations in sample. The overlap percentages for BPD and BD mentioned by Gunderson were derived from carefully analyzed samples: he referred to those, among others, of Fyer et al. (1988), Gunderson et al. (1999), and Zanarini et al. (1988). But there are other samples composed of borderline patients from generally higher socioeconomic class and from cultural backgrounds where parental neglect, brutality, and incest are quite rare: patients (females in particular) from these more protected settings may show a heightened overlap-ratio with BD—there being little else to account for their (dual) pathology apart from risk genes for bipolarity. Patients with BPD from settings of an opposite sort, where adverse environmental factors (sexual molestation, especially) are present to a marked degree, appear to develop their borderline clinical picture primarily from the early traumata; a family history of BDs may be quite uncommon in such samples—where even Gunderson's somewhat conservative percentages would seem much too high. Perhaps most common are BPD patients who occupy a middle ground in the nature/nurture debate—in whom the brain systems regulating impulsivity and emotion may partly have their origin in the "nature" side, but in whom the full-blown picture of BPD is pushed into clinical recognizability by adverse environmental factors (Pally 2002). I have, for example, served as consultant to a hospital unit devoted to BPD patients in Brisbane, Australia—where the family history of BDs was negligible, but where a history of incest and parental brutality was near universal (Stone et al. 1988). Clinicians attached to centers with sample differences of this kind are prone to develop hypotheses about the origins of BPD that were indeed "correct" for their clinic or hospital—but widely divergent from the hypotheses generated by clinicians from other centers. It was sample differences of this sort that, I believe, helped to account for the often acrimonious disputes about the degree to which BD, PTSD, incest, or dissociative identity disorder should be awarded "pride of place" in the etiological hierarchy of BPD. This disputation was common, if not inevitable, in the era antedating the neurophysiological studies of the past two decades: studies which have begun to shed at least a little light into the inner workings of the brain in BPD and in a number of other conditions that either overlap with, or could be confused with, BPD.

Before we address the data from these neurophysiological studies, we should pay attention to another stream of data, stemming from recent studies in evolutionary psychiatry—for it is these that help us to understand the erstwhile puzzling but long-known lopsided sex distribution among BPD patients. They are mostly female (Oglodek 2011). Women with BPD are more prone to experience premenstrual dysphoric disorder (with its temporarily heightened depressed mood, irritability, anxiety, and lability of affect—as outlined in DSM-V 2013). In some women, not previously regarded as "borderline", the symptoms may be particularly intense and may be accompanied by suicidal feelings or self-injurious behaviors, along with the "inordinate anger" that suggest to the clinician the presence of BPD. The syndrome tends to be a more regular feature, and more pronounced in its intensity, in women with BDs (such as Bipolar-II)—each condition alerting the clinician to the possible presence of the other. One will, to be sure, encounter other women with certain gynecological conditions (e.g., polycystic ovary) who experience severe premenstrual dysphoria (sometimes with endometriosis as well), who do not show the additional features of either BPD or BD. But mood disorders characterized primarily by depression are more common in women—in a way that appears to have implications for our evolution as a species. Annette Schirmer in her comprehensive chapter on sex and emotion (2013) summarizes the sex differences in this way: "For some emotions, men show stronger subjective feelings, cognitive, and/or behavioral effects than women (e.g., anger and contempt), whereas for other (emotions) we find the opposite (e.g., sadness, fear, disgust)," (p. 605) (the latter emotions being more often stronger in women). She adds: "…territorial behavior, in humans and other primates, is more strongly developed in males. Thus, emotional responses that facilitate aggression (anger, contempt) may have been of greater value to men…Conversely, early female typical tasks such as food gathering and child care were less confrontation and dangerous…Women present more often than men with disorders of prosocial emotions. That is, they are more likely than men to suffer from intense and prolonged feelings of fear and sadness" (ibid, p. 605). This sex-difference factor, coupled with the far greater vulnerability of daughters, compares with sons, to childhood sexual molestation by older-generation relatives (which can promote the later development of BPD symptomatology) helps one understand why many of the neurophysiological studies of BPD are based preponderantly on female subjects (Stone 1990b).

2 Neurophysiological Studies in the Borderline Domain

Since the last quarter of the last century, the biological aspects of psychiatry have become increasingly important. Electroencephalography (EEG) had already been in use for a long time. As for BPD, electrophysiological studies, until recently, offered only modest help in our understanding of the relevant brain changes (Boutros et al. 2003).

But newer techniques have now been brought more and more into use, such as Single Photon Emission Computed Tomography (SPECT) and Magnetic Resonance Imaging (MRI). These and related techniques have helped to establish a deeper understanding of brain correlates in many of the major psychiatric disorders. They have also shown promise in enabling us to base diagnostic distinctions along more objectifiable lines, rather than having to depend solely on the external criteria discernible to the clinician. In traditional psychiatry, that is, only the patterns of *thought-emotion-behavior* were available for making diagnostic decisions—patterns that often show confusing degrees of overlap between one supposedly "distinct" entity and another.

The use of MRI for medical purposes began in the 1970s (Goldstein and Price 2004) but has blossomed in the last 15 years. The spotlight in this article is on BPD, though MRI and other imaging studies relative to conditions that often overlap of co-occur are also included.

3 MRI and Related Findings in BPD

One of the earliest MRI studies of BPD reported smaller frontal lobe volumes in the patients, though the authors (Lyoo et al. 1998) mentioned that findings were inconsistent in other reports. Electroencephalographic studies of event-related potentials (ERPs) have demonstrated abnormalities in fluctuations 300 milliseconds after the presentation of certain (auditory or visual) "events"—the P300 response—in various psychiatric disorders, including BPD. But the P300 changes are not very specific, since they are noted in such other conditions as panic-disorder, substance abuse, schizophrenia, and PTSD (Kuperberg 2004). Viewed from one perspective, it is not surprising that unanimity was not found in the neuroimaging and EEG studies. BPD has long been recognized as heterogeneous from an etiological standpoint, better viewed as an array of dimensions (i.e., as a syndrome) rather than as a specific disorder (Stone 1980; Schmahl et al. 2002). We have alluded earlier to a number of routes that may converge into the BPD syndrome: genetic factors, parental neglect or brutality, early sexual abuse, and serious traumata of other sorts. There are also cases where maternal abuse of illicit drugs in the first trimester of pregnancy, very low birth weight or fetal hypoxia at delivery can also predispose to a clinical picture of BPD later on. Yet despite all this heterogeneity, there is widespread consensus that the essential clinical features of BPD are *impulsivity* and *emotional dysregulation*. These abnormalities occur in BPD, as Hughes et al. (2012) mention, almost invariably within an interpersonal context. But this consensus about the key features has spawned an outpouring of studies dedicated to discovering what brain peculiarities may underlie impulsivity and emotional dysregulation.

Many investigators of brain changes in BPD have drawn attention to the kind of *frontolimbic dysfunction* Hughes regarded as the predominant neural substrate underlying the personality disorder. The "fronto" aspect has been characterized in general as a diminished "top-down" control of affective responses; specifically, because of decreased responsiveness of certain midline areas of the prefrontal cortex (New et al. 2008). As these authors mention, besides neuroanatomical abnormalities there may also be a neuroendocrine factor, such as reduced serotonin availability—with a resultant dysregulation in the form of emotional *disinhibition*. In agreement with these impressions are the observations of Dell'Osso et al. (2010), how noted alterations in the serotonin system, but also in dopaminergic and glutaminergic systems, appear to play a role in the impulse dyscontrol and aggressivity in borderline patients. In a more recent study, Kamphausen et al. (2013) pinpoints *ventromedial prefrontal cortex* (vmPFC) dysfunction as the "top-down" element and, in their fMRI analysis of female BPD patients exposed to visual "threat" stimuli, a prolonged *amygdala* response, as the "bottom-up" component (Herpertz et al. 2001). The patients also showed an increased connectivity between the amygdala and the vmPFC. In a similar study higher connectivity was noted, during an fMRI "fear-scan", between the amygdala and the rostral portion of the *anterior cingulate cortex* (ACC) (Cullen et al. 2011). Borderline patients were shown to make more mistakes on fMRI than did the controls in a task involving distinguishing emotional from neutral faces in other areas as well, such as the insula, amygdala, and fusiform gyrus (Guitart-Masip et al. 2009; Koenigsberg et al. 2009). Abnormalities of this sort were seen as contributing to the heightened sensitivity in BPD to negative emotion, with consequent social disturbances: particularly, the tendency in borderline patients to become too angry too quickly in interpersonal situations others handle more calmly (Domes et al. 2009). Similar difficulties in suppressing their reaction to negative emotion was noted also in an ERP study of borderline patients (Marissen et al. 2010). In another study, reduced gray matter in female BPD patients was noted in the dorsolateral prefrontal cortex (dl-PFC) (bilaterally) and in the left orbitofrontal cortex (OFC); the prefrontal cortical changes did not, however, appear specific to BPD, insofar as similar changes were observed in a control group of other psychiatric disorders (Brunner et al. 2010), and also in a still wider array of psychiatric disorders including panic disorders and other Cluster-B personality disorders (Jackowski et al. 2012). White-matter abnormalities have also been implicated: BPD patients were shown to have abnormalities in the long association bundles connecting the association cortex with the hippocampus and thalamus—of a sort that appeared to play a role in the disruption of emotional regulation in BPD (Maier-Hein et al. 2014). In general, it is the OFC that plays a major role in top-down inhibitory control via "reverse-learning"—where maladaptive impulses and choices are suppressed in favor of more adaptive/socially appropriate choices (Jentsch 2012; Jentsch et al. 2002). This has relevance to BPD, but also to abuse of certain drugs such as cocaine and methamphetamine—which cause blockade of the dopamine-related D-2 receptors and impairment of the inhibitory control otherwise exercised by the OFC. Abuse of such drugs is common in BPD, aggravating a problem in top-down control typical of BPD psychopathology even in the absence of drug-abuse.

Neuroanatomically, size appears to matter, as in the study of Ruocco et al. (2012), in whose meta-analysis of MRI research in BPD–volume reductions of about 11 %

were noted in the amygdala, with comparable reductions also in the hippocampus. Similar reduction in amygdalar volume (from 11 to 17 %) was noted by Tebartz van Elst et al. (2007), who also found an *increased creatine concentration* in the left amygdala.The latter correlated with the patient's anxiety level, and might provide another clue to the emotional dysregulation in BPD—this time in the form of a neurochemical abnormality. Volume reduction in amygdala and hippocampus has been seen as correlates of other "bottom-up" (i.., limbic) abnormalities in BPD that lead to impulsivity and heightened aggression associated with this disorder (Nunes et al. 2009). Berdahl (2010) has advocated that we include even deeper—in effect, sublimbic—regions in the circuitry relevant to BPD: a network involving not only the Anterior Cingulate and ventromedial prefrontal cortex (ACC/vmPFC) and the amygdala, but also the brain-stem center—the periaqueductal gray. Panksepp and Biven (2012) have underlined the importance of the periaqueductal gray as constituting the first portal of entry for incoming stimuli affecting the emotions in humans and in other animals, including the negative emotions of fear, rage, and panic/grief. Unregulated feelings of the PANIC/GRIEF system may, in their view, underlie the stormy social relationships, depression, and avoidance of abandonment that plague the patients we label as BPD (Panksepp and Biven 2012, p. 75).

A number of authors have drawn attention to the ironical situation of reduced amygdala volume in BPD patients, yet *hyperactivity* in the amygdala's responses when confronted with emotion-related stimuli (Stein 2009; Siever and Weinstein 2009; O'Neill and Frodl 2012). Allele differences in the 5-hydroxytryptamine-1a receptor (5-HTR-1a) gene may account for some of the disagreement in the literature about amygdala-size: BPD patients with the G allele had smaller amygdala sizes than did those with the C/C genotype, and may be more prone to the impulsive and aggressive behavior that characterizes BPD (Zetzsche et al. 2007). Presumably, however, it is ultimately dysfunction in the top-down centers that should be held responsible for the dysregulation of impulse and affect in BPD (Soloff et al. 2008), since the amygdala (and perhaps before that—the periaqueductal gray) are the earlier recipients of stimuli carrying negative emotional valence, thence broadcast to the higher centers for evaluation and reaction.

Another peculiarity noted in many BPD patients, besides their emotional overreactivity etc., is a comparative insensitivity to pain. This, too, may answer to abnormalities in cortico-limbic centers: Kluetsch et al. (2012) noted that painful stimulation is handled differently in normals than in borderline patients. In their study of 25 women with BPD, almost all of whom (23) had a history of self-harm, showed altered pain-processing in regions (such as the cingulate- and left dorsolateral prefrontal cortices) involved in cognitive and affective evaluation of pain. This paradoxical reaction may underlie the tendency in many BPD patients to self-cut: they are less sensitive than other people to the sheer *physical* pain but able to use the (for them, milder) physical pain to distract them from the often overwhelming *psychological* pain of their everyday life.

Although contemporary research on brain changes associated with BPD has relied on MRI, Several groups using EEG have also made notable contributions. Brain activation as assessed by EEG-vigilance, for example, was noted to be lower in a sample of BPD patients (compared with OCD patients); the lowered vigilance has, in turn, been associated with the impulsivity and sensation-seeking manifested by many borderline patients (Hegerl et al. 2008). In another study EEG was used along with thyrotropin-releasing hormone (TSH), neurological soft-signs, and dexamethasone suppression in a search for associations between otherwise seemingly unrelated variables (De la Fuente et al. 2011). EEG and TSH emerged as the variables that influenced most of the others, in their Bayesian network model, raising the hope that such measures might strengthen subsequent diagnostic criterion-sets for BPD.

4 Borderline Personality Disorder and Aggression

Although violence *per se* is not among the DSM criteria for BPD, many BPD patients manifest the "items": inappropriate intense anger, impulsivity, and transient paranoid ideation under stress. These attributes may culminate in outbursts of violence. In less dramatic instances the violence may be limited to punching a lover or mate, or to smashing glassware. But in the forensic hospital where I work, most of the female patients carry the BPD diagnosis (often with "antisocial" comorbidity), and were remanded to the hospital following acts of greater violence: assault, arson, or murder. The same "top-down" and "bottom-up" abnormalities mentioned earlier are usually operative in such case: hyper-responsivity in the amygdala, and concomitant failure of the "braking system" in the prefrontal cortex (Siever 2008). The predisposition to violence may be aggravated by insufficient availability of serotonin, upon which the cortical braking system is partly dependent (Brendel et al. 2005; Siever 2008). These findings are mirrored in the important work by Coccaro and his colleagues on impulsive aggression and on the syndrome of Intermittent Explosive Disorder (IED): disorders noted frequently in persons comorbid for both BPD and antisocial personality disorder (ASPD), such as the female forensic patients just mentioned (Coccaro et al. 2007, 2011).

5 Borderline Personality Disorder and Childhood Abuse

A group of Québec researchers interested in BPD, aware of the importance of building a bridge between neurophysiological data and psychological material, brain and mind being two sides of the same coin, have drawn attention to the way in which childhood abuse can aggravate, or perhaps bring about de novo, the executive and frontal dysregulation that underlay the BPD syndrome (Bouchard et al. 2010). Severe and prolonged childhood abuse (especially physical and sexual) has been implicated in epigenetic changes—where otherwise silent genes become activated (here: in response to the abuse) but without any actual change in the sequence of DNA (Lewin 2008, p. 819). Such changes, as part of the body's mechanism in coping with the abuse, may take on the kind of permanence as though the child had inherited the genome transformations that developed as a consequence. Minzenberg et al. (2008) found a linkage between BPD patients with an abuse history and executive dysfunction; specifically, attachment-avoidance that correlated with temporo-limbic dysfunction (whether brought about, or made worse by the abuse). The avoidance apparently served in such patients as a compensatory mechanism whereby they could sidestep the kinds of interpersonal stresses that would otherwise reawaken the abnormal frontal lobe processing to which the earlier abuse had predisposed them. The various brain regions involved were outlined in an earlier fMRI study that dealt with facial emotion processing in BPD (Minzenberg et al. 2007).

6 Borderline Personality Disorder and Dissociative States

From a neurophysiological perspective, dissociative disorders, which often accompany BPD, have been connected with abnormalities in the parietal lobe. In a study out of Göttingen in Germany, young women with BPD who had been the victims of childhood sexual and physical abuse were shown, via structural MRI, to have a 9 % smaller volume in the right-precuneus area of the parietal lobe, and a 13 % increased volume in the left post-central gyrus of the superior lobe. The latter finding was correlated with the clinical conditions of dissociative amnesia and dissociative identity disorder (akin to the former "multiple personality") (Irle et al. 2007). A different brain area was implicated in another study: abnormalities in the function of the OFC appeared linked to the impulsivity, over-reaction to negative emotion, and to difficulty in retrieving autobiographical memories in BPD patients. The latter type of impairment was correlated with the *dissociative* symptoms that frequently occur in BPD (Poletti 2009). These brain areas are involved in memory, such that abnormalities in size or function might predispose to the varieties of dissociative and related memory disturbances seen in BPD, especially in patients who had been subjected to early abuse. Issues concerning causation versus correlation remain to be further elucidated. Irle et al. speculated that some BPD patients might have a neurodevelopmental defect of the right cerebral hemisphere that could render them more susceptible to the effects of early abuse. If so, this would suggest that being born with such a defect might heighten the vulnerability to abuse during childhood, as opposed to a situation where the early abuse somehow caused volume changes in key brain areas.

7 Borderline Personality Disorder and Post-Traumatic Stress Disorder

Women (and to a much smaller extent, men) who had been exposed to severe sexual abuse, especially by an older male relative, may develop the clinical picture of post-traumatic stress disorder (PTSD), with its chronic anxiety, flashback memories, nightmares, and startle-responses. Some of course will show dissociative symptoms or depersonalization as well. Irle et al. (2005) noted in women with both BPD and PTSD a 17 % smaller hippocampal size. They also noted a leftward asymmetry in the parietal cortex, which correlated with a greater vulnerability to schizoid traits and to psychotic symptoms. In a similar study hippocampal volumes were reduced in BPD patients in general, but particularly so in those with concomitant PTSD (Rodrigues et al. 2011). Also in the combined BPD with PTSD women, marked reduction (34 %) was found in the amygdala size: a greater reduction than was noted in the BPD patients who had experienced trauma, but did not show the PTSD picture—where the reduction is size was about 22 % (Weninger et al. 2011). The question about amygdalar size remains controversial, inasmuch as a Brazilian group, in their meta-analysis, noted smaller amygdala size in BPD where PTSD was not an accompaniment; they suggested that the reduced amygdalar volume in BPD might not be explainable as a consequence of concomitant PTSD (de-Ameida et al. 2012). In their observation about stress, in general, and its effect on limbic structures, Wingenfild et al. (2010) viewed that stress exerted damaging effects on the hippocampus, which had special relevance to BPD. These authors underlined the importance of studying further the hypothalamus-pituitary-adrenal (HPA) axis and its vicissitudes vis-à-vis BPD patients who have endured marked stresses, whether from abuse or neglect, in their formative years. The role of the HPA axis in BPD has been studied extensively by (Teicher et al. 2003), who point out that the major brain-structural consequences of stress related to childhood traumata concern not only the neocortex, amygdala, and hippocampus, but also the corpus callosum (CC) (where reduced size has been noted in its mid-portion—in children with a history of PTSD). Regarding hippocampal changes, severe early stress may be more associated with dissociative symptoms than simply withy declarative memory (Stein 1997), and thus have particular relevance to BPD.

On the neuropeptide side, Prossin et al. (2010) have shown greater regional mu-opioid non-displaceable binding potential (BP-ND) via PET scan in female patients with BPD, when compared with normal subjects. Brain regions involved included the amygdala, caudate, N accumbens, and OFC. Negative emotion challenges (sadness induction) led to greater reductions in BP-ND in the BPD patients, especially in left-sided regions. The authors did not focus on BPD with comorbid PTSD. The differences in response to negative emotion and the accompanying stress did, however, appear related to some of the typical stress-related phenomena in BPD patients. In their commentary on such stresses as rejection and abandonment, Stanley and Siever (2010) drew attention to how the reactions elicited by these stresses in BPD patients (viz., impulsive and self-destructive/suicidal behaviors) suggest a malfunction of psychological systems oriented to attachment and affiliation. This in turn lends an importance, for enhancing our understanding BPD, to certain neuropeptides that play a role in these interpersonal actions; specifically, opioids (in pain-related phenomena), oxytocin (in affiliative responses), and vasopressin (in homeostasis and memory formation).

8 Bipolar Disorder: Neurophysiological Aspects and Relation to BPD

Compared with the etiologically heterogeneous BPD, BDs represent a more unified nosologic construct, since genetic factors appear to play the primary role in their origin. Confusion and controversy in the domain of bipolar and borderline conditions stem from the observation of similarities in symptom presentation, similarities in neurophysiological underpinnings, and the fact, as mentioned in Gunderson's work cited earlier, that an impressive proportion of persons diagnosed in adolescence or early adult life with BPD are seen later to show the characteristics of a BD, and *mutatis mutandis*, persons diagnosed as "bipolar" in their teens frequently meet, later on, criteria for BPD.

Some of the similarities are outlined by Antoniadis et al. (2012), who point to the main clinical features of BD; namely, impulsivity and affective instability—the same as found in BPD. Alterations in the limbic system have been found on MRI in both, though amygdala size has been reported as smaller in BPD; larger, in BD. Heritability, clearer significant in BD, has been found in some studies of BPD, but there do not appear to be genes specific in any way for the disorder. Environmental factors, meanwhile, appear to be more important in BPD than in BD. At the clinical level, Benazzi (2006a, b) takes the position, in the dispute whether BPD is a bipolar "spectrum" condition or is a separate entity, that the DSM-IV (1994) definition of BPD may be conflating two sets of unrelated features: an emotional instability related more to BD (especially to the milder form of Bipolar-II disorder), and an impulsivity dimension more applicable to BPD. In Bipolar-II patients cyclothymic temperament (an inherited quality) and borderline traits (short of meeting full criteria for BPD) clustered more with Bipolar-II Disorder than with Major Depressive Disorder. Among the BPD traits, lability of affect (*unstable mood*), unstable interpersonal relationships, identity weakness (*unstable self-image*), and chronic anger sorted in factor analysis—more with Bipolar-II, but impulsivity did not (Benazzi 2006a). Coulston (2012) was also struck by the clinical similarity between BD and BPD, the presence of one predicting the (sooner or later) presence of the other. In his view, childhood trauma predisposed to both conditions, and also to rapid cycling (in BD) or to the analogous quickly fluctuating "mood lability" in traumatized BPD patients. But in BPD the mood changes tend to be briefer and vary between anxiety, on the one side, versus anger and depression, on the other. This in contrast with the rapid cycling in BD, where the shifts tend to vary between elation and sadness (Fiedorowicz and Black 2010). Mackinnon and Pies (2006) also comment on the similarity between the rapid cycling in BD and the affective instability/mood lability in BPD, adding that anticonvulsant medications are regularly helpful in BD and often so in BPD as well. They regard this as pointing to a biological overlap, though do not buttress their argument with data from neurophysiological research.

What neurophysiological data do exist are also equivocal regarding the question: are BD and BPD separate conditions—or two sides of the same coin. Rossi et al. (2012), using an MRI technique, studied 26 mostly female BPD patients and

15 mostly male BD patients. The BPD group showed smaller hippocampal volumes bilaterally; in the Bipolar patients, there was smaller right-hippocampal volume (of the right dentate gyrus that lies between the fimbria of the hippocampus and the hippocampal gyrus). The authors speculated that these volumetric differences might be related to the clinical phenomenology of each disorder. In a subsequent paper (Rossi et al. 2013) where gray and white matters were compared in the two conditions, it was noted that gray-matter density changes were more diffuse and severe in BD than in BPD. Each disorder had specific regions of abnormality involving both cortical and subcortical structures in BD; in BPD—mainly fronto-limbic regions. Despite areas of overlap in gray matter changes, the topography in those changes appeared more consistent with a "separate-conditions" hypothesis. The separate-conditions hypothesis finds support in a psychological study of set-shifting and reversal learning in BPD (Barker et al. 2014). BPD patients in this study did not show significant deficits in extra-dimensional shift (EDS) or in reversal learning; this appeared to distinguish them from bipolar subjects—who did demonstrate deficits in tests of EDS. MRI suggests that performances on reversal learning reflects OFC functioning, whereas EDS relies on prefrontal cortex function. Since deficits in these tasks were not found in the BPD patients, the authors suggested that, in contrast to the bipolar patients, the limbic system was the primary locus of pathology in BPD (Barker et al. 2014, p. 9).

In some patients there is a convergence clinically between the symptoms and traits of BPD and *Bipolar Depression.* Patients with (depressive) mood disorder are known to have a higher prevalence of BPD—the combined condition showing marked instability of mood and poorer response to medication (Radaelli et al. 2012). When MRI data were obtained from patients with Bipolar Depression alone and with those diagnosed with both conditions, the Depression + BPD group showed a lower activation of the dlPFC than in the Depression-only patients. Emotional dysregualtion appeared greater in the combined group. In contrast, MRI data from another research group found an important abnormality in the ACC (Brodmann Area #24), whose function is ordinarily to assess the salience of emotional information and to help regulate emotional responses. This gray matter (but not white) in this area was shown to be smaller in patients with BPD adolescents with Major Depression (Goodman et al. 2011). Still, there are many areas of overlap on the biologic side between BPD and Major Depression: amygdala hyper-reactivity, smaller ACC volume, and diminished serotonergic function, such that the data that might help firm up the similarities and also make more meaningful distinctions between the two are not as yet available (Goodman et al. 2010).

9 The Spotlight on Bipolar-I Disorder

The mood dysregulation characteristic of BD has been ascribed to dysfunction in the prefrontal cortex, leading to inhibitory dyscontrol (Anticevic et al. 2012). In their study, based on BD patients with psychosis, there was reduced medial prefrontal cortex (mPFC) connectivity within the prefrontal cortex as a whole, and also reduced connectivity between the amygdala and the dl-PFC. The mPFC is considered the key region for emotional regulation. Dysconnectivity was also noted even in remitted bipolar patients, suggesting that this abnormality might constitute a risk factor for the phasic (mood-fluctuating) features of BD. The memory deficits sometimes associated with BD were linked, in another study, to the lowered hippocampal volume, as documented above by a number of other investigators (Chepenik et al. 2012). Improvement in memory was noted in some BD patients following treatment with antidepressant medications, though whether the memory improvement was associated with morphological changes in the hippocampus (such as regaining its volume) was not assessed. The hyper-reactivity of the amygdala to emotional stimuli (looking at sad faces, for example, in fMRI experiments) was noted in BD and linked to deficient activation in the dl-PFC (Garrett et al. 2012)—as a factor in the emotional dysregulation in BD—comparable, however, to what has been noted in BPD as well. Amygdalar hyperactivity to emotional facial expressions has been found to be particularly marked in children and adolescents (aged 7–18) with BD (Kim 2012). Yet in a study of Bipolar-II patients with depression, Vizueta and colleagues (2012) found amygdalar *hypo*-activity (which they felt might be state-dependent), though *OFC hypoactivation* was similar to what has been observed in Bipolar-I—which might therefore warrant consideration as a trait-marker of BDs in general.

Hajek et al. (2012, 2013a) have shown in an fMRI study that BD patients underactivated the right inferior frontal gyrus (R-IFG) relative to a control group, irrespective of current mood state. This suggested that the impaired response inhibition to emotional stimuli (in effect: poorer top-down control) in BD may be a biological marker for the condition. Oddly, the authors noted that the IFG was significantly larger not only in the BD patients, but also in their family members (with or without the disorder). Usually, abnormal function has been associated with smaller-, rather than with larger volumes in the affected regions. Further along in the course of bipolar illness, however, they noted that the IFG became smaller than normal—possibly because of the neurotoxic effects of the illness on gray matter (cf. also: Hajek et al. 2013a). But lithium-treated BD patients eventually were shown to have normal R-IFG volume, despite having had the condition for a long time (Hajek et al. 2013b). Hippocampal volumes in BD, smaller in volume at the outset, were also noted to appear normal in volume following two years or more of lithium treatment (Hayek et al. 2013c). As for other higher centers, a London-based group found evidence in their meta-analysis–of *reduced* gray matter in BD in the right ventral prefrontal cortex (r VPRC), as well as in the temporal cortex and insula (Selvaraj et al. 2012). Malhi et al. (2013), in a review concerning the effects of lithium, mentioned that lithium may exert its beneficial effects in BD via helping to preserve or even increase the volume of brain structures involved in emotional regulation.

In an effort to develop a consensus concerning the evolution of BD and key brain areas involved, a workgroup organized by the University of Cincinnati Department of Psychiatry met to discuss their research. Their impression was that in the early development of BD there was disruption of brain networks that modulate emotional responses. Because of what may be excessive prefrontal pruning during adolescence, one finds decreased connectivity among ventral prefrontal networks and the relevant limbic centers, especially the amygdala. These changes appear to prepare the path for manic symptomatology (Strakowski et al. (2012). In a study devoted to finding neurophysiological signs that might be helpful in distinguishing BD from schizophrenia in a reliable way, Whalley et al. (2012) examined the fMRI literature, from which they found evidence that over-activation in the medial temporal lobe in emotion and memory tasks—occurred in BD more so than in schizophrenia. As to the distinction between, versus the similarity of, BD and schizophrenia, at least from the standpoint of pharmacological response, it is of interest that in a recent Danish study, clozapine proved useful not just in schizophrenia, but also in refractory BD (Nielsen et al. 2012). Vacheron-Trystram et al. (2004) found clozapine more useful in BD even than in schizophrenia.

10 Attention-Deficit Disorder, BPD, and BD: Similarities and Differences

The conceptual and, to an increasing extent, neurophysiological overlap between BPD and the spectrum of BDs—in a significant percentage of cases—is also pertinent to Attention-Deficit Hyperactivity Disorder (ADHD). This is because, in the same way that BPD is over-represented in samples of BD and vice versa, ADHD is over-represented in a similar way. Granted that ADHD tends to be over-diagnosed in the US, nevertheless young persons with the genuine disorder are more likely, as they enter adulthood, to be diagnosed also with either BPD or BD or both (Makris et al. 2012). Similarly, in the background of adults with BPD or BD, there is a higher proportion of ADHD histories than would be expected in the general population (Faraone et al. 2012). (Philipsen 2006), besides mentioning the clinical interconnection between ADHD and BPD, expressed the view, in the light of recent neuroimaging studies, that ADHD and BPD may be not so much two distinct disorders, but rather a manifestation of two aspects or dimensions of one (underlying) disorder. In a like manner, Rüsch et al. (2007) examined a group of women with BPD who were comorbid also for ADHD. In their imaging study the women with BPD had a narrower *isthmus* of the CC (i.e., the portion where the anterior parts of the CC: rostrum, genu, and body and the posterior part: the splenium—are fused during embryogenesis). A history of childhood sexual abuse was associated with a thinner posterior body of the CC, indicating a possible loss of some of the 190,000,000 axons that make up the CC and subserve interhemispheric connectivity. This may account for some of the deficits in the combined BPD/ADHD disorder. In the study of Posner et al. (2011) the focus was on the amygdala and the lateral prefrontal cortex (LPFC) in their fMRI assessment (using subliminal presentation of fearful faces) of adolescents with ADHD. Their strategy was based on the connection of the amygdala and the LPFC in monitoring emotional reactivity. They found that activity in the right amygdala was greater in adolescents with ADHD than in the control group, along with greater connectivity between the amygdala and the LPFC. Parenthetically, they noted that stimulants (a common treatment for ADHD) had a normalizing effect on the activity of the right amygdala and on its connections with the LPFC. ADHD is associated with deficits in cortical inhibition, as has been noted in BPD, but also in Gilles de la Tourette syndrome. Barnow et al. (2009) used transcranial magnetic stimulation (TMS) to assess whether BPD patients showed decreased cortical inhibition, or else—increased cortical excitation. They controlled their result for ADHD symptomatology and still noted an association between BPD and deficits in cortical inhibition.

11 Discussion

There is fairly good agreement among investigators concerning the neurophysiological alterations in BPD. The brain-regions chiefly implicated in BPD, along with neurophysiological correlates noted in other disorders: Bipolar, Major Depressive, and ADHD—are summarized in Table 1.

The various studies, relying primarily on MRI, invoke fronto-limbic malfunction as the most general way of addressing the impulsivity and emotional dysregulation that characterize BPD. Because of the widespread recognition that BPD, BDs (whether fluctuating between manic and depressive episodes, or predominantly depressive), and Attention-Deficit/Hyperactivity Disorder often co-occur in various combinations and sequences, imaging researchers who focus on one disorder often include patients with the related conditions as well. The very fact that patients with all three disorders are well-known to clinicians has spurred interest in finding possible commonalities at the deeper level of brain physiology. Eagerness to explore this area has been heightened further following the recent discovery by the Cross-Disorder Group of the Psychiatric Genomics Consortium (2013), that five psychiatric disorders viewed as distinct from a *clinical* perspective nevertheless share certain *genetic* features in common; namely, single nucleotide polymorphisms (SNPs) in two genes involved in calcium-channel activity.

Alessandro Serretti and Chiara Fabbri (2013) and colleagues from the Human Genome Project contributed importantly to this work, which has been reviewed recently by Smoller (2013). The five disorder: Autism, Attention Deficit-Hyperactivity Disorder, BD, Major Depressive Disorder, and Schizophrenia include three of relevance to the topic of BPD. Serretti and Fabbri have argued that there is abundant pleiotropy in human complex disorders, such that the same genetic variant may play a role in several diseases that—to the clinical eye—have appeared separate and unrelated.

Table 1 Neurophysiological abnormalities noted in brain regions in BPD and related conditions

Disorder	Cortical/subcortical regions	Limbic regions
BPD	Prefrontal cortex, orbitofrontal cortex, dorsolateral PFC (including decreased gray matter), ventromedical PFC, parietal lobe, right hemisphere, corpus callosum, insula, N accumbens	Amygdala size (decreased), amygdala reactivity (increased) hippocampus size (decreased)
Bipolar I, II	Prefrontal cortex, dorsolateral PFC, middle prefrontal cortex, right inferior frontal gyrus, medial temporal lobe, right ventral prefrontal cortex	Amygdala size (decreased), amygdala reactivity (decreased; though increased in bipolar-II), hippocampal size (decreased)
Major depression	Dorsolateral prefrontal cortex, anterior cingulate cortex	Amygdala reactivity (increased)
ADHD	Lateral prefrontal cortex, corpus callosum	Amygdala reactivity (increased)

Note The abnormalities observed in the cortical regions usually involved smaller than normal volumes, though in Bipolar-I the R Inferior Frontal Gyrus was larger than normal initially, but smaller—as the illness progressed (Hajek et al. 2013a)

The new research will enable us to move beyond a nosology based on description of signs and symptoms, toward a classification based progressively more on fundamental causes. An effort was recently made in regard to BPD by Calati et al. (2013): she and her colleagues looked for serotonergic polymorphisms, but did not find a direct role in BPD for the three genetic polymorphisms of interest. Perhaps this is less surprising—to the extent that BPD, as noted above, is a markedly heterogeneous clinical syndrome based more on adverse environmental factors than on putative genetic factors. There is in all likelihood a *subset* of BPD cases where genetic factors play a major, not to say, a determinative, role in predisposing to the development (usually discernible at puberty) of the borderline syndrome (à la DSM). Persons with clear-cut and severe BD, who in addition have a family history of bipolar disorders—but who have *no history* of neglect, abuse (whether sexual, physical, or verbal), perinatal complications, drug abuse, or head injury—would constitute the most concentrated pool of patients for the assessment of a genetic linkage to BD, and also for whatever gene polymorphisms may be a part of the picture. There are also social-class and cultural factors to take into account. Persons from economically poorer backgrounds are much more likely to have experienced childhood physical abuse than their better-off counterparts (Straus and Gelles 1992). Incest histories are common in certain cultural settings; rare, in others. In many samples of BPD patients, including those devoted to MRI and fMRI studies, these factors are not separated out, thus complicating any search for specific gene peculiarities. Given the difficulties inherent in carrying out MRI analyses in infants or very young children, it is not easy to determine whether the brain-changes associated with bipolar disorder were already detectable at birth (and later paved the way for development of BPD), or whether they were epiphenomena of adverse environmental factors. We still tend to be divided into two camps: the "lumpers" and the "splitters": diagnostic lumpers who might argue that BPD, BD, and ADHD are three continents of the same nosologic planet, and the splitters, who claim that they are diagnostically, and perhaps even genetically, separable. As for the BPD question, the time is ripe for further genetic analysis, based on patient-samples that have been more scrupulously homogenized: borderline patients with no family history and no diagnostic indications of bipolarity, versus bipolar patients with a strong family history of BD and no signs at all of environmental adversity. Further neurophysiological and genetic analysis of such groups will help resolve many of the as yet unanswered questions in this domain.

References

Akiskal HS (1981) Subaffective disorders: dysthymic, cyclothymic, and bipolar-II disorders in the 'borderline' states. Psychiatr Clin North Am 4(1):25–46

Anticevic A, Brumbaugh MS, Winkler AM, Lombardo LE, Barrett J, Corlett PR, Kober H, Gruber J, Repovs G, Cole MW, Krystal JH, Pearlson GD, Glahn DC (2012) Global prefrontal and fronto-amygdala dysconnectivity in bipolar-I disorder with psychosis history. Biol Psychiatry. doi:10.1016/j.biopsych.2012.07.031

Antoniadis D, Samakouri M, Livaditis M (2012) The association of bipolar spectrum disorders and borderline personality disorder. Psychiatr Q 83:449–465

Barker V, Pope M, Smith S, Brown V, Hall J (2014) Set shifting and reversal learning in borderline personality disorder. Pers Ment Health 8:1–13

Barnow S, Völker KA, Möller B, Freyberger HJ, Spitzer C, Grabe HJ, Daskalakis ZJ (2009) Neurophysiological correlates of borderline personality disorder: a transcranial magnetic stimulation study. Biol Psychiatry 65:313–318

Benazzi F (2006a) Does temperamental instability support a continuity between bipolar-II disorder and major depressive disorder? Eur Psychiatry 21:274–279

Benazzi F (2006b) Borderline personality-bipolar spectrum relationship. Prog Neuropsychopharmacol Biol Psychiatry 30:68–74

Berdahl CH (2010) A neural network of borderline personality disorder. Neural Netw 23:177–188

Bouchard S, Lemelin S, Dubé C, Giguère JF (2010) Dysregulation of the executive system and theory of mind: clinical interest of a neuroscientific conception of BPD. Santé Ment Québec 35:227–251

Boutros NN, Torello M, McGlashan TH (2003) Electrophysiological aberrations in borderline personality disorder: state of the evidence. J Neuropsychiatry Clin Neurosci 15:145–154

Brendel GR, Stern E, Silbersweig DA (2005) Defining the neurocircuitry of borderline personality disorder: functional neuroimaging approaches. Dev Psychopathol 17:1197–1206

Brunner R, Henze R, Parzer P, Kramer J, Feigl N, Lutz K, Essig M, Resch F, Stieltjes B (2010) Reduced prefrontal and orbitofrontal gray matter in female adolescents with borderline personality disorder: is it disorder specific? Neuroimage 49:114–120

Calati R, Gresier F, Balestri M, Serretti A (2013) Genetic modulation of borderline personality disorder: systematic review and meta-analysis. J Psychiatr Res 47:1275–1287

Chepenik LG, Wang F, Sencer L, Spann M, Kalmar JH, Womer F, Kale-Edmiston E, Pittman B, Blumberg HP (2012) Structure-function associations in hippocampus in bipolar disorder. Biol Psychol 90:18–22

Coccaro EF, McCloskey MS, Fitzgerald DA, Phan KL (2007) Amygdala and orbitofrontal reactivity to social threat in individuals with impulsive aggression. Biol Psychiatry 62:168–178

Coccaro EF, Sripada CS, Yanowitch RN, Phan KL (2011) Corticolimbic function in impulsive aggressive behavior. Biol Psychiatry 69:1153–1159

Coulston CM, Tanious M, Mulder RT, Porter RJ, Mahli GS (2012) Borderling onbipolar: the overlap between borderline personality disorder and bipolarity. Aust NZ J Psychiatry 46:506–521

Cross-Disorder Group of the Psychiatric Genomics Consortium (2013) Nat Genet 45:984–994

Cullen KR, Vizueta N, Thomas KM, Han GJ, Kim LO, Camchong J, Mueller BA, Bell CH, Heller MD, Schultz SC (2011) Amygdala functional connectivity in young women with borderline personality disorder. Brain Connect 1:61–71

de Almeida CP, Wenzel A, de Carvalho CS, Powell VB, Araújo-Neto C, Quarantini LC, Oliveira IR (2012) Amygdalar volume in borderline personality disorder with and without comorbid post-traumatic stress disorder: a meta-analysis. CNS Spectr 17:70–75

De la Fuente JM, Bengoetxea E, Navarro F, Bobes J, Alarcon RD (2011) Interconnection between biological abnormalities in borderline personality disorder: use of the Bayesian networks model. Psychiatry Res 186:315–319

Dell'Osso B, Berlin HA, Serati M, Altamura AC (2010) Neuropsychobiological aspects, comorbidity patterns and dimensional models in borderline personality disorder. Neuropsychobiology 61:169–179

Diagnostic and statistical manual of psychiatric disorders [DSM-III] (1980), 3rd edn. American Psychiatric, Washington DC

Diagnostic and statistical manual of psychiatric disorders [DSM-IV] (1994), 4th edn. American Psychiatric, Washington DC

Diagnostic and statistical manual of psychiatric disorders [DSM-V] (2013), 5th edn. American Psychiatric, Washington DC

Domes G, Schulze L, Herpertz SC (2009) Emotion recognition in borderline personality disorder: a review of the literature. J Pers Disord 23:6–19

Faraone SV, Biederman J, Wozniak J (2012) Examining the comorbidity between attention deficit hyperactivity disorder and bipolar I disorder: a meta-analysis. Am J Psychiatry 169:1256–1266

Fiedorowicz JG, Black DW (2010) Borderline, bipolar or both? Curr Psychiatry 9:21–31

Fyer MR, Frances AJ, Sullivan T et al (1988) Comorbidity of borderline personality disorder. Arch Gen Psychiatry 45:348–352

Garrett AS, Reiss AL, Howe ME, Kelley RG, Singh MK, Adleman NE, Karchsmsky A, Chang KD (2012) Abnormal amygdala and prefrontal cortex activation to facial expressions in pediatric bipolar disorder. J Am Acad Child Adolesc Psychiatry 51:821–831

Goldstein MA, Price BH (2004) Magnetic resonance imaging. In: Dougherty DD, Rauch SL, Rosenblum JF (eds) Essentials of neuroimaging for clinical practice. American Psychiatric, Washington, DC, pp 21–73

Goodman M, EA Hazlett, Avedon JB, Siever DR, Chu KW, New AS (2011) Anterior cingulated volume reduction in adolescents with borderline personality disorder and co-morbid major depression. J Psychiatr Res 45:803–807

Goodman M, New AS, Triebwasser J, Collins KA, Siever LJ (2010) Phenotype, endophenotype, and genotype comparisons between borderline personality disorder and major depressive disorder. J Pers Disord 24:38–59

Guitart-Masip M, Pascual JC, Carmona S, Hoekzema E, Bergé D, Pérez V, Soler J, Soliva JC, Rovira M, Bulbena A, Villaroya O (2009) Neurla correlates of impaired emotional discrimination in borderline personality disorder: an fMRI study. Prog Neuropsychopharm Biol Psychiatry 33:1537–1545

Gunderson JG (2001) Borderline personality disorder: a clinical guide. American Psychiatric, Washington, DC

Gunderson JG, Singer MT (1975) Defining borderline patients: an overview. Am J Psychiatry 132:1–10

Gunderson JG, Triebwasser J, Phillips KA (1999) Personality and vulnerability to affective disorders. In: Cloninger CR (ed) Personality and psychopathology. American Psychiatric, Washington, DC, pp 3–32

Gunderson JG, Weinberg I, Daversa MT, Kueppenbender KD, Zanarini MC, Shea MT, Skodol AE, Sanislow CA, Yen S, Morey LC, Grilo CM, McGlashan TH, Stout RL, Dick I (2006) Descriptive and longitudinal observations on the relationship of borderline personality disorder and bipolar disorder. Am J Psychiatry 163:1173–1178

Hajek T (2012) Brain-region size may be long-sought biomarker. Psychiatric News, (reported by J Arehart-Treichel). Accessed 17–19 Sept 2012

Hajek T, Alda M, Hajek E, Ivanoff J (2013a) Functional neuroanatomy of response inhibition in bipolar disorders: combined voxel-based and cognitive performance meta-analysis. J Psychiatr Res 47:1955–1966

Hajek T, Cullis J, Novak T, Kopecek M, Blagdon R, Propper L, Stopkova P, Duffy A, Hoschl C, Uher R, Paus T, Young LT, Alda M (2013b) Brain structural signature of familial predisposition for bipolar disorder: replicable evidence for involvement of the right inferior frontal gyrus. Biol Psychiatry 73:144–152

Hajek T, Cullis J, Novak T, Kopecek M, Höschl C, Blagdon R, O'Donovan C, Bauer M, Young LT, Macqueen G, Alda M (2013c) Hippocampal volumes in bipolar disorders: opposing effects of illness burden and lithium treatment. Bipolar Disord 14:261–270

Hegerl U, Stein M, Mulert C, Mergl R, Olbrich S, Dichgans E, Rujescu D, Pogarell O (2008) EEG-vigilance differences between patients with borderline personality disorder, patients with obsessive-compulsive disorder and healthy controls. Eur Arch Psychiatry Clin Neurosci 258:137–143

Herpertz SC, Dietrich TM, Wenning B, Krings T, Erberich SG, Willmes K, Thron A, Sass H (2001) Evidence of abnormal amygdala functioning in borderline personality disorder: a functional MRI study. Biol Psychiatry 50:292–298

Hughes AE, Crowell SE, Uyeji L, Coan JA (2012) A developmental neuroscience of borderline pathology: emotional dysregulation and social baseline theory. J Abnorm Child Psychol 20:21–33

Irle E, Lange C, Sachsse U (2005) Reduced size and abnormal symmetry of parietal cortex in women with borderline personality disorder. Biol Psychiatry 57:173–182

Irle E, Lange C, Weninger G, Sachsse U (2007) Size abnormalities in the superior parietal cortices are related to dissociation in borderline personality disorder. Psychiatry Res 156:139–149

Jankowski AP, Araújo Filho GM, Almeida AG, Araújo CM, Reis M, Nery F, Batista IR, Silva I, Lacerda AL (2012) The involvement of the orbitofrontal cortex in psychiatric disorders: an update of neuroimaging findings. Rev Bras Psiquiatr 24:207–212

Jentsch JD (2012) Lecture on inhibitory control. New York State Psychiatric Institute, New York

Jentsch JD, Olausson P, De La Garza R, Taylor JR (2002) Impairments in reversal learning and response perseveration after repeated, intermittent cocaine administration to monkeys. Neuropsychopharmacology 26:183–190

Kamphausen S, Schröder P, Maier S, Bader K, Feige B, Kaller CP, Glauche V, Ohlendorf S, Tebartz van Elst L, Klöppel S, Jacob GA, Silbersweig D, Lieb K, Tüshcer O (2013) Medical prefrontal dysfunction and prolonged amygdala response during instructed fear processing in borderline personality disorder. World J Biol Psychiatry 14:307–318

Kernberg OF (1967) Borderline personality organization. J Am Psychoanal Assoc 15:641–685

Kim P, Thomas LA, Rosen BH, Moscicki AM, Brotman MA, Zarate CA Jr, Blair RJ, Pine DS, Leibenluft E (2012) Differing amygdala responses to facial expressions in children and adults with bipolar disorder. Am J Psychiatry 169:642–649

Kluetsch RC, Schmahl C, Niedfeld I, Densmore M, Calhoun VD, Daniels J, Kraus A, Ludaescher P, Bohus M, Lanius RA (2012) Alterations in default mode network connectivity during pain processing in borderline personality disorder. Arch Gen Psychiatr 69:993–1002

Koenigsberg HW, Siever LJ, Lee H, Pizzarello S, New AS, Goodman M, Cheng H, Flory J, Prohovnik I (2009) Neural correlates of emotion processing in borderline personality disorder. Psychiatry Res 172:192–199

Kohut H (1971) Analysis of the self. International Universities Press, New York

Kraepelin E (1905) Einfürung in die psychiatrische Klinik, 2nd edn. Leipzig, Barth

Kraepelin E (1921) Manic-depressive insanity and paranoia. E. & S. Livingstone, Edinburgh

Kuperberg GR (2004) Electroencephalography, event-related potentials and magnetoencephalography. In: Dougherty DD, Rauch SL, Rosenblum JF (eds) Essentials of neuroimaging for clinical practice. American Psychiatric, Washington DC, p 117–127

Lewin B (2008) Genes IX. Jones & Bartlett, Boston

Lyoo IK, Han MH, Cho DY (1998) A brain MRI study in subjects with borderline personality disorder. J Affect Disord 50:235–243

Mackinnon DF, Pies R (2006) Affective instability as rapid cycling: theoretical and clinical implications for borderline personality and bipolar spectrum disorders. Bipolar Disord 8:1–14

Mahli GS, Tanious M, Das P et al (2013) Potential mechanisms of action of lithium in bipolar disorder: current understanding. CNS Drugs 27:135–153

Maier-Hein KH, Brunner R, Lutz K, Henze R, Parzer P, Feigl N, Kramer J, Meinzer HP, Resch F, Stieltzes B (2014) Disorder-specific white matter alterations in adolescent borderline personality disorder. Biol Psychiatry 75:81–88

Makris N, Seidman LJ, Brown A et al (2012) Further understanding of the comorbidity between attention-deficit/hyperactivity disorder and bipolar disorder in adults: an MRI study of cortical thickness. Psychiatry Res 202:1–11

Marissen MA, Meuleman L, Franken IH (2010) Altered emotional information processing in borderline personality disorder: an electrophysiological study. Psychiatry Res 181:226–232

Minzenberg MJ, Fan J, New AS, Tang CY, Siever LJ (2007) Fronto-limbic dysfunction in response to facial emotion in borderline personality disorder: an event-related fMRI study. Psychiatry Res 155:231–243

Minzenberg MJ, Poole JH, Vinogradov S (2008) A neurocognitive model of borderline personality disorder: effects of childhood sexual abuse and relationship to adult social attachment disturbance. Dev Psychopathol 20:341–368

New AS, Goodman M, Triebwasser J, Siever L (2008) Recent advances in the biological study of personality disorders. Psychiatr Clin North Am 31:441–461

Nielsen J, Kane JM, Correll CU (2012) Real-world effectiveness of clozapine in patients with bipolar disorder: results from a two-year mirror-image study. Bipolar Disord 14:863–869

Nunes PM, Wenzel A, Borges KT, Porto CR, Caminha RM, de Oliveira IR (2009) Volumes of the hippocampus and amygdala in patients with borderline personality disorder: a meta-analysis. J Pers Disord 23:333–345

Oglodek E (2011) Emotional dysregulation in borderline personality disorder. Pol Merkur Lekarski 30:160–163

O'Neill A, Frodl T (2012) Brain structure and function in borderline personality disorder. Brain Struct Funct 217:767–782

Pally R (2002) The neurobiology of borderline personality disorder: the synergy of "nature and nurture". J Psychiatric Pract 8:133–142

Panksepp J, Biven L (2012) The archaeology of mind. WW Norton, New York

Philipsen A (2006) Differential diagnosis and comorbidity of attention-deficit/hyperactivity disorder (ADHD) and borderline personality disorder (BPD) in adults. Eur Arch Psychiatry Clin Neurosci 256(1):i42–i46

Poletti M (2009) Neurocognitive functioning in borderline personality disorder. Riv Psichiatr 44:374–383

Posner J, Nagel BJ, Maia TV, Mechling A, Oh M, Wang Z, Peterson BS (2011) Abnormal amygdalar activiation and connectivity in adolescents with attention-deficit/hyperactivity disorder. J Am Acad Child Adolesc Psychiatry 50:828–837

Prossin AR, Love TM, Koeppe RA, Zubieta JK, Silk KR (2010) Dysregulation of regional endogenous opioid function in borderline personality disorder. Am J Psychiatry 167:925–933

Radaelli D, Poletti S, Dallaspezia S, Colombo C, Smeraldi E, Benedetti F (2012) Neural responses to emotional stimuli in comorbid borderline personality disorder and bipolar depression. Psychiatry Res 203:61–66
Rodrigues E, Wenzel A, Ribeiro MP et al (2011) Hippocampal volume in borderline personality disorder with and without comorbid posttraumatic stress disorder: a meta-analysis. Eur Psychiatry 26:452–456
Rossi R, Lanfredi M, Pievani M, Beneduce R, Rillosi L, Giannakopoulos P, Thompson PM, Rossi G, Frisoni GB (2012) Volumetric and topographic differences in hippocampal subdivisions in borderline personality disorder and bipolar disorders. Psychiatry Res 203:132–138
Rossi R, Pievani M, Lorenzi M et al (2013) Structural brain features of borderline personality and bipolar disorders. Psychiatry Res 213:83–91
Ruocco AC, Amirthavasagam S, Zakzanis KK (2012) Amygdala and hippocampal volume reductions as candidate endophenotypes for borderline personality disorder: a meta-analysis of magnetic resonance imaging studies. Psychiatry Res 201:245–252
RüschN Luders E, Lieb K, Zahn R, Ebert D, Thompson PM, Toga AW, van Elst LT (2007) Corpus callosum abnormalities in women with borderline personality disorder and comorbid attention deficit hyperactivity disorder. J Psychiatry Neurosci 32:417–422
Schirmer A (2013) Sex and emotion. In: Armony J, Vuilleumier P (eds) The Cambridge handbook of human affective neuroscience. Cambridge University Press, New York, pp 591–610
Schmahl CG, McGlashan TH, Bremner JD (2002) Neurobiological correlates of borderline personality disorder. Psychopharmacol Bull 36:69–87
Selvaraj S, Arnone D, Job D, Stanfield A, Farrow TF, Nugent AC, Scherk H, Gruber O, Chen X, Sachdev PS, Dickstein DP, Malhi GS, Ha TH, Ha K, Phillips ML, McIntosh AM (2012) Grey matter differences in bipolar disorder: a meta-analysis of voxel-based morphometry studies. Bipolar Disord 14:135–145
Serretti A, Fabbri C (2013) Shared genetics among major psychiatric disorders. Lancet 381:1339–1341
Siever LJ (2008) Neurobiology of aggression and violence. Am J Psychiatry 165:429–442
Siever LJ, Weinstein LN (2009) The neurobiology of personality disorders: implications for psychoanalysis. J Am Psychoanal Assoc 57:361–398
Smoller JW (2013) Disorders and borders: psychiatric genetics and nosology. Am J Med Genet B Neuropsychiatry Genet. doi:10.1002/ajmg.b.32174
Soloff P, Nutche J, Goradia D, Diwadkar V (2008) Structural brain abnormalities in borderline personality disorder: a voxel-based morphometry study. Psychiatry Res 164:223–236
Stanley B, Siever LJ (2010) The interpersonal dimension of borderline personality disorder: toward a neuropeptide model. Am J Psychiatry 167:24–39
Stein D (2009) Borderline personality disorder: toward integration. CNS Spectr 14:352–356
Stein M (1997) Hippocampal volume in women victimized by childhood sexual abuse. Psychol Med 27:951–959
Stern A (1938) Psychoanalytic investigation and therapy in the borderline group of neuroses. Psychoanal Q 7:467–489
Stone MH (1980) The borderline syndromes. McGraw Hill, New York
Stone MH (1981) Borderline syndromes: a consideration of subtypes and an overview. Psychiatric Clin N Am 4(1):3–24
Stone MH (1990a) The fate of borderline patients: successful outcome and psychiatric practice. Guilford, New York
Stone MH (1990b) Incest in borderlines. In: Kluft R (ed) Incest-related syndromes of adult psychopathology. American Psychiatric, Washington DC, pp 183–204
Stone MH, Unwin A, Beacham B, Swenson C (1988) Incest in borderlines: its frequency and impact. Int J Fam Psychiatry 9:277–293
Strakowski SM, Adler CM, Almeida J, Altschuler LL, Blumberg HP, Chang KD, DelBello MP, Frangou S, McIntosh A, Phillips ML, Sussman JE, Townsend JD (2012) The functional neuroanatomy of bipolar disorder: a consensus model. Bipolar Disord 14:313–325

Straus MA, Gelles RJ (1992) Physical violence in American families, 2nd edn. Transaction, New Brunswick

Tebartz van Elst L, Ludaescher P, Thiel T, Büchert M, Hesslinger B, Bohus M, Rüsch N, Hennig J, Ebert D, Lieb K (2007) Evidence of disturbed amygdalar energy metabolism in patients with borderline personality disorder. Neurosci Lett 417:36–41

Teicher MH, Anderson SL, Polcari A, Anderson CM, Nvalta CP, Kim DM (2003) The neurobiological consequences of early stress and childhood maltreatment. Neurosci Biobehav Rev 27:33–44

Vacheron-Trystram MN, Braitman A, Cheref S, Auffray L (2004) Antipsychotics in bipolar disorders. Encéphale 30:417–424

Vizueta N, Rudie JD, Townsend JD, Torrisi S, Moody TD, Bookheimer SY, Altschuler LL (2012) Regional fMRI hypoactivation and altered functional connectivity during emotion processing in nonmedicated depressed patients with bipolar-II disorder. Am J Psychiatry 169:831–840

Weninger G, Lange C, Sachsse U, Irle E (2011) Reduced amygdala and hippocampal size in trauma-exposed women with borderline personality disorder and without posttraumatic stress disorder. J Psychiatry Neurosci 34:383–388

Whalley HC, Papmeyer M, Sprooten E, Lawrie SM, Sussman JE, McIntosh AM (2012) Review of functional magnetic resonance imaging studies comparing bipolar disorder with schizophrenia. Bipolar Disord 14:411–431

Wingenfeld K, Spitzer C, Rullkötter N, Löwe B (2010) Borderline personality disorder: hypothalamus-pituitary-adrenal axis and findings from neuroimaging studies. Psychoneuroendocrinology 35:154–170

Zanarini MC, Frankenburg FR, Dubo ED et al (1988) Axis I comorbidity of borderline personality disorder. Am J Psychiatry 155:1733–1739

Zetzsche T, Preuss UW, Bondy B, Frodl T, Zill P, Schmitt G, Koutsouleris N, Rujescu D, Born C, Reiser M, Möller H-J, Meisenzahl EM (2007) 5-HT-1areceprot gene C-1019G polymorphism and amygdala volume in borderline personality disorder. Genes Brain Behav 7:306–313

Zilboorg Z (1941) Ambulatoiry schizophrenia. Psychiatry 4:149–155

Physiological Correlates of Bipolar Spectrum Disorders and their Treatment

Tim Outhred, Andrew H. Kemp and Gin S. Malhi

Abstract Bipolar spectrum disorders (BSDs) are associated with great personal and socioeconomic burden, with patients often facing a delay in detection, misdiagnosis when detected, and a trial-and-error approach to finding the most appropriate treatment. Therefore, improvement in the assessment and management of patients with BSDs is critical. Should valid physiological measures for BSDs be identified and implemented, significant clinical improvements are likely to be realized. This chapter reviews the physiological correlates of BSDs and treatment,

For submission to *Psychophysiology in Psychiatry and Psychopharmacology* Current Topics in Behavioral Neuroscience

T. Outhred · A. H. Kemp · G. S. Malhi
Discipline of Psychiatry, Sydney Medical School, University of Sydney,
Royal North Shore Hospital, Sydney, NSW 2065, Australia

T. Outhred · A. H. Kemp
SCAN Research and Teaching Unit, School of Psychology, University of Sydney,
Sydney, NSW 2008, Australia

A. H. Kemp · G. S. Malhi
CADE Clinic, Department of Psychiatry, Royal North Shore Hospital, Sydney,
NSW 2065, Australia

A. H. Kemp · G. S. Malhi
Advanced Research and Clinical Highfield Imaging (ARCHI),
University of Sydney Royal North Shore Hospital, Sydney,
NSW 2065, Australia

A. H. Kemp
Centro de Pesquisa Clínica e Epidemiológica, Hospital Universitário,
Universidade de São Paulo, Av Lineu Prestes
2565, São Paulo SP 05508-000, Brazil

G. S. Malhi (✉)
CADE Clinic, Level 3, Royal North Shore Hospital, Main Hospital Building,
St Leonards, NSW 2065, Australia
e-mail: gin.malhi@sydney.edu.au

Curr Topics Behav Neurosci (2014) 21: 47–102

DOI: 10.1007/7854_2014_297

Published Online: 22 May 2014

and in doing so, examines the neuroimaging, electroencephalogram, and event-related potential, and peripheral physiological correlates that both characterize and differentiate BSDs and their response to treatment. Key correlates of BSDs involve underlying disturbances in prefrontal and limbic network neural activity, early neural processing, and within the autonomic nervous system. These changes appear to be mood-related and can be normalized with treatment. We adopt an "embodied" perspective and propose a novel, working framework that takes into account embodied psychophysiological mechanisms in which the physiological correlates of BSD are integrated. This approach may in time provide the objective physiological measures needed to improve assessment and decision making when treating patients with BSDs. Future research with integrative, multimodal measures is likely to yield potential applications for physiological measures of BSD that correlate closely with diagnosis and treatment.

Keywords Bipolar disorder · Bipolar spectrum disorders · Physiology · Psychophysiology · Treatment · Heart rate and its variability · EEG · FMRI · GSR · Embodied cognition

Contents

1 Introduction

Determining the physiological correlates of psychiatric disorders and assessing the potential clinical utility of physiological measures are major pursuits in biological psychiatry research (Insel et al. 2010, 2013). The impetus for such research is the promise of objective measures for assessment and predicting response to treatments, which would improve diagnostic validity and inform treatment selection (Insel et al. 2010, 2013). To facilitate this goal, the National Institute of Mental Health has proposed Research Domain Criteria (or RDoC), which provide researchers with a novel framework in which research can be presented (Morris and Cuthbert 2012). In this context, physiological measures provide an important tool through which diagnosis could be improved and treatment options tailored. Currently, the assessment and management of bipolar spectrum disorders (BSDs) often entails misdiagnosis following a significant delay in detection (Hirschfeld et al. 2003; Suppes et al. 2001), the inability to predict course (Crowe et al. 2012; Malhi et al. 2012, 2013), trial and error in treatment selection (Malhi et al. 2009; Sachs 2013), and a failure to implement long-term management strategies (Keck 2006; Malhi et al. 2009). Therefore, improvement in the assessment and treatment of those with BSDs is critical given the great personal (Bonnín et al. 2012; Merikangas et al. 2007) and socioeconomic burden associated with BSDs (Merikangas et al. 2007). Critically, these disorders are associated with the highest risk of suicide of any mental disorder (Nock et al. 2009), highlighting the need for early and accurate detection with improved diagnosis and a more personalized approach to effective treatment. Initiation of successful treatment early in the course of the disorder will undoubtedly reduce morbidity (Baldessarini et al. 2003; Post et al. 2010) and improve treatment outcomes (Berk et al. 2011; Ketter et al. 2006; Malhi 2012). Given the potential for these needs to be met with the translation of physiological measures into clinical practice (Morris and Cuthbert 2012), identification of physiological markers that could be employed in assessment and treatment selection remains an ambitious but worthy goal.

BSDs represent a cluster of disorders characterized by extreme changes in mood (Malhi et al. 2012). Depression, mania, hypomania, and euthymia (periods of remission) are phases of illness that are subjectively experienced by patients, and objectively determined by clinicians (Tohen et al. 2009). These include cyclothymia, Bipolar I disorder, Bipolar II disorder, and Bipolar Disorder Not Elsewhere Classified (NEC) and are partitioned from major depression on the basis of cycling into mood elevation such as (hypo)mania. The spectrum is conceptualized as increasing in severity and burden from Bipolar Disorder NEC, Cyclothymia, Bipolar II Disorder, through to Bipolar I Disorder (Merikangas et al. 2007), but in reality, this does not always hold true. Currently, there are no physiological tests that can be employed to assist with the detection, assessment, and diagnosis of BSDs. The physiological correlates implicated in cognitive and emotional disturbances underlying BSDs and the different phases of illness have thus far been investigated using neuroimaging and peripheral physiology techniques. Studies have revealed disturbances in prefrontal and limbic network neutral activity (see Strakowski et al. 2012), neural activity states and early neural processing (see Degabriele and Lagopoulos 2009), and within the autonomic nervous system (ANS; Gruber et al. 2011; Lee et al. 2012), respectively. Understandably, most studies have considered these neural and autonomic activation characteristics separately. Hence we now consider the neural and autonomic characteristics of BSDs in the context of an embodied (see Craig 2009; Niedenthal 2007; Price et al. 2011) disturbance affecting both the brain and body so as to develop a novel framework (see Fig. 1) in which the physiological correlates of the disorder can be investigated.

In addition to characterizing the physiological correlates of the BSDs, we further characterize the correlates—or markers—of treatment effect, and predictors of response to treatment. Patients with BSDs are typically prescribed pharmacological treatment (including lithium, antipsychotics [e.g., olanzapine, risperidone, quetiapine], and anticonvulsants [e.g., valproate, carbazepine lamotrigine)] either alone or in combination) but first-line treatment is often ineffective (Malhi et al. 2009) reflecting the fact that information gained from clinical assessment alone is insufficient for planning and implementing treatment. If physiological measures could anticipate treatment efficacy, then the trial and error involved in first-line treatment strategies may be diminished or subsided altogether. Therefore, the need to identify potential physiological markers of BSDs and to ascertain their clinical utility is imperative.

2 Measuring the Physiological Correlates of Bipolar Spectrum Disorders

Studies on BSDs have most often utilized neuroimaging techniques such as functional magnetic resonance imaging (fMRI), positron emission tomography (PET), and single-photon emission computed tomography (SPECT) and findings from studies using these neuroimaging methods have been largely consistent (Strakowski et al. 2005). The next most employed measures have been electroencephalography (EEG) and event-related potential (ERP). While neuroimaging methods enable localization of regional responses to cognitive and emotional tasks across the whole brain (Friston et al. 1998), EEG and ERP have higher temporal resolution such that fluctuations in neural activity states (Davidson 1998, 2004) and early information processing can be examined (Donchin and Coles 1988), respectively. In terms of peripheral physiological measures, cardiovascular measures including heart rate (HR) and heart rate variability (HRV) gathered using electrocardiography (ECG), and galvanic skin response (GSR) have been widely employed. Within these, HR is a measure of overall ANS arousal (see Duschek et al. 2013; Lopes and White 2006) under tonic inhibitory control by the parasympathetic nervous system (PNS; Saul 1990; Thayer et al. 2009), high frequency HRV measures PNS activity (see Duschek et al. 2013; Lopes and White 2006), and galvanic skin response (GSR) measures reflect sympathetic nervous system (SNS) activity (see Dawson et al. 2007). Notably, only GSR has been employed in combination with other measures for multimodal investigations. Building toward a more embodied perspective of BSDs, we delineate, examine, and then integrate the cognitive neuropsychological (brain; fMRI and EEG, and ERP) "neurocorrelates" and the peripheral physiological (body; HR and HRV, and GSR) correlates.

A methodological issue that has been discussed in most physiological studies of BSDs is heterogeneity with respect to previous and current treatment effects, clinical course, and comorbidities. Though it is often not feasible to exclude or stratify patients on all these bases, future research should report and, where possible, take these factors into account. Another self-evident issue that has challenged researchers is the difficulty in recruiting and testing manic bipolar patients in laboratory settings, explaining why there are relatively fewer studies of this phase of illness (Small et al. 1999). With technological advances, future research will likely take advantage of less invasive, ambulatory sensors including smartphone (Heathers 2013) and sensorized clothing (e.g., Mariani et al. 2012; Quintana et al. 2012; Siegel 2013), which will enable the collection of longitudinal physiological data across illness and treatment phases within BSD patients.

3 Neurocorrelates of Bipolar Spectrum Disorders and Their Treatment

3.1 Functional Neuroimaging Correlates

3.1.1 Characterization and Differentiation

Characterization

Characterization of the functional neuroanatomy of BSDs has been extensive, systematic, and consistent for the last two decades (see Table 1). The results and conclusions obtained across neuroimaging modalities—including fMRI, PET, and SPECT—have been largely consistent and non-modality specific (see Strakowski et al. 2005). Overall, BSDs display prefrontal cortex (PFC) hypoactivity and limbic hyperactivity during emotional and cognitive tasks, and these findings correlate with trait and state emotional lability and mood disturbances in BSDs (see Strakowski et al. 2012). In addition, bipolar disorder is characterized by dysfunctional connectivity among ventral prefrontal networks and limbic brain regions, particularly the amygdala (Blond et al. 2012; Chen et al. 2011; Houenou et al. 2011; Strakowski et al. 2012; Townsend et al. 2012) indicating both difficulty in regulating mood alongside a dysfunction of emotion processing. Impaired PFC regulation subsequently leads to a loss of neurological emotional homeostasis, emotional lability, and mood disturbances (Strakowski et al. 2012). It is posited that a disruption of frontal regulatory networks allows for extreme mood states, switching among mood states, and mixed states (Strakowski et al. 2012). These abnormalities have been conceptualized as dysfunction within oscillatory mechanisms, which perhaps worsen over time, and result in the many manifestations of the illness (Schneider et al. 2012).

Interestingly, bipolar patients have consistently decreased frontal activation across the ventrolateral PFC (VLPFC), a region critical for emotional processing and mood regulation (Blond et al. 2012; Chen et al. 2011; Houenou et al. 2011;

Table 1 Meta-review of fMRI assessment in BSD characterization and differentiation

Study	Subjects	Medication	Stimuli	Char/Diff	Principle findings	Interpretation
Chen et al. (2011)	fMRI meta-analysis of 65 adult studies Bipolar patients, $n = 1,040$ Controls, $n = 1,074$	Yes, no test of effects	Emotional tasks Cognitive tasks	✔/✔	Bipolar patients had decreased IFG and putamen and increased parahippocampus, hippocampus, and amygdala and basal ganglia activation, relative to controls. Decreased IFG activity was seen in both cognitive and emotional processing, while increased limbic activation was seen in emotional processing. Inferior frontal activity was decreased in manic but not in euthymic and depressed states. Limbic activation increases were not associated with mood states	Bipolar disorder can be characterized by abnormal frontal-limbic activation Manic state can be differentiated from depressed and euthymic states by decreased IFG activity Bipolar disorder fMRI studies are largely consistent

(continued)

Table 1 (continued)

Study	Subjects	Medication	Stimuli	Char/Diff	Principle findings	Interpretation
Houenou et al. (2011)	fMRI meta-analysis of 13 emotion processing studies Bipolar versus control Bipolar patients, $n = 130$ Controls, $n = 141$ Euthymic versus control Bipolar patients, $n = 68$ Controls, $n = 73$ Manic versus control Bipolar patients, $n = 35$ Controls, $n = 40$	Yes, no test of effects	Emotion processing tasks Explicit and implicit affect recognition tasks Emotional go/no-go Emotional Stroop task Emotional Sternberg memory task Emotional face-matching paradigm	✔/✔	Bipolar patients had increased parahippocampus, amygdala, caudate, thalamus, and right MFG activation, and decreased right IFG, right precuneus, right MFG, right thalamus and right cerebellum activation, relative to controls Euthymic patients had increased L parahippocampus and amygdala activation, and decreased right precuneus, right thalamus and right cerebellum activation, compared to controls Manic patients had increased parahippocampus, amygdala, L thalamus, and right MFG activation, and decreased right IFG activation, compared to controls	Bipolar disorder can be characterized by dysfunction in the emotional networks involved in voluntary regulation and cognitive control of emotion. However, the specificity of these network characteristics to bipolar disorder versus major depressive disorder is unknown

(continued)

Table 1 (continued)

Study	Subjects	Medication	Stimuli	Char/ Diff	Principle findings	Interpretation
Townsend et al. (2012)	Adult bipolar patients—mania, depression, euthymia	Yes, tested for effects. Some studies test for effects No effects Few studies have used medication-free subjects, and have results consistent with other studies	Emotion activation tasks Emotion processing tasks Emotion regulation task Stimuli: Faces Pictures Emotional Go/No-go words Emotional Stroop Emotional distracters	✔/✔	Amygdala activation varies as a function of mood state. Hyperactivity in mania, valence-related activation in depression, and normal activation in euthymia VLPFC is hypoactivated across mood states	Emotional dysregulation and lability in mania and depression may reflect disruption of a frontal-limbic functional network. There is not enough evidence for classifying state using fMRI

(continued)

Table 1 (continued)

Study	Subjects	Medication	Stimuli	Char/Diff	Principle findings	Interpretation
Blond et al. (2012)	Adult bipolar patients—mania, depression Euthymia Adolescent samples	Yes, tested for effects Despite medication differences within and between studies, there is still convergence of findings	Resting state Emotion activation tasks with Faces or Pictures Go/No-go tasks Stroop Working memory	✔/✔	Dysfunctional connectivity between the amygdala and the ventral PFC, insula, and temporopolar cortex have been shown in bipolar disorder across mood states, suggesting that they may be trait features of the disorder that are developmental	Bipolar disorder may be characterized by dysfunction in the amygdala–anterior paralimbic neural system. Connectivity within the system is highlighted key to state changes, and may be key to bipolar disorder development. These abnormalities may distinguish bipolar from major depressive disorder. There has been no work to attempt to classify based on this circuit

(continued)

Table 1 (continued)

Study	Subjects	Medication	Stimuli	Char/ Diff	Principle findings	Interpretation
Schneider et al. (2012)	Adult bipolar patients Adolescent samples "At risk" adolescent samples	Yes, some studies test for medication effects. Some effects may be related to medication exposure	*N*-back Parametric sentence completion Stroop tasks Emotional processing tasks with faces or pictures Working memory Go/No-go tasks Reward tasks Dominant hand motor task	✔/×	Studies suggest a pattern of abnormalities in neural development early in the appearance of bipolar disorder that gives way to progressive neuropathic changes at least influenced by the course of illness leading to an iterative process in which functional changes drive clinical symptoms and are in turn exacerbated by the consequences of these symptoms	The few studies characterizing developmental aspects of functional neuroanatomy involved in bipolar disorder are limited There have been no longitudinal studies of functional abnormalities in bipolar disorder

(continued)

Table 1 (continued)

Study	Subjects	Medication	Stimuli	Char/ Diff	Principle findings	Interpretation
Whalley et al. (2012)	Adult bipolar patients compared against Adults schizophrenia patients	Yes, no test of medication effects	Emotional prosody Reward task Associative memory task Emotional memory Verbal fluency Verbal tasks Working memory Comprehension task Auditory oddball, ICA, and temporal network Resting state, default mode	✔/✔	Over-activation in the medial temporal lobe and associated structures was found in bipolar relative to schizophrenia patients in tasks involving emotion or memory. Differences between the disorders in prefrontal regions were not as consistent. Differentiation of diagnosis with fMRI is less accurate in bipolar than schizophrenia patients. Few studies that report symptom associations implicate limbic regions with manic symptoms	Limited number of studies to show ability to classify bipolar and schizophrenia using fMRI

(continued)

Table 1 (continued)

Study	Subjects	Medication	Stimuli	Char/ Diff	Principle findings	Interpretation
Strakowski et al. (2012)	A consensus view on the functional neuroanatomy bipolar disorder based on reviews conducted by separate groups	Bipolar disorder is characterized by dysfunctional connectivity among ventral prefrontal networks and limbic brain regions, particularly the amygdala. Dysfunction is associated the abnormal development of the component brain regions. Dysregulation of the limbic brain then leads to loss of emotional homeostasis, resulting in mood instability. Due to disruption of networks that can restore this homeostasis, extreme mood states, switching among mood states, and mixed states occur as different unregulated systems oscillate in the absence of homeostatic control. During euthymia, recovery of prefrontal control, along with compensation from other brain regions, temporarily restores homeostasis. However, the underlying functional abnormalities leave the bipolar disorder individual at risk for disruption of this fragile homeostasis under even minor stress				With more evidence, fMRI may be able to discriminate bipolar disorder states and other diagnoses based on the function of these networks
Hajek et al. (2013)	fMRI meta-analysis of 30 response inhibition studies Bipolar patients, $n = 635$ Controls, $n = 667$	Yes, no test of effects in most of the studies	Response inhibition tasks Go/no go Stop signal response task Stroop test Hayling Sentence Completion task	✔/✔	Bipolar patients had decreased R IFG activation regardless of current mood state, relative to controls. Euthymic patients had increased L superior temporal and right MFG and decreased basal ganglia activation. Manic patients had increased bilateral basal ganglia activation and reduced right inferior and MFG activation. Changes in euthymia were associated with normal cognitive performance, whereas manic patients had more errors during response inhibition	Bipolar disorder is characterized by decreased response inhibition and associated R IFG hypoactivation. Euthymic patients subjects may compensate with hyperactivations of adjacent cortical areas, yielding comparable performance in inhibitory functions

Note Char/Diff = the review differentiates between studies that attempt to (1) characterize BSDs (✔/×) or (2) characterize and, differentiate BSDs or differentiate BSDs from other diagnoses (✔/✔); *L* left; *R* right

Strakowski et al. 2012; Townsend and Altshuler 2012), and consistently decreased inferior frontal gryus (IFG) activity, specifically the right IFG (R IFG), a region associated with regulatory inhibition (Hajek et al. 2013). Furthermore, bipolar patients have also increased activity within limbic regions including parahippocampus, hippocampus, and amygdala and basal ganglia (Blond et al. 2012; Chen et al. 2011; Houenou et al. 2011; Strakowski et al. 2012; Townsend and Altshuler 2012), which may underpin abnormalities of primary emotion processing. Hence, decreased IFG activity in bipolar disorder which is seen during both cognitive and emotional processing, and increased limbic activation that is seen during emotional processing (Chen et al. 2011; Hajek et al. 2013; Houenou et al. 2011; Strakowski et al. 2012) may reflect trait-based correlates of BSDs.

With respect to the manic phase of bipolar disorder, IFG activity is decreased in mania but not in euthymic or depressed states, and limbic activation increases are not associated with mood states (Chen et al. 2011; Houenou et al. 2011; Strakowski et al. 2012). However, amygdala activation varies as a function of mood state and the valence of the emotional stimuli: hyperactivity to emotional stimuli in mania; hyperactivity to negative stimuli and hypoactivity to positive stimuli in depression, and normalized activations in euthymia (Townsend and Altshuler 2012). However, it is important to note that many of these findings are preliminary and may be contingent on additional factors such as the tasks used and disorder phenotype, but they do suggest that manic and depressed phases of bipolar disorder can be differentiated on the basis of altered IFG activity and valence-mood congruent activation of the amygdala (Strakowski et al. 2012; Townsend and Altshuler 2012). With respect to response inhibition, the manic phase is associated with reduced performance, associated with decreased R IFG and medial frontal gyrus (MFG) activation and increased bilateral basal ganglia activation (Hajek et al. 2013; Houenou et al. 2011; Strakowski et al. 2012). In the euthymic phase, response inhibition is not dysfunctional, although activity in left superior temporal and right MFG is increased and basal ganglia activation decreased (Hajek et al. 2013; Houenou et al. 2011; Strakowski et al. 2012). Therefore, euthymic patients compensate for reduced inhibitory IFG activity with increased activation of adjacent cortical areas, thereby yielding normalized inhibitory functions (Hajek et al. 2013). During euthymia, recovery of frontal control, along with compensation from other brain regions, temporarily restores neurological emotional homeostasis (Strakowski et al. 2012). However, the underlying functional abnormalities in the VLPFC networks leave the risk for emotional and cognitive disruption, leading to manic, depressed, or mixed phases, even under minor stress (Strakowski et al. 2012).

Differentiation

Over-activation in the medial temporal lobe during tasks involving emotion or memory may differentiate patients with bipolar disorder from patients with schizophrenia (Whalley et al. 2012). However, differential diagnosis with fMRI has been less accurate with bipolar disorder than schizophrenia (Whalley et al. 2012).

While promising preliminary findings have been reported suggesting that BSDs may be distinguished from unipolar major depression in small samples (Diler et al. 2013; Grotegerd et al. 2013; Marchand et al. 2013), large-scale studies are needed to determine the sensitivity and specificity of these findings.

3.1.2 Treatment

Effective treatment would be expected to normalize the state-based and trait-based VLPFC-limbic network disturbances correlated with BSDs using fMRI. In comparison to the fMRI correlates characterizing and differentiating BSDs, fMRI correlates of treatment are understudied. In the last decade, eight controlled studies examining the impacts of treatment administration in BSDs on cognitive and emotional stimuli have been published, with each demonstrating some sort of normalization effect (Table 2). Lithium appears to have prophylactic effects on cognition after 14 days' treatment and acts on frontal regions in the euthymic phase of BSDs with little impact during the depressed phase (Silverstone et al. 2005). After 12 weeks of lamotrigine administration during the euthymic phase, there are increases in the prefrontal and cingulate regions, thereby normalizing the activity of circuitry involved in emotion regulation (Haldane et al. 2008; Jogia et al. 2008). In the depressed phase, 8 weeks of lamotrigine administration reduces amygdala reactivity to negative stimuli, with greater reductions in reactivity being correlated with reductions in depression symptoms after 8 weeks (Chang et al. 2008). When patients are given a 4-week course of antipsychotics and then a 14-week course of lamotrigine, decreases in mania symptoms following treatment are associated with increased VLPFC and dorsolateral (DLPFC) activity during cognitive-emotional (Pavuluri et al. 2010b) and response inhibition tasks (Pavuluri et al. 2010a). In subsyndromal patients, there are no consistent differences after 12 weeks of valproate treatment (Chang et al. 2009). Finally, a study investigating the effect of psychotherapy showed normalization of IFG hypoactivity after 12 weekly sessions (Favre et al. 2013); however, it was difficult to differentiate the effect of psychotherapy from improvement with the medication patients were already receiving (Favre et al. 2013).

In sum, there appears to be normalization of the cognitive and emotional neural networks implicated in BSDs with treatment when patients present in the depressed or manic phases of illness. Additionally, prophylactic treatment appears to affect these networks. However, the majority of investigations thus far have small sample sizes (less than 20 patients) and many did not have a control group for comparison (e.g., Chang et al. 2008; Haldane et al. 2008). Additionally, some studies investigated adolescents (e.g., Chang et al. 2008, 2009; Pavuluri et al. 2010a, b), who are likely to have fundamentally different responses compared to adults. Nevertheless, there are promising findings from this relatively new line of research that should encourage future research with larger samples, across different treatments. In doing so, clinically useful fMRI treatment markers for predicting treatment response and treatment monitoring may be determined.

Table 2 Literature review findings of fMRI assessment in BSD treatment

Study	Modality measure	Subjects	Treatment	Principal findings	Interpretation	Comment
Silverstone et al. (2005)	fMRI 1.5 Tesla BOLD Working memory task Word generation task	5 bipolar disorder patients—depressed Bipolar II, $n = 3$ Taking other medication, $n = 2$ Five bipolar disorder—euthymic Bipolar II, $n = 4$ Taking antidepressants, $n = 2$	Lithium, 14 days	After lithium, euthymic bipolar patients had decreased L precentral, Broca's aera, and supplementary motor area activation during the word generation task. No change in the depressed group. No group effect for the working memory task	Lithium has effects on brain activation that are task, region, and state-dependent	
Haldane et al. (2008)	fMRI 1.5 Tesla BOLD *N*-back task Angry Facial Affect Recognition Task	12 bipolar disorder—euthymic Titrated off medications	Lamotrigine, 12 weeks	After lamotrigine, patients had increased activation in the superior and medial PFC and cingulate gyrus during both tasks	Lamotrigine may impact the neural circuits involved in memory and emotion regulation	No control group to show that practice effects did not cause the changes
Jogia et al. (2008)	fMRI 1.5 Tesla BOLD Sad facial affect recognition task	12 bipolar disorder—euthymic Titrated off medications 12 controls	Lamotrigine, 12 weeks	At baseline, patients had increased temporal and dorsomedial and right VLPFC, and dorsal cingulate gyrus, relative to controls. After lamorigine, patients had reduced temporal and increased prefrontal activation	Lamotrigine may normalize the dysfunctional circuits involved in affect recognition	No treated control group to show that, perhaps differential, practice effects in the patients did not cause the changes

(continued)

Table 2 (continued)

Study	Modalitymeasure	Subjects	Treatment	Principal findings	Interpretation	Comment
Chang et al. (2008)	fMRI 3 Tesla BOLD Emotional pictures task Negative Neutral	8 bipolar patients—depressed, adolescents (aged 13–17) Bipolar I, Bipolar II, Bipolar NOS, Previous medication was tapered and discontinued for 2–4 week washout. Other mood stabilizers, antipsychotics, and ADHD treatments with no dose changes within 1 month. Did not test for effects	Lamotrigine, 8 weeks	After lamotrigine, clinical improvement in depression was correlated with decreased right amygdala activation with treatment. After lamotrigine, depression symptoms were positively correlated with bilateral amygdala activation. DLPFC activation was not correlated with clinical improvement in depression	Lamotrigine reduces reactivity to negative stimuli, which is related to clinical improvement in depression	No control group to show that practice effects did not cause the changes
Chang et al. (2009)	fMRI 3 Tesla BOLD Emotional pictures task Negative Positive Neutral (also sMRI, spectroscopy)	Six subsyndromal children, with parent with bipolar disorder (aged 9–18) Two patients previously medicated Antidepressant, $n = 1$ Stimulant, $n = 1$ Five controls	Valproate semisodium, 12 weeks	There were no differences in amygdala or DLPFC activation between groups at baseline or after valproate treatment. Clinical improvement in depression was positively correlated with decreased DLPFC activation	Clinical change may predate neurobiological change that was detectable by the methods employed	Small sample. Perhaps there is large heterogeneity in subsyndromal groups

(continued)

Table 2 (continued)

Study	Modalitymeasure	Subjects	Treatment	Principal findings	Interpretation	Comment
Pavuluri et al. (2010b)	fMRI 3 Tesla BOLD Affective color matching task	17 bipolar disorder patients—manic, hypomanic, or mixed, adolescent (aged 10–18) Unmedicated 14 controls	Initially treated with atypical psychotics, four weeks Lamotrigine, 14 weeks	After treatment, the VLPFC and DLPFC hypoactivity between patients and controls was partially normalized. In the patient group, mania symptom improvement was correlated with the increased VLPFC activity	Pharmacotherapy results in amelioration of cognitive and affective circuitry dysfunction, which improves symptoms of mania	Cannot differentiate the impacts of the two different medications. No treated control group to show that, perhaps differential, practice effects in the patients did not cause the changes
Pavuluri et al. (2010a)	fMRI 3 Tesla BOLD Go/No-go Task	17 bipolar disorder patients—manic, hypomanic, or mixed, adolescent (aged 10–18) Unmedicated 14 controls	Initially treated with atypical psychotics, four weeks Lamotrigine, 14 weeks	After treatment, the prefrontal hypoactivity between patients and controls was partially normalized. In the patient group, mania symptom improvement was correlated with the increased VLPFC activity	Pharmacotherapy normalizes dysfunction in cognitive circuitry function supporting voluntary behavioral inhibition	Cannot differentiate the impacts of the two different medications. No treated control group to show that, perhaps differential, practice effects in the patients did not cause the changes
Favre et al. (2013)	fMRI 3 Tesla BOLD Word-Face emotional Stroop Task	16 bipolar patients—euthymic Bipolar I, $n = 11$ Bipolar II, $n = 5$ Lithium, $n = 9$ Anticonvulsants, $n = 10$ Antidepressant, $n = 8$ Atypical antipsychotics, $n = 1$ Drug-free, $n = 1$ Medication was stable between phases. Did not test for effects 16 controls	Psychoeducation therapy, including stress and symptoms management, weekly session for 12 weeks with a psychiatrist and a psychologist	After treatment, the inferior frontal gyrus hypoactivity between patients and controls was partially normalized	Treatment improved dysfunctional cognitive control and emotional regulation mechanisms	Specificity of psychoeducation-related improvement, particularly in relation to medication effects, should be evaluated by comparing it to non-structured interventions

Note L left; *R* right

3.2 EEG and ERP Correlates

3.2.1 Characterization and Differentiation

Within the EEG and ERP literature, there has been one systematic review on the correlates that characterize and differentiate BSDs (Degabriele and Lagopoulos 2009). Here, we updated this review (Table 3). There have been 22 studies examining the BSD characteristics with EEG and ERP measures. Overall, there appears to be measurable correlates in frequency band, ERP component, and sleep EEG characteristics (see Degabriele and Lagopoulos 2009), and network properties (Kam et al. 2013; Kim et al. 2013) that can characterize and differentiate BSD phases and BSDs from unipolar depression and schizophrenia. Single electrode EEG (see Iacono et al. 1983), clinical EEGs (see Cook et al. 1986; Small et al. 1999), and associating EEG data with neuroanatomical abnormalities from computerized tomography images (see Dewan et al. 1988) are yet to produce in measurable correlates characterizing and differentiating BSDs.

BSD studies employing EEG data show that differential power at specific frequency bands, which are associated with different activity states, are correlated with traits and states of BSDs. Specifically, differential activity in the alpha band between the frontal lobe hemispheres, frontal alpha asymmetry—an index associated with behavioral motivation (Davidson 1998, 2004)—correlates with BSD phases. In the depressed phase, increased right-dominant, withdrawal-related, frontal alpha asymmetry, relative to controls, is characteristic at rest (Nusslock et al. 2012). Additionally, bipolar disorder patients with decreased functional network integration and decreased optimal balance of network segregation in functional fronto-central and centro-parietal networks had higher depression scores (see Kim et al. 2013). In the manic phase, various frequency characteristics can be observed during rest (see Clementz et al. 1994; Kano et al. 1992), with increased 'busy thinking' related, beta activity correlating with increased mania symptoms (Kam et al. 2013). Increased left-dominant, goal striving frontal alpha asymmetry, relative to controls, appears to be characteristic of mania (Harmon-Jones et al. 2008; Nusslock et al. 2012), opposing the activity characterizing the depressed phase. In the hypomanic phase, increased left-dominant frontal alpha asymmetry at rest is also observed (Harmon-Jones et al. 2008; Nusslock et al. 2012), an effect that also correlates with hypomanic personality (Peterson and Harmon-Jones 2008; Wyczesany et al. 2010). In the euthymic phase, bipolar patients appear to have more normalized frequency characteristics and frontal alpha asymmetry relative to controls (Nusslock et al. 2012), though some residual frequency characteristics remain suggesting some trait-based cognitive dysfunction (El-Badri et al. 2001).

ERP components, such as the commonly reported P300, provide an opportunity to determine whether early information-processing is impaired in patient samples (see Degabriele and Lagopoulos 2009; Donchin and Coles 1988; Kemp et al. 2009). In studies with bipolar patients in no specific phase, results show ERP

Table 3 Literature review findings of EEG assessment in BSD characterization and differentiation

Study	Modality measure	Subjects	Medication	Stimuli	Char/ Diff	Principal findings	Interpretation	Comment
Iacono et al. (1983)	Single Au electrode EEG at Cz site Phi, Alpha, Kappa (also cardiac measures and GSR, reported in Tables 1 and 7, respectively)	24 bipolar affective disorder—euthymic 26 Unipolar 46 controls	Yes, tested for effects Imipramine only, not lithium for at least one month, $n = 13$ Lithium only, not imipramine for at least one month, $n = 6$ Imipramine + Lithium = 26 Drug-free for at least 3 months, $n = 5$	Resting state, 5 min, eyes closed	✔/✔	No patient group differences. No impact of medication	Euthymic bipolar patients cannot be differentiated from other groups using EEG	No spatial resolution from one Cz electrode
Cook et al. (1986)	16-channel EEG "EEG abnormalities"	23 bipolar patients with 'abnormal' EEGs 23 bipolar patients with 'normal' EEGs	Unmedicated, 48 h washout period	Clinical EEG	✔/×	"EEG abnormalities" were related to less family history of bipolar disorder	Aetiology of bipolar disorder is unclear	Perhaps clinical EEG abnormalities may not characterize bipolar disorder
Dewan et al. (1988)	EEG Details not reported	26 Euthymic bipolar patients Nine with CT abnormalities 17 without CT abnormalities	Yes, no test of effects Lithium, $n = 24$ Some on carbamazepine and antipsychotics	Two periods of eyes opening and closing Hyperventilation, 3 min	✔/×	EEG could not differentiate between bipolar patients with or without CT abnormalities	CT abnormalities may have no clinical relevance to bipolar disorder	

(continued)

Table 3 (continued)

Study	Modality measure	Subjects	Medication	Stimuli	Char/ Diff	Principal findings	Interpretation	Comment
Muir et al. (1991)	Single Ag/AgCl electrode at Cz site P300	99 bipolar patients—mixed phases 48 MDD patients 32 in-patient controls 213 controls	Yes, tests of effects Bipolar patients Drug-free, $n = 28$ Neuropeptics, $n = 34$ Antidepressants, $n = 9$ Hypotics, $n = 3$ Lithium, $n = 32$	Covertly counting infrequent tones	✓/✓	P300 latency was greater in bipolar than MDD patients, and controls. No impact of medication on these effects	P300 latency might characterize bipolar disorder, and differentiate bipolar from MDD	No spatial resolution from one Cz electrode
Kano et al. (1992)	16-channel EEG Alpha1 Alpha2 Beta1	7 patients with bipolar disorder—mania 44 controls (also 21 MDD and melancholia. 16 MDD without melancholia)	Yes, no test of effects Across patient sample TCA, $n = 21$ TeCA, $n = 4$ Tranquilizers, $n = 8$ Lithium, $n = 6$ No medication, $n = 6$	Resting state, eyes closed	✓/×	In bipolar patients, Alpha1 was increased in P3 and O1 and decreased at C3. Alpha2 was decreased at F7. Beta2 was increased at F8, relative to controls	Dysfunctional relationships between frontal areas are associated with mood changes	Need to compare patients in different mood states to show that frontal dysfunction is related to mood changes
Clementz et al. (1994)	3-channel EEG Au electrodes at Cz, C3, C4 sites Delta Theta Alpha Beta1 Beta2 Beta3 Total Phi	31 first-episode bipolar patients 35 patient relatives 113 controls 42 relatives 50 first-episode schizophrenia and 55 relatives	Yes, tested for effects Bipolar patients Antipsychotic, $n = 19$ Antidepressant, $n = 4$ Anticholinergic, $n = 9$ Lithium, $n = 12$	Resting state, eyes closed	✓/✓	Bipolar patients had increased delta, increased right alpha, and theta activity compared with controls. Bipolar patients and their relatives had decreased peak alpha frequencies Bipolar patients were differentiable from schizophrenia patients on Cz delta and C4 delta, theta, and alpha	Bipolar patients have a hemispheric dominance in regulation of mood, which is differentiable from schizophrenia	Thorough analysis of medication effects

(continued)

Table 3 (continued)

Study	Modality measure	Subjects	Medication	Stimuli	Char/Diff	Principal findings	Interpretation	Comment
Koles et al. (1994)	8-channel EEG at F7, T3, T4, P3, P4, T6, T4, and F8 sites. Spatial pattern analysis	22 manic patients 33 depressed patients 113 control 31 schizophrenic patients	No. drug washouts	Resting state, eyes open and eyes closed Vocabulary and oral word Fluency Block design	✔/✔	Spatial patterns can differentiate all groups. Bipolar patients had increased left temporal activity and bilateral frontal hyperactivity The depressed group had right-sided hyperactivity and the manic group had left temporal and bilateral frontal hyperactivity	Resting state, spatial patterns rather than temporal, may provide the best diagnostic differentiation using EEG	Shows spatial resolution is important in EEG in BSDs
Souza et al. (1995)	5-channel EEG at Fz, Cz, Pz, T3, T4 sites P300	19 bipolar patients—mixed 27 controls 26 Schizophrenia	Yes, tested effects of medication Bipolar patients Drug-free, $n = 3$ Neuroleptics, $n = 4$. TCAs, $n = 3$ Benzodiazepines, $n = 1$ Lithium, $n = 8$ MAOI, $n = 2$ Carbamazepine, $n = 4$	Covertly counting target tones	✔/✔	Bipolar patients, like schizophrenia patients had increased P300 latency, but bipolar patients had greater P300 amplitude. No effects of medication	P300 may characterize dysfunction in bipolar disorder, but P300 amplitude differentiates bipolar from schizophrenia	
Small et al. (1999)	22-channel EEG Clinical EEG qEEG (previously reported and described in Table 4)	202 bipolar patients—mania	No. washout 10–14 days	Resting state Photic stimulation Hyperventilation Natural or sedative drowsiness Light Sleep Clinical EEG	✔/×	Abnormal EEGs in 16 % of patients Left-lateralized abnormalities were more common than right	Clinical EEG abnormalities are uncommon in bipolar patients	Perhaps clinical EEG is not useful in bipolar disorder. Different states should be examined

(continued)

Table 3 (continued)

Study	Modality measure	Subjects	Medication	Stimuli	Char/ Diff	Principal findings	Interpretation	Comment
El-Badri et al. (2001)	9-channel EEG Delta Theta Alpha Beta	29 euthymic bipolar patients 26 control	Yes, tested for effect of lithium Lithium, $n = 20$ Anticonvulsants, $n = 8$ Antipsychotics, $n = 11$ Antidepressants, $n = 9$ Polypharmacy, $n = 18$ With detailed patient breakdown of specific medications	Resting state, eyes closed and neurocognitive tasks	✔/✗	Bipolar patients had greater power in all wave bands than controls. Greatest differences were for right temporal theta and left occipital beta No effect of lithium Neurocognitive deficits in patients	EEG differences in visuospatial areas may related to neurocognitive deficits in bipolar disorder	
Rao et al. (2002)	Standard poly-somnography EEG	28 adolescent (mean age = 15.4) unipolar patients At 6–8 year follow-up: 26 patients 5 bipolar patients 21 unipolar patients 35 controls	No, 2-week washout.	Poly-somnography study	✔/✔	Patients that converted to bipolar disorder had greater Stage 1 sleep and REM latency and less REM density and duration at baseline than unipolar patients	EEG sleep profile may relate to clinical course of bipolar disorder	Small sample of bipolar-course patients

(continued)

Table 3 (continued)

Study	Modality measure	Subjects	Medication	Stimuli	Char/ Diff	Principal findings	Interpretation	Comment
O'Donnell et al. (2004)	32-channel Ag/AgCl EEG N100 P200 N200 P300	13 bipolar patient—manic or mixed 12 schizophrenia patients 35 controls	Yes, tested for effects Bipolar patients Atypical antipsychotics, $n = 3$ Benzodiazepines, $n = 5$ Unmedicated, $n = 5$ Schizophrenia patients Atypical antipsychotics, $n = 12$	Auditory oddball	✔/✔	Bipolar patients had reduced P300 amplitude and latency relative to controls. Schizophrenia patients had additional N100, P200, and N200 reductions	While both patient groups have attentional and working memory deficits, schizophrenia patients also have disturbances in earlier auditory processing	
Hall et al. (2007)	17-channel EEG P300 P50 Mismatch negativity	Six twin pairs concordant for bipolar disorder 10 twin pairs disconcordant for bipolar disorder 78 control twin pairs	Yes Antipsychotics + mood stabilizers and/or antidepressants, $n = 11$ Antidepressant + Mood stabilizers, $n = 3$ Mood stabilizers, $n = 2$ Antidepressant, $n = 1$ Unmedicated, $n = 5$	Auditory Oddball Conditioning-testing paradigm	✔/×	Bipolar disorder patients had reduced P300 amplitude and decreased P50 suppression. Genetic not environmental factors, accounted for variance in P300 and P50 heritability. MMN and P300 latency were not associated with bipolar disorder	P300 amplitude and P50 suppression ratio may be endophenotypes for bipolar disorder	
Harmon-Jones et al. (2008)	14-channel EEG Alpha frontal asymmetry	41 BSD patients 53 controls	Yes, tested effects. Antidepressant, $n = 7$ Two antidepressants, $n = 1$ Mood stabilizer + anxiolytic, $n = 1$ No information, $n = 2$	Goal-striving task	✔/×	Bipolar patients had greater relative left frontal activity in preparation for winning hard trials. This effect was driven by hypomanic state. No effects of medication	Relative left frontal activity may be involved in mania and triggered by challenging and potentially rewarding events	

(continued)

Table 3 (continued)

Study	Modality measure	Subjects	Medication	Stimuli	Char/ Diff	Principal findings	Interpretation	Comment
Peterson and Harmon-Jones (2008)	27-channel EEG Alpha coherence Hypomanic personality scale	36 controls	Not assessed	Squeeze ball contraction and relaxation	✔/×	During right-hand contractions, increased left mid-frontal (F3-F4) and frontal-central (Fc3-Fc4) coherences were related to increased hypomanic personality scores	Proneness to hypomania may be related to the connectivity of the left motor cortex and the left PFC, priming approach-related motor responses	
Lahera et al. (2009)	3-channel EEG Ag/AgCl electrodes at Fz, Cz, Pz sites. P300	21 euthymic bipolar patients 38 controls	Yes, did not test for effects Mood stabilizers, 29.1 % Mood stabilizers + antipsychotics, 20.8 % Other combinations, 50 %	Auditory oddball	✔/×	No differences between patients and controls on P300 latency and amplitude	Refutes the hypothesis that P300 latency reduction in bipolar disorder is state independent, and than P300 response is driven by symptoms	Did not test for mediation effects. It could be that the medication is normalizing P300
Bestelmeyer et al. (2009)	4-channel EEG Ag/AgCl electrodes at Fz, Cz, Pz, Oz electrode sites P300	19 bipolar disorder patients 21 schizophrenia patients 35 controls	Yes, did not test for effects Lithium, $n = 11$ Carbamazepine, $n = 2$ Other, $n = 6$	Auditory oddball Visual oddball	✔/✔	Auditory both bipolar and schizophrenia patients had decreased P300 amplitude at Pz, relative to controls The bipolar patients could not be differentiated from the schizophrenia patients with P300	Auditory P300 amplitude at Pz may be an endophenotype for psychosis in general rather than specifically for schizophrenia	

(continued)

Table 3 (continued)

Study	Modality measure	Subjects	Medication	Stimuli	Char/Diff	Principal findings	Interpretation	Comment
Wyczesany et al. (2010)	32-channel EEG Alpha1 Alpha2 Beta1 Beta1 Beta3 Laterality coefficients Activation Deactivation Adjective CheckList Energy-Tiredness (ET), Tension-Calmness (TC)	31 hospitalized patients 13 MDD 5 bipolar disorder—mania 13 without mood disorders	Yes, did not test for effects Did not report medications taken	Resting state, eyes open	✔/×	The high-energy group showed right shift of activity in frontocentral and posterior areas visible in alpha and beta range, respectively. High tension was related to right prefrontal dominance and right posterior activation in beta1 band. Also, overall alpha2 decrease and beta2 increase	An increase of right frontocentral cortical activation may be related to conditions of energetic arousal	
Degabriele et al. (2011)	34-channel EEG P100 N170	18 bipolar disorder patients 18 controls	Yes, no test of effects Most patients receiving combination including antipsychotics, antidepressants and anticonvulsants such as sodium valproate or lithium	Emotional face go/no-go inhibition task	✔/×	Bipolar patients had larger P100 amplitudes in response to this happy compared to sad faces. Reaction time was significantly reduced in the happy compared to sad faces. Controls did not have significant differences	Responses to positive emotion is facilitated in bipolar patients. Potentiation of positive stimuli at the early stage of emotion processing may result in bipolar mania	Need to confirm across affective states and medications

(continued)

Table 3 (continued)

Study	Modality measure	Subjects	Medication	Stimuli	Char/ Diff	Principal findings	Interpretation	Comment
Nusslock et al. (2012)	14-channel EEG Alpha Frontal alpha asymmetry at F3/F4 and F7/F8 electrode sites	58 BSD patients depressed, $n = 9$ hypomanic, $n = 21$ euthymic, $n = 21$ no data, $n = 7$ 16 converted to bipolar I after 4.7-year follow-up 59 control	Yes, controlled for medication status Medicated, 55 % Antidepressant, 50 % Lithium/Depakote, 7 % Antipsychotic, 2 % Anxiolytic, 12 %	Eyes open-eyes closed trials	✓/✓	Hypomanic patients had greater relative left F3/F4 than depressed and euthymic patients, and controls. Increased relative left F3/F4 was associated with a greater probability of converting to bipolar I disorder. Increased relative left F3/F4 classified 76 % of individuals with cyclothymia or bipolar II disorder based on whether they did or did not convert to bipolar I disorder	Mid-frontal cortical activity differentiates bipolar affective states and can predict clinical course	Unable to examine conversion to bipolar II disorder from small number of cyclothymia patients
Kim et al. (2013)	29-channel Ag AgCl EEG Synchronization likelihood Delta Theta Alpha Beta Gamma Graph theory functional connectivity analysis	57 bipolar I patients Current episode, $n = 43$ Depressed, $n = 19$ Manic, $n = 10$ Mixed, $n = 14$ Euthymic, $n = 14$ 87 control	Yes, no test of effects 53 medicated Antipsychotic Atypical, $n = 37$ Typical, $n = 8$ Anticonvulsant, $n = 26$ Antidepressant, $n = 24$ Benzodiazepine, $n = 16$ Lithium, $n = 14$ Buspirone, $n = 3$ Stimulant, $n = 2$ Anticholinergic, $n = 4$ No medication, $n = 4$	Resting state, eyes closed	✓/×	Bipolar I patients had decreased mean alpha synchronization, with largest decreases in F4-FC4-Cz-C4 and Cz-C4-CPz connections. Network properties were different in patients relative to controls. Network properties were correlated with depression scores in patients	Bipolar I patients show impaired neural synchronization at rest and a disruption of resting-state functional connectivity n fronto-central and centro-parietal regions	First network-based analysis of bipolar disorder with EEG Need to apply networks to different states and medication status

(continued)

Table 3 (continued)

Study	Modality measure	Subjects	Medication	Stimuli	Char/ Diff	Principal findings	Interpretation	Comment
Kam et al. (2013)	32-channel Ag/ AgCl EEG Delta Theta Alpha1 Alpha2 Beta1 Beta2 Gamma Intra-hemispheric coherence Inter-hemispheric coherence	76 bipolar patients 132 schizophrenia 136 controls	Yes, tested for and then co-varied effects Medicated, $n = 55$ Atypical antipsychotics, $n = 31$ Typical antipsychotics, $n = 6$ Antidepressants, $n = 20$ Anticonvulsants, $n = 27$ Anticholinergics, $n = 2$ Benzodiazepine, $n = 17$ Lithium, $n = 15$ Antiparkinsonians, $n = 1$	Resting state, eyes closed	✔/✔	Bipolar patients had greater beta and gamma power than schizophrenia patients across regions Intra- and Inter-hemispheric coherence was different between bipolar and schizophrenia patients, varying differently across bands Mania symptoms were related to bipolar patients' increase in beta	Bipolar patients had increased high frequency power with few disruptions in neural synchronization. Whereas schizophrenia patients had enhanced synchronization within and across hemispheres	Resting EEG synchronization may differentiate bipolar disorder from schizophrenia

Note Char/Diff = the review differentiates between studies that attempt to (1) characterize BSDs (✔/×) or (2) characterize and, differentiate BSDs or differentiate BSDs from other diagnoses (✔/✔)

component differences characteristic of disturbance in early executive functions (Hall et al. 2007; Muir et al. 1991; O'Donnell et al. 2004; Souza et al. 1995) and accentuation of the early processing of positive stimuli (Degabriele et al. 2011). Furthermore, some ERP components appear to be heritable, endophenotypes for bipolar disorder (Hall et al. 2007). Therefore, early processing deficits appear to be measureable BSD traits.

Clinical course from unipolar to bipolar disorder has been predicted using EEG sleep components (Rao et al. 2002), a finding that may be associated with increasing or differential chronobiological disturbances in BSDs (see Malhi and Kuiper 2013). Bipolar I conversion from cyclothymia and bipolar II is reliably classified by increased manic-related, left-dominant frontal asymmetry at rest (Nusslock et al. 2012). Finally, bipolar disorder can be differentiated from unipolar disorder with specific ERP components (Muir et al. 1991) and network properties (Koles et al. 1994), and from schizophrenia with specific ERP components (O'Donnell et al. 2004; Souza et al. 1995) and network properties (Kam et al. 2013).

In summary, there are identifiable EEG and ERP characteristics that correlate with the states and symptoms of depression and mania that may differentiate the two poles of the illness: withdrawal or negative valence-related right-hemispheric dominance for the depressed phase and approach or positive valence-related left-hemispheric dominance for the manic phase, each related to changes in network properties. In addition, frequency band and early processing disturbances consistently appear to be trait-based characteristics of bipolar disorder, which can differentiate it from unipolar disorder. Furthermore, early processing and network disturbances differentiate bipolar disorder from schizophrenia. Given that classification and predicting clinical course using EEG has been successfully examined, future directions in EEG and ERP research should concern classification of different phases and differential diagnosis using the aforementioned characteristics.

3.3 Treatment

Within the EEG and ERP literature, there has been one review examining correlates of lithium treatment effect (Ikeda and Kato 2003). Here, we updated this review and review other treatments (Table 4). The EEG and ERP characteristics that correlate with the phases and symptoms and trait-based early processing characteristics of BSDs would be expected to normalize with effective treatment. Indeed, all six studies reviewed show that EEG and ERP can detect treatment-related changes with commonly prescribed treatment. Although clinical EEG did not appear to be a useful measure for characterizing BSDs, it appears that existing EEG abnormalities (including, spikes and irregular beta, theta, and slow alpha activity) is a predictor of 3-month lithium treatment non-response (Ikeda et al. 2002). Additionally, clinical EEG and ERP changes with anticonvulsant medication predict treatment response (Gerez and Tello 1992).

Table 4 Literature review findings of EEG assessment in BSD treatment

Study	Modality measure	Subjects	Treatment	Finding	Interpretation	Comment
Small et al. (1989)	22-channel EEG Waking, hyperventilation, photic stimulation and, in some cases, drowsiness and sleep Delta Theta Alpha Beta1 Beta2	14 drug-free mania patients 14 controls 15 drug-free depression patients 15 controls (also 16 schizophrenia, 6 OCD, and controls	Carbamazepine Lithium	EEG changes detected during treatments in mania and depression patients Carbamazepine increased delta activity in the anterior regions, more right-sided increases, and decreased theta; response/non-response could not be differentiated Lithium increased beta1 and increase left in delta, theta, and beta2; treatment response was correlated with increases in delta EEG could not differentiate between disorders	EEG might be useful for treatment monitoring	Cannot determine treatment responses to a particular treatment or treatment combination in this study
Gerez and Tello (1992)	19-channel EEG Resting state, 30 mins P300 Visual evoked potentials P300 Auditory oddball	90 mixed sample of patients, including bipolar disorder patients	Anticonvulsants Treatment as usual	Focal and P300 changes predicted treatment response to anticonvulsants, regardless of diagnosis	EEG might be useful for monitoring anticonvulsant treatment	Cannot determine treatment responses to a particular treatment or treatment combination in this study

(continued)

Table 4 (continued)

Study	Modality measure	Subjects	Treatment	Finding	Interpretation	Comment
Small et al. (1998)	28-channel EEG Slow delta Fast delta Slow theta Fast theta Slow alpha Fast alpha Slow beta Fast beta Total power Eyes closed	37 hospitalized manic patients, newly medicated after washout	Lithium, $n = 5$ or Carbamazepine, $n = 6$ Or Lithium + Carbamazepine, $n = 10$ Lithium + Halopediol, $n = 8$ Lithium + Risperidone, $n = 8$	In the whole sample, non-responders to treatment had higher left fronto-temporal amplitudes (F1, T1, T3, and T5) than responders in the fast delta, theta, and beta bands. Treatment groups were differentiable	Treatment response to commonly prescribed medication for bipolar disorder can be determined from EEG	
Schulz et al. (2000)	8-channel EEG Eyes closed Delta Theta Alpha Beta	12 patients with affective disorders, including bipolar affective disorders (ICD-10, F31) with and without previous medication use Antidepressant, $n = 9$ Neuroleptics, $n = 3$ No medication, $n = 3$	Lithium treatment for 4.4 months	After lithium treatment, relative theta power was increased. Lithium plasma level was correlated with increased theta power. Lithium also decreased relative alpha power in the right centro-parietal region. There was a general slowing of the dominant alpha frequencies	Change in theta power was induced by lithium	No control group. Did not examine the laterality of the effects

(continued)

Table 4 (continued)

Study	Modality measure	Subjects	Treatment	Finding	Interpretation	Comment
Ikeda et al. (2002)	20-channel EEG Clinical EEG	27 bipolar patients Lithium responders, $n = 5$ Lithium non-responders, $n = 17$	Retrospective investigation to impact of lithium treatment response	All of the lithium responders had normal EEGs. Five of the lithium non-responders had abnormal EEGs	Abnormal EEG may be a predictor of lithium treatment non-response in bipolar disorder	
Howells et al. (2012)	6-channel EEG (F3, F4, C3, C4, P3, P4 electrode sites) Resting state, eyes open, and eyes closed Theta Alpha Beta P300 Visual oddball	12 euthymic bipolar patients, medicated Mood stabilizers, $n = 12$ Lithium, $n = 8$ Antipsychotics, $n = 8$ Antidepressants, $n = 2$ Anxiolytics, $n = 1$ 9 controls	Mindfulness-based cognitive therapy, 8 weeks. Bipolar participants only, tested before and after treatment	Bipolar patients had decreased theta and increased beta at F3 and C3 during eyes closed, relative to controls Post treatment, beta was decreased at F4 during eyes closed No significant differences between groups in P300. P300-like wave components were different between groups at F3 and F4. Therapy normalized this difference	Treatment slightly improved attentional readiness, and attenuated activation of non-relevant information processing	Cannot determine treatment responses to a particular treatment or treatment combination in this study. No treated control group

Note Char/Diff = the review differentiates between studies that attempt to (1) characterize BSDs (/) or (2) characterize and, differentiate BSDs or differentiate BSDs from other diagnoses (/); *L* left; *R* right.

On the one hand, studies (Schulz et al. 2000; Small et al. 1989, 1998) that show frequency band component changes with treatment do not interpret these changes in the context of changes in neural activity or cognitive-emotional processing, correlating with normalization of symptoms. Instead, these studies discuss the potential utility of frequency band component changes as a tool to monitor and measure treatment responses. On the other hand, studies (Howells et al. 2012; Schulz et al. 2000; Small et al. 1989, 1998) showing laterality effects suggest that the greater left-dominant frontal approach-related activity characterizing mania may be normalized in treatment responders, relative to treatment non-responders. After 20 weeks of lithium treatment, relative alpha power in the right centro-parietal region is decreased (Schulz et al. 2000). Lithium, carbamazepine, and risperidone treatment non-response are correlated with higher left fronto-temporal amplitudes than responders in the fast delta, theta, and beta bands at baseline (Small et al. 1998). After lithium treatment, beta1 and left delta, theta, and beta2 increase, and treatment response correlates with increases in delta (Small et al. 1989). Lithium plasma level is correlated with increased theta power.

After carbamazepine administration, delta activity in the anterior regions is increased, with more right-sided increases, and theta is decreased (Small et al. 1989). Mindfulness-based cognitive therapy appears to decrease right frontal beta at rest and normalize P300-like ERP components in already medicated euthymic patients (Howells et al. 2012). Although the specificity of changes due to the therapy is uncertain, this was interpreted as improvement in attentional readiness and attenuation of non-relevant information processing (Howells et al. 2012). Although network disturbances appear to characterize and differentiate BSDs (as described in the previous section), the impacts of treatment on these networks are yet to be investigated.

Overall, there appears to be measurable EEG and ERP correlates of general treatment response that normalize phase and symptom-based characteristics of BSDs; however, the specificity of these effects to a particular medication or phase remains uncertain. Future directions would be to examine the specificity of medication effects and consequent treatment responses on EEG, ERP components, and network properties at each illness phase. Additionally, the relationships between frequency band component changes and cognitive-emotional changes with treatment should be determined. These developments would lay the foundations for investigation into the clinical utility of EEG and ERP components as markers of treatment response in BSDs.

4 Peripheral Physiological Correlates of Bipolar Spectrum Disorders and their Treatment

4.1 Cardiovascular Correlates

4.1.1 Characterization and Differentiation

A systematic literature review of cardiovascular correlates in BSD is presented in Table 5. Eight studies have examined cardiovascular measures and suggest that BSDs are associated with a higher heart rate (HR), reflecting increased arousal and reduced PNS function (see Duschek et al. 2013; Lopes and White 2006) can characterize BSDs. This is important as high resting HR is associated with an increased risk of suicide (Lemogne et al. 2011), which may be related to high suicidality in BSDs (Nock et al. 2009). Studies further suggest that lower HRV can characterize BSDs under tonic and phasic conditions. High frequency HRV measures reflects activity within the PNS branch of the ANS (see Duschek et al. 2013; Lopes and White 2006). PNS activity at the heart during emotional responding is associated with engagement of executive PFC control on the limbic system, and thus afferent and efferent brainstem nuclei linked to the heart (Duschek et al. 2013). Studies of BSDs suggest that low HRV are correlated with BSD traits. This is important given that low HRV is associated with poor mental and physical health and psychological flexibility in the face of stress, increasing the risk of cardiovascular disease and mental disorder, and overall morbidity and mortality (see Duschek et al. 2013; Kemp and Quintana 2013).

Studies suggest that euthymic patients have higher HR (Iacono et al. 1983) and lower HRV (Cohen et al. 2003; Lee et al. 2012) at rest than controls, reflecting disturbed capacity to adapt and regulate autonomic arousal. Similarly, in manic (Henry et al. 2010) and subsyndromal (Lee et al. 2012) depressed bipolar patients, there appears to be increased HR and decreased HRV at rest, relative to controls. Furthermore, decreases in HRV appear to be related to both mania (Henry et al. 2010) and depression severity (Lee et al. 2012; Migliorini et al. 2011). In controls at risk for mania, however, there have been findings of increased HRV during emotional films, relative to controls at low risk (Gruber et al. 2008). There has been no comparison between, or within patients, differentiating BSD phases, except for one pilot study showing that HRV seemed to differ from controls when a patient was in a depressed state, rather than in a euthymic state (see Migliorini et al. 2011). Studies that compare bipolar to unipolar (Iacono et al. 1983) and schizophrenia (Henry et al. 2010) patients have not revealed any differential findings. However, many studies have encountered measurement problems—including poor consistency of findings across different HRV measures and short recording times—when compared against established guidelines (Task Force of the European Society of Cardiology the North American Society of Pacing Electrophysiology [Task Force] 1996). These issues have been highlighted in the "Comment" column in Table 5. Therefore, the results of these studies should be

Table 5 Literature review findings of cardiovascular assessment in BSD characterization and differentiation

Study	Modality measure	Subjects	Medication	Stimuli	Char/ Diff	Principal findings	Interpretation	Comment
Iacono et al. (1983)	HR (cardiotachometer coupler R&L arm electrodes) Phasic HR acceleration Phasic HR deceleration (Main measure was GSR; also EEG)	24 Bipolar Affective Disorder—euthymic 26 Unipolar 46 controls	Yes, tested for effects Imipramine only, not lithium for at least one month, $n = 13$ Lithium only, not imipramine for at least one month, n = 6 Imipramine + Lithium = 26 Drug-free for at least three months, $n = 5$	Resting state with controlled breathing, 2 min Dishabituation Auditory Task, for phase HR changes	✔/✔	The affective disorder groups had higher tonic HR and phasic acceleration than controls	Affective disorders have greater cardiovascular reactivity, due to a tonic difference; however, medication status (imipramine) may explain group differences	While a wide range of physiological factors determines heart rate, the heart is under tonic inhibitory control by the PNS
Cohen et al. (2003)	Holter monitor ECG HR HRV (SDNN, VLF, LF, HF, HFn, LF/HF, TP) QTc	39 Bipolar disorder patients—euthymic 39 controls	Yes, tested for effects Lithium only, $n = 18$ Lithium + other, $n = 11$ Other, $n = 10$	Resting state, no controlled breathing, no time period specified	✔/×	Euthymic patients had lower SDNN,TP, LF/HF, and higher HF%. No difference on HR, QTc, HF, LF, VLF. No differences between patients on different treatments, with or without lithium	Euthymic bipolar patients had decreased HRV	Inconsistent results. Inconclusive due to no specification of the time period

(continued)

Table 5 (continued)

Study	Modality measure	Subjects	Medication	Stimuli	Char/ Diff	Principal findings	Interpretation	Comment
Todder et al. (2005)	Holter monitor ECG HRV (MED, LLE, Shenon entropy, SD1, SD2, SD1/ SD2)	32 Bipolar disorder patients—euthymic 24 controls	Yes, did not test for medication effects Lithium, $n = 28$ Valporic acid, $n = 4$	Resting state, no controlled breathing, no time period specified	✔/×	No differences	No difference between euthymic bipolar and controls on non-linear HRV measures	Inconclusive due to no specification of the time period
Gruber et al. (2008)	Ambulatory monitor HR HRV (RMSSD) (also GSR, facial expressions,)	36 High risk mania 54 low risk mania	Yes, tested for effects. Antidepressants in the high risk group, $n = 2$. No medication use in the low risk group	Baseline (90 s) Emotional Films (positive, negative, neutral; 50–170 s)	✔/×	At baseline, high risk mania subjects had higher RMSSD than low risk subjects. No difference for HR or GSR High risk mania subject had higher RMSSD regardless of valence. No differences for HR or GSR	Tonic differences between those at risk for mania and those who are at low risk	Inconclusive due to short time period at baseline, and the short, variable, time periods for the emotional images

(continued)

Table 5 (continued)

Study	Modality measure	Subjects	Medication	Stimuli	Char/ Diff	Principal findings	Interpretation	Comment
Henry et al. (2010)	Sensorized with 3 lead ECG HR HRV (SDNN, RMSSD, pNN50, LF/HF, LFn, HFn, SampEn, entropy h)	23 Bipolar disorder patients—acute manic 23 controls 14 schizophrenia patients	Yes, tested for effects Bipolar patients on Antipsychotic + mood stabilizer, $n = 17$ Antipsychotic only, $n = 4$ Mood stabilizer only, $n = 1$ On risperidone, $n = 11$ On valproate, $n = 9$ On lithium, $n = 7$ Not medicated, $n = 1$	Resting state, 5 min	✔/×	Acute manic bipolar patients had greater HR, lower RMSSD, pNN50, HFn, LF/HF, SampEn Entropy h than controls. No difference on SDNN, LFn LF/HF, but no other measure, was positively correlated with scores on mania scale No differences between bipolar and schizophrenia patients	Decreased HRV in acute mania. Severity of acute mania is associated with decreased HRV	Decrease in HRV in acute mania appears to be valid; however, conclusions regarding severity and HRV are in doubt due to inconsistent results. However, it is the study with the least methodological issues. Guidelines prescribe reporting of raw LF and HF power

(continued)

Table 5 (continued)

Study	Modality measure	Subjects	Medication	Stimuli	Char/Diff	Principal findings	Interpretation	Comment
Migliorini et al. (2011)	Sensorized with 3 lead ECG HR HRV (MEANNN,SDNN, RMSSD, VLF, LF, HF, LF/HF, LFn, HFn, SampEn, Lempel–Ziv, DFA a1, DFA a2, 1/f slope) (also three axis accelerometer)	1 bipolar patient—depressed and euthymic, over time Eight controls	The patient had ECT	Sleep studies over 4 nights	✔/×	MEANNN, SDNN, RMSSD was decreased in the bipolar patient, relative to controls. Lempel–Ziv and SampEn seem to correlate with depression level	In the bipolar patient, HRV seemed to differ when they were in a depressed state, rather than euthymic. When they were euthymic, they had similar values as the controls	Pilot, proof-of-concept study into how HRV measures may be used for home monitoring of changes in clinical state. Needs to be replicated

(continued)

Table 5 (continued)

Study	Modality measure	Subjects	Medication	Stimuli	Char/Diff	Principal findings	Interpretation	Comment
Gruber et al. (2011)	2 lead ECG HR HRV (HF) (also GSR, facial expressions,)	26 bipolar disorder patients—euthymic Bipolar I, $n = 24$. Bipolar II, $n = 2$ 23 controls	Yes, poorly reported tests of effects Lithium, $n = 4$ Anticonvulsants, $n = 14$, Antidepressants, $n = 19$ Neuroleptics, $n = 13$ Benzodiazepines ($n = 4$), Stimulants, $n = 1$ Sedative-hypnotics, $n = 1$	Resting baseline 60 s Positive emotional film, 150 or 181 s Resting baseline 60 s Positive emotional memory task. 60 s	✔/×	Bipolar patients had decreased HR during the positive memory No differences on HR for the positive film Bipolar patients had greater HRV in with the film and memory than controls	HRV may be a biomarker for extreme positive emotion in bipolar disorder	Inconclusive results. The recording periods are too short according to guidelines, and are variable. No connection was made between high positive emotion in the bipolar patients and high HRV Medication also may have affected the results

(continued)

Table 5 (continued)

Study	Modality measure	Subjects	Medication	Stimuli	Char/Diff	Principal findings	Interpretation	Comment
Jung-Sun et al. (2012)	2 limb lead ECG HR HRV (HRV index, SDNN, RMSSD, pNN50, TP, VLF, LF, HF)	33 bipolar patients—subsyndromal depressive phase 59 controls	Yes, only tested effect of lithium versus no lithium Lithium only, $n = 3$ Lamotrigine only, $n = 3$ Antipsychotic only, $n = 16$ Polypharmacy, $n = 16$ Lithium, $n = 22$ No Lithium, $n = 10$	Resting state, 5 min	✔/×	Patients had less SDNN, pNN50, Log TP, and VLF. No differences on HR, LF, HF, or RMSSD	The subsyndromal depressive state is associated with reduced HRV, relative to controls. Increased severity is associated with reduced HRV. HRV is a biomarker of clinically remitted depressive state	VLF cannot be reliably measured in a 5 min period. Log TP and SDNN are equivalent measures. HRV index is a cruder, more summative, geometric measure of variability, which is not related to vagal tone specifically. Regardless, established standards prescribe at least 20 mins of recording for geometric HRV Polypharmacy patients may have included those on TCAs

Note HRV (HRV measures) = the different time and frequency domain, and nonlinear HRV measures employed in the studies (Lopes and White 2006); Char/Diff = the review differentiates between studies that attempt to (1) characterize BSDs (✔/×) or (2) characterize and, differentiate BSDs or differentiate BSDs from other diagnoses (✔/✔)

interpreted with caution. Given the resurgence of interest in this type of research, future directions for research into cardiovascular measures in the characterization and differentiation of BSDs should replicate findings across phases of illness between, and within, patients and other diagnostic groups, in accordance with established guidelines (Task Force 1996). Furthermore, given that cardiovascular measures are some of the least costly, time-consuming, invasive, and most mobile (see Heathers 2013; Mariani et al. 2012; Quintana et al. 2012) of commonly employed measures, the potential clinical utility of HR and HRV in BSD assessment and monitoring should be explored.

4.1.2 Treatment

A review of the cardiovascular literature focusing on the treatment of BSDs is presented in Table 6. The cardiovascular findings of the increased HR and decreased HRV characterizing BSD phases would expect to be normalized with effective treatment. For example, lithium is known to decrease depolarization at the sinoatrial node of the heart (Chong et al. 2001); thus, it could be predicted that lithium administration increases HRV, in the context of the wider literature (Duschek et al. 2013; see Lopes and White 2006). However, no study has directly investigated this possibility. Decreased depolarization at the sinoatrial node could be a direct effect of lithium or a more indirect, neural-ANS integration effect.

The impacts of commonly prescribed pharmacological treatment have been documented as manipulation checks in studies that characterize BSDs using cardiovascular measures. Additionally, there are some other potentially relevant findings from pharmacological HRV studies. Both these types of findings are provided in Table 6. Tricyclic antidepressants are known to decrease HRV in unipolar patients (Kemp et al. 2010) and in a mixed sample of bipolar and unipolar patients (Paclt et al. 2003). Promisingly, there appears to be some effect of lithium on HRV (Henry et al. 2010), consistent with decreased depolarization at the sinoatrial node. This finding should be followed up in trials that investigate the impact of lithium, along with other commonly prescribed medications, on tonic and phasic changes in HR and HRV across BSD phases and in healthy controls. The potential clinical utility of these HR and HRV measures for treatment monitoring in BSDs should be explored. This is important given cardiovascular measures provide insights into suicidality (Lemogne et al. 2011), health and wellbeing (Duschek et al. 2013; Kemp and Quintana 2013), and morbidity and mortality (Kemp and Quintana 2013).

Table 6 Literature review findings of cardiovascular assessment in BSD treatment

Study	Modality measure	Subjects	Finding	Interpretation	Comment
No trials of commonly prescribed treatment on cardiovascular measures in BSD					
Lithium is known to decrease depolarization at the sinoatrial node (Chong et al. 2001). Thus, it is likely that lithium administration increases HRV					
Examples of relevant exploratory results that have been reported					
Iacono et al. (1983)	Cardiotachometer coupler R&L arm electrodes Tonic HR Phasic HR acceleration Phasic HR deceleration (Main measure was GSR; also EEG)	Mixed bipolar and unipolar sample. Imipramine only, not lithium for at least one month, $n = 13$ Lithium only, not imipramine for at least one month, $n = 6$ Imipramine + Lithium = 26 Drug-free for at least 3 months, $n = 5$	Medication status impacted tonic HR and phasic HR acceleration There were no differences between patients taking lithium and those who were not Patients taking imipramine had increased tonic HR and smaller phasic HR deceleration	Tricyclics may have some effect on tonic heart rate and/or cardiac reactivity	Suggests that use of tricyclics may impact cardiovascular tone and reactivity, and should be taken into account when comparing patient groups; consistent with Kemp et al. (Kemp et al. 2010) (below)
Jung-Sun et al. (2012)	2 limb lead ECG HR HRV (HRV index, SDNN, RMSSD, pNN50, TP, VLF, LF, HF)	22 bipolar patients on lithium 10 bipolar patients not on lithium	Reduced LF in patients on lithium, no differences on any other measure	No effect of lithium on HR or HRV due to inconsistent results	Inconsistent results, yet fits with (Chong 2001wf) (above)
Cohen et al. (2003)	Holter monitor ECG HR HRV (SDNN, VLF, LF, HF, HFn, LF/HF, TP) QTc	Euthymic bipolar patients Lithium only, $n = 18$ Lithium + Other, $n = 11$ Other, $n = 10$	No differences between treatments, with or without lithium	Medications have no effects	The “Other” medication category is undefined, so result is inconclusive

(continued)

Table 6 (continued)

Study	Modality measure	Subjects	Finding	Interpretation	Comment
Paclt et al. (2003)	ECG HR RR (also PQRST characteristics, vectorcardiogram, and body surface potential maps)	Bipolar and depressed patients, mixed sample Patients taking dosulepine only, $n = 43$ Patients taking lithium only, $n = 30$ Patients taking citalopram only, $n = 40$ Controls, $n = 21$	HR increased and RR was decreased in patients treated with dosulepine, relative to controls	Tricyclics decrease HRV in a mixed sample	There may be an effect of diagnosis. Cannot differentiate the effects between the bipolar and depressed patients
Henry et al. (2010)	Sensorized with three lead ECG HR HRV (SDNN, RMSSD, pNN50, LF/HF, LFn, HFn, SampEn, entropy h)	23 acute manic bipolar patients Antipsychotic + mood stabilizer, $n = 17$ Antipsychotic only, $n = 4$ Mood stabilizer only, $n = 1$ On risperidone, $n = 11$ On valproate, $n = 9$ On lithium, $n = 7$ Not medicated, $n = 1$ 23 Controls	No different effects of risperidone or valproate on HRV. No effects of antipsychotic medication alone compared to combination of antipsychotic and mood stabilizer. Patients treated with lithium had decreased LF/HF ratio compared to all other groups	Lithium may blunt sympathetic activation	No consistent results. Inconclusive
Kemp et al. (2010)	Meta-analysis of six treatment studies. Mixed methods and analysis techniques HRV (TD, RSA, HF)	Not a study on bipolar patients Meta-analysis of 186 major depressive disorder patients treated with tricyclic, selective noradrenaline inhibitors, and selective serotonin reuptake inhibitors 407 controls	No effect of commonly prescribed antidepressants (SSRIs, SNRIs) in depressed patients, but TCAs decrease HRV	Tricyclic antidepressants decrease HRV	

Note HRV (HRV measures) = the different time and frequency domain, and nonlinear HRV measures employed in the studies (Lopes and White 2006)

4.2 GSR and Multimodal Correlates

4.2.1 Characterization and Differentiation

Here, we present the first review of multimodal methods to study the physiological correlates of BSDs (Table 7). If BSDs can be characterized and differentiated by embodied dysfunction in brain-body integrated systems, multimodal studies would be expected to show this dysfunction. GSR has been employed only in the context of multimodal physiology measurement. In contrast to the parasympathetic HRV measures, GSR is used as a measure of tonic SNS activation at rest or during experimental tasks, whereby task-related phasic changes in skin conductance can be compared from a baseline (see Dawson et al. 2007 for full explanation). Increased magnitude of GSRs—associated with the strength of SNS activation—but not the number of responses has been shown in BSDs relative to controls (Iacono et al. 1983; Malhi et al. 2005). Given that studies employing GSR phasic responses in response to brief emotional stimuli could not make this differentiation (Gruber et al. 2008, 2011), GSR magnitude appears to be a trait marker of bipolar disorder. Though simultaneous GSR, EEG, and HR and HRV measures (Iacono et al. 1983)—along with measurement of positive facial emotion expression (Gruber et al. 2008, 2011)—have been utilized, results from these simultaneous measures have been largely inconsistent within studies and thus have not provided complementary information. However, this may be due to methodology (see "Comment" column in Table 7), including lack of spatial resolution using single electrode EEG (e.g., Iacono et al. 1983) and short recording times for HRV measures (e.g., Gruber et al. 2008, 2011).

With the improvement of MR-compatible GSR systems and techniques, preliminary research on BSD with simultaneous fMRI and GSR measurement in BSDs has been conducted (e.g., Malhi et al. 2005). This work suggests that bipolar patients may have cognitive deficits related to arousal and appraisal of emotional stimuli given simultaneous VLPFC hypoactivity and increased SNS activity during an emotional stroop task (Malhi et al. 2005). The future research should examine whether simultaneous and integrative neurological and peripheral physiological measurement provides further insight into BSDs. Should integrative measurements be employed, the neural responses when the central and peripheral ANSs (both PNS and SNS) are concurrently active could be measured (see Gray et al. 2009), and thus embodiment of cognitive and emotional processes can be observed.

With GSR-fMRI integration, for example, neural activity during periods of SNS activity can be partitioned (Gray et al. 2009). This is promising given that PFC activity—the disturbances of which are implicated in BSDs—is positively associated with GSR amplitude (Critchley et al. 2000). Future work will employ these methods given that BSDs may be characterized by a lack of inhibition of the VLPFC on increased amygdala activity which relates to increased SNS activity, which then may be related to the disengagement of the PFC regions involved in appraisal during emotional stimuli, in accordance with aforementioned neural-ANS

Table 7 GSR and multimodal assessment in BSD characterization and differentiation

Study	Modality measure	Subjects	Medication	Stimuli	Char/ Diff	Principle findings	Interpretation	Comment
Iacono et al. (1983)	Single Au electrode EEG at Cz site Phi, Alpha, Kappa GSR HR (cardiotachometer coupler R & L arm electrodes) Phasic HR acceleration Phasic HR deceleration	24 Bipolar Affective Disorder—in remission 26 Unipolar 46 controls	Yes, tested for effects Imipramine only, not lithium for at least one month, $n = 13$ Lithium only, not imipramine for at least one month, $n = 6$ Imipramine + Lithium = 26 Drug-free for at least three months, $n = 5$	Resting state with controlled breathing (2 min) Dishabituation Auditory Task, for phase HR changes	✔/✔	On average, electrodermal activity was decreased across patient groups for each task Five bipolar patients and one unipolar patient had increased activity There were no differences on EEG. HR findings might be explained by medication effects	Decreased electrodermal activity may serve as a marker of affective disorder	No spatial resolution from one Cz electrode HR and phasic HR measures do not reflect activity of the PNS or SNS, only indices of arousal and changes in arousal
Malhi et al. (2005)	fMRI 3 Tesla BOLD GSR	12 Bipolar disorder patients—euthymic 12 controls	Yes, did not test for effects Lithium only, $n = 3$ Valporate only, $n = 4$ Lithium + Valporate, $n = 1$ Drug-free, $n = 4$	Emotional stroop	✔/×	Bipolar patients had greater electrodermal arousal, relative to controls. Mean GSR responses between groups were equal across valence Bipolar patients had decreased amygdala and VLPFC activity	Cognitive deficit in bipolar disorder might be a due to arousal differences	Concurrent, rather than integrative, illustration of SNS-related arousal differences PNS activity may also play a role

(continued)

Table 7 (continued)

Study	Modality measure	Subjects	Medication	Stimuli	Char/ Diff	Principle findings	Interpretation	Comment
Gruber et al. (2008)	GSR HR HRV Facial expressions	36 High risk mania 54 low risk mania	Yes, tested for effects Antidepressants in the high risk group, $n = 2$. No medication use in the low risk group	Baseline (90 s) Emotional films (positive, negative, neutral; 50–170 s)	✔/×	High-risk group reported elevated positive emotion in response to all film clips, but were equally facially expressive High-risk group had higher resting vagal tone than did the low-risk group High-risk group exhibited higher HRV across all film clips No group differences were found for HR or GSR	Mania is associated with elevated tonic positive emotion	Inconclusive results. The recording periods are too short according to guidelines, and are variable No integration between high positive emotion in the bipolar patients and HRV or GSR
Gruber et al. (2011)	GSR HR HRV Facial expressions	26 bipolar patients Bipolar I, $n = 24$. Bipolar II, $n = 2$ 23 control	Yes, under-reported tests of effects Lithium, $n = 4$ Anticonvulsants, $n = 14$, Antidepressants, $n = 19$ Neuroleptics, $n = 13$ Benzodiazepines ($n = 4$), Stimulants, $n = 1$ Sedative-hypnotics, $n = 1$	Resting baseline 60 s Positive emotional film, 150 or 181 s Resting baseline 60 s Positive emotional memory task. 60 s	✔/×	Bipolar patients had greater HRV in with the film and memory than controls Bipolar patients had increased positive facial expression and decreased HR during the positive memory No differences on HR, GSR, affect, or facial expressions for the positive film	HRV may be a biomarker for extreme positive emotion in bipolar disorder	Inconclusive results. The recording periods are too short according to guidelines, and are variable No integration between high positive emotion in the bipolar patients and HRV or GSR Medication also may have affected the results

Note Char/Diff = the review differentiates between studies that attempt to (1) characterize BSDs (✔/×) or (2) characterize and, differentiate BSDs or differentiate BSDs from other diagnoses (✔/×); *L* left; *R* right integration accounts (see Damasio 1996; Porges 2007; Thayer et al. 2009) and preliminary findings (e.g., Malhi et al. 2005). In addition to neural-SNS integration, dysfunctional prefrontal engagement of the inhibitive PNS might be involved in BSDs. Thus, the future work in BSDs should explore the integration of the PNS with integrative measurement (e.g., fMRI-HRV measurement; see Thayer et al. 2012). Finally, in order to provide a complete account of the neutral-autonomic integration, longitudinal integrative measurement (e.g., fMRI-GSR-HRV; EEG-GSR-HRV) should be employed. A more complete embodied account and multimodal measurement techniques may provide objective measures of the mechanisms through which emotional lability and mood disturbance occurs in BSDs. Consideration of an embodied psychophysiological mechanism, not just examining individual physiological correlates, may provide the clinically useful, objective physiological measures needed to characterize and differentiate BSDs.

4.2.2 Treatment

Although there have been studies that employ multimodal physiology to characterize BSDs, there are yet to be multimodal investigations of treatment effects and responses. Additionally, treatment studies employing GSR are yet to be conducted. Research has already employed simultaneous fMRI-GSR measurement (e.g., Malhi et al. 2005); thus, determining whether treatment normalizes these correlates is a likely next step. In doing so, researchers may be able to illustrate the manner in which treatment impacts the embodied neural-SNS integration of cognitive and emotional stimuli, which in turn may be related to improvement in depressed and manic symptoms, and prophylaxis in the euthymic phase.

5 An Embodied Framework for the Psychophysiology of Bipolar Spectrum Disorders

Here, we interpret the physiological correlates of BSDs in line with embodied, neural-autonomic integration perspectives (see Craig 2009; Niedenthal 2007), particularly with respect to the crucial role for the body in emotion, motivation, and cognition (see Price et al. 2011). Characterization of BSD as an embodied disturbance is gaining some attention with consideration of molecular biological correlates across circadian, homeostatic, and stress systems (see Malhi et al. 2012; Malhi and Kuiper 2013). A previous review of BSDs (Green et al. 2007) has discussed the potential impact of the neurological dysfunction on autonomic arousal systems; however, the exact disturbance remains to be determined. Initial support for an embodied perspective on BSD psychophysiological correlates originates from preliminary multimodal investigations (e.g., Malhi et al. 2005). These studies may provide the objective physiological measures needed to characterize and differentiate, and make treatment decisions in patients with BSDs.

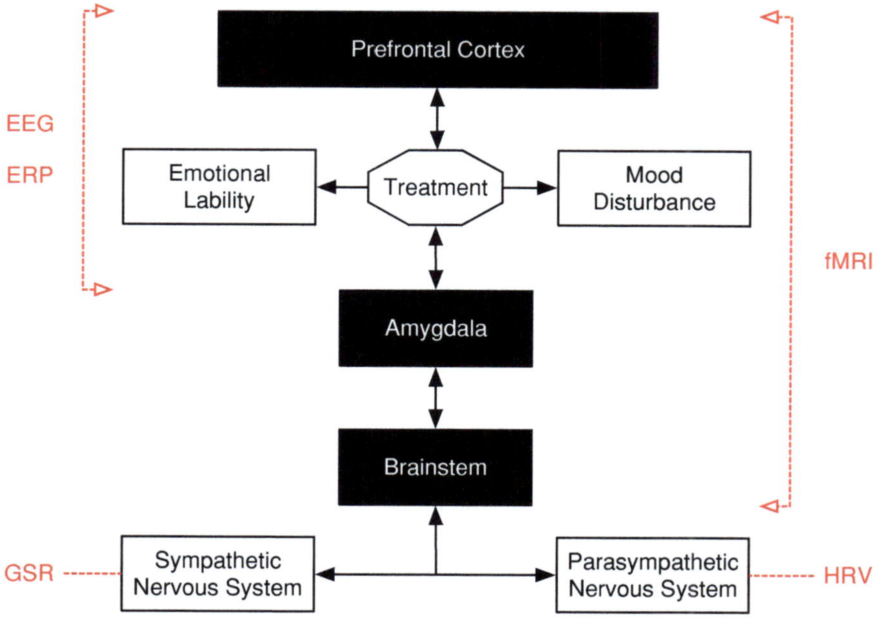

Fig. 1 Embodied framework for the psychophysiology of bipolar spectrum disorders (*BSDs*). This schematic provides a working framework, which enables the characterization of BSDs in an embodied manner. It details the dysfunctional neural-autonomic integration of cognitive and emotional processes and responses, which result in emotional lability and mood disturbances. Within this framework, this dysfunction would be normalized by effective treatment. With respect to noninvasive multimodal measurement of the physiological correlates, cortical neural activity can be measured (but not limited to the measures reviewed; see in red) using electroencephalogram (*EEG*) and functional magnetic resonance imaging (*fMRI*), the temporal information processing stream from the cortex with event-related potential (*ERP*), subcortical neural activity with fMRI, sympathetic nervous system activity with galvanic skin response (*GSR*), and parasympathetic nervous system activity with high frequency heart rate variability (*HRV*)

After having provided overviews of the cardiovascular, EEG and ERP, fMRI, and multimodal literatures that characterize and differentiate BSDs, we incorporated these areas into a simplified, working embodied framework for the physiology of BSDs (see Fig. 1). Accordingly, we suggest that BSDs may be considered as a cluster of disorders characterized by dysfunctional embodied cognitive and emotional neural integration with the ANS, resulting in emotional lability and mood disturbances. In a simplified illustration, dysfunctional VLPFC connectivity between the amygdala impacts the neural-autonomic integration of SNS and PNS activity, through the brainstem, resulting in the core cognitive and emotional dysfunctions of BSDs, which is the target of treatment. Dysfunctional neural-autonomic integration will lead to decreased control over primary visceromotor activity and decreased flexibility for adaptation to stress: SNS hyperactivation and PNS hypoactivation. Dysregulated primary viscerosensory feedback from the periphery will then be inadequately regulated by hypoactivation of the medial frontal regions involved in the appraisal of the stimuli, resulting in emotional lability and mood disturbance characteristic of BSDs. It is perhaps through bidirectional projections between brain and body that underpin core cognitive and emotional dysfunction in BSDs: VLPFC may fail to regulate the amygdala effectively, increasing SNS activity through visceral efferent pathways, leading to a state of ANS rigidity, which subsequently impacts on the brain through afferent feedback, resulting in the core dysfunctions. An embodied approach to better understanding BSDs may, in time, provide clinically useful physiological measures needed to improve assessment and make treatment decisions in patients with BSDs.

5.1 Novel Predictions Within the Embodied Framework

Within the simple, embodied framework for the physiology of BSDs, novel predictions regarding BSD psychophysiology arise. Within BSD patients during the euthymic phase in comparison to other phases, increased VLPFC activity would be associated with higher PNS and low SNS activity (e.g., greater HRV; lower GSR). However, in comparison to healthy controls, the characteristic decreased VLPFC activity would be associated with lower PNS and higher SNS activity (e.g., lower HRV; greater GSR). We further predict that neural and peripheral responses in response to positive, approach-related stimuli will differ from those in response to negative, withdrawal-related stimuli, and that these responses during mania will differ from those in the depressed phase. We predict that manic phase will be associated with decreased VLPFC and R IFG activity, in addition to increased left-dominant alpha asymmetry, lower PNS and higher SNS activity (e.g., lower HRV; higher GSR), and in turn medial PFC disengagement involved in appraisal of goal-oriented positive stimuli, relative to negative stimuli. By contrast, during the depressed phase, we predict the opposite: decreased VLPFC and R IFG, increased right-dominant alpha asymmetry, low PNS and higher SNS activity (e.g., lower HRV; higher GSR), and in turn medial PFC disengagement from appraisal of withdrawal-oriented negative stimuli, relative to positive stimuli. We predict that effective treatments will act on neural-autonomic integration, thereby normalizing these differences.

6 Conclusions and Future Directions

In conclusion, there are measurable physiological correlates of BSDs and their responses to treatment. Of particular promise, discernible fMRI and EEG and ERP correlates that characterize and differentiate BSDs, and responses to treatment in BSD patients are beginning to emerge. However, the ability to use these correlates to aid classification of BSDs and to improve treatment selection and prediction of response requires further validation. Additionally, cardiovascular correlates of BSDs and treatment response are still in the initial stages of investigation, but again early findings hold promise especially when considered within the context of the wider literature regarding autonomic disturbances and their relationship with suicidality, and longer term morbidity and mortality. Other than the aforementioned future directions specific to each physiological modality, a number of general future initiatives for research are recommended.

From having developed a simplified integrative embodied framework of BSDs, a major future direction for research into the physiological correlates of BSDs will be to employ simultaneous recording techniques in order to determine dysfunction in the neural-autonomic integration of cognitive and emotional responses. Such an approach will aid our understanding of the adverse impact of BSDs on brain and body function and facilitate the characterization of treatment targets. Building on previous work on neural-autonomic integration in BSDs (e.g., Malhi et al. 2005), future work could partial-out the inhibitory and excitatory neural activation associated with the activity of both the inhibitory PNS and excitatory SNS, respectively. Such work would illustrate the manner in which BSD patients attend to, process, respond to, regulate responses, and recover from cognitive and emotional stimuli, accounting for the bidirectional nature of the central and peripheral ANSs. In doing so, the differential cognitive and emotional inhibition and activation changes associated with the manic and depressed phases may be better integrated and understood. Further, longitudinal and simultaneous multimodal physiological assessment is likely to better differentiate phases of illness, characterize clinical trajectory, and provide insights into the chronobiological changes associated with phase and course in BSDs (see Malhi et al. 2012; Malhi and Kuiper 2013).

With the development of a simplified integrative framework reflecting the current state of the literature, future work may be guided toward examining the embodied, cognitive, and emotional dysfunction that is associated with BSD emotional lability and mood disturbance in accordance with the aforementioned novel predictions. Taking an embodied account of psychophysiological mechanisms, and not just examining the physiological correlates of the dysfunctional parts, is likely to yield more clinical meaningful and objective physiological measures that will in turn improve assessment and therapeutic decision making in patients with BSDs. Given complexity of assessment and the multitude of treatment considerations associated with BSDs, and the fact that they continue to exert an overwhelming burden because of their prevalence and poor response to treatment, there remains a critical need to continue with such endeavors.

7 Conflict of Interest

The authors Tim Outhred and Andrew H. Kemp are supported by an Australian Postgraduate Award and an International Research Professorship from the University of São Paulo, respectively. Gin S. Malhi has received research support from

AstraZeneca, Eli Lilly, Organon, Pfizer, Servier, and Wyeth. He has been a speaker for AstraZeneca, Eli Lilly, Janssen Cilag, Lundbeck, Pfizer, Ranbaxy, Servier, and Wyeth. He has been a consultant for AstraZeneca, Eli Lilly, Janssen Cilag, Lundbeck, and Servier.

Acknowledgments The authors acknowledge the assistance of James A. J. Heathers with regard to insight into some of the heart rate variability research presented.

References

Baldessarini RJ, Tondo L, Hennen J (2003) Treatment-latency and previous episodes: relationships to pretreatment morbidity and response to maintenance treatment in bipolar I and II disorders. Bipolar Disord 5:169–179. doi:10.1034/j.1399-5618.2003.00030.x

Berk M, Brnabic A, Dodd S, Kelin K, Tohen M, Malhi GS, Berk L, Conus P, McGorry PD (2011) Does stage of illness impact treatment response in bipolar disorder? Empirical treatment data and their implication for the staging model and early intervention. Bipolar Disord 13:87–98. doi:10.1111/j.1399-5618.2011.00889.x

Bestelmeyer PEG, Phillips LH, Crombie C, Benson P, St Clair D (2009) The P300 as a possible endophenotype for schizophrenia and bipolar disorder: Evidence from twin and patient studies. Psychiatry Res 169:212–219. doi:10.1016/j.psychres.2008.06.035

Blond BN, Fredericks CA, Blumberg HP (2012) Functional neuroanatomy of bipolar disorder: structure, function, and connectivity in an amygdala-anterior paralimbic neural system. Bipolar Disord 14:340–355. doi:10.1111/j.1399-5618.2012.01015.x

Bonnín CM, Sanchez-Moreno J, Martínez-Arán A, Solé B, Reinares M, Rosa AR, Goikolea JM, Benabarre A, Ayuso-Mateos JL, Ferrer M, Vieta E, Torrent C (2012) Subthreshold symptoms in bipolar disorder: Impact on neurocognition, quality of life and disability. J Affect Disord 136:650–659. doi:10.1016/j.jad.2011.10.012

Chang K, Karchemskiy A, Kelley R, Howe M, Garrett A, Adleman N, Reiss A (2009) Effect of divalproex on brain morphometry, chemistry, and function in youth at high-risk for bipolar disorder: a pilot study. J Child Adolesc Psychopharmacol 19:51–59. doi:10.1089/cap.2008.060

Chang KD, Wagner C, Garrett A, Howe M, Reiss A (2008) A preliminary functional magnetic resonance imaging study of prefrontal-amygdalar activation changes in adolescents with bipolar depression treated with lamotrigine. Bipolar Disord 10:426–431. doi:10.1111/j.1399-5618.2007.00576.x

Chen C-H, Suckling J, Lennox BR, Ooi C, Bullmore ET (2011) A quantitative meta-analysis of fMRI studies in bipolar disorder. Bipolar Disord 13:1–15. doi:10.1111/j.1399-5618.2011.00893.x

Chong SA, Mythily Mahendran R (2001) Cardiac effects of psychotropic drugs. Ann Acad Med Singap 30:625–631

Clementz BA, Sponheim SR, Iacono WG, Beiser M (1994) Resting EEG in first-episode schizophrenia patients, bipolar psychosis patients, and their first-degree relatives. Psychophysiol 31:486–494. doi:10.1111/j.1469-8986.1994.tb01052.x

Cohen H, Kaplan Z, Kotler M, Mittelman I, Osher Y, Bersudsky Y (2003) Impaired heart rate variability in euthymic bipolar patients. Bipolar Disord 5:138–143. doi:10.1034/j.1399-5618.2003.00027.x

Cook BL, Shukla S, Hoff AL (1986) EEG abnormalities in bipolar affective disorder. J Affect Disord 11:147–149

Craig ADB (2009) How do you feel–now? The anterior insula and human awareness. Nat Rev Neurosci 10:59–70. doi:10.1038/nrn2555

Critchley HD, Elliott R, Mathias CJ, Dolan RJ (2000) Neural activity relating to generation and representation of galvanic skin conductance responses: a functional magnetic resonance imaging study. J Neurosci 20:3033–3040

Crowe M, Inder M, Carlyle D, Wilson L, Whitehead L, Panckhurst A, O'Brien A, Joyce P (2012) Feeling out of control: a qualitative analysis of the impact of bipolar disorder. J Psychiatr Ment Health Nurs 19:294–302. doi:10.1111/j.1365-2850.2011.01786.x

Damasio AR (1996) The somatic marker hypothesis and the possible functions of the prefrontal cortex. Philos Trans R Soc Lond B Biol Sci 351:1413–1420. doi:10.1098/rstb.1996.0125

Davidson RJ (1998) Affective style and affective disorders: perspectives from affective neuroscience. PCEM 12:307–330. doi:10.1080/026999398379628

Davidson RJ (2004) What does the prefrontal cortex "do" in affect: perspectives on frontal EEG asymmetry research. Biol Psychol 67:219–233. doi:10.1016/j.biopsycho.2004.03.008

Dawson ME, Schell AM, Filion DL (2007) The electrodermal system. In: Cacioppo JT, Tassinary LG, Berntson GG (eds) Handbook of psychophysiology. Cambridge University Press, Cambridge, pp 159–181

Degabriele R, Lagopoulos J (2009) A review of EEG and ERP studies in bipolar disorder. Acta Neuropsychiatrica 21:58–66. doi:10.1111/j.1601-5215.2009.00359.x

Degabriele R, Lagopoulos J, Malhi G (2011) Neural correlates of emotional face processing in bipolar disorder: an event-related potential study. J Affect Disord 133:212–220. doi:10.1016/j.jad.2011.03.033

Dewan MJ, Haldipur CV, Boucher MF, Ramachandran T, Major LF (1988) Bipolar affective disorder. II. EEG, neuropsychological, and clinical correlates of CT abnormality. Acta Psychiatr Scand 77:677–682

Diler RS, de Almeida JRC, Ladouceur C, Birmaher B, Axelson D, Phillips M (2013) Neural activity to intense positive versus negative stimuli can help differentiate bipolar disorder from unipolar major depressive disorder in depressed adolescents: a pilot fMRI study. Psychiatry Res. doi:10.1016/j.pscychresns.2013.06.013

Donchin E, Coles MGH (1988) Event-related potentials and cognition: a critique of the context updating hypothesis and an alternative interpretation of P3. Behav Brain Sci 11:357–374. doi:10.1017/S0140525X00058027

Duschek S, Werner NS, Reyes del Paso GA (2013) The behavioral impact of baroreflex function: a review. Psychophysiol 50:1183–1193. doi:10.1111/psyp.12136

El-Badri SM, Ashton CH, Moore PB, Marsh VR, Ferrier IN (2001) Electrophysiological and cognitive function in young euthymic patients with bipolar affective disorder. Bipolar Disord 3:79–87. doi:10.1034/j.1399-5618.2001.030206.x

Favre P, Baciu M, Pichat C, De Pourtalès M-A, Fredembach B, Garçon S, Bougerol T, Polosan M (2013) Modulation of fronto-limbic activity by the psychoeducation in euthymic bipolar patients. Psychiatry Res, A functional MRI study. doi:10.1016/j.pscychresns.2013.07.007

Friston KJ, Fletcher P, Josephs O, Holmes A, Rugg MD, Turner R (1998) Event-related fMRI: characterizing differential responses. NeuroImage 7:30–40. doi:10.1006/nimg.1997.0306

Gerez M, Tello A (1992) Clinical significance of focal topographic changes in the electroencephalogram (EEG) and evoked potentials (EP) of psychiatric patients. Brain Topogr 5:3–10

Gray MA, Minati L, Harrison NA, Gianaros PJ, Napadow V, Critchley HD (2009) Physiological recordings: basic concepts and implementation during functional magnetic resonance imaging. NeuroImage 47:1105–1115. doi:10.1016/j.neuroimage.2009.05.033

Green MJ, Cahill CM, Malhi GS (2007) The cognitive and neurophysiological basis of emotion dysregulation in bipolar disorder. J Affect Disord 103:29–42. doi:10.1016/j.jad.2007.01.024

Grotegerd D, Stuhrmann A, Kugel H, Schmidt S, Redlich R, Zwanzger P, Rauch AV, Heindel W, Zwitserlood P, Arolt V, Suslow T, Dannlowski U (2013) Amygdala excitability to subliminally presented emotional faces distinguishes unipolar and bipolar depression: an fMRI and pattern classification study. Hum Brain Mapp. doi:10.1002/hbm.22380

Gruber J, Dutra S, Eidelman P, Johnson SL, Harvey AG (2011) Emotional and physiological responses to normative and idiographic positive stimuli in bipolar disorder. J Affect Disord 133:437–442. doi:10.1016/j.jad.2011.04.045

Gruber J, Johnson SL, Oveis C, Keltner D (2008) Risk for mania and positive emotional responding: too much of a good thing? Emotion 8:23–33. doi:10.1037/1528-3542.8.1.23
Hajek T, Alda M, Hajek E, Ivanoff J (2013) Functional neuroanatomy of response inhibition in bipolar disorders—combined voxel based and cognitive performance meta-analysis. J Psychiatr Res 47:1955–1966. doi:10.1016/j.jpsychires.2013.08.015
Haldane M, Jogia J, Cobb A, Kozuch E, Kumari V, Frangou S (2008) Changes in brain activation during working memory and facial recognition tasks in patients with bipolar disorder with Lamotrigine monotherapy. Eur Neuropsychopharmacol 18:48–54. doi:10.1016/j.euroneuro.2007.05.009
Hall M-H, Rijsdijk F, Kalidindi S, Schulze K, Kravariti E, Kane F, Sham P, Bramon E, Murray RM (2007) Genetic overlap between bipolar illness and event-related potentials. Psychol Med 37:667–678. doi:10.1017/S003329170600972X
Harmon-Jones E, Abramson LY, Nusslock R, Sigelman JD, Urosevic S, Turonie LD, Alloy LB, Fearn M (2008) Effect of bipolar disorder on left frontal cortical responses to goals differing in valence and task difficulty. Biol Psychiatry 63:693–698. doi:10.1016/j.biopsych.2007.08.004
Heathers JAJ (2013) Smartphone-enabled pulse rate variability: an alternative methodology for the collection of heart rate variability in psychophysiological research. Int J Psychophysiol. doi:10.1016/j.ijpsycho.2013.05.017
Henry BL, Minassian A, Paulus MP, Geyer MA, Perry W (2010) Heart rate variability in bipolar mania and schizophrenia. J Psychiatr Res 44:168–176. doi:10.1016/j.jpsychires.2009.07.011
Hirschfeld RMA, Lewis L, Vornik LA (2003) Perceptions and impact of bipolar disorder: how far have we really come? Results of the national depressive and manic-depressive association 2000 survey of individuals with bipolar disorder. J Clin Psychiatry 64:161–174
Houenou J, Frommberger J, Carde S, Glasbrenner M, Diener C, Leboyer M, Wessa M (2011) Neuroimaging-based markers of bipolar disorder: evidence from two meta-analyses. J Affect Disord 132:344–355. doi:10.1016/j.jad.2011.03.016
Howells FM, Ives-Deliperi VL, Horn NR, Stein DJ (2012) Mindfulness based cognitive therapy improves frontal control in bipolar disorder: a pilot EEG study. BMC Psychiatry 12:15. doi:10.1186/1471-244X-12-15
Iacono WG, Lykken DT, Peloquin LJ, Lumry AE, Valentine RH, Tuason VB (1983) Electrodermal activity in euthymic unipolar and bipolar affective disorders: a possible marker for depression. Arch Gen Psychiatry 40:557. doi:10.1001/archpsyc.1983.01790050083010
Ikeda A, Kato N, Kato T (2002) Possible relationship between electroencephalogram finding and lithium response in bipolar disorder. Prog Neuropsychopharmacol Biol Psychiatry 26:903–907. doi:10.1016/S0278-5846(02)00203-8
Ikeda A, Kato T (2003) Biological predictors of lithium response in bipolar disorder. Psychiatry Clin Neurosci 57:243–250. doi:10.1046/j.1440-1819.2003.01112.x
Insel T, Cuthbert B, Garvey M, Heinssen R, Pine DS, Quinn K, Sanislow C, Wang P (2010) Research domain criteria (RDoC): toward a new classification framework for research on mental disorders. Am J Psychiatry 167:748–751. doi:10.1176/appi.ajp.2010.09091379
Insel TR, Voon V, Nye JS, Brown VJ, Altevogt BM, Bullmore ET, Goodwin GM, Howard RJ, Kupfer DJ, Malloch G, Marston HM, Nutt DJ, Robbins TW, Stahl SM, Tricklebank MD, Williams JH, Sahakian BJ (2013) Innovative solutions to novel drug development in mental health. Neurosci Biobehav Rev. doi:10.1016/j.neubiorev.2013.03.022
Jogia J, Haldane M, Cobb A, Kumari V, Frangou S (2008) Pilot investigation of the changes in cortical activation during facial affect recognition with lamotrigine monotherapy in bipolar disorder. Br J Psychiatry 192:197–201. doi:10.1192/bjp.bp.107.037960
Kam JWY, Bolbecker AR, O'Donnell BF, Hetrick WP, Brenner CA (2013) Resting state EEG power and coherence abnormalities in bipolar disorder and schizophrenia. J Psychiatr Res 47:1893–1901. doi:10.1016/j.jpsychires.2013.09.009
Kano K, Nakamura M, Matsuoka T, Iida H, Nakajima T (1992) The topographical features of EEGs in patients with affective disorders. Electroencephalogr Clin Neurophysiol 83:124–129
Keck PE (2006) Long-term management strategies to achieve optimal function in patients with bipolar disorder. J Clin Psychiatry 67 Suppl 9:19–24– discussion 36–42

Kemp AH, Hopkinson PJ, Hermens DF, Rowe DL, Sumich AL, Clark CR, Drinkenburg W, Abdi N, Penrose R, McFarlane A, Boyce P, Gordon E, Williams LM (2009) Fronto-temporal alterations within the first 200 ms during an attentional task distinguish major depression, non-clinical participants with depressed mood and healthy controls: a potential biomarker? Hum Brain Mapp 30:602–614. doi:10.1002/hbm.20528

Kemp AH, Quintana DS (2013) The Relationship Between Mental and Physical Health: Insights from the Study of Heart Rate Variability. Int J Psychophysiol 1–28. doi: 10.1016/j.ijpsycho.2013.06.018

Kemp AH, Quintana DS, Gray MA, Felmingham KL, Brown K, Gatt JM (2010) Impact of Depression and Antidepressant Treatment on Heart Rate Variability: A Review and Meta-Analysis. Biol Psychiatry 67:1067–1074. doi: 10.1016/j.biopsych.2009.12.012

Ketter TA, Houston JP, Adams DH, Risser RC, Meyers AL, Williamson DJ, Tohen M (2006) Differential efficacy of olanzapine and lithium in preventing manic or mixed recurrence in patients with bipolar I disorder based on number of previous manic or mixed episodes. J Clin Psychiatry 67:95–101

Kim D-J, Bolbecker AR, Howell J, Rass O, Sporns O, Hetrick WP, Breier A, O'Donnell BF (2013) Disturbed resting state EEG synchronization in bipolar disorder: A graph-theoretic analysis. NeuroImage: Clinical 2:414–423. doi: 10.1016/j.nicl.2013.03.007

Koles ZJ, Lind JC, Flor-Henry P (1994) Spatial patterns in the background EEG underlying mental disease in man. Electroencephalogr Clin Neurophysiol 91:319–328

Lahera G, Pedrera A, Cabañes L, Fernandez-Lorente J, Simal P, Montes JM, Saiz-Ruiz J (2009) P300 event-related potential in euthymic patients with bipolar disorder. Prog Neuropsychopharmacol Biol Psychiatry 33:16–19. doi:10.1016/j.pnpbp.2008.09.017

Lee J, Kim B, Hong Y, Joo YH (2012) Heart rate variability in the subsyndromal depressive phase of bipolar disorder. Psychiatry Clin Neurosci 66:361–366. doi:10.1111/j.1440-1819.2012.02335.x

Lemogne C, Thomas F, Consoli SM, Pannier B, Jégo B, Danchin N (2011) Heart rate and completed suicide: evidence from the IPC cohort study. Psychosom Med 73:731–736. doi:10.1097/PSY.0b013e3182365dc7

Lopes P, White J (2006) Heart Rat Variability: Measurement Methods and Practical Implications. In: Maud PJ, C F (eds) Physiological Assessment Of Human Fitness, 2nd ed. Human Kinetics, Champaign, IL, pp 39–49

Malhi GS (2012) Bipolar Antidepressant Agents: Shaken not stirred. Aust N Z J Psychiatry 46:289–292. doi:10.1177/0004867412445131

Malhi GS, Adams D, Cahill CM, Dodd S, Berk M (2009) The management of individuals with bipolar disorder: a review of the evidence and its integration into clinical practice. Drugs 69:2063–2101. doi:10.2165/11318850-000000000-00000

Malhi GS, Bargh DM, Cashman E, Frye MA, Gitlin M (2012) The clinical management of bipolar disorder complexity using a stratified model. Bipolar Disord 14:66–89. doi:10.1111/j.1399-5618.2012.00993.x

Malhi GS, Bargh DM, Coulston CM, Das P, Berk M (2013) Predicting bipolar disorder on the basis of phenomenology: implications for prevention and early intervention. Bipolar Disord. doi:10.1111/bdi.12133

Malhi GS, Kuiper S (2013) Chronobiology of mood disorders. Acta Psychiatr Scand Suppl 2–15. doi: 10.1111/acps.12173

Malhi GS, Lagopoulos J, Sachdev PS, Ivanovski B, Shnier R (2005) An emotional Stroop functional MRI study of euthymic bipolar disorder. Bipolar Disord 7(Suppl 5):58–69. doi:10.1111/j.1399-5618.2005.00255.x

Marchand WR, Lee JN, Johnson S, Gale P, Thatcher J (2013) Differences in functional connectivity in major depression versus bipolar II depression. J Affect Disord 150:527–532. doi:10.1016/j.jad.2013.01.028

Mariani S, Migliorini M, Tacchino G, Gentili C, Bertschy G, Werner S, Bianchi AM (2012) Clinical state assessment in bipolar patients by means of HRV features obtained with a sensorized T-shirt. Conference proceedings: annual international conference of the IEEE

engineering in medicine and biology society IEEE engineering in medicine and biology society conference 2012:2240–2243. doi: 10.1109/EMBC.2012.6346408
Merikangas KR, Akiskal HS, Angst J, Greenberg PE, Hirschfeld RMA, Petukhova M, Kessler RC (2007) Lifetime and 12-month prevalence of bipolar spectrum disorder in the national comorbidity survey replication. Arch Gen Psychiatry 64:543–552. doi:10.1001/archpsyc.64.5.543
Migliorini M, Mendez MO, Bianchi AM (2011) Study of heart rate variability in bipolar disorder: linear and non-linear parameters during sleep. Front Neuroeng 4:22. doi:10.3389/fneng.2011.00022
Morris SE, Cuthbert BN (2012) Research domain criteria: cognitive systems, neural circuits, and dimensions of behavior. Dialogues Clin Neurosci 14:29–37
Muir WJ, St Clair DM, Blackwood DH (1991) Long-latency auditory event-related potentials in schizophrenia and in bipolar and unipolar affective disorder. Psychol Med 21:867–879
Niedenthal PM (2007) Embodying emotion. Science 316:1002–1005. doi:10.1126/science.1136930
Nock MK, Hwang I, Sampson N, Kessler RC, Angermeyer M, Beautrais A, Borges G, Bromet E, Bruffaerts R, de Girolamo G, de Graaf R, Florescu S, Gureje O, Haro JM, Hu C, Huang Y, Karam EG, Kawakami N, Kovess V, Levinson D, Posada-Villa J, Sagar R, Tomov T, Viana MC, Williams DR (2009) Cross-national analysis of the associations among mental disorders and suicidal behavior: findings from the WHO World Mental Health Surveys. PLoS Med 6:e1000123. doi:10.1371/journal.pmed.1000123
Nusslock R, Harmon-Jones E, Alloy LB, Urosevic S, Goldstein K, Abramson LY (2012) Elevated left mid-frontal cortical activity prospectively predicts conversion to bipolar I disorder. J Abnorm Psychol 121:592–601. doi:10.1037/a0028973
O'Donnell BF, Vohs JL, Hetrick WP, Carroll CA, Shekhar A (2004) Auditory event-related potential abnormalities in bipolar disorder and schizophrenia. Int J Psychophysiol 53:45–55. doi:10.1016/j.ijpsycho.2004.02.001
Paclt I, Slavícek J, Dohnalová A, Kitzlerová E, Pisvejcová K (2003) Electrocardiographic dose-dependent changes in prophylactic doses of dosulepine, lithium and citalopram. Physiol Res 52:311–317
Pavuluri MN, Passarotti AM, Harral EM, Sweeney JA (2010a) Enhanced prefrontal function with pharmacotherapy on a response inhibition task in adolescent bipolar disorder. J Clin Psychiatry 71:1526–1534. doi:10.4088/JCP.09m05504yel
Pavuluri MN, Passarotti AM, Parnes SA, Fitzgerald JM, Sweeney JA (2010b) A pharmacological functional magnetic resonance imaging study probing the interface of cognitive and emotional brain systems in pediatric bipolar disorder. J Child Adolesc Psychopharmacol 20:395–406. doi:10.1089/cap.2009.0105
Peterson CK, Harmon-Jones E (2008) Proneness to hypomania predicts EEG coherence between left motor cortex and left prefrontal cortex. Biol Psychol 78:216–219. doi:10.1016/j.biopsycho.2008.01.011
Porges SW (2007) The polyvagal perspective. Biol Psychol 74:116–143. doi:10.1016/j.biopsycho.2006.06.009
Post RM, Leverich GS, Kupka RW, Keck PE, McElroy SL, Altshuler LL, Frye MA, Luckenbaugh DA, Rowe M, Grunze H, Suppes T, Nolen WA (2010) Early-onset bipolar disorder and treatment delay are risk factors for poor outcome in adulthood. J Clin Psychiatry 71:864–872. doi:10.4088/JCP.08m04994yel
Price TF, Peterson CK, Harmon-Jones E (2011) The emotive neuroscience of embodiment. Motiv Emot 36:27–37. doi:10.1007/s11031-011-9258-1
Quintana DS, Heathers JAJ, Kemp AH (2012) On the validity of using the Polar RS800 heart rate monitor for heart rate variability research. Eur J Appl Physiol 112:4179–4180. doi:10.1007/s00421-012-2453-2
Rao U, Dahl RE, Ryan ND, Birmaher B, Williamson DE, Rao R, Kaufman J (2002) Heterogeneity in EEG sleep findings in adolescent depression: unipolar versus bipolar clinical course. J Affect Disord 70:273–280. doi:10.1016/S0165-0327(01)00396-2

Sachs GS (2013) Unmet needs in the assessment and management of bipolar I depression. J Clin Psychiatry 74:e11. doi:10.4088/JCP.12065tx1c

Saul JP (1990) Beat-to-beat variations of heart rate reflect modulation of cardiac autonomic outflow. Physiology 5:32–37

Schneider MR, Delbello MP, McNamara RK, Strakowski SM, Adler CM (2012) Neuroprogression in bipolar disorder. Bipolar Disord 14:356–374. doi:10.1111/j.1399-5618.2012.01024.x

Schulz C, Mavrogiorgou P, Schröter A, Hegerl U, Juckel G (2000) Lithium-induced EEG changes in patients with affective disorders. Neuropsychobiology 42(Suppl 1):33–37

Siegel GJ (2013) Extreme EEG: psychophysiology where and when you'd least expect it. Psychophysiol 50:S16–S17. doi:10.1111/psyp.12112

Silverstone PH, Bell EC, Willson MC, Dave S, Wilman AH (2005) Lithium alters brain activation in bipolar disorder in a task- and state-dependent manner: an fMRI study. Ann Gen Psychiatry 4:14–14. doi:10.1186/1744-859X-4-14

Small J, Milstein V, Kellams J, Miller M, Boyko O, Small T (1989) EEG topography in psychiatric diagnosis and drug treatment. Ann Clin Psychiatry 1:7–17. doi:10.3109/10401238909149858

Small JG, Milstein V, Malloy FW, Klapper MH, Golay SJ, Medlock CE (1998) Topographic EEG studies of mania. Clin EEG Neurosci 29:59–66. doi:10.1177/155005949802900203

Small JG, Milstein V, Malloy FW, Medlock CE, Klapper MH (1999) Clinical and quantitative EEG studies of mania. J Affect Disord 53:217–224. doi:10.1016/S0165-0327(98)00124-4

Souza VB, Muir WJ, Walker MT, Glabus MF, Roxborough HM, Sharp CW, Dunan JR, Blackwood DH (1995) Auditory P300 event-related potentials and neuropsychological performance in schizophrenia and bipolar affective disorder. Biol Psychiatry 37:300–310. doi:10.1016/0006-3223(94)00131-L

Strakowski SM, Adler CM, Almeida J, Altshuler LL, Blumberg HP, Chang KD, Delbello MP, Frangou S, McIntosh A, Phillips ML, Sussman JE, Townsend JD (2012) The functional neuroanatomy of bipolar disorder: a consensus model. Bipolar Disord 14:313–325. doi:10.1111/j.1399-5618.2012.01022.x

Strakowski SM, Delbello MP, Adler CM (2005) The functional neuroanatomy of bipolar disorder: a review of neuroimaging findings. Mol Psychiatry 10:105–116. doi:10.1038/sj.mp.4001585

Suppes T, Leverich GS, Keck PE, Nolen WA, Denicoff KD, Altshuler LL, McElroy SL, Rush AJ, Kupka R, Frye MA, Bickel M, Post RM (2001) The Stanley Foundation Bipolar Treatment Outcome Network. II. Demographics and illness characteristics of the first 261 patients. J Affect Disord 67:45–59. doi:10.1016/S0165-0327(01)00432-3

Task Force of the European Society of Cardiology the North American Society of Pacing Electrophysiology (1996) Heart rate variability: standards of measurement, physiological interpretation, and clinical use. Circulation 93:1043–1065. doi:10.1161/01.CIR.93.5.1043

Thayer JF, Ahs F, Fredrikson M, Sollers JJ, Wager TD (2012) A meta-analysis of heart rate variability and neuroimaging studies: implications for heart rate variability as a marker of stress and health. Neurosci Biobehav Rev 36:747–756. doi:10.1016/j.neubiorev.2011.11.009

Thayer JF, Hansen AL, Saus-Rose E, Johnsen BH (2009) Heart rate variability, prefrontal neural function, and cognitive performance: the neurovisceral integration perspective on self-regulation, adaptation, and health. Ann Behav Med 37:141–153. doi:10.1007/s12160-009-9101-z

Todder D, Bersudsky Y, Cohen H (2005) Nonlinear analysis of RR interval in euthymic bipolar disorder. Auton Neurosci 117:127–131. doi:10.1016/j.autneu.2004.11.006

Tohen M, Frank E, Bowden CL, Colom F, Ghaemi SN, Yatham LN, Malhi GS, Calabrese JR, Nolen WA, Vieta E, Kapczinski F, Goodwin GM, Suppes T, Sachs GS, Chengappa KR, Grunze H, Mitchell PB, Kanba S, Berk M (2009) The International Society for Bipolar Disorders (ISBD) Task Force report on the nomenclature of course and outcome in bipolar disorders. Bipolar Disord 11:453–473. doi:10.1111/j.1399-5618.2009.00726.x

Townsend J, Altshuler LL (2012) Emotion processing and regulation in bipolar disorder: a review. Bipolar Disord 14:326–339. doi:10.1111/j.1399-5618.2012.01021.x

Townsend JD, Bookheimer SY, Foland-Ross LC, Moody TD, Eisenberger NI, Fischer JS, Cohen MS, Sugar CA, Altshuler LL (2012) Deficits in inferior frontal cortex activation in euthymic bipolar disorder patients during a response inhibition task. Bipolar Disord 14:442–450. doi:10.1111/j.1399-5618.2012.01020.x

Whalley HC, Papmeyer M, Sprooten E, Lawrie SM, Sussmann JE, McIntosh AM (2012) Review of functional magnetic resonance imaging studies comparing bipolar disorder and schizophrenia. Bipolar Disord 14:411–431. doi:10.1111/j.1399-5618.2012.01016.x

Wyczesany M, Kaiser J, Coenen AML (2010) Associations between self-report of emotional state and the EEG patterns in affective disorders patients. J Psychophysiol 24:33–40. doi:10.1027/0269-8803/a000004

Physiological Correlates of Positive Symptoms in Schizophrenia

Silvana Galderisi, Annarita Vignapiano, Armida Mucci and Nash N. Boutros

Abstract Patients with schizophrenia have been hypothesized to have a functional impairment in filtering irrelevant sensory information, which may result in positive symptoms such as hallucinations or delusions. Many evidences suggest that abnormalities in the event-related brain potentials (ERPs), resting state electroencephalography (EEG) and synchronized oscillatory activity of neurons may reflect core pathophysiological mechanisms of schizophrenia. Abnormalities in amplitude and latency of the ERPs reflecting aberrations in gating and difficulties in the detection of changes in auditory stimuli, as well as defects in stimuli evaluation and integration of information are common in patients with schizophrenia. This chapter highlights the findings of electrophysiological studies in schizophrenia dealing with early sensory perception and attention, automatic sensory detection of stimuli changes and cognitive evaluation and integration of information, relevant to the pathophysiological mechanisms underpinning hallucinations and delusions. Results of electrophysiological studies investigating the neural correlates of positive symptoms suggest aberrant intrinsic organization of functional brain networks.

Keywords Schizophrenia · Hallucinations · Delusions · Event-related potentials · Neural oscillations

S. Galderisi · A. Vignapiano · A. Mucci (✉)
Department of Psychiatry, University of Naples SUN, Largo Madonna delle Grazie, 80138 Naples, NA, Italy
e-mail: armida.mucci@gmail.com

S. Galderisi
e-mail: silvana.galderisi@gmail.com

A. Vignapiano
e-mail: vignapiano2004@libero.it

N. N. Boutros
Department of Psychiatry and Neurosciences, University of Missouri Kansas City (UMKC), Kansas, MO, USA
e-mail: boutrosn@umkc.edu

Curr Topics Behav Neurosci (2014) 21: 103–128
DOI: 10.1007/7854_2014_322

Published Online: 12 June 2014

Contents

1 Introduction

Electroencephalography (EEG) was the first physiological method used to study brain information processing organization in schizophrenia. It is a noninvasive neurophysiological technique that allows, by its temporal resolution, to study the fast changing of brain activity during information processing; it is able to record changes of the electrical activity of the brain on a millisecond time scale, the same time window of the neural activity. Unfortunately, the low spatial resolution of the EEG does not allow to assess the corresponding neural sources of electrical potentials with accuracy (van Lutterveld et al. 2011). Neurophysiological correlates of hallucinations and delusions have been detected both in resting state EEG and event-related brain potentials (ERPs) (Kayser and Tenke 2006; van Swam et al. 2013). Furthermore, recent models have proposed that schizophrenia symptoms may be associated with aberrant functional connectivity of multiple neural networks, including those subserving speech perception and production, social cognition and self-referential processing underlying the sense of agency (Wible et al. 2009).

ERP and oscillatory rhythms, compared to other neurobiological indices, allow a detailed analysis of information processing stages: the pre-attentive one, in which redundant information is filtered out by the brain, the perceptive one, modulated by voluntary attention, and later stages in which perceptual information is integrated with the ongoing task and with the internal and external context to guide behavior (Rissling et al. 2010). Furthermore, the use of magnetoencephalography (MEG), which also has a high temporal resolution to record cortical brain activity, is complementary to EEG as it can detect neural sources tangential to the surface of the cortex accurately. In fact, magnetic fields, unlike electrical ones, are minimally influenced by volume conductors, such as brain tissue, skull, and scalp (Hämäläinen et al. 1993).

In this chapter, we focus on ERPs findings related to early sensory perception and attention (P50, N100, and N200), cognitive evaluation (P300), automatic detection of deviant sensory characteristics [mismatch negativity (MMN)], and integration of information at a semantic level (N400), as well as on fast and slow oscillatory activity in order to discuss some key issues concerning pathophysiological models of positive symptoms in schizophrenia.

2 ERP Findings

2.1 P50

The P50 waveform, identified 50 ms after stimulus onset, is a well-known measure of early auditory processing. The attenuation of the P50 response to a second click (S2) delivered within 500 ms from the first click (S1) is usually interpreted as evidence for neuronal sensory gating and was well documented as dysfunctional in schizophrenia (Olincy et al. 2010). The P50 suppression also termed ''P50 gating,'' is conventionally measured as the ratio of S2 to S1 P50 amplitude and is thought to reflect an individual's ability to screen out, or ''gate,'' trivial or repetitive stimuli, in order to protect against information overload (Braff et al. 2007).

Although some earlier studies (Sartorius 1986; Andreasen and Flaum 1991) did not report any significant association between the P50 gating alterations and positive symptoms in subject with schizophrenia, recently Hirano et al. (2010) showed that patients with increased left P50 ratios (i.e., deficit in gating) to the human voice showed more severe auditory hallucinations. Smith et al. (2013) showed a negative correlation between P50 amplitude to S1 and the severity of auditory hallucinations traits, but no relationship of the same parameter with the score on a scale measuring hallucinations during recording. The findings might indicate that P50 gating abnormalities were related to trait but not state severity of hallucinations (Table 1) and predominantly dependent on S1 P50 amplitude reduction. In fact the interrelationship between S1 and S2 response amplitudes in relationship to the assessment of gating is complex (Chang et al. 2011).

2.2 N100

N100 is a middle-latency, negative component of the ERPs, peaking between 80 and 120 ms after stimulus onset, it is investigated by presenting a series of tones and is larger in response to "attended" than "unattended" stimuli. This component, generated in the primary and secondary auditory cortex (Naatanen and Picton 1987), reflects earlier sensory and attentional processing, it is an excellent probe of

Table 1 Electrophysiological correlates of positive symptoms in patients with schizophrenia

Study	Sample	Assessment	Measurement	Findings
P50				
Hirano et al. (2010)	22 Sch 28 HC	SAPS	P50 auditory gating	Higher P50 gating ratio (gating deficit) was related to more severe auditory hallucinations
Smith et al. (2013)	16 Sch 21 HC	PSYRATS	P50 auditory gating	Negative correlations between reduction of P50 amplitude to S1 and the severity of the hallucinations
N100				
Hubl et al. (2007)	7 psychotic patients with AVH	PANSS Oulis AHRS	N100 measured during the simultaneous occurrence of hallucinations	N100 reduced during periods of AVH
Ford et al. (2001b)	15 Sch 15 HC	BPRS	Acoustic and visual stimulation during three conditions: while subjects were silent, when spontaneous inner speech occurred and during directed inner speech	N100 elicited by acoustic stimuli was lower during directed inner speech than during the silent baseline condition in HC but not in Sch
Ford et al. (2001c)	12 Sch 10 HC	BPRS	Acoustic and visual probes during talking aloud, listening to one's speech played back, and silent baseline	Suppression of N100 during speech production in HC and lack of N100 suppression in Sch
Ford and Mathalon (2004)	12 Sch 10 HC	BPRS	Talking and listening task EEG coherence	Missing EEG coherence between frontal and temporal regions during talking in hallucinating patients
Heinks-Maldonado et al. (2007)	20 Sch 17 HC	PANSS SANS SAPS	N100 during speaking or listening	Smaller N100 suppression during speaking was related to more severe hallucinations
Laurent and Baribeau (1992)	7 Sch with thought disorder (FTD+) 7 Sch without FTD 7 HC	Andreasen's Checklist of positive and negative symptoms test for thought disorders	Auditory task	FTD+ correlated with reduced N100 amplitude

(continued)

Table 1 (continued)

Study	Sample	Assessment	Measurement	Findings
Ford et al. (1999)	28 Sch severely ill 29 Sch moderately ill 30 HC	BPRS	Two tone auditory oddball task	No difference of N100 amplitude between severely ill and moderately ill patients
Bruder et al. (2001)	13 Sch paranoid 13 Sch undifferentiated 14 HC	PANSS TDI BPRS	Oddball task using syllables or complex tones	N100 amplitude was not associated with thought disorder
N200				
Bruder et al. (2001)	13 paranoid Sch 13 undifferentiated Sch	PANSS TDI	Phonetic task Tonal task	Greater N200, at fronto-central sites and at left lateral temporo-parietal sites in paranoid Sch
Guillem et al. (2003)	17 Sch with low-scores (RD-) 21 Sch with high scores (RD+) on reality distortion 22 HC	PANSS SANS SAPS	Recognition memory task for faces	Posterior N200 was increased in RD+
Sumich et al. (2006)	20 recent-onset psychosis 20 HC	PANSS	Two tone oddball task	Impaired insight/judgment correlated positively with N200 amplitude
Sumich et al. (2008)	72 HC	Paranormal ideation questionnaire Oxford and Liverpool inventory of feelings and experiences subscale	Two tone oddball task	Unusual experiences were positively associated with left-anterior N200 amplitude
P300				
Ford et al. (1999)	28 severely ill Sch 29 moderately ill Sch 30 HC	BPRS factor	Two tone oddball task	In the combined Sch sample P300 amplitude reduction was associated with the severity of thought disorder

(continued)

Table 1 (continued)

Study	Sample	Assessment	Measurement	Findings
Mathalon et al. (2000b)	36 Sch at multiple time points 34 HC	BPRS	Automatic and effortful auditory and visual paradigm	P3b amplitude tracked BPRS positive symptom severity over time (with larger amplitude associated with symptoms improvement)
Liu et al. (2004)	22 positive symptoms Sch 27 negative symptoms Sch 50 HC	PANSS	Two tone oddball task	Sch with negative symptoms had P300 amplitude decrements and longer latencies with respect to those with positive symptoms
Egan et al. (1994)	16 Sch 16 HC	PSAS	Auditory and visual discrimination tasks	Auditory P300 amplitude correlated inversely with positive symptoms
O'Donnell et al. (1994)	15 Sch 14 HC	SAPS, SANS	Two tone oddball task	Association of P300 amplitude reduction with delusions and thought disorder
Kawasaki et al. (1997)	25 Sch	PANSS	Two tone oddball task	P300 amplitude reduction was associated with positive symptoms
Havermans et al. (1999)	15 AVH 15 non-AVH 17 HC	BPRS	Two tone oddball task	Strong association between P300 and severity of hallucinations
Turetsky et al. (1998)	65 Sch pre and post-treatment with antipsychotic drugs 48 HC	BPRS SANS SAPS	Two tone oddball task	Improvement in delusions was associated with increase in P3b right parietal subcomponent amplitude; improvement in hallucinations was associated with increase of P3a amplitude
Papageorgiou et al. (2004)	13 Sch 16 HC	BPRS	Short memory task	Reduced auditory P300 amplitude during hallucinations
Fisher et al. (2010)	12 AVH 12 non-AVH 12 HC	PSYRATS	Two-tone auditory oddball task	Reduced P3a amplitude associated with predisposition to auditory hallucinations

(continued)

Table 1 (continued)

Study	Sample	Assessment	Measurement	Findings
MMN				
Oades et al. (1997)	12 patients with paranoid symptoms 12 patients without paranoid symptoms 25 HC	DSM III BPRS	Three tone oddball task	Frontal MMN reduced in nonparanoid patients
Schall et al. (1998)	21 Sch 25 HC	SAPS	Visual and auditory discrimination task	Reduced MMN amplitude associated with hallucination severity
Hirayasu et al. (1998)	23 Sch 23 HC	PANSS	Two tone oddball task	Reduced MMN at the left inferior frontal regions was associated with hallucination severity; while MMN reduction at right and left superior frontal regions with negative symptoms
Youn et al. (2002)	15 Sch 15 HC	PANSS	Passive oddball task	Negative correlation of positive scores of PANSS with MMN asymmetry coefficient and left MMN equivalent current dipole
Kasai et al. (2002)	23 Sch 28 HC	PANSS	Three oddball tasks: pure-tone stimuli, phoneme duration and difference between vowels	MMN was lower in the right hemisphere under the phoneme-duration condition and associated with more severe negative symptoms
Fisher et al. (2008a)	12 Hallucinating Sch (HPs) 12 non-Hallucinating Sch (NPs) 12 HC	PANSS PSYRATS	Three oddball tasks: pure-tone stimuli, phoneme duration and difference between vowels	Reduced MMN amplitude associated with increased clarity of AVH in HPs
Fisher et al. (2008b)	12 HPs 12 NPs 12 HC	PANSS PSYRATS	Multi-features MMN paradigm	Reduced MMN to duration deviants in HPs vs. both HC and NPs

(continued)

Table 1 (continued)

Study	Sample	Assessment	Measurement	Findings
Fisher et al. (2012)	12 Sch acutely ill with persistent AHs 15 HC	PANSS PSYRATS	Multi-features MMN paradigm	MMN to several deviants was reduced in Sch; the reduction was associated with severity of AHs. Reduction of MMN to location deviants was associated with perceived location of AHs
N400				
Kostova et al. (2003)	36 Sch 40 HC	PANSS	Sentence-final word task	No correlation of N400 with delusions
Debruille et al. (2007)	14 More-delusional patients 16 Less-delusional patients	BPRS-E	Word-category task	Smaller N400s in more delusional patients
Kiang et al. (2007)	18 Sch 18 HC	SANS SAPS	Word-category task	Smaller amplitude of the N400 associated with SAPS-derived Psychotic factor
Prevost et al. (2011)	34 HC	Schizotypal-personality questionnaire	Word-category task	Delusional-like ideation scores associated with smaller N400 amplitudes
Oscillatory rhythms				
Spencer et al. (2008)	16 first episode Sch 16 first episode affective disorder 33 HC	PANSS	Auditory steady-state EEG responses gamma band	Phase locking factor of 40 Hz harmonic was positively correlated with total positive symptoms in Sch
Spencer et al. (2009)	18 chronic Sch 16 HC	SAPS	40 Hz auditory steady-state EEG	Left hemisphere PLF in Sch was positively correlated with auditory hallucinations

(continued)

Table 1 (continued)

Study	Sample	Assessment	Measurement	Findings
Spencer et al. (2004)	20 Sch 20 HC	PANSS	Gestalt perception task	In Sch, occipital response-locked gamma increase for gestalt stimuli was positively associated with conceptual disorganization on the PANSS, visual hallucinations, delusion of thought withdrawal and global thought disorder scores on the SAPS
Bucci et al. (2007)	10 deficit schizoprenia 13 non deficit schizophrenia 12 HC	SAPS	Three-tone auditory oddball task	In the whole patient sample, evoked gamma power was positively associated with SAPS-derived reality distortion scores (including hallucinations and delusions)
Lee et al. (2008)	25 patients prone to AVH 23 patients without AVH	PANSS	EEG recorded with the eyes alternatively closed and open	Patients prone to AVH showed a significant increased gamma activity in left frontal, temporal, and parietal regions
Ropohl et al. (2004)	1 Sch 13 HC	PANSS	Spontaneous brain activity recorded MEG Dipole density Method	Fast MEG dipole activity in the left auditory cortex associated with hallucinations
Reulbach et al. (2007)	8 AVH 8 Non-AVH 8 HC	PANSS	AVH-button-press task MEG	Presence of dipole activities in the fast frequency range during auditory hallucinations
Ferrarelli et al. (2012)	20 Sch 20 HC	PANSS	High-density electroencephalogram (Hd-EEG) recorded during transcranial magnetic stimulation (TMS)	In HC TMS pulses elicited robust activity in the 25–35 Hz frequency range over frontal electrodes In Sch the frequency of prefrontal oscillatory activity was strongly reduced and was inversely related to positive PANSS score

(continued)

Table 1 (continued)

Study	Sample	Assessment	Measurement	Findings
Fujimoto et al. (2013)	10 Sch 10 HC	PANSS	Auditory oddball task MEG	Imaginary coherence in gamma band between left fronto-parietal and temporal areas showed a positive correlation with hallucinations and delusions in Sch
Koenig et al. (2012)	18 patients with AVH 11 patients without AVH 28 HC	PANSS Oulis AHRS	Auditory stimulation with click tones global measure of phase-locking (GFS)	Global synchronization decreased with stimulation in hallucinating patients
Ford et al. (2007)	24 Sch 25 HC	BPRS SAPS SANS	Subjects uttered "Ah" Prespeech inter-trials coherence analysis	Patients failed to show an increase in frontal β-band inter-trial coherence before speech and the reduced coherence correlated with more severe hallucinations
Resting state EEG				
Lee et al. (2006)	25 patients with AVH 23 patients without AVH	PANSS	Resting state q-EEG	Increased beta frequency oscillations in speech-related areas in patients with AVH
Koenig et al. (1999)	9 neuroleptic-naive, first-episode Sch 18 HC	PANSS	Resting state EEG microstate analysis	One microstate class displayed different field configurations and shorter durations. The duration correlated with paranoid symptomatology
Kindler et al. (2011)	9 AVH	PANSS Oulis AHRS	Resting state EEG microstate analysis	Microstate D was significantly shorter during hallucinations

Sch patients with schizophrenia; *HC* healthy controls; *AVH* auditory verbal hallucination; *AHs* auditory hallucinations; *MEG* magnetoencephalography; *BPRS* brief psychiatric rating scale; *Oulis AHRS* Oulis auditory hallucinations rating scale; *PANSS* positive and negative syndrome scale; *PSAS* psychiatric symptom assessment scale; *PSYRATS* psychotic symptom rating scales; *SANS* scale for the assessment of negative symptoms; *SAPS* scale for the assessment of positive symptoms; *TDI* thought disorder index auditory cortical activity, albeit affected by activity in other areas of the brain (Ford et al. 2012). Its amplitude is strongly dependent on a number of factors such as stimulus intensity, arousal, expectancy, and selective attention (Rosburg et al. 2008).

The N100 amplitude reduction was reported by many studies on schizophrenia (Ford et al. 2001a; Sumich et al. 2006), and several authors investigated its relationship with positive symptoms.

As the N100 is a measure of auditory cortex activity, it can be used to compare auditory cortex activity in the presence and in the absence of hallucinations (van Lutterveld et al. 2011). During periods of auditory hallucinations, in subjects with schizophrenia, Hubl et al. (2007) found smaller N100 amplitudes and a reduced current density sources in the left temporal lobe. These findings could indicate a competition between auditory stimuli and auditory hallucinations for physiological resources in the primary auditory cortex, and that abnormal activation of this brain region during internal verbal production could be a component of auditory hallucinations. These results confirmed previous findings in which left temporal N100 amplitude negatively correlated with positive symptoms including hallucinations (Gallinat et al. 2002; Valkonen-Korhonen et al. 2003). Together, these findings indicated competition between auditory probes and hallucinations for auditory resources, with activation of the primary auditory cortex reflecting the physical acoustic image of verbal thoughts misperceived as voices (Hubl et al. 2007; Ford et al. 2012). As a matter of fact, it has been demonstrated by a series of elegant experiments that during vocalization and inner speech N100 was suppressed with respect to passive listening to speech sounds. This was an evidence of the operation of a corollary discharge from speech production areas to speech reception areas of the brain: an efference copy of the action (vocalization, thoughts) prepared the sensory cortex and competed with the afferent sensory signal (resulting in a suppression of the N100 to that signal), thus "informing" of the own nature of the speech sounds or thoughts.

Ford et al. (2001b, c) hypothesized that a deficit in the corollary discharge could be observed in patients suffering from auditory verbal hallucinations (AVH), who misperceived their own thoughts as externally generated speech sounds. The Authors demonstrated deficits in the corollary discharge, as reflected by the lack of suppression of the N100 during speech production. However, the amount of N100 suppression did not correlate with AVH in this experiment, while neural synchrony before speech onset was related to both subsequent N100 suppression during talking in controls and AVH in patients. The data suggest that the deficit of the corollary discharge is one of the possible mechanisms underlying AVH (Ford et al. 2012). Furthermore, the same authors found that EEG measures of coherence indicated an absence of inter-dependence between frontal speech production and temporal speech reception areas during talking in patients, especially those who hallucinated. On the whole, these data suggest that AVH are related to a failure of a corollary discharge which might be one of the reasons for the misattribution of inner speech to external sources (Ford and Mathalon 2004).

Heinks-Maldonado et al. (2007) conducted an elegant experiment to test the forward model of corollary discharge function: in this model when subjects speak (in the experiment uttering "Ah") the vocal motor system sends the efference copy to the auditory cortex so that when the subjects hear their self-produced speech sound the auditory cortex is suppressed and the subject recognize his own verbal production. In the experiment, they also altered the auditory feedback so that during vocalization (again uttering "Ah") the subjects heard their speech sound unaltered or shifted by 2-semitones or substituted by unaltered or shifted alien voice. As a control condition they also passively listened to the same speech sounds (unaltered or shifted self-voice or unaltered or shifted alien voice). In healthy controls N100 was suppressed only when receiving as feedback the unaltered self-voice during vocalization. Hallucinating patients (HPs) did not show the attenuation of N100 amplitude to unaltered self-voice. The lack of N100 suppression in the left hemisphere correlated with the Hallucination score on the Scale for the Assessment of Positive Symptoms (SAPS). In non-hallucinators, the pattern of suppression was heterogeneous with some subjects suppressing and others not suppressing the N100 to unaltered self-voice. This could reflect differences in hallucination history, suggesting that N100 suppression deficit might be related to a "hallucination trait" (Ford et al. 2012). Summarizing the findings of several studies support a role for the N100 (as well as for M100, the corresponding MEG component) in the study of abnormalities of the corollary discharge and AVH (Table 1). Although very often AVH are correlated with delusions, the association of N100 abnormalities with delusions are less clear (Ford et al. 2012).

A relationship between N100 abnormalities and thought disorder was rarely investigated and inconsistently reported (Laurent and Baribeau 1992; Ford et al. 1999; Bruder et al. 2001).

2.3 N200

The N200 is a negative deflection occurring at about 200 ms post-stimulus. It is thought to reflect top–down regulatory processes involved in automatic detection and recognition of novel stimuli (Tarbi et al. 2011). Some studies showed that in subjects with recent-onset psychosis, left fronto-central N200 amplitude enhancement was associated with poor insight/judgment, which could exacerbate positive symptoms (Sumich et al. 2006); however no relationship between N200 amplitude and reality distortion was observed. Only in paranoid patients, Bruder et al. (2001) found a greater N200, at fronto-central sites and at left lateral temporo-parietal sites, during a phonetic task. These findings were consistent with neuropsychological evidence of greater verbal abilities and left hemisphere dominance in paranoid than nonparanoid schizophrenia. The results were also in line with Guillem et al. (2003) that showed, in patients with high score on Reality Distortion, an increased posterior N200. These Authors hypothesized that N200 increase could reflect the over-activity related to emotionally significant stimuli during a state of reality distortion. Sumich et al. (2008) found in a nonclinical population a relationship between unusual experiences and N200 amplitude increase over left-anterior regions. As the unusual experiences were also associated with N100 decrease the Authors suggest that a poor bottom-up sensory processing, indexed by N100 reduction, may give greater control over perception to top–down processes, as indexed by increased N200 amplitude. The enhanced N200 could reflect shared mechanisms between psychotic symptoms in patients and unusual experiences in healthy subjects, such as inhibition of sensory input or auditory hypersensitivity (Hooley and Campbell 2002), a phenomenon that is observed in subjects with high scores for magical ideation (Dubal and Viaud-Delmon 2008). N200 data suggest that reality distortion is the consequence of several dysfunctions occurring in the cascade of cognitive and neural mechanisms involved in information processing (Table 1).

2.4 P300

The P300 is a positive wave that occurs 300 ms after rare or task relevant stimuli that are counted, detected or otherwise processed (Johnson 1986). The P300 amplitude is thought to reflect aspects of conscious processing of relevant stimuli. A target stimulus elicits the maximal P300 (called "P3b") with centro-parietal distribution that has been associated to controlled information processing, context updating, and response-related decisional stages (Campanella and Guerit 2009). While an infrequent distractor, novel or otherwise salient stimulus, with no task relevance, generates a robust fronto-centrally maximal P300 (called "P3a"), that reflects a shift in attention and the processing of novelty (Ford et al. 2012; van Swam et al. 2013).

P300 amplitude reduction in patients with schizophrenia versus healthy controls, initially reported by Roth and Cannon (1972), has been consistently replicated, independently from medication status, in both first-episode and chronic patients (Galderisi et al. 2009) and proposed as a potential endophenotype (Ishii et al. 2012). It has not been clarified, however, what kinds of clinical symptoms can result from the disturbance in information processing related to this P300 abnormality. Several studies have reported associations between P300 amplitude and the severity of negative symptoms (Pfefferbaum et al. 1989; Strik et al. 1993; Preuss et al. 2010; Kim et al. 2014), fewer have reported associations between P300 amplitude and positive symptoms (Juckel et al. 1996; Jeon and Polich 2003; Higashima et al. 2003). Ford et al. (1999) analyzed P300 in two groups of patients with schizophrenia that had a different degree of clinical severity. A marked reduction of P300 amplitude was found in patients with the most severe clinical picture; however, a relationship between P300 amplitude and Thought Disorder was found only when patients were combined. However, other studies have failed to find significant correlations between P300 amplitude and Thought Disorder or positive symptom cluster of schizophrenia (Liu et al. 2004; Bruder et al. 2001).

One reason for the inconsistent association between P300 and positive symptoms could be a restricted range of symptom severity values. Mathalon et al. (2000a) argued that inconsistent results in cross-sectional samples might be due to the trait and state influences on P300 amplitude variance and that multiple observations could be a better way to demonstrate a relationship between P300 and symptoms. Studies using longitudinal data have demonstrated significant increases in P300 amplitude with symptoms improvement (Schall et al. 1998; Turetsky et al. 1998) but most of these studies included only two time points, limiting their power to detect correlations between changes in P300 and clinical state. Mathalon et al. (2000b) carried out a longitudinal study with multiple observations discovering that P300 amplitude reduction tracked brief psychiatric rating scale (BPRS) total and positive symptom severity over time. Only auditory P3b amplitude reduction tracked positive symptoms severity. In their opinion, this could suggest that delusions and hallucinations interfered with fundamental information-processing functions, such as effortful attention to target stimuli, deducting neural resources generally available for processing environmental stimuli.

Some studies focused on the hypothesis that P300 amplitude reduction over the left temporal region, positive symptoms, and left temporal lobe anatomic abnormalities represented a cohesive process in chronic patients with schizophrenia (McCarley et al. 1991; O'Donnell et al. 1994; Egan et al. 1994) suggesting that the neuronal systems underlying the alterations of P300 could also mediate positive symptoms. McCarley et al. (1991) in their review reported a link between severity of thought disorder and total SAPS score and left temporal P300 amplitude. O'Donnell et al. (1994) reported correlations of P300 amplitude reduction with SAPS delusions and thought disorder subscores. Also Egan et al. (1994) discovered weaker but significant correlations between auditory P300 and measures of left temporal lobe structures and an inverse correlation between auditory P300 amplitude and positive symptoms. Kawasaki et al. (1997) failed to replicate the association between auditory P300 amplitude and left temporal lobe volume but confirmed that the P300 amplitude was correlated with the severity of positive thought disorder.

Turetsky et al. (1998) showed that, although the P300 deficits were relatively stable over time, a P3b right parietal subcomponent increase was associated with an improvement of delusions, a P3b left temporal subcomponent increase with an improvement of negative symptoms and BPRS total scores, whereas P3a increase correlated with auditory hallucinations reduction. This pattern of results suggest that psychotic symptoms interfere with processes associated with both P3a and P3b by diverting limited attentional resources from both automatically orienting and effortful directing attention to auditory stimuli.

As most patients with schizophrenia who experience auditory hallucinations and delusions also experience other symptoms, such as disorganization and negative symptoms, heterogeneous P3 findings may be related to diversity in symptomatology (van Lutterveld et al. 2011).

Authors investigating the relationship between P300 and AVH, showed that the amplitude of the P300 was inversely correlated to AVH severity, indicating an additional state dependency of P300 alterations (Havermans et al. 1999; Turetsky et al. 1998; Papageorgiou et al. 2004).

Fisher et al. (2008a) proposed that patients with AVH have an impairment in the processing of speech sounds that could be related to a competition between external and internal auditory stimuli, and deficits in attributing significance to incoming stimuli. They carried out a study comparing P3a amplitude, as an index of involuntary shift to auditory changes and novelty processing between hallucinating (HPs) and non-hallucinating schizophrenia (NPs) patients (Fisher et al. 2010). Only in HPs was found a decreased P3a. Furthermore, they found a negative correlation in HPs between P3a amplitude at central area and scores on the PSYRATS, which measures AVH as a trait, suggesting that the predisposition to auditory hallucinations is associated with reduced P3a amplitude. The finding may also provide additional evidence of AVH competing with incoming external stimuli for neural resources (Ford et al. 2009) (Table 1).

2.5 MMN

MMN is an ERP component that reflects a largely automatic and preattentive form of sensory processing. It is considered automatic because it does not require any behavioral response and can be elicited in the absence of explicit instructions to attend to the auditory stream (Lyytinen et al. 1992). MMN can be elicited by any auditory event that is deviant from the preceding events in a sequence: its elicitation indicates that a sequence was learned and that an auditory change was detected (Ford et al. 2012). The maximum of the amplitude difference between standard and deviant stimuli is located at fronto-central regions and MMN generators are located bilaterally in the left and right supratemporal lobes (Naatanen et al. 2007).

Marked MMN abnormalities are found in schizophrenia and appear to be robust and reliable especially for duration-deviant stimuli (Umbricht et al. 2003).

Some studies have investigated relationship between MMN and delusions with inconsistent results (Oades et al. 1997; Liu et al. 2007) but Authors only compared paranoid and nonparanoid patients.

Many studies focused on auditory hallucinations and their association with MMN deficits as indices of frontal and temporal lobe dysfunction (Schall et al. 1998). Three studies showed a significant negative correlation between severity of hallucinations and MMN abnormalities (amplitude or current source density reduction) at left hemispheric sites (Hirayasu et al. 1998; Youn et al. 2002; Fisher et al. 2011a, b). The correlation at left fronto temporal sites could reflect contributions of superior temporal gyrus sources (Hirayasu et al. 1998) and the increased asymmetry in patients with more severe positive symptoms (left < right) suggested that positive symptoms might be associated with abnormalities of the left temporal structures for speech (Youn et al. 2002). The correlation between left hemisphere MMN and positive symptomatology was not replicate in the study of Kasai et al. (2002) in which MMN was lower in the right hemisphere under the phoneme-duration condition and this was significantly associated with more severe negative symptoms: The discrepancies may be at least partly explained by the different methodologies (Fisher et al. 2008).

Fisher et al. (2008a) with the same paradigms used by Kasai et al. (2002) compared MMN amplitudes and latencies, for speech and nonspeech stimuli, between HPs and NPs. There was no difference between HPs and NPs on amplitude of MMN to the different stimulus conditions. However, in the HP subgroup, MMN amplitude in frontal area was negatively associated with AHs clarity (but not the duration or loudness), suggesting a relationship between AHs and preconscious auditory stimulus detection. Auditory hallucinations could compete with incoming external stimuli for finite resources in the auditory cortex; this was confirmed in a subsequent study by Fisher et al. (2008b), who used a multifeature MMN paradigm and found smaller MMNs in HPs to duration deviants vs. both healthy controls and NPs. In a more recent study, Fisher et al. (2012) applied the same experimental task to evaluate the MMN during an acute psychotic episode with severe degree of hallucinations. Patients showed deficits in MMN amplitude across several deviant types, which negatively correlated with state, trait, and frequency measures of AHs. Furthermore, the MMN to location deviants correlated with perceived location of AHs. On the whole, these data demonstrate that MMN may be a useful noninvasive tool for probing relationships between hallucinations and neural dysfunctions in schizophrenia (Table 1).

2.6 N400

The N400 is a negative deflection, manifesting 400 ms following stimulus presentation, sensitive to modulations of meaning at the semantic/associative level. It is widely spread across the scalp, usually with a maximum at centro-parietal electrode sites. This component has been taken to index semantic integration processes (Hagoort and Levelt 2009). Its amplitude is inversely proportional to the extent to which the eliciting stimulus fits semantic expectancies (Kutas and Hillyard 1984).

Schizophrenia patients exhibit N400 semantic priming deficits, suggesting impairment in using meaningful context to activate related concepts (Kiang et al. 2014), presumably resulting from impaired semantic memory associative networks (Metzler et al. 2013). The failure of the integration of the new information could explain the persistence of the delusions despite the awareness of contradictory information (Misra and Holcomb 2003).

Some studies investigated the association between N400 amplitude, as an index of information integration, and presence or severity of delusions, reporting discrepant findings. Kostova et al. (2003) revealed a correlation with formal thought disorder but not with delusions. Debruille et al. (2007) reported that patients who were more delusional displayed smaller N400 amplitude to category discrepant target words than less delusional patients, confirming the hypothesis that subjects with delusional beliefs do not properly process information that disconfirm any idea or belief. Kiang et al. (2007) showed a correlation between SAPS-derived Psychotic factor and reduction of the N400. They speculated that this relationship might reflect an association of psychotic symptoms with either decreased context use or broader spread of activation, suggesting that delusions could be related to a smaller than normal difference in processing of typical versus atypical members of a category, at least at a semantic level. Prevost et al. (2011) hypothesized that smaller N400 amplitude could be related to paranoid feelings. Using the same semantic task of Debruille et al. (2007) they induced paranoid feelings in healthy controls and showed that greater delusional-like ideation scores were associated with smaller N400 amplitudes. These findings suggested that the induction of paranoid feelings promoted a cognitive strategy similar to that displayed by patients (Debruille et al. 2007; Kiang et al. 2007), suggesting that inaccurate beliefs or ideas, including delusions, are associated with a weakness of semantic processes when participants experience paranoid feelings (Table 1).

3 Neural Oscillations in Schizophrenia

Synchronous rhythms represent a core mechanism for dynamic temporal coordination of neural activity in distributed brain networks (Wang 2010). Neural oscillations at low- (theta, alpha) and high-(beta/gamma) frequency ranges facilitate the transient formation of large-scale networks that represent the neural substrates of cognitive processes (Varela et al. 2001). Lower EEG frequencies preferentially establish synchronization over long distances, whereas oscillations in the beta and gamma range create synchronization with great precision in more circumscribed local cortical networks (Uhlhaas 2013). Consistent observations are that EEG fluctuations in the beta and gamma bands are abnormal in schizophrenia as a result of impaired interplay among many distributed cortical areas and their connections (Uhlhaas and Singer 2010). Many studies attempted to directly assess the correlation between synchrony, as assessed by power, amplitude or phase synchronization of evoked and induced beta- and gamma-band activity, and positive symptoms of schizophrenia (Lee et al. 2006; Spencer 2009). Spencer et al. (2004) found that occipital response-locked gamma increase was significantly correlated with conceptual disorganization on the positive and negative syndrome scale (PANSS), visual hallucinations, delusion of thought withdrawal and global thought disorder scores on the SAPS. Bucci et al. (2007) found a positive correlation between reality distortion (sum of delusions and hallucinations subscales scores on SAPS) and evoked gamma power, which is thought to be associated to early sensory and attentional processes. Spencer et al. (2008) found that the overall hallucination history of first-episode schizophrenia patients was positively correlated with the phase locking of the 40 Hz harmonic of 20 Hz auditory steady-state response. Subsequently, they found that 40 Hz source evoked power and phase locking factor (PLF) in the left auditory cortex were reduced in subjects with schizophrenia; PLF positively correlated with auditory hallucinations (Spencer et al. 2009). Other authors (Lee et al. 2008) found in patients prone to AVH a significant increased gamma activity in left frontal, temporal, and parietal regions. Using MEG investigations in patients with schizophrenia with auditory hallucinations, Ropohl et al. (2004) described an increase in fast dipole activity over the left temporal lobe during auditory hallucinations and Reulbach et al. (2007) found a significant increase of dipole activity in the fast frequency range during AVH in left frontal and temporal regions. It has been speculated that increased fast oscillations during AVH might index enhanced fronto-temporal synchronization, in main speech production and perception areas. These findings were in line with results of a recent study by Fujimoto et al. (2013) showing that the low gamma band imaginary coherence between left occipital and right fronto-parietal lobes had a positive correlation with hallucinations and delusions. Koenig et al. (2012), using a global measure of phase-locking (GFS), observed a left-lateralized decrease of 40 Hz EEG activity during stimulation only in the hallucinators suggesting a disruption of the normally observed increased synchronization in the cortical networks.

The pattern of spontaneously occurring gamma-band oscillations may differ from that associated with entrainment of neural activity through transcranial magnetic stimulation (TMS). Ferrarelli et al. (2012) applied TMS over four cortical areas and analyzed stimulus-evoked EEG-activity. In controls, TMS pulses elicited robust activity in the 25- to 35-Hz frequency range over frontal electrodes. In patients, the peak frequency of evoked oscillations over frontal leads was characterized by a reduction of ~ 10 Hz and the prefrontal natural frequency was inversely related with delusions PANSS sub-scores, suggesting a deficit of the underlying prefrontal cortical/thalamocortical circuitry. In contrast to gamma-band activity, the role of beta-band oscillations has been less explored. Ford et al. (2007) suggested that the synchronized beta-band activity could reflect a forward model which dampened auditory responsiveness to self-generated speech, and in patients this mechanism could be impaired. The reduced pre-speech inter-trial coherence negatively correlated with hallucination severity. In HPs, a significant increase of beta frequency oscillations in speech-related areas was demonstrated (Lee et al. 2006), and a reduced beta-band synchrony in the pre-speech condition was found (Ford et al. 2007).

The local increase of neural oscillations seen in schizophrenia patients with positive symptoms, accompanied by a deficit in the precise synchronization of these oscillations between cortical areas might be a neurophysiological correlate of impairment in corollary discharge (Uhlhaas et al. 2008) (Table 1).

4 Resting State EEG

Splitting EEG-resting state data into sub second time epochs is possible to investigate microstates, (Lehmann et al. 1987) thought to be manifestations of momentary global functional states of the brain. Distinct scalp fields are generated by differently activated neural networks, thus altered microstates correspond to altered brain functions (Lehmann et al. 2010).

In patients with schizophrenia, consistent differences to normative data were reported in resting-state EEG microstates, with shorter microstate with fronto-central distribution (Koenig et al. 1999; Lehmann et al. 2005). This shortening was correlated to paranoid symptomatology (Koenig et al. 1999). Kindler et al. (2011) investigated various microstate markers and compared periods with and without auditory hallucinations for understanding if the alterations reflected positive psychotic traits or present psychopathology. The fronto-central microstate was significantly shorter in periods with hallucinations, suggesting that premature termination of this microstate might facilitate the misattribution of self-generated inner speech to external sources or inadequate processing of context information (Table 1).

5 Conclusions

This chapter provided an overview of electrophysiological signs of aberrant intrinsic organization of multiple functional networks in schizophrenia, which correlate with the severity of positive symptoms, including hallucinations and delusions and help in the refinement of pathophysiological hypotheses concerning these complex phenomena.

In current diagnostic classification systems, the diagnostic categories used are not based on neurobiological knowledge but rather on phenotypes, course, prognosis, and therapy; therefore, it is not surprising that electrophysiological studies in patients with schizophrenia demonstrate inconsistent and conflicting findings. Patients with schizophrenia who experience delusions and/or auditory hallucinations do experience also other symptoms such as disorganization or avolition, dimensions which have different pathophysiology and electrophysiological correlates (van Lutterveld et al. 2011). The dimensional or symptom capture approaches demonstrated a more homogeneous pattern of electrophysiological findings (Fisher et al. 2011a; Ford et al. 2012; van Swam et al. 2013; Boutros et al. 2014, this volume). Several studies have fostered hypothesis on hallucinations, formal thought disorders and delusions and demonstrated an association of these phenomena with aberrant intrinsic organization of functional brain networks. The safety, noninvasiveness, and limited costs of electrophysiological methods allow to study healthy subjects and/or unaffected relatives and to carry out longitudinal investigations in patients to test hypotheses and verify trait and state influences on the reported abnormalities, as well as degree of heritability to identify endophenotypes for genetic studies. The standardization of methods might allow the design of large multicenter studies, instrumental in the progress of research knowledge in heterogeneous disorders.

References

Andreasen NC, Flaum M (1991) Schizophrenia: the characteristic symptoms. Schizophr Bull 17(1):27–49

Boutros NN, Mucci A, Vignapiano A, Galderisi S (2014) Electrophysiological abberations associated with negative symptoms in schizophrenia. Curr Top Behav Neurosci. doi:10.1007/7854_2014_303

Braff DL, Light GA, Swerdlow NR (2007) Prepulse inhibition and P50 suppression are both deficient but not correlated in schizophrenia patients. Biol Psychiatry 61(10):1204–1207. doi:10.1016/j.biopsych.2006.08.015

Bruder GE, Kayser J, Tenke CE, Friedman M, Malaspina D, Gorman JM (2001) Event-related potentials in schizophrenia during tonal and phonetic oddball tasks: relations to diagnostic subtype, symptom features and verbal memory. Biol Psychiatry 50(6):447–452

Bucci P, Mucci A, Merlotti E, Volpe U, Galderisi S (2007) Induced gamma activity and event-related coherence in schizophrenia. Clin EEG Neurosci 38(2):96–104

Campanella S, Guerit JM (2009) How clinical neurophysiology may contribute to the understanding of a psychiatric disease such as schizophrenia. Neurophysiol Clin (Clinical neurophysiology) 39(1):31–39. doi:10.1016/j.neucli.2008.12.002

Chang WP, Arfken C, Boutros NN (2011) Probing the relative contribution of the first and second responses to sensory gating indices: a meta-analysis. Psychophysiology 48(7):180–192. doi:10.1111/j.1469-8986.2010.01168.x

Debruille JB, Kumar N, Saheb D, Chintoh A, Gharghi D, Lionnet C, King S (2007) Delusions and processing of discrepant information: an event-related brain potential study. Schizophr Res 89(1–3):261–277. doi:10.1016/j.schres.2006.07.014

Dubal S, Viaud-Delmon I (2008) Magical ideation and hyperacusis. Cortex (a journal devoted to the study of the nervous system and behavior) 44(10):1379–1386. doi:10.1016/j.cortex.2007.06.008

Egan MF, Duncan CC, Suddath RL, Kirch DG, Mirsky AF, Wyatt RJ (1994) Event-related potential abnormalities correlate with structural brain alterations and clinical features in patients with chronic schizophrenia. Schizophr Res 11(3):259–271

Ferrarelli F, Sarasso S, Guller Y, Riedner BA, Peterson MJ, Bellesi M, Massimini M, Postle BR, Tononi G (2012) Reduced natural oscillatory frequency of frontal thalamocortical circuits in schizophrenia. Arch Gen Psychiatry 69(8):766–774. doi:10.1001/archgenpsychiatry.2012.147

Fisher DJ, Labelle A, Knott VJ (2008a) Auditory hallucinations and the mismatch negativity: processing speech and non-speech sounds in schizophrenia. Int J Psychophysiol (official journal of the International Organization of Psychophysiology) 70(1):3–15. doi:10.1016/j.ijpsycho.2008.04.001

Fisher DJ, Labelle A, Knott VJ (2008b) The right profile: mismatch negativity in schizophrenia with and without auditory hallucinations as measured by a multi-feature paradigm. Clin Neurophysiol (official journal of the International Federation of Clinical Neurophysiology) 119(4):909–921. doi:10.1016/j.clinph.2007.12.005

Fisher DJ, Labelle A, Knott VJ (2010) Auditory hallucinations and the P3a: attention-switching to speech in schizophrenia. Biol Psychol 85(3):417–423. doi:10.1016/j.biopsycho.2010.09.003

Fisher DJ, Grant B, Smith DM, Borracci G, Labelle A, Knott VJ (2011a) Effects of auditory hallucinations on the mismatch negativity (MMN) in schizophrenia as measured by a modified 'optimal' multi-feature paradigm. Int J Psychophysiol (official journal of the International Organization of Psychophysiology) 81(3):245–251. doi:10.1016/j.ijpsycho.2011.06.018

Fisher DJ, Grant B, Smith DM, Knott VJ (2011b) Effects of deviant probability on the 'optimal' multi-feature mismatch negativity (MMN) paradigm. Int J Psychophysiol (official journal of the International Organization of Psychophysiology) 79(2):311–315. doi:10.1016/j.ijpsycho.2010.11.006

Fisher DJ, Labelle A, Knott VJ (2012) Alterations of mismatch negativity (MMN) in schizophrenia patients with auditory hallucinations experiencing acute exacerbation of illness. Schizophr Res 139:237–245

Ford JM, Mathalon DH (2004) Electrophysiological evidence of corollary discharge dysfunction in schizophrenia during talking and thinking. J Psychiatr Res 38(1):37–46

Ford JM, Mathalon DH, Marsh L, Faustman WO, Harris D, Hoff AL, Beal M, Pfefferbaum A (1999) P300 amplitude is related to clinical state in severely and moderately ill patients with schizophrenia. Biol Psychiatry 46(1):94–101

Ford JM, Mathalon DH, Kalba S, Marsh L, Pfefferbaum A (2001a) N1 and P300 abnormalities in patients with schizophrenia, epilepsy, and epilepsy with schizophrenialike features. Biol Psychiatry 49(10):848–860

Ford JM, Mathalon DH, Kalba S, Whitfield S, Faustman WO, Roth WT (2001b) Cortical responsiveness during inner speech in schizophrenia: an event-related potential study. Am J Psychiatry 158(11):1914–1916

Ford JM, Mathalon DH, Kalba S, Whitfield S, Faustman WO, Roth WT (2001c) Cortical responsiveness during talking and listening in schizophrenia: an event-related brain potential study. Biol Psychiatry 50(7):540–549

Ford JM, Roach BJ, Faustman WO, Mathalon DH (2007) Synch before you speak: auditory hallucinations in schizophrenia. Am J Psychiatry 164(3):458–466

Ford JM, Roach BJ, Jorgensen KW, Turner JA, Brown GG, Notestine R, Bischoff-Grethe A, Greve D, Wible C, Lauriello J, Belger A, Mueller BA, Calhoun V, Preda A, Keator D, O'Leary DS, Lim KO, Glover G, Potkin SG, Fbirn, Mathalon DH (2009) Tuning into the voices: a multisite FMRI study of auditory hallucinations. Schizophr Bull 35(1):58–66. doi:10.1093/schbul/sbn140

Ford JM, Dierks T, Fisher DJ, Herrmann CS, Hubl D, Kindler J, Koenig T, Mathalon DH, Spencer KM, Strik W, van Lutterveld R (2012) Neurophysiological studies of auditory verbal hallucinations. Schizophr Bull 38(4):715–723. doi:10.1093/schbul/sbs009

Fujimoto T, Okumura E, Takeuchi K, Kodabashi A, Otsubo T, Nakamura K, Kamiya S, Higashi Y, Yuji T, Honda K, Shimooki S, Tamura T (2013) Dysfunctional cortical connectivity during the auditory oddball task in patients with schizophrenia. Open Neuroimaging J 7:15–26. doi:10.2174/1874440001307010015

Galderisi S, Mucci A, Volpe U, Boutros N (2009) Evidence-based medicine and electrophysiology in schizophrenia. Clin EEG Neurosci 40(2):62–77

Gallinat J, Mulert C, Bajbouj M, Herrmann WM, Schunter J, Senkowski D, Moukhtieva R, Kronfeldt D, Winterer G (2002) Frontal and temporal dysfunction of auditory stimulus processing in schizophrenia. NeuroImage 17(1):110–127

Guillem F, Bicu M, Pampoulova T, Hooper R, Bloom D, Wolf MA, Messier J, Desautels R, Todorov C, Lalonde P, Debruille JB (2003) The cognitive and anatomo-functional basis of reality distortion in schizophrenia: a view from memory event-related potentials. Psychiatry Res 117(2):137–158

Hagoort P, Levelt WJ (2009) Neuroscience. The speaking brain. Science 326(5951):372–373. doi:10.1126/science.1181675

Hämäläinen M, Hari R, Ilmoniemi R, Knuutila J, Lounasmaa O (1993) Magnetoencephalography—theory, instrumentation, and applications to non invasive studies of the working human brain. Rev Mod Phys 65:413–497

Havermans R, Honig A, Vuurman EF, Krabbendam L, Wilmink J, Lamers T, Verheecke CJ, Jolles J, Romme MA, van Praag HM (1999) A controlled study of temporal lobe structure volumes and P300 responses in schizophrenic patients with persistent auditory hallucinations. Schizophr Res 38(2–3):151–158

Heinks-Maldonado TH, Mathalon DH, Houde JF, Gray M, Faustman WO, Ford JM (2007) Relationship of imprecise corollary discharge in schizophrenia to auditory hallucinations. Arch Gen Psychiatry 64(3):286–296. doi:10.1001/archpsyc.64.3.286

Higashima M, Nagasawa T, Kawasaki Y, Oka T, Sakai N, Tsukada T, Koshino Y (2003) Auditory P300 amplitude as a state marker for positive symptoms in schizophrenia: cross-sectional and retrospective longitudinal studies. Schizophr Res 59(2–3):147–157

Hirano Y, Hirano S, Maekawa T, Obayashi C, Oribe N, Monji A, Kasai K, Kanba S, Onitsuka T (2010) Auditory gating deficit to human voices in schizophrenia: a MEG study. Schizophr Res 117(1):61–67. doi:10.1016/j.schres.2009.09.003

Hirayasu Y, Potts GF, O'Donnell BF, Kwon JS, Arakaki H, Akdag SJ, Levitt JJ, Shenton ME, McCarley RW (1998) Auditory mismatch negativity in schizophrenia: topographic evaluation with a high-density recording montage. Am J Psychiatry 155(9):1281–1284

Hooley JM, Campbell C (2002) Control and controllability: beliefs and behaviour in high and low expressed emotion relatives. Psychol Med 32(6):1091–1099

Hubl D, Koenig T, Strik WK, Garcia LM, Dierks T (2007) Competition for neuronal resources: how hallucinations make themselves heard. Br J Psychiatry (the journal of mental science) 190:57–62. doi:10.1192/bjp.bp.106.022954

Ishii R, Takahashi H, Kurimoto R, Aoki Y, Ikeda S, Hata M, Ikezawa K, Canuet L, Nakahachi T, Iwase M, Takeda M (2012) Endophenotypes in schizophrenia: a review of electrophysiological studies. Seishin shinkeigaku zasshi (Psychiatria et neurologia Japonica) 114(6):629–646

Jeon YW, Polich J (2003) Meta-analysis of P300 and schizophrenia: patients, paradigms, and practical implications. Psychophysiology 40(5):684–701

Johnson R Jr (1986) A triarchic model of P300 amplitude. Psychophysiology 23(4):367–384

Juckel G, Muller-Schubert A, Gaebel W, Hegerl U (1996) Residual symptoms and P300 in schizophrenic outpatients. Psychiatry Res 65(1):23–32

Kasai K, Nakagome K, Itoh K, Koshida I, Hata A, Iwanami A, Fukuda M, Kato N (2002) Impaired cortical network for preattentive detection of change in speech sounds in schizophrenia: a high-resolution event-related potential study. Am J Psychiatry 159(4):546–553

Kawasaki Y, Maeda Y, Higashima M, Nagasawa T, Koshino Y, Suzuki M, Ide Y (1997) Reduced auditory P300 amplitude, medial temporal volume reduction and psychopathology in schizophrenia. Schizophr Res 26(2–3):107–115

Kayser J, Tenke CE (2006) Principal components analysis of Laplacian waveforms as a generic method for identifying ERP generator patterns: I. Evaluation with auditory oddball tasks. Clin Neurophysiol (official journal of the International Federation of Clinical Neurophysiology) 117(2):348–368. doi:10.1016/j.clinph.2005.08.034

Kiang M, Kutas M, Light GA, Braff DL (2007) Electrophysiological insights into conceptual disorganization in schizophrenia. Schizophr Res 92(1–3):225–236. doi:10.1016/j.schres.2007.02.001

Kiang M, Christensen BK, Zipursky RB (2014) Event-related brain potential study of semantic priming in unaffected first-degree relatives of schizophrenia patients. Schizophr Res 153(1–3):78–86. doi:10.1016/j.schres.2014.01.001

Kim DW, Shim M, Kim JI, Im CH, Lee SH (2014) Source activation of P300 correlates with negative symptom severity in patients with schizophrenia. Brain Topogr 27(2):307–317. doi:10.1007/s10548-013-0306-x

Kindler J, Hubl D, Strik WK, Dierks T, Koenig T (2011) Resting-state EEG in schizophrenia: auditory verbal hallucinations are related to shortening of specific microstates. Clin Neurophysiol (official journal of the International Federation of Clinical Neurophysiology) 122(6):1179–1182. doi:10.1016/j.clinph.2010.10.042

Koenig T, Lehmann D, Merlo MC, Kochi K, Hell D, Koukkou M (1999) A deviant EEG brain microstate in acute, neuroleptic-naive schizophrenics at rest. Eur Arch Psychiatry Clin Neurosci 249(4):205–211

Koenig T, van Swam C, Dierks T, Hubl D (2012) Is gamma band EEG synchronization reduced during auditory driving in schizophrenia patients with auditory verbal hallucinations? Schizophr Res 141(2–3):266–270. doi:10.1016/j.schres.2012.07.016

Kostova M, Passerieux C, Laurent JP, Hardy-Bayle MC (2003) An electrophysiologic study: can semantic context processes be mobilized in patients with thought-disordered schizophrenia? Can J Psychiatry (Revue canadienne de psychiatrie) 48(9):615–623

Kutas M, Hillyard SA (1984) Brain potentials during reading reflect word expectancy and semantic association. Nature 307(5947):161–163

Laurent JP, Baribeau J (1992) AERPs and clinical evolution of thought disorders over five years. Int J Psychophysiol (official journal of the International Organization of Psychophysiology) 13(3):271–282

Lee SH, Wynn JK, Green MF, Kim H, Lee KJ, Nam M, Park JK, Chung YC (2006) Quantitative EEG and low resolution electromagnetic tomography (LORETA) imaging of patients with persistent auditory hallucinations. Schizophr Res 83(2–3):111–119. doi:10.1016/j.schres.2005.11.025

Lee SH, Choo JS, Im WY, Chae JH (2008) Nonlinear analysis of electroencephalogram in schizophrenia patients with persistent auditory hallucination. Psychiatry Investig 5(2):115–120. doi:10.4306/pi.2008.5.2.115

Lehmann D, Ozaki H, Pal I (1987) EEG alpha map series: brain micro-states by space-oriented adaptive segmentation. Electroencephalogr Clin Neurophysiol 67(3):271–288

Lehmann D, Faber PL, Galderisi S, Herrmann WM, Kinoshita T, Koukkou M, Mucci A, Pascual-Marqui RD, Saito N, Wackermann J, Winterer G, Koenig T (2005) EEG microstate duration and syntax in acute, medication-naive, first-episode schizophrenia: a multi-center study. Psychiatry Res 138(2):141–156. doi:10.1016/j.pscychresns.2004.05.007

Lehmann D, Pascual-Marqui RD, Strik WK, Koenig T (2010) Core networks for visual-concrete and abstract thought content: a brain electric microstate analysis. NeuroImage 49(1):1073–1079. doi:10.1016/j.neuroimage.2009.07.054

Liu Y, Shen X, Zhu Y, Xu Y, Cai W, Shen M, Yu R, Wang W (2007) Mismatch negativity in paranoid, schizotypal, and antisocial personality disorders. Neurophysiol Clin (Clinical neurophysiology) 37(2):89–96. doi:10.1016/j.neucli.2007.03.001

Liu Z, Tam WC, Xue Z, Yao S, Wu D (2004) Positive and negative symptom profile schizophrenia and abnormalities in the P300 component of the event-related potential: a longitudinal controlled study. Psychiatry Res 132(2):131–139. doi:10.1016/j.pscychresns.2004.03.003

Lyytinen H, Blomberg AP, Naatanen R (1992) Event-related potentials and autonomic responses to a change in unattended auditory stimuli. Psychophysiology 29(5):523–534

Mathalon DH, Ford JM, Pfefferbaum A (2000a) Trait and state aspects of P300 amplitude reduction in schizophrenia: a retrospective longitudinal study. Biol Psychiatry 47(5):434–449

Mathalon DH, Ford JM, Rosenbloom M, Pfefferbaum A (2000b) P300 reduction and prolongation with illness duration in schizophrenia. Biol Psychiatry 47(5):413–427

McCarley RW, Faux SF, Shenton ME, Nestor PG, Holinger DP (1991) Is there P300 asymmetry in schizophrenia? Arch Gen Psychiatry 48(4):380–383

Metzler S, Theodoridou A, Aleksandrowicz A, Muller M, Obermann C, Kawohl W, Heekeren K (2013) Evaluation of trait adjectives and ego pathology in schizophrenia: an N400 study. Psychiatry Res. doi:10.1016/j.psychres.2013.12.025

Misra M, Holcomb PJ (2003) Event-related potential indices of masked repetition priming. Psychophysiology 40(1):115–130

Naatanen R, Picton T (1987) The N1 wave of the human electric and magnetic response to sound: a review and an analysis of the component structure. Psychophysiology 24(4):375–425

Naatanen R, Paavilainen P, Rinne T, Alho K (2007) The mismatch negativity (MMN) in basic research of central auditory processing: a review. Clin Neurophysiol (official journal of the International Federation of Clinical Neurophysiology) 118(12):2544–2590. doi:10.1016/j.clinph.2007.04.026

O'Donnell BF, Hokama H, McCarley RW, Smith RS, Salisbury DF, Mondrow E, Nestor PG, Shenton ME (1994) Auditory ERPs to non-target stimuli in schizophrenia: relationship to probability, task-demands, and target ERPs. Int J Psychophysiol (official journal of the International Organization of Psychophysiology) 17(3):219–231

Oades RD, Dittmann-Balcar A, Zerbin D, Grzella I (1997) Impaired attention-dependent augmentation of MMN in nonparanoid vs paranoid schizophrenic patients: a comparison with obsessive-compulsive disorder and healthy subjects. Biol Psychiatry 41(12):1196–1210

Olincy A, Braff DL, Adler LE, Cadenhead KS, Calkins ME, Dobie DJ, Green MF, Greenwood TA, Gur RE, Gur RC, Light GA, Mintz J, Nuechterlein KH, Radant AD, Schork NJ, Seidman LJ, Siever LJ, Silverman JM, Stone WS, Swerdlow NR, Tsuang DW, Tsuang MT, Turetsky BI, Wagner BD, Freedman R (2010) Inhibition of the P50 cerebral evoked response to repeated auditory stimuli: results from the consortium on genetics of schizophrenia. Schizophr Res 119(1–3):175–182. doi:10.1016/j.schres.2010.03.004

Papageorgiou C, Oulis P, Vasios C, Kontopantelis E, Uzunoglu N, Rabavilas A, Christodoulou GN (2004) P300 alterations in schizophrenic patients experiencing auditory hallucinations. Eur Neuropsychopharmacol (the journal of the European College of Neuropsychopharmacology) 14(3):227–236. doi:10.1016/S0924-977X(03)00147-0

Pfefferbaum A, Ford JM, White PM, Roth WT (1989) P3 in schizophrenia is affected by stimulus modality, response requirements, medication status, and negative symptoms. Arch General Psychiatry 46(11):1035–1044

Preuss UW, Zetzsche T, Pogarell O, Mulert C, Frodl T, Muller D, Schmidt G, Born C, Reiser M, Moller HJ, Hegerl U, Meisenzahl EM (2010) Anterior cingulum volumetry, auditory P300 in schizophrenia with negative symptoms. Psychiatry Res 183(2):133–139. doi:10.1016/j.pscychresns.2010.05.008

Prevost M, Rodier M, Lionnet C, Brodeur M, King S, Debruille JB (2011) Paranoid induction reduces N400 s of healthy subjects with delusional-like ideation. Psychophysiology 48(7):937–949. doi:10.1111/j.1469-8986.2010.01160.x

Reulbach U, Bleich S, Maihofner C, Kornhuber J, Sperling W (2007) Specific and unspecific auditory hallucinations in patients with schizophrenia: a magnetoencephalographic study. Neuropsychobiology 55(2):89–95. doi:10.1159/000103907

Rissling AJ, Makeig S, Braff DL, Light GA (2010) Neurophysiologic markers of abnormal brain activity in schizophrenia. Curr Psychiatry Rep 12(6):572–578. doi:10.1007/s11920-010-0149-z

Ropohl A, Sperling W, Elstner S, Tomandl B, Reulbach U, Kaltenhauser M, Kornhuber J, Maihofner C (2004) Cortical activity associated with auditory hallucinations. NeuroReport 15(3):523–526

Rosburg T, Boutros NN, Ford JM (2008) Reduced auditory evoked potential component N100 in schizophrenia—a critical review. Psychiatry Res 161(3):259–274. doi:10.1016/j.psychres.2008.03.017

Roth WT, Cannon EH (1972) Some features of the auditory evoked response in schizophrenics. Arch Gen Psychiatry 27(4):466–471

Sartorius N (1986) Psychiatry in the 21st century. Anales de la Real Academia Nacional de Medicina 103(4):539–552

Schall U, Catts SV, Chaturvedi S, Liebert B, Redenbach J, Karayanidis F, Ward PB (1998) The effect of clozapine therapy on frontal lobe dysfunction in schizophrenia: neuropsychology and event-related potential measures. Int J Neuropsychopharmacol (official scientific journal of the Collegium Internationale Neuropsychopharmacologicum) 1(1):19–29. doi:10.1017/S146114579800100X

Smith DM, Grant B, Fisher DJ, Borracci G, Labelle A, Knott VJ (2013) Auditory verbal hallucinations in schizophrenia correlate with P50 gating. Clin Neurophysiol (official journal of the International Federation of Clinical Neurophysiology) 124(7):1329–1335. doi:10.1016/j.clinph.2013.02.004

Spencer KM (2009) The functional consequences of cortical circuit abnormalities on gamma oscillations in schizophrenia: insights from computational modeling. Frontiers Human Neurosci 3:33. doi:10.3389/neuro.09.033.2009

Spencer KM, Nestor PG, Perlmutter R, Niznikiewicz MA, Klump MC, Frumin M, Shenton ME, McCarley RW (2004) Neural synchrony indexes disordered perception and cognition in schizophrenia. Proc Nat Acad Sci U.S.A 101(49):17288–17293. doi:10.1073/pnas.0406074101

Spencer KM, Salisbury DF, Shenton ME, McCarley RW (2008) Gamma-band auditory steady-state responses are impaired in first episode psychosis. Biol Psychiatry 64(5):369–375. doi:10.1016/j.biopsych.2008.02.021

Spencer KM, Niznikiewicz MA, Nestor PG, Shenton ME, McCarley RW (2009) Left auditory cortex gamma synchronization and auditory hallucination symptoms in schizophrenia. BMC Neurosci 10:85. doi:10.1186/1471-2202-10-85

Strik WK, Dierks T, Maurer K (1993) Amplitudes of auditory P300 in remitted and residual schizophrenics: correlations with clinical features. Neuropsychobiology 27(1):54–60. doi:10.1159/000118953

Sumich A, Harris A, Flynn G, Whitford T, Tunstall N, Kumari V, Brammer M, Gordon E, Williams LM (2006) Event-related potential correlates of depression, insight and negative symptoms in males with recent-onset psychosis. Clin Neurophysiol (official journal of the International Federation of Clinical Neurophysiology) 117(8):1715–1727. doi:10.1016/j.clinph.2006.04.017

Sumich A, Kumari V, Gordon E, Tunstall N, Brammer M (2008) Event-related potential correlates of paranormal ideation and unusual experiences. Cortex (a journal devoted to the study of the nervous system and behavior) 44(10):1342–1352. doi:10.1016/j.cortex.2007.10.012

Tarbi EC, Sun X, Holcomb PJ, Daffner KR (2011) Surprise? Early visual novelty processing is not modulated by attention. Psychophysiology 48(5):624–632. doi:10.1111/j.1469-8986.2010.01129.x

Turetsky BI, Colbath EA, Gur RE (1998) P300 subcomponent abnormalities in schizophrenia: I. Physiological evidence for gender and subtype specific differences in regional pathology. Biol Psychiatry 43(2):84–96. doi:10.1016/S0006-3223(97)00258-8

Uhlhaas PJ (2013) Dysconnectivity, large-scale networks and neuronal dynamics in schizophrenia. Curr Opin Neurobiol 23(2):283–290. doi:10.1016/j.conb.2012.11.004

Uhlhaas PJ, Singer W (2010) Abnormal neural oscillations and synchrony in schizophrenia. Nat Rev Neurosci 11(2):100–113. doi:10.1038/nrn2774

Uhlhaas PJ, Haenschel C, Nikolic D, Singer W (2008) The role of oscillations and synchrony in cortical networks and their putative relevance for the pathophysiology of schizophrenia. Schizophr Bull 34(5):927–943. doi:10.1093/schbul/sbn062

Umbricht D, Koller R, Schmid L, Skrabo A, Grubel C, Huber T, Stassen H (2003) How specific are deficits in mismatch negativity generation to schizophrenia? Biol Psychiatry 53(12):1120–1131

Valkonen-Korhonen M, Purhonen M, Tarkka IM, Sipila P, Partanen J, Karhu J, Lehtonen J (2003) Altered auditory processing in acutely psychotic never-medicated first-episode patients. Brain Res Cogn Brain Res 17(3):747–758

van Lutterveld R, Sommer IE, Ford JM (2011) The neurophysiology of auditory hallucinations—a historical and contemporary review. Front Psychiatry 16(2):28

van Swam C, Dierks T, Hubl D (2013) Electrophysiological exploration of hallucinations (EEG, MEG). In: Jardri R, Cachia A, Thomas P, Pins D (eds) The neuroscience of hallucinations. Springer Science+Business Media, New York

Varela F, Lachaux JP, Rodriguez E, Martinerie J (2001) The brainweb: phase synchronization and large-scale integration. Nat Rev Neurosci 2(4):229–239. doi:10.1038/35067550

Wang XJ (2010) Neurophysiological and computational principles of cortical rhythms in cognition. Physiol Rev 90(3):1195–1268. doi:10.1152/physrev.00035.2008

Wible CG, Preus AP, Hashimoto R (2009) A cognitive neuroscience view of schizophrenic symptoms: abnormal activation of a system for social perception and communication. Brain Imaging Behav 3(1):85–110. doi:10.1007/s11682-008-9052-1

Youn T, Hae-Jeong P, Jae-Jin K (2002) Altered hemispheric asymmetry and positive symptoms in schizophrenia: equivalent current dipole of auditory mismatch negativity. Schizophr Res 59:253–260

Electrophysiological Aberrations Associated with Negative Symptoms in Schizophrenia

Nash N. Boutros, Armida Mucci, Annarita Vignapiano and Silvana Galderisi

Abstract Clinical heterogeneity is a confound common to all of schizophrenia research. Deficit schizophrenia has been proposed as a homogeneous disease entity within the schizophrenia syndrome. The use of the Schedule for the Deficit Syndrome (SDS) has allowed the definition of a subgroup dominated by persistent and primary negative symptoms. While a number of studies have appeared over the years examining the electrophysiological correlates of the cluster of negative symptoms in schizophrenia, only a few studies have actually focused on the Deficit Syndrome (DS). In this chapter, electrophysiological investigations utilizing EEG, Evoked Potentials (EPs), polysomnography (PSG), or magnetoencephalography (MEG) to probe "negative symptoms," or "Deficit Syndrome" are reviewed. While this line of research is evidently in its infancy, two significant trends emerge. First, spectral EEG studies link increased slow wave activity during wakefulness to the prevalence of negative symptoms. Second, sleep studies point to an association between decrease in slow wave sleep and prevalence of negative symptoms. Several studies also indicate a relationship of negative symptoms with reduced alpha activity. A host of other abnormalities including sensory gating and P300 attenuation are less consistently reported. Three studies specifically

N. N. Boutros (✉)
Department of Psychiatry and Neurosciences, University of Missouri Kansas City (UMKC), 1000 East 24th Street, Kansas City, MO 64108, USA
e-mail: boutrosn@umkc.edu

A. Mucci · A. Vignapiano · S. Galderisi
Department of Psychiatry, University of Naples SUN, Largo Madonna delle Grazie, 80138 Naples, Italy
e-mail: Armida.mucci@gmail.com

A. Vignapiano
e-mail: vignapiano2004@libero.it

S. Galderisi
e-mail: Silvana.Galderisi@gmail.com

Curr Topics Behav Neurosci (2014) 21: 129–156

DOI: 10.1007/7854_2014_303

Published Online: 27 March 2014

addressed electrophysiology of the DS. Two of the three studies provided evidence suggesting that the DS may be a separate disease entity and not simply a severe form of schizophrenia.

Keywords Deficit syndrome · Negative symptoms · Schizophrenia · Evoked potentials (EPs) · Electroencephalography (EEG) · Polysomnography (PSG)

Contents

1 Introduction

Enduring negative symptoms in patients with chronic schizophrenia correlate with decreased functional outcome (Tandon et al. 2000; Buchanan 2007; Green 2006; Kirkpatrick et al. 2006) and are refractory to common pharmacological interventions (Buchanan 2007; Buchanan et al. 2007; Galderisi and Maj 2009). It has long been recognized that negative symptoms can be secondary to medication effects, co-morbid mood disorders, or even resulting from significant cognitive disorganization. The development of the Schedule for the Deficit Syndrome (SDS) (Kirkpatrick et al. 1989) has allowed the definition of a subgroup of schizophrenia patients dominated by clusters of primary and enduring negative symptoms (Kirkpatrick et al. 1989). This subgroup of patients (deficit schizophrenia, DS) arguably suffers from the most severe and costly form of all of the subgroups of schizophrenia (Galderisi and Maj 2009; Kirkpatrick and Galderisi 2008).

Functional or anatomical probing of individual psychiatric symptoms or symptom clusters within a psychiatric syndrome is a relatively recent endeavor. Schizophrenia symptomatology varies widely affecting the cognitive, affective, and reality testing domains. Three sets of symptoms received particular attention: positive, negative, and cognition-related symptoms. Based on the fundamental differences and likely brain structures involved in the generation of these symptom clusters, an assumption can be made that the underlying physiopathologies mediating these symptoms are different. Neuropathological findings associated with schizophrenia in general have implicated most of brain structures but with significant emphasis on frontal, temporal, and thalamic regions. While the complete understanding of the pathophysiology of the DS is not possible with the current level of knowledge, most studies suggest pervasive bottom-up deficiencies that may lead to cascading information processing problems. This tentative conclusion is based on the observation that information processing abnormalities gleaned from evoked potentials (EPs) studies tend to occur earlier in the sequence of EPs and starting during periods that are commonly considered as "pre-attentive."

Significant efforts are now being expended in attempting to ascertain whether DS is a subgroup of schizophrenia with identified abnormalities proving to be only quantitative and not qualitatively different from rest of schizophrenia subgroups. On the other hand, it is suggested that if identified abnormalities closely associated with the DS are qualitatively different then it would be possible to postulate DS as a separate psychopathological entity. The clear demarcation of the DS and its early identification could lead to better treatment and rehabilitation efforts.

Electrophysiological investigations, including electroencephalography (EEG), magnetoencephalography (MEG), evoked potentials (EPs), and sleep studies represent investigative methodology that can assess chronology at a millisecond by millisecond level of the information processing cascade in the intact behaving human. Research utilizing this methodology has pointed out to a number of highly replicable physiological aberrations in individuals suffering from schizophrenia. None the less, no biological abnormality has proven to be diagnostic or even present in a significant majority of patients. One reason commonly advanced for such is the agreed upon heterogeneity of the syndrome. By identifying the aberrations that are more specifically linked to a particular symptom (e.g., hallucination) or to a symptom cluster (positive, negative or cognitive), advances can be made toward better understanding of the neural circuitries subserving different symptom clusters, as well as the disorder as a whole. Defining the electrophysiological changes most closely linked to negative symptoms may allow guided research to probe specific negative symptoms like avolition and emotional blunting.

Physiological probing of brain function and dysfunction uses two fundamentally different technologies; functional (as contrasted to structural) neuroimaging and electrophysiology. While neuroimaging has superior spatial resolution, electrophysiology enjoys a superior temporal resolution. It is widely agreed upon that the two methodologies are complementary (Boutros et al. 2011) and efforts are underway for the simultaneous recording of both kinds of brain activity. In this review, we focus on electrophysiological investigations probing negative symptoms or the deficit syndrome and attempt to leverage the existing electrophysiology literature in schizophrenia in an attempt to investigate the possibility of a unifying hypothesis for the pathophysiology of negative symptom in schizophrenia patients. A testable hypothesis is whether positive symptoms represent a more localized dysfunction in the frontotemporal-thalamo-striatal-cortical circuitry while negative symptoms (and more fully the DS) may be secondary to a more widespread cortical-thalamo-striatal-cortical dysfunction. It is also quite possible that the nature of involvement of the frontal cortex is different in groups of schizophrenia patients with predominance of negative, positive, or cognitive dysfunctions.

2 EEG Changes in Association with Negative Symptoms

EEG abnormalities in schizophrenia have been noted since the early days of electroencephalography. The emergence of the ability to analyze EEG signals with the aid of the computer allowed the intense and detailed interrogation of the rich electrophysiological data. An increase of slow activity in the delta and theta range (1–8 Hz) was consistently reported in schizophrenia (Galderisi and Maj 2009). Spectral EEG studies have linked increased slow wave activity to the prevalence of negative symptoms (Boutros et al. 2008a).

Table 1 lists studies that directly addressed the correlation between negative symptoms and/or the deficit syndrome EEG changes. Table 1 also highlights the varying EEG methodologies used in probing negative symptoms (and schizophrenia in general) (Boutros et al. 2008a). Totally, 5 of the 11 studies using spectral analysis point to an increased slow activity (mainly theta rhythms of 4–8 Hz frequency) in association with negative symptoms. A relatively recent study supported the observation of a significant correlation between increased theta activity and negative symptoms (Venables et al. 2009). In this study the strongest correlations were obtained from frontal, central, and occipital regions while correlation between EEG data obtained from temporal regions and negative symptoms barely reached significance. This observation suggests that the increased theta seen in association with negative symptoms may be more represented in certain brain regions and may have significant implications to the eventual identification of the specific circuitry underlying the development of these symptoms (Venables et al. 2009). These findings also underline the possibility that the pathophysiology associated with negative symptoms may be widespread. The computational role of theta oscillation has been linked to periodic modulation of synaptic transmission and plasticity (Lengyel et al. 2005). Persistent theta neural activity lasting for seconds after transient stimulation has been observed in several brain regions. This activity has been taken to be indicative of the integration of inputs on long time scales (Huhn et al. 2005).

As far as delta band activities are concerned, some studies have found enhanced delta band activity in the prefrontal cortex in neuroleptic-naïve subjects with schizophrenia (Pascual-Marqui et al. 1999) or a generalized increase in delta band activity, most prominently in the anterior cingulate gyrus and temporal lobes (Mientus et al. 2002). Itoh et al. (2011) conducted the first study to assess, in first episode drug-naïve patients, if delta band activity would be increased in brain

Table 1 EEG studies examining correlation with negative symptoms

Study	Sample	Assessment	Measurement	Findings
Williamson et al. 1989[a]	20 Sch 20 HC	DSM-IIIR SCID	Spectral EEG resting and during WCST	Neg Sx correlate with smaller increase in beta during WCST
Kessler et al. (1991)	18 Sch (med free) 13 HC.	DSM-IIIR Diagnostic Interview Schedule	Spectral EEG with auditory emotionally salient and control stimuli	Residual/Undifferentiated subgroup with predominant neg Sx showed more beta1 and less alpha at temporal sites and more beta1 and beta2 at frontal sites versus controls; they had different lateralization patterns in the delta band after emotionally salient stimuli
Merrin and Floyd (1996)[b]	17 Sch (med free (14 days)	DSM-IIIR by the SADS, BPRS	Spectral EEG	Reduced alpha power, decreased alpha coherence between hemispheres associated with neg Sx
Sponheim et al. (1997)	28 Sch (winter born) 81 Sc3h (non-winter born) 18 non-sch psychosis	PSE, DSM III	Nonwinter born CS and non-schizophrenia	This paper strongly documents heterogeneity within schizophrenia groups. Psychotic patients had increased low frequency and decreased alpha power. Winter born Sch and non-schizophrenia psychotic patients had no power abnormalities
Sponheim et al. (2000)[b]	112 Sch, 78 psychotic not sch, 107 controls	PSE DSM-IV	EEG spectral, relative, only Cz data examined	Low alpha—high slow activity factor scores associated with negative Sx in Sch patients
Knott et al. (2000)[a]	17 Sch, free of neuroleptics	DSM-IIIR, clinical, PANSS	EEG spectral analysis, intra, and interhemispheric coherence	No clear correlate of neg Sx pre clozapine Tr. Greater interhemispheric theta and beta asymmetry predicted good response in both pos and neg Sx. Greater intrahemispheric delta asymmetry predicted response of neg Sx

(continued)

Table 1 (continued)

Study	Sample	Assessment	Measurement	Findings
Winterer et al. (2000)	Two trainer (33 Sch, 49 Con), and a test set (32 Sch, 49 HC)	ICD-10 clinical interview	EEG spectral resting and activated and auditory evoked responses	Patients with predominantly negative symptoms were more readily classifiable based on frontally pronounced delta activity and decreased power of the N100/P200 evoked response
Strelets et al. (2002)[b]	Only male subjects. Pos Sx Sch 16 Neg Sx Sch 20, 16 HC	DSM-IIIR clinical interview	High frequency oscillation connectivity	Decreased connectivity between frontal regions in neg Sx patients
Fehr et al. (2003)[a]	Sch 30 (9 unmedicated),17 HC	DSM-IV Clin Int, PANSS	MEG, spectral analysis	Relative % of slow activity in temporal regions corr with neg Sx
Manchanda et al. (2008)[a]	117 first episode psychosis	SCID-IV, SAPS, SANS	Standard (visually inspected) EEG (sEEG)	Pretreatment sEEG abnormalities predict poorer response of both positive and neg Sx
Venables et al. (2009)[a]	Sch 48 (medicated), 61 1st-degree relatives	25-item BPRS	Spectral analysis	Increased theta activity during eyes closed condition in Sch
Itoh et al. (2011)	Sch 17 first episode drug-naïve 17 HC	SCID, SAPS, SANS	Delta band activity, using LORETA current density	LORETA values for delta band activity in left inferior temporal gyrus, right middle frontal gyrus, right superior frontal gyrus, right inferior frontal gyrus, and right parahippocampal gyrus. were negatively correlated with negative, but not positive symptoms
Noh et al. (2013)	15 Sch with chronic schizophrenia 17 HC	DSM-IV PANSS	MEG, spectral analysis	CLGFcircuits in the cortical functional network are related to the abnormal synchronization and correlated to the negative symptom

Applicability to the DS [a] Possibly applicable
[b] *Likely applicable* description clear
[c] Description inadequate for a determination
Sx Symptoms, *PSE* Present State Exam, *sEEG* Standard visually inspected EEG

areas with a relevant relationship with the pathophysiology of schizophrenia and its correlation with the symptomatology of schizophrenia (e.g., positive and negative symptoms). Low Resolution Brain Electromagnetic Tomography (LORETA) was used to generate current source density images of delta, theta, alpha, and beta activities and localize the difference in EEG activity between patients and healthy controls. Current density for delta band activity was greater for patients in the left inferior temporal gyrus, right middle frontal gyrus, right superior frontal gyrus, right inferior frontal gyrus, and right parahippocampal gyrus. Current density values for delta band at these brain regions showed a negative correlation with the Scale for Assessment of Negative Symptoms (SANS) total score and no significant correlation with Scale for Assessment of Positive Symptoms (SAPS) score. The results of this study indicated that increased slow activity in frontal regions is not associated with positive symptoms and is more marked in subjects with fewer negative symptoms. In the light of recent conceptualization of increased delta band over frontal regions as an electrophysiological sign of reduced attention to the external input, the authors hypothesized that subjects with increased frontal delta may be able to reduce external input, cope better with positive symptoms and have less secondary negative symptoms. It is thus clear that further research is needed to more fully understand the interrelationship between increased slow wave activity, particularly in the frontal regions, and schizophrenia symptomatology.

A less consistently reported abnormality is altered reactivity and/or predominance of faster rhythms (beta activity). Noh et al. (2013) analyzed magnetoencephalography (MEG) data in chronic patients with schizophrenia and found reduced gamma power and increased beta band synchronization. Using an index of intra and interregional cortical functional connectivity (the amount of coupled local and global feedback (CLGF) circuits) they found that dysfunctional connectivity was related to the abnormal synchronization of the beta band and to negative symptoms.

Another important approach of describing dynamical systems is nonlinear analysis, called chaos theory, which allows researchers to determine changes in the dimensionality of the system over time (Kotini and Anninos 2002). A higher rate of phase-state transitions (i.e., higher dimensionality) characterizes more complex systems. Earlier work revealed that nonlinearity scores were significantly lower during awake state in schizophrenia patients compared to control subjects suggesting that there may be diminished interplay between different generators of the various EEG rhythms (Keshavan et al. 2004). Decreased nonlinear complexity correlated with neurocognitive deficits. Other groups found an increase in the dimensional complexity in schizophrenia patients (Koukkou et al. 1993; Rockstroh et al. 1997). Kotini and Anninos (2002) when using MEG to examine nonlinearity in schizophrenia patients also reported lower dimensional complexity. One possible contributor to the discrepancy is heterogeneity of study samples. A readily testable hypothesis would be that decreased dimensionality would correlate with increased cognitive deficit and possibly negative symptoms.

Yet another recently evolving methodology is to stimulate the cortex using transcranial magnetic stimulation (TMS) and examine the resultant EEG effects. While yet to be applied specifically to probe the deficit syndrome, Guller et al. (2012) used this methodology to test the hypothesis that direct physiological stimulation of the cortex will produce an abnormal thalamic response in individuals with schizophrenia. They stimulated the precentral gyrus with single pulse TMS (spTMS) and measured the response to this pulse in synaptically connected regions (thalamus, medial superior frontal cortex, insula) using concurrent functional magnetic resonance imaging. Compared with healthy subjects, patients with schizophrenia showed a reduced response to spTMS in the thalamus, medial superior frontal cortex, and in the insula. Functional connectivity analyses revealed weaker thalamus–medial superior frontal cortex and thalamus–insula connectivity in patients with schizophrenia compared with control subjects. Subsequently, the same group utilized the same technology probing EEG effects to assess the natural frequency of the posterior parietal, motor, premotor, and prefrontal cortices in patients with schizophrenia and healthy control subjects. High-density EEG recordings during TMS of four cortical areas were performed. Several TMS-evoked EEG oscillation parameters, including synchronization, amplitude, and natural frequency, were compared between the schizophrenia and control groups. Patients with schizophrenia showed a slowing in the natural frequency of the frontal/prefrontal regions compared with control subjects (from an average of a 2-Hz decrease for the motor area to an almost 10-Hz decrease for the prefrontal cortex). The prefrontal natural frequency of individuals with schizophrenia was slower than in any healthy comparison subject and correlated with both positive and negative syndrome scale scores suggesting that these abnormalities may be common between patients dominated by positive or negative symptoms and further implicates thalamic abnormalities in schizophrenia. It could be speculated that frontothalamo-striatal-frontal circuit dysfunctions are common to all schizophrenia patients and perhaps contributing to the common coexistence of the varying manifestations of the disorder. According to our hypothesis advanced above, the development of a DS picture may require involvement of other cortical regions beyond frontal and perhaps temporal regions.

3 Evoked Potentials (EP) Changes Associated with Negative Symptoms

Disturbances in sensory processing have been suggested as significant contributors to the deficit state in both clinical and neurophysiological studies (Turetsky et al. 1998; Mucci et al. 2007; Li et al. 2013). Two studies have thus far demonstrated qualitative differences between DS and NDS (nondeficit) patients utilizing EP measures (Turetsky et al. 1998; Mucci et al. 2007). Thus, surrogate markers of sensory information processing are promising tools for probing the pathophysiology of the DS as well as delineating differences between DS and NDS physiological aberrations. Table 2 lists studies that directly examined the correlation between negative symptoms and evoked potential changes.

3.1 *Mid-Latency Auditory Evoked Responses and Sensory Gating*

The mid-latency auditory evoked responses (MLAERs) have been extensively used to study information processing both in mental health and disease. Numerous studies have shown two particular components to be abnormally reduced in amplitude in patients with schizophrenia. These two components are the N100 (a negative component seen approximately 100 ms following an auditory stimulus) and the P50 (a positive component seen about 50 ms after the presentation of an auditory stimulus) (Erwin et al. 1991). The N100 component has been extensively examined in schizophrenia patients. The majority of studies report decreased amplitude of the N100 which is not readily attributable to medication effects (Rosburg et al. 2008). N100 amplitude is commonly (but not invariably) found to be reduced in patients with schizophrenia (Rosburg et al. 2008) and in unaffected first-degree relatives of the same patients (Ahveninen et al. 2006). The relationship between N100 amplitude and specific schizophrenia symptoms remains unclear (Rosburg et al. 2008).

The P50 MLAER has also been used extensively to examine the phenomenon of amplitude attenuation with stimulus repetition. The term "sensory gating" has been linked with studies of sensory inhibition utilizing the P50 MLAER in a paired stimulus paradigm (PSP). Abnormal sensory gating has been proposed as a fundamental mechanism by which psychotic symptoms evolve (Freedman et al. 1991). In a PSP, two identical stimuli (S1 and S2) are delivered with short interstimulus interval of 500 ms and a longer interpair interval of 8–10 s (Adler et al. 1982). A sensory gating deficit has been repeatedly demonstrated in schizophrenia patients (Adler et al. 1982; Boutros et al. 1991). Meta-analysis of the P50 gating deficit in groups of nonselective schizophrenia, patients found the effect size to be more modest than earlier reports suggested (Patterson et al. 2008). Heterogeneity of both methodology and composition of patient groups were suggested as possible causes for variation in results. A number of studies suggest that the gating deficit, particularly of the N100 component, may be more strongly found in association with negative symptoms (Boutros et al. 2009).

The gating deficit as assessed by the P50 component was examined in relationship to negative symptoms and was not found to correlate with any symptom cluster in earlier studies (Adler et al. 1990; Boutros et al. 2004). Subsequent investigations of gating of the P50 MLAER component reported a significant correlation with negative symptoms (Ringel et al. 2004; Louchart-de la Chapelle et al. 2005; Arnfred 2006). More recently Santos et al. (2010) investigated P50

Table 2 EPs

Study	Sample	Measurement	Assessment	Findings
Pfefferbaum et al. (1989)	31 Sch 37 HC	P300 amplitude	BPRS	P3 smaller in patients with a pos correlation with neg Sx in unmedicated patients
Adler et al. (1990)[a]	20 medicated (9 predominantly negative) Sch, 12 control	P50 gating		No correlation of P50 gating and negative symptoms
Boutros et al. (1991)[a]	13 medicated nonparanoid, 13 paranoid, Sch, and 13 control	P50 amplitude and gating	BPRS	P50 amp and gating decreased in nonparanoid patients
Boutros et al. (1993)[a]	13 paranoid Sch 11 nonparanoid (all unmedicated) 11HC	P50 amplitude	DSM-3-R clinical interview	P50 amp decreased in nonparanoid patients
Turetsky et al. (1998)[b]	65 Sch (30 un-medicated), 48 HC	P300	BPRS, SANS, SAPS	Asymmetrical P3 reduction only in NDS while DS patients had reduction in r-Parietal
Baldeweg et al. (2002)	Medicated Sch (N=14), control (N=14)	Duration MMN	ICD-10 (clinical interview, Manchester Scale, Digit span, verbal fluency	No correlations with symptoms clusters
Jeon and Polich (2003)	Literature review	P300	NA	No overall correlation with neg Sx but a strong negative correlation with paranoid symptoms
Boutros et al. (2004)	Medicated Sch (atypical neuroleptics) 23 cont, 23	P50 gating	PANSS	No relationship between P50 gating and neg Sx
Ringel (2004)	34 Sch (medicated) 12 HC	P50 gating	PANSS	P50 gating corr + with neg Sx

(continued)

Table 2 (continued)

Study	Sample	Measurement	Assessment	Findings
Louchart-de La Chapell (2005)	81 Sch (medicated) 88 HC	P50 gating and P50 latency	SAPS and SANS	P50 gating dec in gen but more in assoc with neg Sx
Thoma et al. (2005)[b]	20 Sch (medicated)	P50 EP and MEG gating	SCID for DSM-IV, SANS, PANSS	No relationship between P50 (EP) gating and neg Sx. Right hemisphere M50 (MEG) gating ratios positively correlated with neg Sx (worse gating-more neg Sx)
Arnfred (2006)[b]	17 Sch spectrum (unmedicated), 24 HC	P50 EP and P50 Gating	SANS and SAPS	Difference wave calculated by subtracting S2 wave from S1 wave. Decreased P50 amplitude and difference wave in patients with severe negative Sx
Toyomaki et al. (2008)[c]	23 Sch (medicated)	Duration MMN	SCID and WCST	Strong correlation between MMN amplitude and executive functioning
Mucci et al. (2007)[b]	Sch (medicated) (DS = 20), non-DS = 20, and HC = 20	N100 and P300	SANS/SAPS and EBPRS	Double dissociation:N100 small in DS and P300 small in non-DS
Boutros et al. (2009)[a]	45 Sch (medicated), 49 HC	P50, N100, P200 gating	PANSS	N100 gating deficit correlates with negative symptoms
Santos et al. 2010[b]	60 DS, 60 NDS, 60 HC	P50 gating	SDS	No correlation to negative or positive symptoms but patients in general had worse gating which corr with poor outcome
Olincy et al. (2010)	181 Sch (medicated), HC 333	P50 gating	SCID and SANS/PANS	P50 gating decreased in group in general but no corr with clinical clusters

(continued)

Table 2 (continued)

Study	Sample	Measurement	Assessment	Findings
Smucny (2013)	21 Sch 23 HC	Sensory gating at the beta and gamma frequencies	DSM-IV BPRS SANS	Poor beta gating was associated with negative symptoms
Wynn et al. (2010)	34 Sch 36 HC	CNV SPN	SCID I BPRS SANS TEPS Social anhedonia scale Physical anhedonia scale	Patients demonstrated generally lower CNV and SPN across pleasant, neutral, and unpleasant picture conditions without a relationship with clinical symptomatology and anhedonia
Horan et al. (2012)	35 Sch 33 HC	FN ERN	SCID I BPRS	Patients and controls demonstrated comparable FN differentiation between reward and nonreward condition, and no correlation between negative symptoms and FN
Li et al. (2013)	21 DS, 38 NDS, 50 HC	P300 (P3a and P3b), MMN, P50 amp and gating, CNV		DS patients showed delayed point A in CNV which correlated with poorer Global assessment of Functioning scale but not with any individual negative symptom
Wysse (2013)	13 neg Sx Sch and 13 HC	LDAEP N1/P2		LDAEP in the right hemisphere in patients was related to higher scores in scales assessing negative symptoms

[a] Possibly applicable
[b] *Likely applicable* description clear
[c] Description inadequate for a determination
LDAEP Loudness dependence of auditory evoked potentials; *CNV* Contingent negative variation

gating in patients with DS and in those with nondeficit schizophrenia (NDS). These authors did not find differences in P50 gating between the two groups of patients, but found an association between the abnormality of P50 gating and poor functional outcome. These contradictory findings suggest that the relationship between primary, enduring negative symptoms and the P50 gating deficit are not simple or straightforward and may depend strongly on the specific composition of the study sample. Given the documented high prevalence of smoking among schizophrenia patients it is also important to control for smoking status of the patient and the time elapsed between the last cigarette and P50 testing as transient improvement in P50 gating following nicotine/cigarette smoking has been demonstrated (Leonard et al. 2007). It is also possible that the effect size of the observation is not robust enough to be detected in smaller samples or in the presence of a high noise to signal ratio as is common in P50 studies due to the relative small amplitude of the component (Boutros et al. 2009). Furthermore, the specific scale or instrument used to assess the negative symptoms (or for that matter any symptom cluster) may also influence the correlations identified (Boutros et al. 2009). An alternative hypothesis is that a mediating variable (not as yet identified), cross-correlated with negative symptoms and gating deficits (e.g., poor outcome), is responsible for the association between P50 gating deficit and negative symptoms. Further studies are needed to clarify the issue.

Evidence for a frontal cortex role in mediating both auditory and somatosensory habituation has been provided by MEG studies (Bowyer et al. 2007; Weiland et al. 2008), EEG studies (Korzyukov et al. 2007; Garcia-Rill et al. 2008) and from direct brain recordings (Grunwald et al. 2003; Boutros et al. 2008a, b). Physiological data have indicated that the inability to suppress irrelevant inputs, coupled with difficulty in novelty detection (which may be related and indexed here by the MMN), impairs the coding at the beginning and ending of discrete events. Information is stored with incorrect spatiotemporal tags (Knight et al. 1995). A number of studies documented a working memory deficit in schizophrenia patients (Goldman-Rakic 1999). It has further been postulated that it is the inability to gate irrelevant information during the recall period, that causes the deficit to become manifest. Chao and Knight (1995) tested this hypothesis in a number of patients with dorsolateral prefrontal cortex (DLPFC), temporal–parietal junction (TPJ), or posterior hippocampus lesions. They showed that patients with prefrontal lesions were significantly impaired by distractors at all delays while patients with temporo-parietal lesions performed similar to control subjects. Individuals with DLPFC lesions exhibited increased amplitudes of the mid-latency auditory evoked responses (MLAERs) beginning 25–35 ms following auditory stimulation (Knight et al. 1989). It is possible to speculate that the preattentive gating abnormality (mediated by the P50) may be common to all schizophrenias while the later occurring (or early attentive) gating abnormalities (reflected by either the N100 or the P200 EPs) may be more linked to negative symptoms (also a readily testable hypothesis).

Smucny et al. (2013) most recently used time–frequency analysis to compare sensory gating at the beta (15–26 Hz) and gamma (30–50 Hz) frequencies between schizophrenia patients and healthy controls. Relative to controls, patients showed impaired gating of total beta and gamma power. Poor beta gating was associated with negative symptoms. Time–frequency analysis of beta and gamma gating may thus be a translational method of assessing the genetic basis of gating deficits in schizophrenia.

3.2 Mismatch Negativity

The mismatch negativity (MMN) is an early EP (with latency around 150 ms from the stimuli onset) related to automatic probing of auditory sensory traces of a repetitious stimulus by a deviant one elicited in an oddball paradigm (Näätänen et al. 1978; Sams et al. 1985). MMN is considered a preattentive process and thus mainly a bottom-up function. While MMN has been shown to be diminished in a number of studies of schizophrenia patients, not all studies find similar deviations (some find duration and others find frequency deviation MMN alterations). We propose that one possible explanation of the variance in the findings is related to inclusion of different patient samples and not paying specific attention to negative versus positive symptoms as well as not paying attention to enduring primary negative versus secondary negative symptoms. The MMN has not yet been specifically investigated in well-characterized DS patients. Based on the assumption that DS patients suffer from an early (preattentive and bottom-up) sensory processing deficit, we fully expect that future MMN investigation will reveal abnormalities of this component in DS patients. At the time of preparing this chapter, only a relationship between diminished frequency MMN and executive dysfunction has been reported (Toyomaki et al. 2008).

3.3 The P300

The P300 response is a positive deflection appearing around 300 ms from stimuli onset in response to rare novel or deviant stimuli in an auditory oddball paradigm. Its amplitude is proportional with the amount of attentional resources allocated to the task and memory performance (Polich 2007). The P300 as well as the MMN have been found to be abnormally small (and sometimes delayed) in schizophrenia populations (Javitt et al. 1993; Näätänen and Kähkönen 2009; Jeon and Polich 2003; Gjini et al. 2010). The P300 response represents processing of information at more advanced cognitive levels, such as a shift of attention, context updating, or orienting to a relatively novel or deviant stimulus (Polich and Kok 1995). Examination of the P300 is of particular importance as evidence suggests that it may be relatively spared in DS patients (Mucci et al. 2007; Jeon and Polish 2003). It is of importance to note that a meta-analysis of the P300 ERP in schizophrenia found a correlation with positive but not negative symptoms (Jeon and Polish 2003).

Mucci et al. (2007) were able to corroborate this finding in a group of well-characterized DS patients. Confirmation of this important observation would significantly contribute to the delineation of the neural circuitry implicated in DS. Engagement of frontal attention mechanisms during evaluation of novel incoming stimuli produces the P3a response with a midline centrofrontal maximum as compared to the classic P300 (i.e., P3b) activity to attended target stimuli showing a midline parietal maximum and has been related to context updating operations and subsequent memory storage (Polich 2007). A recent fMRI study (Wolf et al. 2008) showed a negative correlation of novelty-induced BOLD signal in the left frontal cortex with SANS scores, indicating lower levels of activation in patients with more severe negative symptoms. No significant correlations were identified for novel-induced BOLD signal and positive symptoms; also no significant correlations for target-induced BOLD with either positive or negative symptoms. Based on the limited available literature examining the correlations between P3 and negative symptoms we predict that DS patients will exhibit deficit in the ability to respond to novelty reflected by generating lower amplitude P3a components, but preserved target P300 (P3b).

Kim et al. (2013) analyzed four event-related potential (ERP) components (P100, N170, N250, and P300) and their source activities using EEG data acquired from 23 schizophrenia patients while they were presented with facial emotion picture stimuli. Correlations between positive and negative syndrome scale (PANSS) scores and source activations during facial emotion processing were calculated to identify the brain areas affected by symptom scores. Analysis demonstrated that PANSS positive scores were negatively correlated with major areas of the left temporal lobule for early ERP components (P100, N170) and with the right middle frontal lobule for a later component (N250), which indicates that positive symptoms affect both early face processing and facial emotion processing. On the other hand, PANSS negative scores were negatively correlated with several clustered regions, including the left fusiform gyrus (at P100), most of which are not overlapped with regions showing correlations with PANSS positive scores. These recent results suggest that positive and negative symptoms affect independent brain regions during facial emotion processing.

3.4 *Error-Related and Correct-Response Negativity (ERN and CRN)*

Deficits in self monitoring are a core feature of cognitive dysfunction in schizophrenia, and may be the basis for disturbances of self and lack of insight, ultimately impacting social functioning. DS patients are characterized by a significant lack of insight. However, the functional and structural neural correlates of such deficits in self monitoring are not well understood (Araki et al. 2013). This group of researchers investigated this issue using measurements of neurophysiological and structural brain indices, i.e., error-related and correct-response negativity (ERN and CRN) ERPs, and gray matter volume of the anterior cingulate cortex (ACC), and tested whether the association between these indices is altered in patients with schizophrenia compared to healthy controls. The two groups did not differ in ERN amplitude. In contrast, schizophrenia patients showed significantly larger CRN amplitudes than did healthy subjects. Although the two groups did not significantly differ in gray matter volume of the ACC subregions, a significant negative correlation was found between ERN amplitudes at the frontocentral electrodes and absolute gray matter volumes of the left region of ACC only in healthy controls. These results suggest a disruption of function–structure coupling of the brain regions subserving self monitoring in schizophrenia. A direct comparison between DS and non-DS schizophrenia patients would be informative in this regards.

3.5 Anticipatory Components: Contingent Negative Variation (CNV) and Stimulus-Preceding Negativity (SPN)

The negative symptoms construct has been refined with the modern conceptualization identifying 5 core component that have been shown to separate into two domains: 1) a motivational dimension consisting of anhedonia, avolition, and asocialty and 2) a diminished expressivity dimension consisting of restricted affect and alogia (Strauss et al. 2013). In schizophrenia reward deficits involve more than just anhedonia (Foussias and Remington 2010). In some cases, while the ability to experience pleasure may even remain intact, patients with schizophrenia more consistently demonstrate an impaired anticipation of reward, not immediately available (Gold et al. 2008) and impairments in several other facets of reward processing involved in motivation (Strauss et al. 2013).

ERP components studied in relation to anticipation of reward include the contingent negative variation (CNV), a slow negative brain wave shown to reflect the anticipation of or orienting to the upcoming stimulus and response preparation, and has been related to preparatory attention, motivation, and response readiness (Walter et al. 1964), and the stimulus-preceding negativity (SPN), a negative ERP detected in paradigms that involve anticipation of feedback about the correctness of prior performance or stimuli that are motivationally significant (van Boxtel and Bocker 2004). Schevernels et al. (2014) showed that reward anticipation was linked to early cue processing components, as well as the early and later parts of the CNV. Wynn et al. (2010) investigated anticipatory deficits in subjects with schizophrenia in a paradigm involving a cued motor response (CNV) and no motor response (SPN) and the relationships of these ERP components abnormalities with self-reported trait anhedonia or anticipatory pleasure and clinically rated negative symptoms. Patients demonstrated generally lower CNV and SPN across pleasant, neutral, and unpleasant conditions; SANS total score did not correlate with the ERP variables; higher trait anhedonia was related at a trend-level to lower overall SPN.

Other facets of motivation found to be impaired in schizophrenia include difficulties using internal representations of emotional experiences, previous rewards, and motivational goals to drive current and future behavior, deficits that have major clinical significance in terms of functional capacity (Barch and Dowd 2010). The feedback-related negativity (FRN) is an ERP component that has been localized to the ACC (Nieuwenhuis et al. 2005) and has been hypothesized to reflect the function of a performance monitoring/evaluative system that rapidly assesses the motivational impact and/or salience of environmental feedback (Segalowitz et al. 2010). Fitting this view, the FRN has been found to be sensitive to both errors in reward prediction, negative valence of emotional stimuli (Pfabigan et al. 2011) and the magnitude of outcome (Holroyd and Krigolson 2007; Holroyd et al. 2004). In their study, Horan et al. (2012) tried to clarify the scope of feedback processing impairments in schizophrenia analyzing FRN during a simple monetary gambling task. The authors also investigated the relationships between FRN abnormalities and negative symptoms, assessed using the BPRS. They showed that patients and controls demonstrated comparable FRN differentiation between reward and nonreward feedback and higher positive symptoms were associated with greater differences between FRN to positive and negative feedback. There were no significant or trend-level correlations for negative symptoms. In their opinion the use of the BPRS may have limited ability to detect an association between negative symptoms and reward processing because the BPRS negative symptom subscale focuses on expressive symptoms (e.g., blunted affect) whereas experience-related symptoms (e.g., avolition, asociality) have a stronger theoretical link to feedback and reward processing (Blanchard et al. 2011).

3.6 Loudness Dependence of Auditory Evoked Potentials (LDAEP)

There is an evidence that alterations of serotonin (5-HT) system functioning also play a crucial role in the pathophysiology of disabling negative symptoms. From post mortem and genetic studies on patients with negative symptoms a 5-HT dysfunction is documented. In addition, atypical neuroleptics and some antidepressants improve negative symptoms via serotonergic action. So far, no research has been done to directly clarify the association between the serotonergic functioning and the extent of negative symptoms. Wysse et al. (2013) based on the above, examined the status of brain 5-HT level in negative symptoms in schizophrenia by means of the loudness dependence of auditory evoked potentials (LDAEP). The LDAEP provides a well-established and noninvasive in vivo marker of the central 5-HT activity. They investigated 13 patients with schizophrenia with predominant negative symptoms treated with atypical neuroleptics and 13 healthy controls. The LDAEP of the N1/P2 component was evaluated by dipole source analysis and single electrode estimation at Cz. Psychopathological parameters, nicotine use, and medication were assessed to control for additional influencing factors. Schizophrenia patients showed significantly higher LDAEP in both hemispheres than controls. Furthermore, the LDAEP in the right hemisphere in patients was related to higher scores in scales assessing negative symptoms. A relationship with positive symptoms was not found. These data might suggest a diminished central serotonergic neurotransmission in patients with predominant negative symptoms.

It should be noted that only three papers specifically addressed ERPs in the deficit syndrome (Turetsky et al. 1998; Mucci et al. 2007; Li et al. 2013). Two studies provided data suggestive that the deficit syndrome is not simply a severe form of schizophrenia but more likely a separate clinical entity. Turetsky et al. (1998) while not employing the SDS, used the BPRS, SANS, and SAPS to address the criteria for the deficit syndrome. They examined the P300 component and found two patterns that cannot be seen as different grades of the same process. Patients with NDS showed the greatest reduction over the left temporal regions while the deficit subgroup showed the greatest reduction over the right parietal region. The second study specifically addressing DS, using the SDS, found a double dissociation where only NDS patients exhibited the asymmetrical left temporal P300 amplitude deficiency, while DS patients exhibited a decreased amplitude of the N100 (Mucci et al. 2007). Most recently, Li et al. (2013) compared a number of EPs between 21 patients with DS, 38 patients with NDS schizophrenia and 50 healthy control subjects. They included the P300 (both P3a and P3b), MMN, amplitudes and gating of the P50 MLAER, and the contingent negative variation (CNV). To our knowledge this is the only report examining the CNV in relationship to negative symptoms schizophrenia. The authors reported a great deal of similarity between the two schizophrenia groups. As compared to the healthy controls both patient groups exhibited delayed N1, N2, MMN, and P3a latencies as well as reduced N1, N2, and CNV amplitudes. Similarly, P50 gating was decreased in both groups. Only NDS, when compared to controls, showed delayed latency of P3b. Only DS patients showed delayed point A in CNV which correlated with poorer Global Assessment of Functioning scale but not with any individual negative symptom. These recent data suggest a significant overlap between the DS and NDS syndromes but with some distinctive characteristics that need further explorations in larger and perhaps unmedicated patients (Li et al. 2013).

4 Sleep Changes in Association with Negative Symptoms

Table 3 lists studies that directly reported on the association between sleep architectural changes and negative schizophrenia symptoms.

Decreased delta sleep in association with negative symptoms has been a relatively consistent finding in schizophrenia patients (Tandon et al. 2000). Of the

Table 3 Sleep

Study	Sample	Measurement	Assessment	Results
Ganguli et al. (1987)[a]	8 Sch (drug-naïve), 16 HC	Standard sleep analysis	BPRS, Wing negative symptoms scale	An inverse relationship between slow wave sleep and negative Sx
van Kammen et al. 1988[a]	10 Sch (unmedicat ed)	Standard sleep analysis	Scale for negative symptoms, Bunney-Hamburg global assessment scale for psychosis (83)	An inverse relationship between slow wave sleep and negative Sx
Tandon et al. 1989	10Sch and 10HC	Standard Sleep Analysis	BPRS, SANS, HAMD	Decreased SWS and Inverse correlation between SWS and severity of neg Sx
Tandon et al. (1992)	20 drug free and 20 drug-naive Sch and 15 HC	Standard sleep analysis	BPRS, SANS and 17-item HAMD	REM latency was inversely correlated with the severity of negative symptoms and unrelated to depressive symptoms. SWS did not differ between the groups and was unrelated to any clinical parameter
Neylan et al. (1992)[a]	18 Sch (haloperidol), repeat drug free (N = 9)	Standard sleep analysis	Bunny-Hamburg scale, SADS	An inverse relationship between slow wave sleep and negative Sx
Keshavan et al. 1995[a]	24Sch, Delusional disorder (N = 5), Un-medicated	Standard and automated sleep analysis	BPRS, SANS, SAPS, SCID	An inverse relationship between slow wave sleep and negative Sx
Kajimura et al. (1996)	Sch (N = 6) and 6 archival HC	Delta count during all night sleep	DSM-IV and BPRS	Decreased SWS and delta count particularly high amplitude significant negative correlation between the half-delta wave count and negative Sx
Kato et al. (1999)[c]	Sch (N = 7)	Standard sleep analysis	DSM-IV clinical	An inverse relationship between slow wave sleep and negative Sx
Tandon et al. (2000)[b]	Sch (drug free) (N = 60)	Standard sleep analysis	SADs and RDC	An inverse relationship between REM latency and slow wave sleep and negative symptoms
Müller et al. (2004)[b]	Sch (drug free, N = 10)	Standard sleep analysis	DSM-IV (clinical), PANSS	Decreased SWS and REM percentage

(continued)

Table 3 (continued)

Study	Sample	Measurement	Assessment	Results
Sekimoto et al. (2007)	11 Sch, 12 HC	Frontal delta wave count during sleep	BPRS	Delta count in frontal regions inversely correlated with BPRS negative symptoms scores
Poulin et al. (2008)[b]	Sch (first episode, drug naive) (N = 10), control (N = 30)	Spectral analysis	DSM-IV clinical and BPRS	Magnitude of absolute alpha correlated positively with negative symptoms and negatively with positive symptoms
Sekimoto et al. (2011)	17 Sch and 18 HC	Period-amplitude analysis of all night sleep	BPRS	Inverse correlation between neg Sx and delta wave counts in all brain regions
Yetkin et al. (2011)	13 Sch and 13 HC	Standard sleep analysis	BPRS, SANS, and SAPS	Sch had less total sleep time, less sleep efficiency, but no decrease in SWS. SWS was inversely correlated with formal thought disorder

[a] Possibly applicable
[b] *Likely applicable* description clear
[c] Description inadequate for a determination
BPRS Brief Psychiatric Rating Scale; *SADS* Schedule for Affective Disorders and Schizophrenia; *SANS* Scale for Negative Symptoms; *SAPS* Scale for Positive Symptoms; SCID Standardized Clinical Interview for DSM

13 studies identified, 10 reported decreased slow wave sleep (SWS) in association with negative symptoms. Of these 10 papers, seven found a significant negative correlation between the severity of negative symptoms and percent of SWS (i.e., with increased severity of negative symptoms, less SWS is noted).

Recent sleep studies depend heavily on computer-based analysis. Sekimoto et al. (2007) utilized a specialized program (the Medilog Sleep Analyzing Computer) to perform period-amplitude analysis of the delta wave count (delta half-wave analysis). They reported that schizophrenia patients in general exhibited a significant decrease in the percent of stage 2 sleep and a marked decrease of SWS. The delta half-wave count in the bilateral frontal regions was inversely correlated with BPRS negative symptoms scores. Poulin et al. (2008) reported a negative correlation between sleep absolute alpha activity and negative symptoms scores. Sarkar et al. (2010) reported a significant decrease in SWS, stages one and two as well as total sleep time in general but with no correlation with negative symptoms. On the other hand, Yetkin et al. (2011) found no evidence of a SWS abnormality in a group of 13 male schizophrenia patients (undifferentiated type).

Relatively more recently, Sekimoto et al. (2011) reported a significant inverse correlation between negative symptoms scores and delta wave counts in all regions examined (frontal, central, parietal, occipital). Utilizing self-report (Pittsburgh Sleep Quality Index), Lunsford-Avery et al. (2013) observed a strong association between negative symptoms severity (and not positive symptoms) and increased sleep dysfunction (no PSG data obtained).

5 Conclusions

Progress in the investigation of negative symptoms pathophysiology has been hindered by intrinsic heterogeneity of this symptom cluster. Primary and enduring negative symptoms are recognized as distinct disease processes with respect to broadly defined negative symptoms. Recent conceptualizations and factor analytic studies of both primary and broadly defined negative symptoms support two distinct domains: avolition and impaired emotional expression, probably related to separate pathophysiological mechanisms (Cohen et al. 2007). These sources of heterogeneity have seldom been considered in the electrophysiological literature. However, the field is rapidly changing and studies of the deficit syndrome as well as investigations of the separate domains of negative symptoms are rapidly increasing.

Two prominent/reasonably consistent electrophysiological correlates of negative symptoms emerge through this review: increased slow frequencies during wakefulness (as assessed by awake spectral EEG) and decreased slow wave sleep during night time recording.

While the largest group of studies identified were those related to EP/ERP findings, a consistent or strong trend was difficult to identify perhaps due to the tendency of different research groups to examine one particular EP component like

P300, MMN, or sensory gating. Studies examining a number of EP components simultaneously are sparse (Li et al. 2013). Despite the small number of studies and the varying EPs examined, the literature points to a deficit in the sensory gating of the mid-latency evoked responses reported from more than one laboratory (Boutros et al. 2009; Ringel et al. 2004; Louchart-de la Chapelle et al. 2005). Furthermore, a decreased amplitude of the N100 response has also been linked to the deficit syndrome (Mucci et al. 2007). It is of importance to note that a meta-analysis of the P300 ERP in schizophrenia found a correlation with positive but not negative symptoms (Jeon and Polich 2003). Mucci et al. (2007) were able to corroborate this finding in a group of well-characterized DS patients.

As is a direct reflection of the research expertise represented in a particular laboratory, most research groups tend to probe one physiological variable (e.g., ERP, EEG, or sleep). Multicomponent electrophysiological studies will help further define the abnormalities detected. For example, if an EEG or EP abnormality disappears during sleep this would influence the understanding of its pathophysiology as compared to the abnormality being persistent during sleep. If an EP abnormality is linked to a particular EEG state, this finding would be of significance for the identification of the neural circuitry mediating this abnormality.

The current level of knowledge is inadequate to propose a unifying theory of the pathophysiology underlying the described anomalies associated with the DS. However, most studies suggest pervasive bottom-up deficiencies that may lead to cascading information processing problems. This tentative conclusion is based on the observation that information processing abnormalities gleaned from EP studies (the only investigative methodology that can assess chronology at a millisecond by millisecond level in the intact behaving human), tend to occur earlier in the sequence of EPs and starting during periods that are commonly considered as "pre-attentive." Sleep deviations suggest a serious abnormality of the restorative deep sleep stages and the awake EEG abnormality suggest difficulty generating the faster frequencies, reflecting decreased ability to generate efficient smaller neuronal ensembles to deal with more focused or effective information processing.

A unifying hypothesis of the deviant oscillations during wakefulness (increased slow activity) and decreased slow wave sleep (SWS), could point to a dysfunction involving thalamocortical circuitry (Pinault 2011; Sekimoto et al. 2011). Thalamocortical circuits exhibit two fundamentally different modes of operation across the sleep-wake cycle: a state of tonic activation (desynchrony) during waking and REM sleep and a state of rhythmic synchronized activity during SWS (Steriade and Llinas 1988). Thalamic relay receives significant input from a number of brain stem structures and thus are subject to changes with a number of ascending neurotransmitter inputs. Furthermore, thalamic relay neurons also send collateral projections to the thalamic reticular nucleus which are reciprocally inhibitory (GABAergic) with thalamic nuclei (Siegel 2011). The thalamus (and more specifically the reticular nucleus) has been proposed to be important for the function of sensory gating which has been repeatedly shown to be deficient in schizophrenia and may be more specifically associated with negative symptoms (Krause et al. 2003). Kirkpatrick and Buchanan (1990) proposed a neural circuit that may be at the heart of the DS. Components of this circuit include the amygdala, periamygdalar cortex, and parts of the prefrontal cortex. A number of thalamic nuclei (including the anterior, midline, mediodorsal, lateral anterior, and lateral dorsal as well as the intralaminar) have extensive connections with all these components (Clarke et al. 2010). The involvement of the thalamus is also supported by observation of worsening of somatosensory gating in patients with thalamic strokes with recovery of the function over time (Stain et al. 2002). Recent reviews suggest a central role for thalamic abnormalities in the generation of schizophrenia symptomatology (Byne and Hazlett 2009). Recent studies suggested an association between DS and genuine movement disorders (unrelated to antipsychotic treatment) pointing to cortico-striatal-thalamic circuits (Peralta et al. 2014). It could be hypothesized that abnormalities of the latter circuits might be related to avolition, while those concerning the limbic-thalamic circuits to the emotional expression domain of the negative symptoms.

Based on the above, it can be stated that while research on the electrophysiological correlates of the deficit syndrome and enduring negative symptoms remains minimal, available data strongly support the need and likely profitability of this line of investigation. Most notably is the absence of studies where EPs, EEGs and sleep studies are performed in the same individuals in order to examine the correlation and interrelationships among these deviations.

Acknowledgements No support was received for the development of this chapter.

References

Adler LE, Pachtman E, Franks RD, Pecevich M, Waldo MC, Freedman R (1982) Neurophysiological evidence for a defect in neuronal mechanisms Involved in sensory gating in schizophrenia. Biol Psychiatry 17:639–654

Adler LE, Waldo MC, Tatcher A, Cawthra E, Baker N, Freedman R (1990) Lack of relationship of auditory gating defects to negative symptoms in schizophrenia. Schizophr Res 3:131–138

Ahveninen J, Jääskeläinen IP, Osipova D, Huttunen MO, Ilmoniemi RJ, Kaprio J, Lönnqvist J, Manninen M, Pakarinen S, Therman S, Näätänen R, Cannon TD (2006) Inherited auditory-cortical dysfunction in twin pairs discordant for schizophrenia. Biol Psychiatry 60:612–620

Araki T, Niznikiewicz M, Kawashima T, Nestor PG, Shenton ME, McCarley RW (2013) Disruption of function-structure coupling in brain regions sub-serving self monitoring in schizophrenia. Schizophr Res 146:336–343

Arnfred S (2006) Exploration of auditory P50 gating in schizophrenia by way of difference wave. Behav Brain Funct 2:6

Baldeweg AK, Gruzelier JH, Hirrsch SR (2002) Impairment in frontal but not temporal components of mismatch negativity in schizophrenia. Int J Psychophys 43:111–122

Barch DM, Dowd EC (2010) Goal representations and motivational drive in schizophrenia: the role of prefrontal-striatal interactions. Schizophr Bull 36:919–934

Blanchard JJ, Kring AM, Horan WP, Gur R (2011) Toward the next generation of negative symptom assessments: the collaboration to advance negative symptom assessment in schizophrenia. Schizophr Bull 37(2):291–299

Boutros NN, Zouridakis G. Overall J. (1991) Replication and extension of P50 findings in schizophrenia. Clinical EEG. 22(1):40-45

Boutros NN, Arfken C, Galderisi S, Warrick J, Pratt G, Iacono W (2008a) The status of EEG abnormality as a diagnostic test for schizophrenia. Schizophr Res 99:225–237

Boutros NN, Brockhaus-Dumke A, Gjini K, Vedeniapin A, Elfakhani M, Burroughs S, Keshavan M (2009) Sensory-gating deficit of the N100 mid-latency auditory evoked potential in medicated schizophrenia patients. Schizophr Res 113:339–346

Boutros NN, Gjini K, Arfken CL (2011) Advances in electrophysiology in the diagnosis of behavioral disorders. Expert Opin Med Diagn 5:441–452

Boutros NN, Korzyukov O, Jansen B (2004) Sensory-gating deficits during the mid-latency phase of information processing in medicated schizophrenia patients. Psychiatry Res 126:203–215

Boutros NN, Mears R, Pflieger PE (2008b) Sensory gating in the human hippocampal and rhinal regions: regional differences. Hippocampus 18:310–316

Boutros NN, Rustin T, Zouridakis G, Peabody CA, Warner MD (1993) The P50 auditory evoked responses and subtypes of schizophrenia. Psychiatry Res 47:243–254

Bowyer SM, Boutros N, Korzyukov O, Tepley N (2007) Supra-modal role for frontal cortex in sensory gating. Int Congr Ser 1600:367–370

Buchanan RW (2007) Persistent negative symptoms: an overview. Schizophr Bull 33:1013–1022

Buchanan RW, Javitt DC, Marder SR, Schooler NR, Gold JM, McMahon RP, Heresco-Levy U, Carpenter WT (2007) The cognitive and negative symptoms in schizophrenia trial (CONSIST): the efficacy of glutamatergic agents for negative symptoms and cognitive impairments. Am J Psychiatry 164:1593–1602

Byne W, Hazlett EA, Buchsbaum MS (2009) The thalamus and schizophrenia:current status of research. Acta Neuropathol 117:347–368

Chao LL, Knight RT (1995) Human prefrontal lesions increase distractability to irrelevant sensory inputs. NeuroReport 6:1605–1610

Clarke DL, Boutros NN, Mendez MF (2010) The brain and behavior: an introduction to behavioral neuroanatomy, 3rd edn. Cambridge, New York pp 156–166

Cohen AS, Saperstein AM, Gold JM, Kirkpatrick B, Carpenter WT Jr, Buchanan RW (2007) Neuropsychology of the deficit syndrome: new data and meta-analysis of findings to date. Schizophr Bul 33:1201–1212

Erwin RJ, Mawhinney-Hee M, Gur RC, Gur RE (1991) Midlatency auditory evoked responses in schizophrenia. Biol Psychiatry 30:430–442

Fehr T, Kissler J, Wienbruch C, Moratti S, Elbert T, Watzl H, Rockstroh B (2003) Source distribution of neuromagnetic slow-wave activity in schizophrenic patients–effects of activation. Schizophr Res 63:63–71

Foussias G, Remington G (2010) Negative symptoms in schizophrenia: avolition and Occam's razor. Schizophr Bull 36:359–369

Freedman R, Waldo M, Bickford-Wimer P, Nagamoto H (1991) Elementary neuronal dysfunctions in schizophrenia. Schizophr Res 4:233–243

Galderisi S, Maj M (2009) Deficit schizophrenia: an overview of clinical, biological and treatment aspects. Eur Psychiatry 24:493–500

Ganguli R, Reynolds CF 3rd, Kupfer DJ (1987) Electroencephalographic sleep in young, never-medicated schizophrenics. A comparison with delusional and nondelusional depressives and with healthy controls. Arch Gen Psychiatry 44:36–44

Garcia-Rill E, Moran K, Garcia J, Findley WM, Walton K, Strotman B et al (2008) Magnetic sources of the M50 response are localized to frontal cortex. Clin Neurophysiol 119(2):388–398

Gjini K, Arfken C, Boutros NN (2010) Relationships between sensory "gating out" and sensory "gating in" of auditory evoked potentials in schizophrenia: a pilot study. Schizophr Res 121:139–145

Gm Winterer, Dorn K, Frick C (2000) Frontal dysfunction in schizophrenia: a newelectrophysiological classifier for research and clinical applications. Eur Arch Psychiatry Clin Neurosci 250:207–214

Gold JM (2008) Reward processing in schizophrenia: a deficit in the representation of value. Schizophr Bull 34:835–847

Goldman-Rakic PS (1999) The physiological approach: functional architecture of working memory and disordered cognition in schizophrenia. Biol Psychiatry 1(46):650–661

Green MF. (2006) Cognitive impairment and functional outcome in schizophrenia and bipolar disorder. J Clin Psychiatry. 67 Suppl 9:3-8; discussion 36-42

Grunwald T, Boutros NN, Pezer N, von Oertzen T, Fernandez G, Schaller C, Elger CE (2003) Neural substrates of Sensory gating within the human brain. Biol Psychiatry 53:511–519

Guller Y, Ferrarelli F, Shackman AJ (2012) Probing thalamic integrity in schizophrenia using concurrent transcranial magnetic stimulation and functional magnetic resonance imaging. Arch Gen Psychiatry 69:662–671

Holroyd CB, Krigolson OE (2007) Reward prediction error signals associated with a modified time estimation task. Psychophysiology 44:913–917

Holroyd CB, Larsen JT, Cohen JD (2004) Context dependence of the event-related brain potential associated with reward and punishment. Psychophysiology 41:245–253

Horan WP, Foti D, Hajcak G, Wynn JK, Green MF (2012) Impaired neural response to internal but not external feedback in schizophrenia. Psychol Med 42:1637–1647

Huhn Z, Orbán G, Erdi P, Lengyel M (2005) Theta oscillation-coupled dendritic spiking integrates inputs on a long time scale. Hippocampus 15(7):950–962

Itoh T, Sumiyoshi T, Higuchi Y (2011) LORETA analysis of three-dimensional distribution of delta band activity in schizophrenia: relation to negative symptoms. Neurosci Res 70:442–448

Javitt DC, Doneshka P, Zylberman I, Ritter W, Vaughan HG Jr (1993) Impairment of early cortical processing in schizophrenia: an event-related potential confirmation study. Biol Psychiatry 33:513–519

Jeon YW, Polich J (2003) Meta-analysis of P300 and schizophrenia: patients, paradigms, and practical implications. Psychophysiology 40:684–701

Kajimura N, Kato M, Okuma T (1996) Relationship between delta activity during all-night sleep and negative symptoms in schizophrenia: a preliminary study. Biol Psychiatry 39:451–454

Kato M, Kajimura N, Okuma T, Sekimoto M, Watanabe T, Yamadera H, Takahashi K (1999) Association between delta waves during sleep and negative symptoms in schizophrenia. Pharmaco-eeg studies by using structurally different hypnotics. Neuropsychobiology 39:165–172

Keshavan MS, Cashmere JD, Miewald J, Yergani VK (2004) Decreased nonlinear complexity and chaos during sleep in first episode schizophrenia: a preliminary report. Schizophr Res 71:263–272

Keshavan MS, Miewald J, Haas G et al (1995) Slow-wave sleep and symptomatology in schizophrenia and related psychotic disorders. J Psychiatr Res 29:303–314

Kessler C, Kling A (1991) EEG power variation in schizophrenic subgroups: effects of emotionally salient stimuli. Biol Psychiatry 30:335–348

Kim DW, Kim HS, Lee SH (2013) Positive and negative symptom scores are correlated with activation in different brain regions during facial emotion perception in schizophrenia patients: a voxel-based sLORETA source activity study. Schizophr Res 151:165–174

Kirkpatrick B, Buchanan RW (1990) The neural basis of the deficit syndrome of schizophrenia. J Nerv Ment Dis 178:545–555

Kirkpatrick B, Buchanan RW, McKenney PD, Alphs LD, Carpenter WT Jr (1989) The Schedule for the deficit syndrome: an instrument for research in schizophrenia. Psychiatry Res 30:119–123

Kirkpatrick B, Fenton WS, Carpenter WT, Marder SR (2006) The NIMH MATRICS consensus statement on negative symptoms. Schizophr Bull 32:214–219

Kirkpatrick B, Galderisi S (2008) Deficit schizophrenia: an update. World Psychiatry. 7:143–147

Knight RT, Grabowecky MF, Scabini D (1995) Role of human prefrontal cortex in attention control. In: Jasper HH, Riggio S, Goldman-Rakic PS (eds) Epilepsy and the functional anatomy of the frontal lobe. Raven Press, New York, pp 21–36

Knight RT, Scabini D, Woods DL (1989) Prefrontal cortex gating of auditory transmission in humans. Brain Res 504:338–342
Knott VJ (2000) Quantitative EEG methods and measures in human psychopharmacological research. Hum Psychopharmacol 15:479–498
Korzyukov OA, Pflieger ME, Wagner M, Bowyer SM, Rosburg T, Sundaresan K et al (2007) Generators of the intracranial P50 response in auditory sensory gating. NeuroImage 35(2):814–26
Kotini A, Anninos P (2002) Detection of non-linearity in schizophrenic patients using magnetoencephalography. Brain Topogr 15:107–113
Koukkou M, Lehmann D, Wackermann J, Dvorak I, Henggeler B (1993) Dimensional complexity of EEG brain mechanisms in untreated schizophrenia. Biol Psychiatry 33:397–407
Krause M, Hofffmann WE, Hajos M (2003) Auditory sensory gating in hippocampus and reticular thalamic neurons in anesthetized rats. Biol Psychiatry 53:244–253
Lengyel M, Huhn Z, Erdi P (2005) Computational theories on the function of theta oscillations. Biol Cybern 92(6):393–408
Leonard S, Mexal S, Freedman R (2007) Smoking, genetics and schizophrenia: evidence for self medication. J Dual Diagn 3(3–4):43–59
Li Z, Zheng B, Deng W (2013) Multi-components of evoked-brain potentials in deficit and nondeficit schizophrenia. Asia Pac Psychiatry 5:69–79
Louchart-de la Chapelle S, Levillain D (2005) P50 inhibitory gating deficit is correlated with the negative symptomatology of schizophrenia. Psychiatry Res 136:27–34
Lunsford-Avery JR, Orr JM, Gupta T (2013) Sleep dysfunction and thalamic abnormalities in adolescents at ultra high-risk for psychosis. Schizophr Res 151:148–153
Manchanda R, Norman R, Malla A (2008) EEG abnormalities and 3-year outcome in first episode psychosis. Acta Psychiatr Scand 117:277–282
Merrin EL, Floyd TC (1996) Negative symptoms and EEG alpha in schizophrenia: a replication. Schizophr Res 19:151–161
Mientus S, Gallinat J, Wuebben Y, Pascual-Marqui RD, Mulert C, Frick K, Dorn H, Herrmann WM, Winterer G (2002) Cortical hypoactivation during resting EEG in schizophrenics but not in depressives and schizotypal subjects as revealed by low resolution electromagnetic tomography (LORETA). Psychiatr Res 116:95–111
Mucci A, Galderisi S, Kirkpatrick B, Bucci P, Volpe U, Merlotti E, Centanaro F, Catapano F, Maj M (2007) Double dissociation of N1 and P3 abnormalities in deficit and nondeficit schizophrenia. Schizophr Res 92:252–261
Müller MJ, Rossbach W, Mann K (2004) Subchronic effects of olanzapine on sleep EEG in schizophrenic patients with predominantly negative symptoms. Pharmacopsychiatry 37: 157–162
Näätänen R, Gaillard AWK, Mäntysalo S (1978) Early selective-attention effect on evoked potential reinterpreted. Acta Psychol 42:313–329
Näätänen R, Kähkönen S (2009) Central auditory dysfunction in schizophrenia as revealed by the mismatch negativity (MMN) and its magnetic equivalent MMNm: a review. Int J Neuropsychopharmacol 12:125–135
Neylan TC, van Kammen DP, Kelley ME (1992) Sleep in schizophrenic patients on and off haloperidol therapy. Clinically stable versus relapsed patients. Arch Gen Psychiatry 49: 643–649
Nieuwenhuis S, Slagter HA, von Geusau A, Heslenfeld DJ, Holroyd CB (2005) Knowing good from bad: differential activation of human cortical areas by positive and negative outcomes. Eur J Neurosci 21:3161–3168
Noh K, Shin KS, Shin D (2013) Impaired coupling of local and global functional feedbacks underlies abnormal synchronization and negative symptoms of schizophrenia. BMC Syst Biol 10(7):30
Olincy A, Braff DL, Adler LE, Cadenhead KS, Calkins ME, Dobie DJ, Green MF, Greenwood TA, Gur RE, Gur RC, Light GA, Mintz J, Nuechterlein KH, Radant AD, Schork NJ, Seidman LJ, Siever LJ, Silverman JM, Stone WS, Swerdlow NR, Tsuang DW, Tsuang MT, Turetsky

BI, Wagner BD, Freedman R (2010) Inhibition of the P50 cerebral evoked response to repeated auditory stimuli: results from the consortium on genetics of schizophrenia. Schizophr Res 119:175–182

Pascual-Marqui RD, Lehmann D, Koenig T, Kochi K, Merlo MC, Hell D, Koukkou M (1999) Low resolution brain electromagnetic tomography (LORETA) functional imaging in acute, neuroleptic-naive, first-episode, productive schizophrenia. Psychiatr Res 90:169–179

Patterson JV, Hetrick WP, Boutros NN, Jin Y, Potkin S, Sandman C, Bunney WE (2008) P50 sensory gating ratios in schizophrenics and controls: a review and data analysis. Psychiatry Res 158:226–247

Peralta V, Moreno-Izco L, Sanchez-Torres A, García de Jalón E, Campos MS, Cuesta MJ (2014) Characterization of the deficit syndrome in drug-naive schizophrenia patients: the role of spontaneous movement disorders and neurological soft signs. Schizophr Bull 40:214–24

Pfabigan DM, Alexopoulos J, Bauer H (2011) Manipulation of feedback expectancy and valence induces negative and positive reward prediction error signals manifest in event-related brain potentials. Psychophysiology 48:656–664

Pfefferbaum A, Ford MJ, White PM, Roth WT (1989) P3 in schizophrenia is affected by stimulus modality, response requirements, medication status, and negative symptoms. Arch Gen Psychiatry 46:1035–1044

Pinault D (2011) Dysfunctional thalamic-related networks in schizophrenia. Schizophr Bull 37:238–243

Polich J (2007) Updating P300: an integrative theory of P3a and P3b. Clin Neurophysiol 118:2128–2148

Polich J, Kok A (1995) Cognitive and biological determinants of P300: an integrative review. Biol Psychol 41:103–146

Poulin J, Stip E, Godbout R (2008) REM sleep EEG spectral analysis in patients with first-episode schizophrenia. J Psychiatr Res 42:1086–1093

Ringel TM, Heidrich A, Jacob CP, Pfuhlmann B, Stoeber G, Fallgatter AJ (2004) Sensory gating deficit in a subtype of chronic schizophrenic patients. Psychiatry Res 125:237–245

Rockstroh B, Watzl H, Kowalik ZJ, Cohen R, Sterr A, Müller M, Elbert T (1997) Dynamical aspects of the EEG in different psychopathological states in an interview situation: a pilot study. Schizophr Res 28:77–85

Rosburg T, Boutros NN, Ford JM (2008) Reduced auditory evoked potential component N100 in schizophrenia: a critical review. Psychiatry Res 161:259–274

Sams M, Paavilainen P, Alho K (1985) Auditory frequency discrimination and event-related potentials. Electroencephalogr Clin Neurophysiol 62:437–448

Santos JL, Sánchez-Morla EM, Aparicio A, García-Jiménez MA, Villanueva C, Martínez-Vizcaíno V, Arango C (2010) P50 gating in deficit and non deficit schizophrenia. Schizophr Res 119:183–190

Sarkar S, Katshu MZ, Nizamie SH (2010) Slow wave sleep deficits as a trait marker in patients with schizophrenia. Schizophr Res 124:127–133

Schevernels H, Krebs RM, Santens P, Woldorff MG, Boehler CN (2014) Task preparation processes related to reward prediction precede those related to task-difficulty expectation. Neuroimage 84:639–647

Segalowitz SJ, Santesso DL, Chantziantoniou DK (2010) Retest reliability of medial frontal negativities during performance monitoring. Psychophysiology 47:260–270

Sekimoto M, Kato M, Watanabe T, Kajimura N, Takahashi K (2007) Reduced frontal asymmetry of delta waves during all-night sleep in schizophrenia. Schizophr Bull 33:1307–1311

Sekimoto M, Kato M, Watanabe T, Kajimura N, Takahashi K (2011) Cortical regional differences of delta waves during all-night sleep in schizophrenia. Schizophr Res 126:284–290

Siegel JM (2011) Sleep mechanisms and phylogeny. In: Kryger MH, Roth T, Dement WC (eds) Principles and practice of sleep medicine, 5th edn. Elsevier, St. Lious, MO, pp 76–91

Smucny J, Wylie K, Rojas D (2013) Evidence for gamma and beta sensory gating deficits as translational endophenotypes for schizophrenia. Psychiatry Res 214:169–174

Sponheim SR, Clementz BA, Iacono WG (2000) Clinical and biological concomitants of resting state EEG power abnormalities in schizophrenia. Biol Psychiatry 48:1088–1097

Sponheim SR, Iacono WG, Clementz BA, Beiser M (1997) Season of birth and electroencephalogram power abnormalities in schizophrenia. Biol Psychiatry 41:1020–1027

Stains WR, Black SE, Graham SJ (2002) Somatosensory gating and recovery from stroke involving the thalamus. Stroke 33:2642–2651

Steriade M, Llinás RR (1988) The functional states of the thalamus and the associated neuronal interplay. Physiol Rev 68:649–742

Strauss GP, Waltz JA, Gold JM (2013) A review of reward processing and motivational impairment in schizophrenia. Schizophr Bull PMID: 24375459 [Epub ahead of print]

Strelets VB, Novototsky-Vlasov VY, Golikova JV (2002) Cortical connectivity in high frequency beta-rhythm in schizophrenics with positive and negative symptoms. Int J Psychophysiol 44:101–115

Tandon R, DeQuardo JR, Taylor SF (2000) Psychotic phasic and deficit enduring subtypes of negative symptoms: biological markers and relationship to outcome. Schizophr Res 45:191–201

Tandon R, Shipley JE, Eiser AS, Greden JF (1989) Association between abnormal REM sleep and negative symptoms in schizophrenia. Psychiatry Res 27:359–361

Tandon R, Shipley JE, Taylor S, Greden JF, Eiser A, DeQuardo J, Goodson J (1992) Electroencephalographic sleep abnormalities in schizophrenia. Relationship to positive/negative symptoms and prior neuroleptic treatment. Arch Gen Psychiatry 49:185–194

Thoma RJ, Hanlon FM, Moses SN, Ricker D, Huang M, Edgar C, Irwin J, Torres F, Weisend MP, Adler LE, Miller GA, Canive JM (2005) M50 sensory gating predicts negative symptoms in schizophrenia. Schizophr Res 73:311–318

Toyomaki A, Kusumi I, Matsuyama T (2008) Tone duration mismatch negativity deficits predict impairment of executive function in schizophrenia. Prog Neuropsychopharmacol Biol Psychiatry 32:95–99

Turetsky B, Colbath EA, Gur RE (1998) P300 subcomponents abnormalities in schizophrenia I. Physiological evidence for gender and subtype specific regional pathology. Biol Psychiatry 43:84–96

van Boxtel GJM, Bocker KBE (2004) Cortical measures of anticipation. J Psychophysiol 18:61–75

van Kammen DP, van Kammen WB, Peters J, Goetz K, Neylan T (1988) Decreased slow-wave sleep and enlarged lateral ventricles in schizophrenia. Neuropsychopharmacology 1:265–271

Venables NC, Bernat EM, Sponheim SR (2009) Genetic and disorder-specific aspects of resting state EEG abnormalities in schizophrenia. Schizophr Bull 35:826–839

Walter WG, Cooper R, Aldridge VJ (1964) Contingent negative variation: an electric sign of sensori-motor association and expectancy in the human brain. Nature 203:380–384

Weiland BJ, Boutros NN, Moran JM, Tepley N, Bowyer SM (2008) Evidence for a frontal cortex role in mediating both auditory and somatosensory habituation: a MEG study. NeuroImage 42:827–835

Williamson PC, Kutcher SP, Cooper PW (1989) Psychological, topographic EEG, and CT scan correlates of frontal lobe function in schizophrenia. Psychiatry Res 29:137–149

Wolf DH, Turetsky BI, Loughead J, Elliott MA, Pratiwadi R, Gur RE et al (2008) Auditory Oddball fMRI in schizophrenia: association of negative symptoms with regional hypoactivation to novel distractors. Brain Imaging Behav 2(2):132–145

Wynn J, Horan P, Kring AM (2010) Impaired anticipatory event-related potentials in schizophrenia. Int J Psychophysiol 77:141–149

Wysse C, Hitz K, Hengartner MP, Theodoridou A, Obermann C, Uhl I, Roser P, Grünblatt E, Seifritz E, Juckel G, Kawohl W (2013) The loudness dependence of auditory evoked potentials (LDAEP) as an indicator of serotonergic dysfunction in patients with predominant schizophrenic negative symptoms. PLoS ONE 12(8):e68650. doi:10.1371/journal.pone.0068650

Yetkin S, Aydın H, Özgen F (2011) Sleep architecture in schizophrenia patients. Turk Psikiyatri Derg. Spring 22(1):1–9 (Article in Turkish)

Psychophysiology in the Study of Psychological Trauma: Where Are We Now and Where Do We Need to Be?

D.T. Acheson, M.A. Geyer and V.B. Risbrough

Abstract Posttraumatic stress disorder (PTSD) is a major public health concern, which has been seeing increased recent attention partly due to the wars in Iraq and Afghanistan. Historically, research attempting to understand the etiology and treatment of PTSD has made frequent use of psychophysiological measures of arousal as they provide a number of advantages in providing objective, non-self-report outcomes that are closely related to proposed neurobiological mechanisms and provide opportunity for cross-species translation. Further, the ongoing shift in classification of psychiatric illness based on symptom clusters to specific biological, physiological, and behavioral constructs, as outlined in the US National Institute of Mental Health (NIMH) Research Domain Criteria project (RDoC), promises that psychophysiological research will continue to play a prominent role in research on trauma-related illnesses. This review focuses on the current state of the knowledge regarding psychophysiological measures and PTSD with a focus on physiological markers associated with current PTSD symptoms, as well as markers of constructs thought to be relevant to PTSD symptomatology (safety signal learning, fear extinction), and psychophysiological markers of risk for developing PTSD following trauma. Future directions and issues for the psychophysiological study of trauma including traumatic brain injury (TBI), treatment outcome studies, and new wearable physiological monitoring technologies are also discussed.

Keywords Psychophysiology · PTSD · Startle · Heart rate variability · Electrodermal response · TBI

D.T. Acheson · M.A. Geyer · V.B. Risbrough (✉)
Department of Psychiatry, University of California San Diego,
9500 Gilman Dr. Mail Code 0804, La Jolla, CA 92093-0804, USA
e-mail: vrisbrough@ucsd.edu

V.B. Risbrough
Center for Excellence in Stress and Mental Health, VA San Diego Healthcare System,
San Diego, USA

M.A. Geyer
Mental Illness Research, Education and Clinical Center, VA San Diego Healthcare System,
San Diego, USA

Curr Topics Behav Neurosci (2014) 21: 157–183
DOI: 10.1007/7854_2014_346

Published Online: 27 August 2014

Contents

1 Introduction

Posttraumatic stress disorder (PTSD) is a major public health concern with lifetime prevalence rates in the USA estimated to be 6.8–12.2 %, and 12-month prevalence rates estimated to be 3.5 % (Breslau 2009). Due to the wars in Iraq and Afghanistan, PTSD has received significant attention in the past 10–13 years, in terms of both popular media coverage and funds directed toward its research. This attention is warranted, given that rates of PTSD have increased in service members by 656 % since 2001 and the cost to the US Department of Defense (DoD) for treating these service members doubled between 2007 and 2012 (Blakeley and Jansen 2013 Congressional Research Service Report). In addition, it is important to note that PTSD affects more than just combat veterans and occurs in civilians following physical and sexual assaults, forced captivity, muggings/robberies, motor vehicle accidents, natural disaster, and life-threatening illness among other events (Breslau 2009). The DSM-IV classification of PTSD consisted in exposure to the traumatic event, as well as 3 clusters of symptoms: re-experiencing, avoidance and numbing, and hyperarousal. With the recent publication of DSM-5, the definition has expanded into 4 symptom clusters: intrusion, avoidance, negative alterations in cognitions and mood, and alterations in arousal and reactivity. This expansion recognizes broader, more heterogeneous symptom expressions (such as dysphoria and anger) while allowing for more dynamic changes in arousal and reactivity. Current treatments for PTSD are mainly psychotherapy based (e.g., exposure therapy and cognitive therapy). Pharmacological treatments, such as serotonin-selective and serotonin-norepinephrine reuptake inhibitors (SSRI/SNRIs), have also achieved modest efficacy (Committee on treatment of posttraumatic stress disorder IoMotNA 2007).

There is a clear need for the development of novel preventive and therapeutic treatment strategies for PTSD via increased understanding of etiological and maintaining factors of the disorder (Baker et al. 2009). To this end, there is a new focus on utilization of biological, physiological, and behavioral tools to enable a "paradigm shift" from sole reliance on self-report measures to assess symptom status and diagnosis for psychiatric disorders such as PTSD. The US National Institute of Mental Health (NIMH) Research Domain Criteria project (RDoC) represents a framework for research in this area, with an emphasis on developing a diagnostic classification scheme based upon valid observable markers of common biological processes across the range of currently identified diagnostic categories. The negative valence system (NVS) domain suggested by the NIMH contains the constructs of acute threat of "fear," potential harm or "anxiety," and sustained threat. The 2011 NVS working group meeting identified many of the physiological measures reviewed below as important research tools for understanding these constructs. Psychophysiological measures may have utility as static markers of these constructs, as well as dynamic markers of change enabling the elucidation of the roles of learning and memory processes in the expression of these constructs. Thus, psychophysiological measures are poised to play an important role in the future understanding of mental illness generally, and traumatic stress-related disorder characterized by negative valence states more specifically.

Psychophysiological outcome measures have a number of advantages in neuropsychiatric research. (1) Psychophysiological measures provide objective, non-self-report outcomes and thus are less subject to bias by the subject and/or researcher. (2) Physiological measures are quantifiable. (3) Compared to self-report symptom scales, physiological measures may represent more discrete symptom domains that probe specific neurobiological pathways enabling mechanistic study of neurobiological abnormalities underlying symptoms. (4) Physiological measures enable cross-species translation to examine causal mechanisms of psychophysiological abnormalities linked to trauma exposure that cannot be achieved with self-report measures. The current manuscript will review the current state of knowledge on psychophysiological outcomes in PTSD with attention to their use as markers of current symptoms as well as markers of PTSD-related processes. We will also discuss these variables in terms of their sensitivity and selectivity for PTSD symptoms versus other anxiety and mood disorders and comorbid disorders such as traumatic brain injury (TBI). Further, we will discuss potential future avenues for integrating psychophysiology into emerging areas of PTSD research. We have limited our review to relatively common psychophysiological measures of arousal/threat, including cardiovascular, electromyographic, and electrodermal measures.

2 Psychophysiological Markers of Current PTSD Symptoms

2.1 Cardiovascular Activity

Baseline: Current conceptualizations of PTSD, reflected in the Diagnostic and Statistical Manual of the American Psychiatric Association (DSM 5; APA 2014) criteria, recognize that PTSD has a complex phenomenology expressed not just as fear-based hyperarousal, but also as anhedonic and dysphoric emotional states. In contrast, earlier conceptualizations of the disorder, reflected in DSM-III through IV criteria, placed a larger emphasis on fear-related arousal. Given the past emphasis on arousal-related symptoms, research has long focused on identifying and understanding the psychophysiological basis of elevated arousal. Though studies have assessed the construct of arousal across a number of psychophysiological measures, an extensive body of work has focused on the cardiovascular system. Cardiovascular physiology is a convenient domain to focus on since it can be measured relatively easily using a number of different methods and equipment typically present in an emergency department or urgent care clinic. Further, some elements of cardiac physiology can be interpreted as a readout of sympathetic/parasympathetic balance, which has long been theorized to be disrupted in PTSD (see below).

Blanchard et al. (1982) observed that Vietnam veterans with PTSD had higher resting baseline heart rate (HR) and blood pressure (BP) than Vietnam veterans without PTSD. These initial observations were later largely confirmed in a meta-analysis by Buckley and Kaloupek (2001), which reviewed 34 studies of resting cardiovascular activity in PTSD conducted up to that time. This meta-analysis found support for elevated resting HR and diastolic blood pressure (BP), though systolic BP levels were similar across PTSD subjects and healthy controls. A more recent meta-analysis of psychophysiological studies in PTSD (Pole 2007) reviewed 55 studies conducted until that time and also supported increased resting HR in PTSD relative to healthy controls. However, elevations in systolic and diastolic BP were only present under relaxed criteria for statistical significance.

While the evidence for altered cardiovascular activity at rest in PTSD appears fairly strong, some researchers have suggested a more nuanced relationship. First, some studies (i.e., Shalev et al. 1992) have failed to find HR differences in new-onset PTSD. Further, Buckley and Kaloupek (2001) showed a greater effect size for HR in patients with chronic PTSD (>13 years). Taken together, these findings suggest that elevated HR may be a consequence of physiological changes driven by long-term PTSD. Second, studies monitoring cardiovascular activity over 24-h periods have suggested that HR and BP may fluctuate widely across the day, complicating previous studies (Muraoka et al. 1998; Buckley et al. 2004). One study using 24-h HR monitoring did, however, confirm increased HR in veterans with PTSD, with more pronounced effects during the night, perhaps related to the sleep disturbances commonly associated with PTSD (Agorastos et al. 2013). Third, there is disagreement among researchers regarding whether resting state activity is actually being measured in these studies, or if what is actually being captured are cardiovascular responses to a stressful situation/challenge induced by the testing environment (see below; Zoladz and Diamond 2013). Other studies suggest that PTSD subjects are hyperresponsive to stress or threat across a number of physiological markers, including HR, startle, and skin conductance (see below). Further, increased HR is not specific to PTSD, but is also reported in panic disorder and depression (Cohen et al. 2000; Blechert et al. 2007; Kamphuis et al. 2007).

An additional marker of resting-state cardiovascular activity that is altered in PTSD is heart rate variability (HRV). HRV is a measure of the variation in time between heart beats, which indicates autonomic flexibility (the higher the variation, the more flexibility). HRV is most accurately measured via electrocardiogram; however, photoplethysmography is also utilized. HRV is measured as time-domain variables (e.g., changes in the standard deviation of beat–beat interval) and frequency domains using power spectral density analysis methods. Frequency components are thought to represent sympathetic and parasympathetic control over HR, with the high frequency domain (HF; 0.15–0.4 Hz) representing parasympathetic or vagal tone, while the low frequency (LF; 0.04–0.15 Hz) is comprised of both parasympathetic and some sympathetic elements (see Heathers 2014; Berntson et al. 1997 for review). Finally, respiratory sinus arrhythmia, HRV due to respiration, is another measure of vagal control of autonomic activity. Reduced HRV is associated with mortality and cardiovascular symptoms in patients with PTSD, highlighting the clinical importance of these measures (Kubzansky et al. 2007). There is growing evidence that both LF and HF are reduced in PTSD patients, which may be suggestive of an imbalance between sympathetic and parasympathetic drive on cardiovascular output (Cohen et al. 2000; Blechert et al. 2007; Jovanovic et al. 2009), though exceptions have been reported (Sahar et al. 2001). In a recent twin study of combat-related PTSD in Vietnam era veterans, Shah et al. (2013) found that HRV abnormalities (lower LF and HF) were present only in the twin with PTSD, suggesting that reduced HRV is an acquired consequence of the disorder. They also suggested that HRV abnormalities were not present in subjects with remitted PTSD, suggesting HRV reductions are indicative of symptom state. We have recently shown that HRV reductions (reduced HF) are also associated with new-onset PTSD symptoms in active duty marines who served in Iraq/Afghanistan, suggesting that reduced HRV is not related to age or chronicity of PTSD (Minassian et al. 2014). These studies have also shown that reductions in HRV in these populations are not due to depression or TBI, nor are they related to degree of combat exposure or deployment history per se (Shah et al. 2013; Minassian et al. 2014). Finally, reduced HRV is reported in untreated subjects (Minassian et al. 2014; Chang et al. 2013), indicating that this phenotype is not due simply to medication side effects. Although HRV measures appear to be sensitive to PTSD symptoms, they are not specific to PTSD. Indeed, reduced HRV, in particular HF, may be a more general marker of anxiety disorders (Pittig et al. 2013) or even mental illness, as it is reduced across multiple disorders including anxiety, depression, bipolar disorder, and schizophrenia (Moon et al. 2013). It is possible that multiple mechanisms underlie the reductions in HRV across these diverse patient groups, or that reductions in HRV are due to the higher stress or allostatic load experienced by those with neuropsychiatric illness (McEwen 2000).

Response to Challenge: In contrast to resting-state cardiovascular markers, several studies have assessed cardiovascular activity in response to challenges from either loud acoustic stimuli (startle) or trauma-related cues. A large body of literature documents larger HR reactivity to startling sounds in PTSD patients (Pallmeyer et al. 1986; Shalev et al. 1992; Orr et al. 2002). Pole (2007) investigated 10 studies measuring HR response to loud acoustic stimuli and found that elevated HR response was among the most robust effects found using this paradigm. Pitman et al. (2006) examined elevated HR reactivity to sudden loud tone presentation in a twin sample of Vietnam veterans. They found elevated HR reactivity only in the twin with PTSD, indicating HR response is an acquired consequence of the disorder rather than a predisposing trait.

HR response to trauma-related reminder cues has also been examined, which may probe biological mechanisms relevant to fear memory processes. These studies typically involve either "standardized" cues, such as combat sounds (Liberzon et al. 1999) that are held constant across the sample being studied, or "ideographic" cues which are tailored to be specific to each subject's traumatic experience. Pole (2007) reviewed 16 studies investigating HR response to standardized trauma cues and another 22 investigating HR response to ideographic trauma-related cues. Elevated HR response to standardized cues in PTSD emerged as one of the more robust effects in these paradigms. Support for increased HR responses to ideographic trauma cues was also found, though less robust than that for standardized cues. Recent studies have also supported these findings in both standardized (Adenauer et al. 2010; Suendermann et al. 2010; Ehlers et al. 2010) and ideographic trauma cues (Barkay et al. 2012). Barkay et al. (2012) have investigated the neurobiological correlates of this effect using PET imaging and found correlations between HR and rCBF in the orbitofrontal, precentral, and occipital regions of the cortex only in patients with PTSD and not in trauma-exposed non-PTSD subjects. These findings are suggestive that increased HR responses to trauma reminders may overlap in neural substrates (orbitofrontal cortex) with the reduced ability to inhibit fear responses (Shin et al. 2006). In PTSD, there are correlations between HR response to trauma and norepinephrine concentrations in cerebrospinal fluid (Geracioti et al. 2008), suggesting that central noradrenergic hypersignaling could play a role in this phenotype. It is unclear whether increased HR or other cardiovascular abnormalities are ameliorated by treatment, however, despite the use of noradrenergic reuptake inhibitors (Hoge et al. 2012) as well as clinical trials of the alpha 1 receptor antagonist prazosin (Raskind et al. 2013). Whether increases in HR are an epiphenomenon of increased centrally mediated fear responses, or are a core feature of PTSD pathology is unclear. One intriguing recent finding suggests that inhibitors of angiotensin I signaling, commonly given for hypertension, are associated with fewer PTSD symptoms in a cross-sectional sample of highly traumatized civilian populations (Khoury et al. 2012). Other common hypertension medications were not associated with fewer symptoms, suggesting that the angiotensin pathway may play a role in PTSD-related pathology. Thus, more research is clearly needed to further elucidate pathways involved in elevated cardiovascular responses in PTSD.

Summary of Cardiovascular Markers of PTSD Symptom State: Cardiovascular physiology is an active and important area of research in PTSD, especially given reported links between PTSD and increased incidence of cardiovascular disease (Wentworth et al. 2013). While there is strong evidence that resting-state cardiovascular activity, as well as HR response to standardized and ideographic trauma cues, is altered in PTSD, this is still an active area of research that is not without controversy. Specifically, the degree to which the testing situation contributes to findings of elevated HR in PTSD is unclear. The extent to which elevated HR is a feature of core PTSD pathology versus simply a consequence of chronic stress is also unknown. Some studies have suggested that HR soon after trauma may predict development of PTSD, suggestive of HR being a proxy for biological risk factors for PTSD (see below). However, a recent study suggests that HR is not altered in relatively "recent" PTSD cases after combat (Minassian et al. 2014), arguing against elevated HR as a risk factor. HR increases are also not specific to PTSD, but are increased in other anxiety disorders more generally. Research investigating the time course and neurobiological correlates of altered cardiovascular activity in PTSD is needed to further clarify these issues.

Many questions still remain for the association of HRV with PTSD symptoms. Although twin studies suggest that altered HRV is specific to PTSD symptom state, prospective studies are needed to confirm HRV measures as symptom dependent or markers of risk for PTSD (Baker et al. 2012). Similarly, although there is some evidence from cross-sectional analysis in small samples for symptom remission to be associated with normalization of HRV (Shah et al. 2013), longitudinal treatment studies are required to best address this question. The biological mechanisms responsible for HRV reductions in PTSD are also unclear. However, dysregulated sympathetic output (e.g., via increased noradrenergic tone, Geracioti et al. 2001, 2008; Pietrzak et al. 2013) and abnormalities in stress and immune systems have been identified as candidate mediators (Risbrough and Stein 2006; Eraly et al. 2014).

2.2 Exaggerated Startle Response

Baseline: The startle response is a sensitive, noninvasive measure of central nervous system activity that is typically accessed via electromyographic (EMG) measurement of strength of contraction of the *orbicularis oculi* muscle controlling eyeblink in response to a sudden acoustic or tactile stimulus (Blumenthal et al. 2005). Exaggerated startle is a symptom of PTSD according to the DSM 5 (APA 2014). Thus, it follows that larger baseline startle responding should be detectable in PTSD. However, evidence for increased startle reactivity under "baseline" conditions in PTSD is mixed, with some studies finding evidence for increased startle in PTSD relative to healthy controls and others finding equivalent startle responses (see Zoladz and Diamond 2013 for a recent review of this literature). There are also some suggestions that increases in baseline startle may only occur in chronic PTSD patients or following certain forms of trauma, such as combat (Grillon and Baas 2003). A significant problem with assessments of "baseline" startle is that it is very difficult to accurately assess this phenomenon. Startle reactivity is extremely plastic, and it is sensitive to many rapid and dynamic modes of inhibition such as habituation and sensorimotor gating, to emotional valence or experimental context, and of course is extremely sensitive to stimulus parameters such as intensity and duration of the startling stimulus, all of which will influence the detection of putative differences.

For example, startle is higher in PTSD patients under low-intensity startle stimuli but not high intensity (Butler et al. 1990), which may reflect a lowering of startle thresholds rather than an exaggeration of startle responses elicited by supra-threshold stimuli. Thus, more robust and reliable startle phenotypes in PTSD and other disorders are measured when comparing startle across multiple stimulus conditions and emotional contexts. Startle has also generally only been explored in terms of magnitude of the response (muscle contraction) compared to controls. However, self-reports of "increased startle" from patients may not simply reflect magnitude, but *the probability* of a response under subthreshold conditions, which has yet to be explored.

In Response to Challenge: Given the inconsistency of baseline startle changes in PTSD, it has been suggested that startle reactivity is higher in PTSD patients only when under threat; thus, this phenomenon is indicative of mechanisms related to increased stress responding rather than disruption of baseline arousal (Grillon and Baas 2003). Grillon et al. (1998) reported normal baseline but increased startle magnitude in Vietnam combat veterans with PTSD during anticipation of experimental electrical shock relative to non-PTSD veterans, demonstrating a higher response in situations of threat or stress in PTSD. Startle is also elevated in response to trauma reminders (imagery, trauma scripts) in PTSD patients (e.g., Cuthbert et al. 2003; McTeague et al. 2010); however, these tasks are relatively unique to individual laboratories and more difficult to generalize across studies. As a whole, these studies suggest that exaggerated startle in PTSD is not indicative of increased arousal at baseline, but is a physiological marker of heightened response to threat and heightened fear responses in the presence of trauma cues. Thus, startle is increasingly used as a quantitative measure of fear responding that complements self-report data on anxiety and stress to identify biological mechanisms underlying PTSD symptoms.

Studies have recently suggested that elevated startle to challenge in PTSD may be subject to gender differences. Kamkwalala et al. (2012) showed that women with PTSD had higher startle in a dark environment relative to a light environment than men and women without PTSD. However, this elevated "dark-enhanced" startle was not present in male subjects with PTSD. Further, dark-enhanced startle has been shown to be associated with pituitary adenylate cyclase-activating polypeptide receptor (PAC1) genotypes in females, a gene that interacts with estrogen and has also been associated with PTSD in females (Ressler et al. 2011). These studies represent a new avenue of PTSD research that is just coming to fruition in utilizing physiological markers as intermediate phenotypes to identify biological pathways related to PSTD risk.

Startle Habituation: Habituation is a non-associative learning process whereby an organism displays a reduction in some innate orienting or defensive response following repeated presentation of a stimulus (Halberstadt and Geyer 2009). Shalev et al. (2000) examined habituation of the startle and electrodermal response to loud acoustic stimulus in a sample of traumatized Israeli civilians tested at 1 week and 1 and 4 months following the traumatic event. Those who developed PTSD began to show reduced habituation in both measures beginning 1 month post-trauma, suggesting that reduced habituation may be an acquired sign of PTSD. The reduced startle habituation finding is confounded, however, as the methodology used to detect startle was flawed, with sample rates that were much too slow (50 Hz) to visualize the very fast on and off rate of a startle response which is typically measured with 1,000 Hz sampling rates. The reduced electrodermal habituation, however, supported earlier findings by this group (Shalev et al. 1992). Other studies had failed to detect reduced startle habituation in PTSD but were compromised by their use of inappropriately slow sampling rates (Pitman et al. 1987, 1993; Orr and Pitman 1993). A more recent study in Croatian combat veterans found that PTSD and control groups did not differ in startle habituation as assessed by quantitative analysis of EMG reduction across trial; however, there was a reduction in PTSD subjects compared to controls when using nonparametric comparisons of a number of subjects who met criteria for habituation (lowest responding at the last trial) (Jovanovic et al. 2009). This study also did not replicate habituation of the electrodermal response, a physiological marker of sympathetic nervous system arousal based on electrical conductivity across the skin due to sweat (see below). Thus, taken together across studies, evidence for differences in startle habituation in PTSD subjects is weak. PTSD subjects may exhibit reduced habituation of fear-potentiated startle during fear association training (Ressler et al. 2011). However, it is unclear whether this effect reflects reduced habituation to startling sounds or increased reactivity to the aversive stimuli used during fear conditioning. Reductions in habituation have been detected in other neuropsychiatric disorders (schizophrenia, panic disorder); thus, it is possible that reductions in habituation of the response may represent a pathology in a subset of patients across disorders, as such a phenotype would have substantial consequences for multiple behavioral functions (Geyer and Braff 1982; Ludewig et al. 2002a, b, 2003, 2005). Habituation is another "intermediate phenotype" that is being used to identify potential gene pathways disrupted in these disorders (Greenwood et al. 2012, 2013).

Prepulse Inhibition of the Startle Response: Prepulse inhibition (PPI), the unlearned suppression of the startle reflex to an intense acoustic stimulus when immediately preceded by a weaker acoustic prepulse, is an operational measure of sensorimotor gating (Geyer et al. 1990; Geyer and Braff 1987). PPI has been shown to be a robust but non-specific biomarker of psychiatric diagnosis. PPI performance is reduced compared to healthy controls in a number of neuropsychiatric disorders including panic disorder, obsessive compulsive disorder, schizophrenia, bipolar disorder, Tourette's disorder, and Huntington's disorder (Braff et al. 2001; Swerdlow et al. 2006; Castellanos et al. 1996; Perry et al. 2001; Ahmari et al. 2012; Ludewig et al. 2002a, b). Many of these disorders are linked to cortico-limbic circuit abnormalities (Kohl et al. 2013). Given the evidence for PTSD to have disruptions in this circuit (Shin et al. 2006), PPI in PTSD subjects has also been examined. However, PPI associations with PTSD are inconsistent, with some studies showing significantly reduced PPI in PTSD patients (Ornitz and Pynoos 1989; Grillon et al. 1996, 1998), while others detected no differences or only marginal differences (Butler et al. 1990; Morgan et al. 1997; Lipschitz et al. 2005; Holstein et al. 2010; Vrana et al. 2013). Thus, additional research is needed to clarify or refute the presence of PPI deficits in PTSD.

Summary of altered startle plasticity in PTSD: Exaggerated startle responding in PTSD patients is seen fairly consistently, most predominantly under conditions of challenge or threat. Pole (2007) conducted a meta-analysis of 20 studies measuring startle response via *orbicularis oculi* EMG both at baseline and after manipulation of contextual threat. This analysis supported a significant increase in startle responses in PTSD; however, this effect was not as robust as elevated cardiovascular responses. Furthermore, increased startle response to threat is also not specific to PTSD, but is also reported in other disorders that are characterized by high physiological arousal and fear (e.g., panic disorder) but not generalized anxiety disorder (Grillon et al. 2009; Grillon 2008). These findings suggest that disorders characterized by exaggerated startle may share an overlapping biological pathway. It is not clear, however, whether these effects are due to increased fear responses per se (e.g., via increased amygdala and/or insula circuit activation), or reduced ability to inhibit or modulate these responses appropriately (e.g., reduced modulation of amygdala output by hippocampal and cortical circuits; see below; Acheson et al. 2012; Klumpers et al. 2007).

Habituation and PPI are both measures of fundamental aspects of information processing that are disrupted in a number of psychiatric disorders and are to some degree heritable (Greenwood et al. 2007). However, there is relatively weak evidence at present for disruptions in PTSD. It is possible that disruption in these processes may indicate one of potentially many biological risk traits for neuropsychiatric disorders. Hence, further understanding of the genetic and neurobiological mechanisms underlying these phenotypes and their relationship to PTSD risk is worth further investigation. Indeed, PTSD is thought to share polygenic risk with other disorders that exhibit information processing deficits, such as bipolar disorder and schizophrenia (Nievergelt et al. in review; Solovieff et al. 2014).

While exaggerated startle per se is not unique to PTSD, it nonetheless represents a powerful method for exploring mechanisms underlying the development of PTSD symptoms. In animals, exaggerated startle phenotypes have long been utilized to test causal hypotheses of potential mechanisms underlying development of anxiety and fear-related behaviors after severe stress, including corticotropin-releasing factor and noradrenergic abnormalities (e.g., Risbrough and Stein 2006; Davis et al. 2010; Grillon et al. 2009). In humans, utilization of startle plasticity as an intermediate phenotype is just now beginning to be exploited (Greenwood et al. 2012). Further, questions of exaggerated startle magnitude versus reduced startle threshold in PTSD remain to be answered (Butler et al. 1990). Finally, surprisingly few pharmacological studies have thus far utilized startle to examine potential biological mechanisms of increased physiological responses in PTSD. Using a pharmacological challenge with the alpha 2 antagonist yohimbine, Morgan et al. (1995) showed that startle reactivity in PTSD patients may be via increased sensitivity to noradrenergic signaling.

2.3 Other Physiological Measures

Electrodermal Level/Response: In addition to HR and startle, researchers have examined electrodermal levels in PTSD both at resting baseline and in response to challenge. Electrodermal response, or the increase in electrical conductivity across the skin due to sweat, is a physiological marker of sympathetic nervous system arousal. A meta-analysis by Pole (2007) looked across 31 studies that measured resting electrodermal levels in subjects with PTSD versus controls and found support for significantly higher levels associated with PTSD, although the effect size was small. Blechert et al. (2007) found that PTSD subjects had higher resting baseline electrodermal level relative to both healthy controls and subjects with panic disorder, suggesting some diagnostic specificity. Resting electrodermal level has historically been reported to be reduced in subjects with depression versus healthy controls (Argyle 1991), further suggesting that this measure may hold some diagnostic specificity.

Electrodermal response to challenge by standardized and ideographic trauma cues has also been examined in relation to PTSD. Pole (2007) looked across 22 studies and found medium effect sizes for elevated electrodermal response to both standard and ideographic cues in PTSD versus controls. Interestingly, Blechert et al. (2007) found blunted electrodermal response in PTSD when subjects were under threat of electrical shock, suggesting that there may be a difference in effect between challenge by reminder cue versus challenge by contextual threat (experimental shock). Similarly, McTeague et al. (2010) found that PTSD subjects with multiple traumas and more severe, chronic PTSD showed blunted defensive responses to ideographic imagery. More recently, Glover et al. (2011) showed overall elevations in fear-potentiated startle in a classical conditioning paradigm in PTSD subjects relative to controls; however, no differences were found in electrodermal responses. It is possible that startle reactivity measures may offer a wider measurable range to detect increased reactivity than skin conductance measures because startle baseline can be controlled by the experimenter (i.e., via adjustments of the intensity of acoustic pulse). Thus, it is possible that startle may be more sensitive to detecting differences in responses even under relatively high arousal states (e.g., under threat). Skin conductance, however, offers other significant advantages over startle, since it does not require a relatively invasive stimulus (e.g., acoustic pulse) for measurement. The passive nature of this measurement has also supported its use as a complementary tool in imaging studies in which subject movement must be severely limited (i.e., startle response movement can disrupt image processing).

Facial EMG: Facial EMG has been used as a physiological measure of emotional response and typically involves measurement of activity in the frontalis, corrugator, and zygomaticus major muscles involved in emotional facial expressions such as smiling and frowning. Pole (2007) found support for increased frontalis and corrugator EMG activity while viewing ideographic trauma cues (12 and 5 studies, respectively). Pole (2007) found no support for altered facial EMG activity at resting baseline, or in response to standardized trauma cues (12 and 6 studies, respectively).

Because these measures are (1) more sensitive to artifact (e.g., non-specific facial and head movements, talking) and (2) are not easily controlled or evoked parametrically compared to reflexive responses such as changes in HR, skin conductance, and startle, they have not been utilized widely. They do not offer cross-species translation nor have well-defined circuits; thus, they may have less utility in understanding biological mechanisms of PTSD.

Summary of Other Physiological Measures Associated with PSTD: Elevated resting-state electrodermal level may be a psychophysiological measure that is relatively specific to PTSD. However, this measure is susceptible to the same methodological difficulties as resting HR or baseline startle response, namely that it is difficult to eliminate contextual factors that may influence stress and thus electrodermal activity. Electrodermal response to challenge presents a complicated picture with findings varying dependent upon both subject-specific and testing protocol variables. There is support for an association between increased facial EMG reactivity specifically in response to idiographic trauma cues; however, the utility of this measure for further biological research is limited.

3 Psychophysiological Markers of PTSD-Relevant Constructs: Fear and Sustained Anxiety

Safety Signal Learning: Safety signal learning is the process by which an individual learns to inhibit a learned fear response in the presence of a cue signaling absence of danger. This process is directly relevant to PTSD phenomenology insofar as PTSD is in part characterized by altered reactivity to trauma-related cues even in "safe" environments. Safety signal learning can be measured by assessing responses to a CS− that is never associated with an aversive event versus a CS+ that is contiguous with an aversive event, or via a specific CS that predicts absence of the aversive event when given in conjunction with the CS+. Using the latter paradigm, Jovanovic et al. (2010) recently tested this process in a sample of trauma-exposed civilians who were healthy, had PTSD, had major depression, or had comorbid PTSD and major depression with fear-potentiated startle as the primary outcome variable. Subjects learned that a cue predicted a blast of air to the throat, but that when this cue was presented along with another cue (the safety signal), the blast of air would not occur. Subjects with PTSD and comorbid PTSD/major depression failed to show inhibition of the potentiated startle response in the presence of the safety cue. Inability of subjects with PTSD to inhibit responding to a safety signal was also confirmed in the former paradigm, a simple CS+/CS− discrimination learning task (Jovanovic et al. 2013). Andero et al. (2013) found associations between the ability to learn to discriminate between the CS+ (danger) and CS− (safety) are impaired in subjects with a single nucleotide polymorphism (SNP) on the opioid receptor 1-like gene which encodes for the amygdala nociception/orphanin FQ receptor involved in pain processing. This SNP was also associated with greater PTSD symptoms, providing further evidence for impaired safety signal processing in PTSD as well as a putative biological pathway for this effect. These results, though preliminary and in need of replication, suggest that failure to learn to distinguish between environmental cues signaling danger versus safety may be an important process that is impaired in PTSD.

Fear Extinction: Fear extinction is the process by which an organism learns that a cue that once signaled threat no longer does so, thus resulting in a progressive reduction in defensive physiological responding in the presence of this cue. Extinction of psychophysiological fear responding has long been considered a putative model of PTSD process due to its similarity to naturalistic recovery from trauma experience. Orr et al. (2000) and Peri et al. (2000) showed that PTSD patients failed to extinguish a conditioned electrodermal response to a cue signaling electrical shock or loud acoustic stimuli, respectively. Subsequent studies using electrodermal responses as the dependent variable have largely supported these original findings (e.g., Wessa and Flor 2007; Blechert et al. 2007). Norrholm et al. (2011) examined fear extinction in PTSD using fear-potentiated startle to a cue signaling an aversive air puff to the throat and found that PTSD patients showed greater potentiated startle in the early and middle portions of extinction training. This finding suggests that enhanced initial fear conditioning produced a greater "fear load" that the PTSD patients had to extinguish. This increased fear responding is also associated with specific symptom clusters of PTSD, re-experiencing (Glover et al. 2011), indicating this paradigm likely probes neural mechanisms of trauma memory.

Not all studies have found evidence for delay of fear extinction learning in PTSD. Milad et al. (2008) found equal levels of extinction performance, as measured by electrodermal response, in combat-related PTSD compared to combat-exposed monozygotic twins without PTSD and controls. However, the PTSD twins failed to recall this fear extinction learning when tested 24 h later. These results suggest that PTSD is not associated with a fear extinction learning deficit, but rather a fear extinction memory deficit. Further, this deficit appears to be an acquired sign of PTSD rather than an inherited trait. This difference in within-session learning results across these studies may be due to the physiological measures of fear used, startle versus skin conductance. The higher magnitude of the startle response to the conditioned cue in PTSD patients is providing a behavioral window to detect reduced/delayed extinction within session, which is not detectable via skin conductance responses (Glover et al. 2011). Taken together, these data suggest overall that there is higher fear responding in PTSD patients, which subsequently takes longer to extinguish fully and is less likely to be fully extinguished upon retesting. Additional research will be needed to determine the time point at which extinction deficits may occur, the most effective method for capturing such deficits, and the specific role these deficits play in PTSD symptomatology.

Summary of Psychophysiological Markers of PTSD-relevant Constructs: Psychophysiological markers have emerged as critical measures of unbiased fear responding to understand fear and anxiety domains disrupted in PTSD. These markers provide quantifiable assessments of autonomic processes that may not be adequately probed by self-report. They have been critical behavioral measures that complement studies of the neural circuits underlying PTSD pathology, such as cortico-hippocampal–amygdala circuit function (Quirk et al. 2006), that can be translated across species for further study of causal factors for PTSD symptoms or PTSD risk. The intriguing preliminary evidence for safety signal learning to be disrupted specifically in PTSD versus depression patients may indicate this is a potential "biomarker" of PTSD, but needs further research and replication. Extinction has shown to be impaired in a number of neuropsychiatric disorders as well as PTSD, including obsessive–compulsive disorder and schizophrenia (Holt et al. 2009; Milad et al. 2013), suggesting that extinction learning may probe common pathological circuits across these disorders. Impairment in these processes is further supported by imaging research showing impaired function and structure of the ventromedial prefrontal/orbitofrontal cortex in PTSD subjects, which are structures known to be central to fear extinction learning and memory (Shin et al. 2006). Recent research suggests involvement of these areas in safety signal learning as well (Jovanovic et al. 2013). Finally, more recently, these paradigms have been utilized in healthy controls or PTSD patients to serve as proof of concept tests for novel treatments for fear-related disorders such as PTSD, with recent or ongoing tests of cannabinoid agonists (Rabinak et al. 2013), oxytocin (Acheson et al. 2013), glucocorticoids (de Quervain et al. 2011), and dopamine agonists (Haaker et al. 2013), among others. It remains to be determined how predictive these paradigms will be for treatment efficacy; however, this is an exciting avenue for PTSD drug discovery.

4 Psychophysiological Markers of Risk for Developing PTSD Following Trauma

Trait Markers: Given that elevated physiological reactivity is a common finding in those with current PTSD, researchers have explored the possibility that this elevated reactivity might serve as a marker of risk prior to or immediately following the traumatic experience. Several studies examined the relationship between HR shortly following trauma and later development of PTSD and found that elevated HR following trauma predicted development of PTSD symptoms (Bryant et al. 2000; Kassam-Adams et al. 2005; Shalev et al. 1998; Zatzick et al. 2005; Kuhn et al. 2006; Gould et al. 2011). Though numerous exceptions have been reported (Blanchard et al. 2002; Buckley et al. 2004; Ehring et al. 2008; Roitman et al. 2013; Price et al. 2014). In a related study, Suendermann et al. (2010) found that HR response to trauma-related images in motor vehicle accident survivors 1 month after trauma predicted PTSD severity at 6 months after trauma. The inconsistency in these findings may be due to the fact that cardiovascular activity assessed immediately post-trauma in the ambulance or emergency department may be subject to too many contextual variables, methodological inconsistencies, or ceiling effects that may limit reproducibility of findings. Newer technology allowing for ambulatory monitoring in the days following trauma (see below) may prove more useful in determining at which time points and under what circumstances post-trauma HR may be most predictive of future PTSD.

While these studies of peri-traumatic HR suggest potential clinical utility as a marker of risk in traumatized individuals, they tell us little about who might be at risk for trauma before the event happens. Toward answering this question, Pitman et al. (2006) examined HR responses to a series of loud tones in Vietnam veterans with PTSD and their non-combat-exposed monozygotic twins. Only the twin with PTSD showed elevated HR response relative to combat-exposed veterans without PTSD and their non-exposed twins, suggesting that elevated HR response is an acquired sign of PTSD rather than a risk factor. However, further longitudinal studies where HR response is measured prior to trauma will be necessary to definitively rule out HR as a prospective marker of risk for PTSD. Pole et al. (2009) measured a number of physiological indices (startle, electrodermal response, HR) in response to startling tones under conditions of varying contextual threat (low, medium, and high threat of electrical shock) in new police academy cadets. These cadets were then later assessed for PTSD symptoms following one year of police work. They found that elevated startle measured by eyeblink EMG (with appropriate sampling rate), elevated electrodermal response, and slower habituation of the electrodermal response predicted PTSD symptom severity, but that HR response did not. Further, the associations between physiological reactivity and PTSD severity varied as a condition of the contextual threat: Greater electrodermal response was associated with PTSD symptom severity under low and high threat, and eyeblink EMG under medium threat was associated with symptom severity. These findings support the hypothesis that increased physiological reactivity to threat may be a useful marker for understanding biological mechanisms of PTSD risk.

Markers of Fear and Anxiety Constructs: Little is known about how abnormalities in safety learning and fear extinction may function as preexisting markers of risk for PTSD. A recent study found that impaired ability to inhibit fear-potentiated startle responding in the presence of a safety cue was associated with PTSD symptoms 2 and 9 months after combat-related trauma (Sijbrandij et al. 2013). These findings suggest that impaired safety signal learning may be important in predicting the maintenance of PTSD symptoms over time. It is not clear, however, whether reductions in safety signal learning predict PTSD prospectively. Investigators have also begun to look at impaired fear extinction processes as risk factors for developing PTSD following trauma. A twin study of combat-related PTSD by Milad et al. (2008) suggested that reduced recall of fear extinction memory is an acquired sign of PTSD rather than a preexisting risk factor. Guthrie and Bryant (2006) examined initial fear extinction learning of an aversively conditioned corrugator EMG response in a sample of firefighter trainees. They found that slower extinction while in training predicted PTSD severity after later exposure to trauma. Lommen et al. (2013) showed similar effects in a sample of Dutch combat veterans, though they only assessed explicit contingency awareness rather than physiological response. Further prospective–longitudinal studies assessing both habituation and extinction prior to trauma are needed to confirm whether or not these are robust markers of PTSD risk.

Summary of Risk Markers: While peri-traumatic physiological response may provide some information regarding who is at risk for developing chronic PTSD, more research is needed to solidify the extant findings and to link elevated physiology following trauma to specific biological changes underlying chronic disorder. Much less is known about using physiological markers to predict risk for PTSD prior to traumatic experience, though the results of Pole et al. (2009) provide promising avenues for future research in this area and suggest the possibility of achieving superior prediction by the integration of multiple psychophysiological domains into a single marker for risk. Knowing who is at risk for PTSD prior to trauma may have utility for screening of soldiers and first responders such as firefighters and police officers. Identification of pretrauma risk factors that are modifiable can inform prevention efforts in these and other populations at high risk for trauma exposure and may also point toward fruitful targets for novel treatment efforts.

5 Future Areas of Application for Psychophysiological Research

Psychophysiological Markers of Treatment Response: Beyond serving as markers of PTSD state or risk for developing the disorder, psychophysiological outcomes may have potential to provide objective markers of treatment response. This utility is particularly relevant as the NIMH now requires treatment studies to include biological and/or physiological markers along with standard symptom scales. To date, however, relatively few studies have made use of physiological outcome measures. To our knowledge, there are no reports of psychophysiological responses in PTSD patients during standard pharmacotherapies, e.g., serotonin reuptake inhibitors. Two recent studies using psychotherapy have included physiological markers. Robinson-Andrew et al. (2014) assessed potentiated startle responding in the presence of trauma-related visual cues in a small number of combat veterans with PTSD before, during, and after either prolonged exposure or "present-centered therapy" treatment. Treatment responders showed a dynamic pattern of increasing and then decreasing startle potentiation across treatment, while non-responders did not change. In another recent study, Rothbaum et al. (2014) compared the effects of d-cycloserine, alprazolam, and placebo on response to 5 sessions of prolonged exposure therapy for PTSD. Outcomes consisted of both self-reported diagnostic assessments as well as potentiated startle response to trauma-related images. The patients receiving d-cycloserine showed significantly lower startle potentiation post-treatment, and magnitude of startle reduction was associated with self-reported treatment response in this group only. However, groups did not differ on self-reported response to the treatment overall. There is no research yet on treatment effects on PTSD-related constructs of fear extinction or safety signal learning.

One earlier area of study where psychophysiological outcomes appeared promising was in predicting potential prophylactic efficacy of propranolol, a beta-adrenergic receptor antagonist. Pitman et al. (2002) originally showed that propranolol given immediately after trauma reduced physiological arousal (HR, electrodermal response, facial EMG) to script-driven traumatic imagery 3 months later, as well as showing a nonsignificant trend toward reduced PTSD symptom severity 1 month following trauma. In a larger study, Hoge et al. (2012) showed mixed results when propranolol or placebo was given to emergency department patients for 19 days following trauma. In "high-medication adherence" subjects, those who took the active drug showed reduced physiological reactivity to trauma imagery across three domains (electrodermal response, HR, lateral frontalis EMG) at 1 month following trauma relative to those who received placebo. However, this difference was not found at 3 months post-trauma, nor was there an effect of treatment on PTSD symptoms. Given the very mixed literature for treatment efficacy of propranolol as a prophylactic treatment for PTSD (Vaiva et al. 2003; Stein et al. 2007; McGhee et al. 2009), the predictive validity of psychophysiological measures for propranolol prevention of PTSD symptoms is inconclusive. Current studies have now shifted to examination of propranolol effects on memory reconsolidation in PTSD patients (www.clinicaltrials.gov), based in part on recent findings that propranolol given immediately after reactivation of the trauma memory via script preparation reduces physiological responding to the same script one week later (Brunet et al. 2009).

Psychophysiological outcomes have also seen limited use in studies investigating potential novel treatments. Jovanovic et al. (2011) showed that dexamethasone treatment reduces fear-potentiated startle in PTSD patients, suggesting that this treatment could reduce physiological symptoms of fear in these patients. These results provide preliminary support for the predictive validity of fear-potentiated startle in PTSD, since glucocorticoid agonists may reduce PTSD symptoms (Aerni et al. 2004; Steckler and Risbrough 2012). An ongoing study is also assessing the efficacy of corticotropin-releasing factor receptor antagonist treatment on both PTSD symptoms and fear-potentiated startle (Dunlop et al. 2014). We expect that more studies will utilize this complementary approach of physiological and self-report measures to assess treatment efficacy in the future.

Overall, psychophysiological outcomes have not been utilized in treatment studies and thus remain largely untested for sensitivity to treatment effects for PTSD. An important caveat is that some studies have shown a pattern of treatment-induced reductions in psychophysiological arousal, but not in self-reported PTSD symptom severity. This pattern of findings suggests several possibilities. First, psychophysiological alterations may not be powerful enough to generalize into symptom change per se (e.g., Hoge et al. 2012). Second, psychophysiological alterations may be one of the several potential mechanisms of change occurring within the same treatment protocol (e.g., Rothbaum et al. 2014). These conclusions suggest that psychophysiological assessment may be used as an objective marker of treatment response and have utility in elucidating mechanism/process of change that may vary across subjects being treated with the same protocol. Further, psychophysiological assessment may have utility for understanding which patients may benefit from among several treatment modalities aimed at the same overt condition (Aikens et al. 2011). More research is required before this approach can be considered a realistic possibility in the near term.

Consideration of Mild Traumatic Brain Injury (mTBI) in Psychophysiological Investigations of Trauma-related Pathology: Many of the traumatic experiences that might result in development of PTSD (motor vehicle accident, physical assault, combat) also involve potential for physical harm. The large numbers of blast-related injuries coming out of the wars in Iraq and Afghanistan (Hoge et al. 2008) have brought into recent focus the potential relationship between mTBI and PTSD. A prospective study of service members deployed in these conflicts suggests a strong association between deployment-related mTBI and post-deployment PTSD symptoms (Yurgil et al. 2014). These findings suggest that mTBI may need to be considered as an important factor in assessing psychophysiological outcomes in PTSD, similar to its potential effects on neurocognitive symptoms in PTSD (Vasterling et al. 2009, 2012). Little research has been conducted on how mTBI affects the physiological markers discussed here, with the exception of HRV. HRV is reduced in some TBI patients, with alterations related to time since injury and injury severity (Keren et al. 2005; Baguley et al. 2006). One study in active duty marines with PTSD suggests that HRV is reduced in PTSD subjects even when controlling for TBI although TBI was also independently associated with reduced HRV (Minassian et al. 2014). Williamson et al. (2013) have suggested that in cases of mTBI-induced damage to white matter tracts involved in emotional behavior (e.g., uncinate fasciculus and the anterior limb of the internal capsule) may cause disruption of top-down control of autonomic nervous system activity reflected in psychophysiological measurements. These forms of disruption could also explain the higher risk for development of PTSD in individuals exhibiting mTBI (Yurgil et al. 2014). Interestingly, recent animal studies have also supported that mild TBI could result in sensitization of fear learning processes (Heldt et al. 2014). Thus, mTBI should be carefully considered in future assessments of PTSD-related physiology, particularly in abnormalities of cortical-mediated inhibitory processes and fear learning constructs, to understand its modulating or mediating role in psychophysiological abnormalities in PTSD.

Wearable Physiological Monitoring Technology: Although the specific physiological abnormalities linked to trauma symptoms are becoming more clear as reviewed above, one of the next steps for the field is to determine whether these measures can translate to clinical applications, such as prediction of symptom development, symptom class, and/or treatment response. Moving these measures to clinical applications faces significant hurdles, one of which is the development of more usable devices that are not dependent on narrow laboratory-specific parameters or expensive and complicated hardware. One potential area for psychophysiology variables in mental health in the future is use of "wearable" devices in subjects that have experienced, or at risk for, trauma (Darwish and Hassanien 2011).

There is a strong push both in private and academic medical sectors to implement wearable devices for a host of medical purposes including diabetes, cardiovascular disease management, cognitive therapy aids, and other lifestyle aids for better wellness. Predictive psychophysiological variables relevant to PTSD phenotypes that may be conducive to wearable technology are measures of physical activity via accelerometers (e.g., Fukukawa et al. 2004), sleep (Suzuki et al. 2014), skin conductance (Rajan et al. 2012), HR and HRV (Billeci et al. 2014), EMG (Grenier et al. 2012), and EEG (Zao et al. 2014). The development of these wearables will enable assessment of dynamics of physiology in naturalist settings, at rest (i.e., sleep) as well as during stress. These devices may help answer the question of which physiological variable, or combination of variables, might be able to predict development of PTSD symptoms after trauma exposure (e.g., after discharge from the ER/hospital). Another question is if physiological markers are sensitive to treatment, and when in the recovery process does this happen (i.e., could these markers serve as early predictors of treatment response?). Many of these variables are not "static," for example, longer-term assessment of sleep variance across multiple nights will enable a much more comprehensive picture than can feasibly be obtained in laboratory settings. Similarly, HRV over long time periods will provide greater fidelity in the assessment of cardiovascular changes after trauma. Some wearable devices may also be utilized in "at-risk" populations, such as rescue service and military personnel, to develop algorithms of risk based on physiological response and recovery after trauma exposure. This approach is currently being examined in the military (Tharion et al. 2013). However, a number of hurdles must be considered in terms of feasibility/practicality of the technology, the data quality, storage capacity, and of course the ethical component of resulting data being used or stored improperly.

One example of current status of technology is assessment of continuous HR. HRV can now be obtained via sophisticated wearable devices (e.g., pulse oximeter introduced into a wrist watch) over long periods of time with little burden to the subject. However, technical challenges must be addressed, including the high sampling rate needed for HRV assessments that can produce power and data storage limitations for continuous monitoring. Data quality is also affected significantly by movement artifact for many of these devices. Thus, despite significant promise, many technical limitations must be addressed before these devices will produce reliable physiological assessments for utility in prediction and intervention.

6 Conclusion

As discussed above, there are now a number of well-validated physiological phenotypes that are reliable across multiple studies/laboratories, including increased and poorly inhibited physiological responses to threat (electrodermal and EMG), as well as altered HRV. We are just now beginning to understand these measures in a larger context of symptom domains, as well as comorbid symptoms (depression, TBI, etc.).

Much more work is needed, however, to refine these phenotypes in terms of specific associations with PTSD symptoms versus other anxiety disorders and comorbid symptoms (depression, TBI). Importantly, many of these phenotypes are now well mapped to circuitry that supports translational research across species for mechanisms driving these phenotypes, which will support development of novel treatment targets. To this end, psychophysiological measures are increasingly being used as complementary measures for integration with both self-report and other biological assessments (e.g., blood-based or genetic markers). We expect much more research in the years to come with these tools for objective assessment of treatment outcome. Finally, in the long term, wearable technology could accelerate the feasibility of these markers as tools to identify risk and symptom development in clinical settings.

Acknowledgments This work was supported by MH074697, MH042228, VA BLR&D Merit Award, and a NARSAD Young Investigator Award.

References

Acheson DT, Gresack JE, Risbrough VB (2012) Hippocampal dysfunction effects on context memory: possible etiology for posttraumatic stress disorder. Neuropharmacology 62:674–685

Acheson DT, Feifel D, de Wilde S, McKinney R, Lohr J, Risbrough VB (2013) The effect of intranasal oxytocin treatment on conditioned fear extinction and recall in a healthy human sample. Psychopharmacology 229:199–208

Adenauer H, Catani C, Keil J, Aichinger H, Neuner F (2010) Is freezing an adaptive reaction to threat? Evidence from heart rate reactivity to emotional pictures in victims of war and torture. Psychophysiology 47:315–322

Aerni A, Traber R, Hock C, Roozendaal B, Schelling G, Papassotiropoulos A, Nitsch RM, Schnyder U, de Quervain DJF (2004) Low-dose cortisol for symptoms of posttraumatic stress disorder. Am J Psychiatry 161:1488–1490

Agorastos A, Boel JA, Heppner PS, Hager T, Moeller-Bertram T, Haji U, Motazedi A, Yanagi MA, Baker DG, Stiedl O (2013) Diminished vagal activity and blunted diurnal variation of heart rate dynamics in posttraumatic stress disorder. Stress 16:300–310

Ahmari SE, Risbrough VB, Geyer MA, Simpson HB (2012) Impaired sensorimotor gating in unmedicated adults with obsessive-compulsive disorder. Neuropsychopharmacology 37:1216–1223

Argyle N (1991) Skin conductance levels in panic disorder and depression. J Nerv Ment Dis 179:563–566

Andero R, Brothers SP, Jovanovic T, Chen YT, Salah-Uddin H, Cameron M, Bannister TD, Almli L, Stevens JS, Bradley B, Binder EB, Wahlestedt C, Ressler KJ (2013) Amygdala-Dependent fear Is regulated by oprl1 in mice and humans with PTSD. Sci Transl Med 5:188ra73

Baguley IJ, Heriseanu RE, Felmingham KL, Cameron ID (2006) Dysautonomia and heart rate variability following severe traumatic brain injury. Brain Inj 20:437–444

Baker DG, Heppner P, Afari N, Nunnink S, Kilmer M, Simmons A, et al. (2009) Trauma exposure, branch of service, and physical injury in relation to mental health among US veterans returning from Iraq and Afghanistan. Mil Med 174:773−778

Baker DG, Nievergelt CM, O'Connor DT (2012) Biomarkers of PTSD: neuropeptides and immune signaling. Neuropharmacology 62:663–673

Barkay G, Freedman N, Lester H, Louzoun Y, Sapoznikov D, Luckenbaugh D, Shalev AY, Chisin RG, Bonne O (2012) Brain activation and heart rate during script-driven traumatic imagery in PTSD: preliminary findings. Psychiatry Res Neuroimaging 204:155–160

Berntson GG, Bigger JT, Eckberg DL, Grossman P, Kaufmann PG, Malik M, Nagaraja HN, Porges SW, Saul JP, Stone PH, van der Molen MW (1997) Heart rate variability: origins, methods, and interpretive caveats. Psychophysiology 34:623–648

Billeci L, Tartarisco G, Brunori E, Crifaci G, Scardigli S, Balocchi R, Pioggia G, Maestro S, Morales MA (2014) The role of wearable sensors and wireless technologies for the assessment of heart rate variability in anorexia nervosa. Eat Weight Disord (Epub ahead of print)

Blakeley K, Jansen DJ (2013) Post-traumatic stress disorder and other mental health problems in the military: oversight issues for congress. Congressional research service report

Blanchard EB, Kolb LC, Pallmeyer TP, Gerardi RJ (1982) A psychophysiological study of post-traumatic stress disorder in Vietnam veterans. Psychiatr Q 54:220–229

Blanchard EB, Hickling EJ, Galovski T, Veazey C (2002) Emergency room vital signs and PTSD in a treatment seeking sample of motor vehicle accident survivors. J Trauma Stress 15:199–204

Blechert J, Michael T, Vriends N, Margraf J, Wilhelm FH (2007) Fear conditioning in posttraumatic stress disorder: Evidence for delayed extinction of autonomic, experiential, and behavioural responses. Behav Res Ther 45:2019–2033

Blumenthal TD, Cuthbert BN, Filion DL, Hackley S, Lipp OV, van Boxtel A (2005) Committee report: guidelines for human startle eyeblink electromyographic studies. Psychophysiology 42:1–15

Braff DL, Geyer MA, Swerdlow NR (2001) Human studies of prepulse inhibition of startle: normal subjects, patient groups, and pharmacological studies. Psychopharmacology 156:234–258

Breslau N (2009) The epidemiology of trauma, PTSD, and other posttrauma disorders. Trauma Violence Abuse 10:198–210

Brunet A, Orr SP, Tremblay J, Robertson K, Nader K, Pitman RK (2009) Effect of post-retrieval propranolol on pscyhophysiologic responding during subsequent script-driven traumatic imagery in post-traumatic stress disorder. J Psychiatr Res 42:503–506

Bryant RA, Harvey AG, Guthrie RM, Moulds ML (2000) A prospective study of psychophysiological arousal, acute stress disorder, and posttraumatic stress disorder. J Abnorm Psychol 109:342–344

Buckley TC, Kaloupek DG (2001) A meta-analytic examination of basal cardiovascular activity in posttraumatic stress disorder. Psychosom Med 63:585–594

Buckley TC, Holohan D, Grei, JL, Bedard M, Suvak M (2004) Twenty-four hour ambulatory assessment of heart rate and blood pressure in chronic PTSD and non-PTSD veterans. J Trauma Stress 17:163–171

Butler RW, Braff DL, Rausch JL, Jenkins MA, Sprock J, Geyer MA (1990) Physiological evidence of exaggerated startle response in a subgroup of Vietnam veterans with combat-related PTSD. Am J Psychiatry 147:1308–1312

Castellanos FX, Fine EJ, Kaysen D, Marsh WL, Rapoport JL, Hallett M (1996) Sensorimotor gating in boys with Tourette's syndrome and ADHD: preliminary results. Biol Psychiatry 39:33–41

Chang HA, Chang CC, Tzeng NS, Kuo TB, Lu RB, Huang SY (2013) Decreased cardiac vagal control in drug-naive patients with posttraumatic stress disorder. Psychiatry Investig 10:121–130

Cohen H, Benjamin J, Geva AB, Matar MA, Kaplan Z, Kotler M (2000) Autonomic dysregulation in panic disorder and in post-traumatic stress disorder: application of power spectrum analysis of heart rate variability at rest and in response to recollection of trauma or panic attacks. Psychiatry Res 96:1–13

Cuthbert BN, Lang PJ, Strauss C, Drobes D, Patrick CJ, Bradley MM (2003) The psychophysiology of anxiety disorder: fear memory imagery. Psychophysiology 40:407–422

Darwish A, Hassanien AE (2011) Wearable and implantable wireless sensor network solutions for healthcare monitoring. Sensors 11:5561–5595

Davis M, Walker DL, Miles L, Grillon C (2010) Phasic vs sustained fear in rats and humans: role of the extended amygdala in fear vs anxiety. Neuropsychopharmacology 35:105–135 de Quervain DJ, Bentz D, Michael T, Bolt OC, Wiederhold BK, Margraf J, Wilhelm FH (2011) Glucocorticoids enhance extinction-based psychotherapy. Proc Natl Acad Sci USA 108:6621–6625

Dunlop BW, Rothbaum BO, Binder EB, Duncan E, Harvey PD, Jovanovic T, Kelley ME, Kinkead B, Kutner M, Iosifescu DV, Mathew SJ, Neylan TC, Kilts CD, Nemeroff CB, Mayberg HS (2014) Evaluation of a corticotropin releasing hormone type 1 receptor antagonist in women with posttraumatic stress disorder: study protocol for a randomized controlled trial. Trials 15:240

Ehlers A, Suendermann O, Boellinghaus I, Vossbeck-Elsebusch A, Gamer M, Briddon E, Martin M, Glucksman E (2010) Heart rate responses to standardized trauma-related pictures in acute posttraumatic stress disorder. Int J Psychophysiol 78:27–34

Ehring T, Ehlers A, Cleare AJ, Glucksman E (2008) Do acute psychological and psychobiological responses to trauma predict subsequent symptom severities of PTSD and depression? Psychiatry Res 161:67–75

Eraly SA, Nievergelt CM, Maihofer AX, Barkauskas DA, Biswas N, Agorastos A, O'Connor DT, Baker DG (2014) Assessment of plasma C-reactive protein as a biomarker of posttraumatic stress disorder risk. JAMA Psychiatry 71:423–431

Fukukawa Y, Nakashima C, Tsuboi S, Kozakai R, Doyo W, Niino N, Ando F, Shimokata H (2004) Age differences in the effect of physical activity on depressive symptoms. Psychol Aging 19:346–351

Geracioti Jr TD, Baker DG, Kasckow JW, Strawn JR, Mulchahey J, Dashevsky BA, Horn PS, Ekhator NN (2008) Effects of trauma-related audiovisual stimulation on cerebrospinal fluid norepinephrine and corticotropin-releasing hormone concentrations in post-traumatic stress disorder. Psychoneuroendocrinology 33:416–424

Geracioti TD, Baker DG, Ekhator NN, West SA, Hill KK, Bruce AB, Schmidt D, Rounds-Kugler B, Yehuda R, Keck PE, Kasckow JW (2001) CSF norepinephrine concentrations in posttraumatic stress disorder. Am J Psychiatry 158:1227–1230

Geyer MA, Braff DL (1982) Habituation of the Blink reflex in normals and schizophrenic patients. Psychophysiology 19:1–6

Geyer MA, Braff DL (1987) Startle habituation and sensorimotor gating in schizophrenia and related animal models. Schizophr Bull 13:643–668

Geyer MA, Swerdlow NR, Mansbach RS, Braff DL (1990) Startle response models of sensorimotor gating and habituation deficits in schizophrenia. Brain Res Bull 25:485–498

Glover EM, Phifer JE, Crain DF, Norrholm SD, Davis M, Bradley B, Ressler KJ, Jovanovic T (2011) Tools for translational neuroscience: PTSD is associated with heightened fear responses using acoustic startle but not skin conductance measures. Depress Anxiety 28:1058–1066

Gould NF, McKibben JB, Hall R, Corry NH, Amoyal NA, Mason ST, McCann UD, Fauerbach JA (2011) Peritraumatic heart rate and posttraumatic stress disorder in patients with severe burns. Clin Psychiatry 72:539–547

Greenwood TA, Braff DL, Light GA, Cadenhead KS, Calkins ME et al (2007) Initial heritability analyses of endophenotypic measures for schizophrenia: the consortium on the genetics of schizophrenia. Arch Gen Psychiatry 64:1242–1250

Greenwood TA, Light GA, Swerdlow NR, Radant AD, Braff DL (2012) Association analysis of 94 candidate genes and schizophrenia-related endophenotypes. PLoS ONE 7:e29630

Greenwood TA, Swerdlow NR, Gur RE, Cadenhead KS, Calkins ME et al (2013) Genome-wide linkage analyses of 12 endophenotypes for schizophrenia from the Consortium on the Genetics of Schizophrenia. Am J Psychiatry 170:521–532

Grenier JG, Millet GY, Peyrot N, Samozino P et al (2012) Effects of extreme-duration heavy load carriage on neuromuscular function and locomotion: a military-based study. PLoS ONE 7: e43586

Grillon C, Morgan CA, Southwick SM, Davis M, Charney DS (1996) Baseline startle amplitude and prepulse inhibition in Vietnam veterans with posttraumatic stress disorder. Psychiatry Res 64:169–178

Grillon C, Morgan CA, Davis M, Southwick SM (1998) Effects of experimental context and explicit threat cues on acoustic startle in Vietnam veterans with posttraumatic stress disorder. Biol Psychiatry 44:1027–1036

Grillon C, Baas J (2003) A review of the modulation of the startle reflex by affective states and its application in psychiatry. Clin Neurophysiol 114:1557–1579

Grillon C (2008) Greater sustained anxiety but not phasic fear in women compared to men. Emotion 8:410–413

Grillon C, Pine DS, Lissek S, Rabin S, Bonne O, Vythillingham M (2009) Increased anxiety during anticipation of unpredictable aversive stimuli in posttraumatic stress disorder but not in generalized anxiety disorder. Biol Psychiatry 66:47–53

Guthrie RM, Bryant RA (2006) Extinction learning before trauma and subsequent posttraumatic stress disorder. Psychosom Med 68:307–311

Haaker J, Gaburro S, Sah A, Gartman N, Lonsdorf TB, Meier K, Singewald N, Pape HC, Morellini F, Kalisch R (2013) Single dose of L-dopa makes extinction memories context-independent and prevents the return of fear. Proc Natl Acad Sci 110:E2423–E2436

Halberstadt AL, Geyer MA (2009) Habituation and sensitization of acoustic startle: opposite influences of dopamine D1 and D2-family receptors. Neurobiol Learn Mem 92:243–248

Heathers JA (2014) Everything Hertz: methodological issues in short-term frequency-domain HRV. Front Physiol 5:177

Heldt SA, Elberger AJ, Deng Y, Guley NH, Del Mar N, Rogers J, Choi GW, Ferrell J, Rex TS, Honig MG, Reiner A (2014) A novel closed-head model of mild traumatic brain injury caused by primary overpressure blast to the cranium produces sustained emotional deficits in mice. Front Neurol 5:2

Hoge EA, McGurk D, Thomas JL, Cox AL, Engel CC, Castro CA (2008) Mild traumatic brain injury in US soldiers returning from Iraq. N Engl J Med 358:453–463

Hoge EA, Worthington JJ, Nagurney JT, Chang Y, Kay EB, Feterowski CM, Katzman AR, Goetz JM, Rosasco ML, Lasko NB, Zusman RM, Pollack MH, Orr SP, Pitman RK (2012) Effect of acute posttrauma propranolol on PTSD outcome and physiological responses during script-driven imagery. CNS Neurosci Ther 18:21–27

Holstein DH, Vollenweider FX, Jancke L, Schopper C, Csomor PA (2010) P50 suppression, prepulse inhibition, and startle reactivity in the same patient cohort suffering from posttraumatic stress disorder. J Affect Disord 126:188–197

Holt DJ, Lebron-Milad K, Milad MR, Rauch SL, Pitman RK, Orr SP, Cassidy BS, Walsh JP, Goff DC (2009) Extinction memory is impaired in schizophrenia. Biol Psychiatry 65:455–463

Jovanovic T, Norrholm SD, Sakoman AJ, Esterajher S, Kozaric-Kovacic D (2009) Altered resting psychophysiology and startle response in Croatian combat veterans with PTSD. Int J Psychophysiol 71:264–268

Jovanovic T, Norrholm SD, Blanding NQ, Davis M, Duncan E, Bradley B, Ressler KJ (2010) Impaired fear inhibition is a biomarker of PTSD but not depression. Depress Anxiety 27:244–251

Jovanovic T, Phifer JE, Sicking K, Weiss T, Norrholm SD, Bradley B, Ressler KJ (2011) Cortisol suppression by dexamethasone reduces exaggerated fear responses in posttraumatic stress disorder. Psychoneuroendocrinology 36:1540–1552

Jovanovic T, Ely T, Fani N, Glover EM, Gutman D, Tone EB, Norrholm SD, Bradley B, Ressler KJ (2013) Reduced neural activation during an inhibition task is associated with impaired fear inhibition in a traumatized civilian sample. Cortex 49:1884–1891

Kamkwalala A, Norrholm SD, Poole JM, Brown A, Donley S, Duncan E, Bradley B, Ressler KJ, Jovanovic T (2012) Dark-enhanced startle responses and heart rate variability in a traumatized civilian sample: putative sex-specific correlates of posttraumatic stress disorder. Psychosom Med 74:153–159

Kamphuis MH, Geerlings MI, Tijhuis MA, Giampaoli S, Nissinen A, Grobbee DE, Kromhout D (2007) Physical inactivity, depression, and risk of cardiovascular mortality. Med Sci Sports Exerc 39:1693–1699

Kassam-Adams N, Garcia-Espana JF, Fein JA, Winston FK (2005) Heart rate and posttraumatic stress in injured children. Arch Gen Psychiatry 62:335–340
Keren O, Yupatov S, Radai MM, Elad-Yarum R, Faraggi D, Abboud S, Ring H, Groswasser Z (2005) Heart rate variability (HRV) of patients with traumatic brain injury (TBI) during the post-insult sub-acute period. Brain Inj 19:605–611
Khoury NM, Marvar PJ, Gillespie CF, Wingo A, Schwartz A, Bradley B, Kramer M, Ressler KJ (2012) The renin-angiotensin pathway in posttraumatic stress disorder: angiotensin-converting enzyme inhibitors and angiotensin receptor blockers are associated with fewer traumatic stress symptoms. J Clin Psychiatry 73:849–855
Klumpers F, Raemaekers MA, Ruigrok AN, Hermans EJ, Kenemans JL, Baas JM (2007) Prefrontal mechanisms of fear reduction after threat onset. Biol Psychiatry 68:1031–1038
Kohl S, Heekeren K, Klosterkötter J, Kuhn J (2013) Prepulse inhibition in psychiatric disorders—Apart from schizophrenia. J Psychiatr Res 47:445–452
Kubzansky LD, Koenen KC, Spiro A, Vokonas PS, Sparrow D (2007) Prospective study of posttraumatic stress disorder symptoms and coronary heart disease in the Normative Aging Study. Arch Gen Psychiatry 64:109–116
Kuhn E, Blanchard EB, Fuse T, Hickling EJ, Broderick J (2006) Heart rate of motor vehicle accident survivors in the emergency department, peritraumatic psychological reactions, ASD, and PTSD severity: a 6-month prospective study. J Trauma Stress 19:735–740
Liberzon I, Taylor SF, Amdur RL, Jung TD, Chamberlain KR, Minoshima S et al (1999) Brain activation in PTSD in response to trauma-related stimuli. Biol Psychiatry 45:817–826
Lipschitz DS, Mayes LM, Rasmusson AM, Anyan W, Billingslea E, Gueorguieva R, Southwick SM (2005) Baseline and modulated acoustic startle responses in adolescent girls with posttraumatic stress disorder. J Am Acad Child Adol Psychiatry 44:807–814
Lommen MJ, Engelhard IM, Sijbrandij M, van den Hout MA, Hermans D (2013) Pre-trauma individual differences in extinction learning predict posttraumatic stress. Behav Res Ther 51:63–67
Ludewig S, Ludewig K, Geyer MA, Hell D, Vollenweider FX (2002) Prepulse inhibition deficits in patients with panic disorder. Depress Anxiety 15:55–60
Ludewig K, Geyer MA, Etzensberger M, Vollenweider FX (2002). Stability of the acoustic startle reflex, prepulse inhibition, and habituation in schizophrenia. Schizophr Res 55:129–137
Ludewig K, Geyer MA, Vollenwweider FX (2003) Deficits in prepulse inhibition and habituation in never-medicated, first-episode schizophrenia. Biol Psychiatry 54:121–128
Ludewig S, Geyer MA, Ramseier M, Vollenweider FX, Rechsteiner E, Cattapan-Ludewig K (2005) Information-processing deficits and cognitive dysfunction in panic disorder. J Psychiatry Neurosci 30:37–43
McEwen BS (2000) Allostasis and allostatic load: implications for neuropsychopharmacology. Neuropsychopharmacology 22:108–124
McGhee LL, Maani CV, Garza TH, Desocio PA, Gaylord KM, Black IH (2009) The effect of propranolol on posttraumatic stress disorder in burned service members. J Burn Care Res 30:92–97
McTeague LM, Lang PJ, Laplante MC, Cuthbert BN, Shumen JR, Bradley MM (2010) Aversive imagery in posttraumatic stress disorder: Trauma recurrence, comorbidity, and physiological reactivity. Biol Psychiatry 15:346–356
Milad MR, Orr SP, Lasko NB, Chang Y, Rauch SL, Pitman RK (2008) Presence and acquired origin of reduced recall for fear extinction in PTSD: results of a twin study. J Psychiatr Res 42:515–520
Milad MR, Furtak SC, Greenberg JL, Keshaviah A, Im JJ, Falkstein MJ, Jenike M, Rauch SL, Wilhelm S (2013) Deficits in conditioned fear extinction in obsessive-compulsive disorder and neurological changes in the fear circuit. JAMA Psychiatry 70:608–618
Minassian A, Geyer MA, Baker DG, Nievergelt CM, O'Connor DT, Risbrough VB, MRS Team (2014) Heart rate variability characteristics in a large group of active-duty Marines and relationship to posttraumatic stress. Psychosom Med 76:292–301

Moon E, Lee SH, Kim DH, Hwang B (2013) Comparative study of heart rate variability in patients with schizophrenia, bipolar disorder, post-traumatic stress disorder, or major depressive disorder. Clin Psychopharmacol Neurosci 11:137–143

Morgan CA, Grillon C, Southwick SM, Nagy LM, Davis M, Krystal JH, Charney DS (1995) Yohimbine facilitated acoustic startle in combat veterans with post-traumatic stress disorder. Psychopharmacology 117:466–471

Morgan CA, Grillon C, Lubin H, Southwick SM (1997) Startle reflex abnormalities in women with sexual assault-related posttraumatic stress disorder. Am J Psychiatry 154:1076–1080

Muraoka MY, Carlson JG, Chemtob CM (1998) Twenty-four hour ambulatory blood pressure and heart rate monitoring in combat-related posttraumatic stress disorder. J Trauma Stress 11:473–484

Norrholm SD, Jovanovic T, Olin IW, Sands LA, Karapanou I, Bradley B, Ressler KJ (2011) Fear extinction in traumatized civilians with posttraumatic stress disorder: relation to symptom severity. Biol Psychiatry 69:556–563

Ornitz EM, Pynoos R (1989) Startle modulation in children with Post-traumatic Stress Disorder. Am J Psychiatry 146:866–870

Orr SP, Pitman RK (1993) Psychophysiologic assessment of attempts to simulate posttraumatic stress disorder. Biol Psychiatry 33:127–129

Orr SP, Metzger LJ, Lasko NB, Macklin ML, Peri T, Pitman RK (2000) De novo conditioning in trauma-exposed individuals with and without posttraumatic stress disorder. J Abnorm Psychol 109:290–298

Orr SP, Metzger LJ, Pitman RK (2002) Psychophysiology of post-traumatic stress disorder. Psychiatr Clin N Am 25:271–293

Pallmeyer TP, Blanchard EB, Kolb LC (1986) The psychophysiology of combat-induced post-traumatic stress disorder in Vietnam veterans. Behav Res Therapy 24:645–652

Peri T, Ben-Shakhar G, Orr SP, Shalev AY (2000) Psychophysiologic assessment of aversive conditioning in posttraumatic stress disorder. Biol Psychiatry 47:512–519

Perry W, Minassian A, Feifel D, Braff DL (2001) Sensorimotor gating deficits in bipolar disorder patients with acute psychotic mania. Biol Psychiatry 50:418–424

Pietrzak RH, Gallezot JD, Ding YS, Henry S, Potenza MN, Southwick SM, Krystal JH, Carson RE, Neumeister A (2013) Association of posttraumatic stress disorder with reduced in vivo norepinephrine availability in the locus coeruleus. JAMA Psychiatry 70:1199–1205

Pitman RK, Orr SP, Forgue DF, Altman B, de Jong JB, Claiborn JM (1987) Psychophysiologic assessment of posttraumatic stress disorder imagery in Vietnam combat veterans. Arch Gen Psychiatry 44:970–975

Pitman RK, Orr SP, Lasko NB (1993) Effects of intranasal vasopressin and oxytocin on physiologic responding during personal combat imagery in Vietnam veterans with posttraumatic stress disorder. Psychiatry Res 48:107

Pitman RK, Sanders KM, Zusman RM, Healy AR, Cheema F, Lasko NB et al (2002) Pilot study of secondary prevention of posttraumatic stress disorder with propranolol. Biol Psychiatry 51:189–192

Pitman RK, Gilbertson MW, Gurvits TV, May FS, Lasko NB, Metzger LJ, Shenton ME, Yehuda R, Orr SP (2006) Clarifying the origin of biological abnormalities in PTSD through the study of identical twins discordant for combat exposure. Ann NY Acad Sci 1071:242–254

Pittig A, Arch JJ, Lam CWR, Craske MG (2013) Heart rate and heart rate variability in panic, social anxiety, obsessive-compulsive, and generalized anxiety disorders at baseline and in response to relaxation and hyperventilation. Int J Psychophysiol 87:19–27

Pole N (2007) The psychophysiology of posttraumatic stress disorder: a meta-analysis. Psychol Bull 133:725–746

Pole N, Neylan TC, Otte C, Henn-Hasse C, Metzler TJ, Marmar CR (2009) Prospective prediction of posttraumatic stress disorder symptoms using fear potentiated auditory startle responses. Biol Psychiatry 65:235–240

Price M, Kearns M, Houry D, Rothbaum BO (2014) Emergency department predictors of posttraumatic stress reduction for trauma-exposed individuals with and without an early intervention. J Consult Clin Psychol 82:336–341

Quirk GJ, Garcia R, Gonzalez-Lima F (2006) Prefrontal mechanisms in extinction of conditioned fear. Biol Psychiary 60:337–343

Rabinak CA, Angstadt M, Sripada CS, Abelson JL, Liberzon I, Milad MR, Phan KL (2013) Cannabinoid facilitation of fear extinction memory recall in humans. Neuropharmacology 64:396–402

Rajan S, Leonard N, Fletcher R, Casarjian B, Casarjian R, Cisse C, Gwadz M (2012) Ambulatory autonomic activity monitoring among at-risk adolescent mothers. J Mob Technol Med 1:25–31

Raskind MA, Peterson K, Williams T, Hoff DJ, Hart K, Holmes H, Homas D, Hill J, Daniels C, Calohan J, Millard SP, Rohde K, O'Connell J, Pritzi D, Feiszil K, Petrie EC, Gross C, Mayer CL, Freed MC, Engel C, Peskind ER (2013) A trial of prazosin for combat trauma PTSD with nightmares in active-duty soldiers returned from Iraq and Afghanistan. Am J Psychiatry 170:1003–1010

Ressler KJ, Mercer KB, Bradley B, Jovanovic T, Mahan A, Kerley K, Norrholm SD, Kilaru V, Smith AK, Myers AJ, Ramirez M, Engel A, Hammack SE, Toufexis D, Braas KM, Binder EB, May V (2011) Post-traumatic stress disorder is associated with PACAP and the PAC1 receptor. Nature 470:492–497

Risbrough VB, Stein MB (2006) Role of corticotropin releasing factor in anxiety disorders: a translational research perspective. Horm Behav 50:550–561

Robison-Andrew EJ, Duval ER, Nelson CB, Echiverri-Cohen A, Giardino N, Defever A, Norrholm SD, Jovanovic T, Rothbaum BO, Liberzon I, Rauch SAM (2014) Changes in trauma-potentiated startle with treatment of posttraumatic stress disorder in combat veterans. J Anxiety Disord 28:358–362

Roitman P, Gilad M, Ankri YL, Shalev AY (2013) Head injury and loss of consciousness raise the likelihood of developing and maintaining PTSD symptoms. J Trauma Stress 26:727–734

Rothbaum BO, Price M, Jovanovic T, Norrholm SD, Gerardi M, Dunlop B, Davis M, Bradley B, Duncan EJ, Rizzo A, Ressler KJ (2014) A randomized, double-blind evaluation of D-cycloserine or alprazolam combined with virtual reality exposure therapy for posttraumatic stress disorder in Iraq and Afghanistan war veterans. Am J Psychiatry 171:640–648

Sahar ME, Shalev AY, Porges SW (2001) Vagal modulation of responses to mental challenge in posttraumatic stress disorder. Biol Psychiatry 49:637–643

Shah AJ, Lampert R, Goldberg J, Veledar E, Bremner JD, Vaccarino V (2013) Posttraumatic stress disorder and impaired autonomic modulation in male twins. Biol Psychiatry 73:1103–1110

Shalev AY, Orr SP, Peri T, Schreiber S, Pitman RK (1992) Physiologic responses to loud tones in Israeli patients with posttraumatic stress disorder. Arch Gen Psychiatry 49:870–875

Shalev AY, Sahar T, Freedman S, Peri T, Glick N, Brandes D, Orr SP, Pitman RK (1998) A prospective study of heart rate response following trauma and the subsequent development of posttraumatic stress disorder. Arch Gen Psychiatry 55:5533–5559

Shalev AY, Peri T, Brandes D, Freedman S, Orr SP, Pitman RK (2000) Auditory startle response in trauma survivors with posttraumatic stress disorder: a prospective study. Am J Psychiatry 157:255–261

Shin LM, Rauch SL, Pitman RK (2006) Amygdala, medial prefrontal cortex, and hippocampal function in PTSD. Ann NY Acad Sci 1071:67–79

Sijbrandij M, Engelhard IM, Lommen MJ, Leer A, Baas JM (2013) Impaired fear inhibition learning predicts the persistence of symptoms of posttraumaticstress disorder (PTSD). J Psychiatr Res 47:1991–1997

Solovieff N, Roberts AL, Ratanatharathorn A, Haloosim M, De Vivo I et al (2014) Genetic Association Analysis of 300 genes identifies a risk haplotype in SLC18A2 for post-traumatic stress disorder in two independent samples. Neuropsychopharmacology 39:1872–1879

Steckler T, Risbrough VB (2012) Pharmacological treatment of PTSD—established and new approaches. Neuropharmacology 62:617−627

Stein MB, Kerridge C, Dimsdale JE, Hoyt DB (2007) Pharmacotherapy to prevent PTSD: results from a randomized controlled proof-of-concept trial in physically injured patients. J Trauma Stress 20:923–932

Suendermann O, Ehlers A, Boelinghaus I, Gamer M, Glucksman E (2010) Early heart rate response to standardized trauma-related pictures predict posttraumatic stress disorder—A prospective study. Psychosom Med 72:301–308

Suzuki G, Tokuno S, Nibuya M, Ishida T, Yamamoto T, Mukai Y, Mitani K, Tsumatori G, Scott D, Shimizu K (2014) Decreased plasma brain-derived neurotrophic factor and vascular endothelial growth factor concentrations during military training. PLoS ONE 9:e89455

Swerdlow NR, Talledo J, Sutherland AN, Nagy D, Shoemaker JM (2006) Anti-psychotic effects on prepulse inhibition in normal 'low gating' humans and rats. Neuropsychopharmacology 31:2011–2021

Tharion WJ, Potter AW, Duhamel CM, Karis AJ, Buller MJ, Hoyt RW (2013) Real-time physiological monitoring while encapsulated in personal protective equipment. J Sport Human Perform

Vaiva G, Ducrocq F, Jezequel K, Averland B, Lestavel P, Brunet A, Marmar CR (2003) Immediate treatment with propranolol decreases posttraumatic stress disorder two months after trauma. Biol Psychiatry 54:947–949

Vasterling JJ, Verfaellie M, Sullivan KD (2009) Mild traumatic brain injury and posttraumatic stress disorder in returning veterans: perspectives from cognitive neuroscience. Clin Psychol Rev 29:674–684

Vasterling JJ, Brailey K, Proctor SP, Kane R, Heeren T, Franz M (2012) Neuropsychological outcomes of mild traumatic brain injury, post-traumatic stress disorder and depression in Iraq-deployed US Army soldiers. Br J Psychiatry 201:186–192

Vrana SR, Calhoun PS, McClernon FJ, Dennis MF, Lee ST, Beckham JC (2013) Effects of smoking on the acoustic startle response and prepulse inhibition in smokers with and without posttraumatic stress disorder. Psychopharmacology 230:477–485

Wentworth BA, Stein MB, Redwine LS, Xue Y, Taub PR, Clopton P, Nayak KR, Maisel AS (2013) Post-traumatic stress disorder: a fast track to premature cardiovascular disease? Cardiol Rev 21:16–22

Wessa M, Flor H (2007) Failure of extinction of fear responses in posttraumatic stress disorder: evidence from second-order conditioning. Am J Psychiatry 164:1684–1692

Williamson JB, Heilman KM, Porges EC, Lamb DG, Porges SW (2013) A possible mechanism for PTSD symptoms in patients with traumatic brain injury: central autonomic network disruption. Front Neuroengineering 6:1–9

Yurgil KA, Barkauskas DA, Vasterling JJ, Nievergelt CM, Larson GE, Schork NJ, Litz BT, Nash WP, Baker DG, MRS Team (2014) Association between traumatic brain injury and risk of posttraumatic stress disorder in active-duty Marines. JAMA Psychiatry 71:149–157

Zao JK, Gan TT, You CK, Chung CE, Wang YT, Rodríguez Méndez SJ, Mullen T, Yu C, Kothe C, Hsiao CT, Chu SL, Shieh CK, Jung TP (2014) Pervasive brain monitoring and data sharing based on multi-tier distributed computing and linked data technology. Front Hum Neurosci 8:370

Zatzick DF, Russo J, Pitman RK, Rivara F, Jurovich G, Roy-Byrne P (2005) Reevaluating the association between emergency department heart rate and the development of posttraumatic stress disorder: a public health approach. Biol Psychiatry 57:91–95

Zoladz PR, Diamond DM (2013) Current status on behavioral and biological markers of PTSD: a search for clarity in a conflicting literature. Neurosci Biobehav Revi 37:860–895

Physiological Aberrations in Panic Disorder

Wenzel Schicho and Oliver Pogarell

Abstract Panic disorder is a frequent and clinically relevant medical entity with a high lifetime prevalence and significant impact on psychosocial stability and function. Regarding the clinical presentation, there are obvious similarities in paroxysmal neurological disorders such as seizures and focal epilepsies. In this context, the detection of EEG abnormalities during the attacks or in asymptomatic intervals, continuously or rhythmical, is of significant interest. Likewise, isolated epileptic discharges (IEDs) are important components of epilepsy. On the other hand, IEDs are also common in non-epileptic psychiatric patients. It is not known exactly which role IEDs play in the genesis of behavioural aberrations. In this chapter, attention is directed towards this issue and its relevance to managing psychiatric patients suffering from panic disorder (PD), as well as understanding the complex relationship between IEDs and the pathophysiology of PD. Two main conclusions are being proposed. First, patients suffering from PD may show a higher rate of unspecific EEG abnormalities and increased beta power, pointing to a state of hyperarousal. Secondly, if first-line treatment of PD fails, the use of anti-epileptic drugs (AEDs) should be considered. There is enough evidence to suggest that IEDs play a significant role in the genesis of PD and that this relationship is far from clear, warranting more research in this field. This article gives an overview of the current literature on the pathophysiology of PD, including studies on altered microstates, coherence imaging and alpha asymmetry.

Keywords Electroencephalography · Epilepsy · Isolated epileptic discharge (IED) · Psychopathology · Panic disorder · Seizure disorder · Synchronicity

---

W. Schicho (✉) · O. Pogarell
Department of Psychiatry and Psychotherapy, University of Munich, Nussbaumstr. 7, 80336 Munich, Germany
e-mail: Wenzel.Schicho@med.uni-muenchen.de

Curr Topics Behav Neurosci (2014) 21: 185–195
DOI: 10.1007/7854_2014_347

Published Online: 18 September 2014

Contents

1 Introduction

With a lifetime prevalence of about 3–4 % (Jacobi et al. 2004), panic disorder (PD) is a rather common medical entity (Hirschfeld 1996; Asnaani et al. 2010; Skapinakis et al. 2011). Episodes of intense fear occurring more than once a week for at least one month are the key feature of PD, but also feelings of derealisation or depersonalisation and miscellaneous autonomic symptoms such as palpitations, nausea, sweating or chest pain are associated with PD (Ham et al. 2005; Hurley et al. 2006). Panic attacks (PA) can result in agoraphobia with severe psychosocial and economic consequences (Boutros et al. 2013a, b). In the clinical routine, it can be difficult to distinguish PA from anxiety attacks in the context of epileptic fits, since there is a significant overlap of these two entities concerning their diagnostic criteria according to DSM-V and ICD-10 (Mintzer and Lopez 2002; Beyenburg et al. 2005). Especially, patients diagnosed with partial seizures originating from temporal structures are likely to present symptoms resembling those of a PD (Adamaszek et al. 2011). Because the limbic system comprises important neural networks involved in the processing of fear and related emotions, it is likely that abnormal synchronous activity induced by epileptic seizures in this area leads to symptoms which are indicative of PA (Hurley et al. 2006; Ray et al. 2007). Patients with seizures or epilepsy suffering solely from so-called ictal fear accompanied by autonomic symptoms can easily be misdiagnosed with PD (Wieser 1983; Young et al. 1995; Sazgar et al. 2003). Kanner (2011) provided a detailed overview of how to clinically distinguish ictal panic, PA and postictal panic. However, he states that apart from clinical observations, EEG, neuroimaging and the assessment of prolactin levels can be of additional help in the diagnostic process (Kanner 2011).

2 Electrophysiology Studies in Anxiety Disorders and Panic Disorder: EEG

Regarding the differential diagnosis PD versus epilepsy, there have been reports on several patients with a PD, showing characteristics of abnormal synchronous cerebral activity or even epileptiform discharges in scalp EEG, but no other traces of an underlying epileptic entity (Jabourian et al. 1992; Gallinat et al. 2003; Avoli et al. 2005). Up to now, clinicians tend to think in a dichotomous way whether a panic attack is of psychiatric origin or a seizure disorder. But there are cases where even extensive workup fails to resolve this dichotomy. It is being discussed that the underlying pathology of PA could be abnormal hyperexcitability which fails to generate potentials large enough to be detected on the scalp (Boutros et al. 2013a, b). Given the relatively high false-negative rate of scalp EEG in detecting epileptic discharges (Boutros 2010), it should be considered that what we see in the standard scalp EEG is only the 'tip of an iceberg' (Boutros et al. 2013a, b).

Weilburg et al. (1993) reported on two patients with atypical PA while their EEGs were being monitored. Focal paroxysms of sharp wave activity appeared in the EEG coinciding with a spontaneous onset of panic attack symptoms in both patients (Weilburg et al. 1993). Consciousness was maintained during these episodes. Later, the same group reported on 15 subjects with atypical PA who met DSM-IIIR criteria for PD and who underwent a routine EEG followed by prolonged ambulatory EEG monitoring using sphenoidal electrodes (Weilburg et al. 1995). They found focal paroxysmal EEG changes consistent with partial seizure activity, which had occurred during PA in 33 % ($N = 5$) of the subjects. It is important to note that multiple attacks were recorded before panic-related EEG changes were demonstrated. Moreover, two of the five subjects with demonstrated EEG abnormalities during PA had perfectly normal baseline EEGs. Weilburg et al. concluded that it would be necessary to monitor the EEG during multiple attacks in order to reveal an association between atypical PA and epileptiform EEG changes. This conclusion is in agreement with a report from Jabourian et al. (1992). They performed 24-h ambulatory EEGs in a population of 300 non-epileptic outpatients with an anxious and depressive pathology and subjects with PA. The recordings revealed a high prevalence of abnormalities in subjects diagnosed with PD. Two groups of 150 medication-free patients each had been selected on the base of DSM-IIIR, one with PA and the other with depressive patients without paroxysmal anxiety (DS). The results showed 62.3 % abnormal records in the PA group, while in the DS group, 74.5 % of the records were normal. Epileptiform abnormalities were four times more frequent in the PA group (80 %) than in the DS group (20 %). MRI allowed the discovery of abnormal cerebral images in 3 patients of the PA group (cyst of the insula, temporal and parietal cryptic angiomas, sequelae of a parietal vasculo-cerebral stroke) (Jabourian et al. 1992). Gallinat and Hegerl (1999) reported on a patient presenting with anxiety and clear PA who intermittently showed bitemporal spikes and spike wave complexes in scalp EEG recordings. The treatment with valproic acid led to a normalisation of the EEG and a remission of psychopathology. They reviewed similar cases in the literature and concluded that at least in a subgroup of patients with PA and PD, their psychopathology could be related to epileptiform activity within the limbic system, but that the confirmation of this association was difficult, since discharges of the limbic system in the depth of the brain often cannot be detected in scalp EEG recordings. Nevertheless, therapeutic trials with antiepileptic drugs (AEDs) such as valproate seemed to be a reasonable approach (Gallinat and Hegerl 1999).

Vaaler et al. (2009) extended the research field from PD to a condition they called acute unstable depressive syndrome (AUDS). They compared EEGs and MRIs of 16 patients diagnosed with AUDS to 16 controls diagnosed with major depressive episodes according to DSM-IV. They also found that short-lasting atypical depressive symptoms seemed to be associated with a high frequency of epileptic and pathologic EEG activity (Vaaler et al. 2009).

3 Resting State EEG: Microstates and Alpha Asymmetry

A normal stimulus processed in a pathological way can lead to PA in patients suffering from PD. By analysing resting state EEGs from both PD patients and healthy controls, it could be elucidated how pre-activated fear-associated networks were involved in this aberrant processing of normal stimuli. Kikuchi et al. (2011) analysed resting state EEGs of 18 drug-naive patients and 18 healthy controls using features of transiently stable brain states, so-called microstates, which correlate significantly with resting state networks assessed by fMRI (Britz et al. 2010; Van de Ville et al. 2010; Kikuchi et al. 2011). It could be observed that some microstate classes differ between healthy controls and PD patients, indicating that some brain functions are already altered during resting state, which may lead to a dysfunctional processing of phobic stimuli (Kikuchi et al. 2011).

Gordon et al. (2010) assessed resting brain laterality in six clinical disorders in order to obtain a more fine-grained image of the utility of alpha asymmetry as a diagnostic and treatment predictive marker. Their results showed that alpha asymmetry may be of more clinical utility as a biomarker for schizophrenia and depression but less so for PD, since the PD group did not show any significant deviance in alpha asymmetry compared to the healthy control group (Gordon et al. 2010).

4 qEEG, MEG and Coherence Analysis

Clark et al. (2009) provided a systematic, evidence-based review of the field of electrophysiology in anxiety disorders, characterising the clinical value of EEG and quantitative EEG (qEEG) in the diagnostic process (Clark et al. 2009). It was summarised that qEEG studies in PD indicate symptom-based abnormalities on basal levels of cortical arousal during waking. In patients not presenting with symptoms of depersonalisation and/or derealisation resting cortical hyperarousal with decreased power in lower frequencies and increased power in the beta range could be detected, while in patients with the aforementioned symptoms, lower frequency power was increased. This could point to a hyperarousal basal state (Clark et al. 2009). Furthermore, it was stated that in patients with PD, habituation seems to be altered, which had already been shown in studies analysing event-related potentials (ERP). It was also mentioned that these alterations could be associated with brainstem pathology and with subsequent early processing of sensory stimuli (Clark et al. 2009).

Engelbregt et al. (2012) reported on a patient suffering from specific phobia similar to agoraphobia with PD. Panic was specifically triggered by leaving the patients domicile, so an EEG was obtained while the patient was provoking the initiation of a panic attack by driving a car out of his hometown. Similar to the aforementioned findings, the EEG showed an increase of frontal beta activity and decrease in frontal–central theta activity during the panic attack (Engelbregt 2012). It is hypothesised that panic could induce changes in functional connectivity patterns in several brain areas including frontal brain, temporal and parietal cortex and subcortical regions, referring to an earlier resting state fMRI study (Dieler et al. 2008; Engelbregt 2012).

Coburn et al. (2006) stated that the qEEG literature on anxiety disorders is small and unimpressive regarding clinical utility, but they also suggested that qEEG may become an important research tool and in the future an important clinical tool (Coburn et al. 2006).

Boutros et al. (2013a, b) reported on a 33-year-old patient with PA and autonomic symptoms such as stomach pain and fainting. The treatment with bupropion, alprazolam, paroxetine and clonazepam did not decrease the PAs; on the contrary, the two latter substances even worsened the patient's complaints. No EEG correlates could be found during the PA. Extensive workup, including qEEG utilising low-resolution electromagnetic tomographie (LORETA), a PET-Scan and magnetencephalography (MEG), pointed to multiple loci of cerebral involvement suggesting the presence of an epileptic process (Boutros et al. 2013a, b). MEG is said to be more sensitive in detecting epileptic discharges in comparison with EEG (Lewine et al. 1999; Lin et al. 2003) and allows the detection of extremely high coherence values in various brain regions in close proximity to each other. Previous studies demonstrated that focal epileptic activity is associated with increased coherence between pathologically altered and other brain regions (Towle et al. 1999; Elisevich et al. 2011). Considering that these findings indicate that in this case of PD there could be an underlying subclinical epileptic process, gabapentin was added to the regimen and led to a decrease in the number of PA (Boutros et al. 2013a, b).

In another study, Boutros et al. (2013a, b) investigated the presence of increased coherence in the limbic frontotemporal regions in patients suffering from PD and was indeed able to show that coherence imaging values were significantly higher in PD patients than in healthy controls (Boutros et al. 2013a, b). These findings could be of importance in choosing the right treatment for the population of patients with PD and increased coherence values, assuming that they would favourably respond to benzodiazepines or AEDs. This is of special importance considering the fact that the treatment strategies available today for PD are only successful in approximately half of the affected subjects (Doyle and Pollack 2004).

5 Panic Disorder and Epilepsy

Toni et al. (1996) compared symptoms of PA to those of temporal lobe complex partial seizures showing a significant overlap (Toni et al. 1996). The study claims to be the first to focus on psychosensorial and related phenomena in PD with agoraphobia and derealisation/depersonalisation (PDA-DD) in comparison with complex partial epilepsy with depersonalisation/derealisation (CPE-DD). The findings show that most features of DD are present in both groups with no significant difference, with the exception for the feeling that one's limbs are unreal (significantly higher in the PDA-DD group) and space or temporal disorientation (significantly higher in the CPE-DD group). It is also stated that hallucination was the main feature to fully discriminate PDA from CPE patients, as hallucinations were not traced in PDA patients (Toni et al. 1996). In conclusion, already in this article it is said that the broad overlap of symptoms of CPE and PDA makes a common neurophysiological substrate of both entities likely, but it has yet to be defined.

Deutsch et al. (2009) reported on a 34-year-old woman with a long history of anxiety attacks together with depersonalisation, derealisation and internal derogatory voices, frightening imagery and suicidal ideation. EEG and MRI results were found to be normal, but temporal lobe epilepsy (TLE) was considered to be a differential diagnosis because the patient did not respond to the first-line treatment of PD. Similar to the aforementioned cases, the patient responded rapidly and dramatically to carbamazepine (Deutsch et al. 2009).

Gerez et al. (2011) reported on two cases diagnosed with PD, who partially responded to first-line treatment such as SSRIs and psychotherapy. After having a panic attack with minimal EEG alterations, both cases developed a clear ictal EEG pattern associated with strong fear content. LORETA analysis revealed increased activity in the right amygdala during panic symptoms, demonstrating the major role of amygdalar hyperactivity in both fear-related conditions (Gerez et al. 2011) and underlining the long debated relationship of PD and TLE (Alvarez-Silva et al. 2006). Again, the most conclusive evidence came from treatment response as both patients were panic-free under carbamazepine.

Two possible anatomical locations seem to be strongly involved in the generation of panic symptoms:

- The amygdala is an almond-sized (about 1-inch long) brain structure which has long been linked to the experience of fear and emotional states in general. Electrical stimulation of the medial temporal structures, and more specifically the amygdala, predictably induces experiential feelings and mainly fear (Gloor et al. 1982). Amygdala source panic is likely to be characterised by a dominance of fear, with other autonomic symptoms following either secondary to the development of epileptic activity or as a psychological reaction to fear (Kalynchuk 2000; Keele 2005). The two cases reported by Gerez et al. (2011) demonstrated the close link between seizures originating from the mesial temporal structures and panic symptoms, with amygdala hyperactivity underlying both phenomena (Gerez et al. 2011). When the amygdala is electrically stimulated in awake humans (usually during epilepsy workup or surgery), fear is the most commonly generated experience (Meletti et al. 2006). The stimulation of even close structures such as the hippocampus (with avoiding an amygdala involvement) leads to far fewer fear reactions. This group also pointed to a gender difference in the reaction to the electrical stimulation of the amygdala, with women more likely to experience fear (Meletti et al. 2006). In addition, the stimulation of both amygdalas (right and left) leads to different responses. The electrical stimulation of the right amygdala induced negative symptoms, especially fear and sadness, while the stimulation of the left amygdala induced either pleasant (e.g. happiness) or unpleasant (e.g. fear, anxiety, sadness) sensations (Lanteaume et al. 2007).

- The insula is located deep within the Sylvian fissure beneath the frontal, parietal and temporal opercula. The insula has been long associated with visceral functions and is responsible for integrating autonomic information. The insula has wide connections with the neocortex, basal ganglia, thalamus and the amygdala among other limbic structures. This wide connectivity explains the varied symptoms of seizures originating from this region (Nguyen et al. 2009). It is important to highlight that epileptic activity emanating from this small and deep region is unlikely to be detected by standard scalp EEG. Insular temporal cortex source PA begin with autonomic symptoms and fear following either secondary spread of epileptic activity or a conditioned response to the experience of sudden and usually unprovoked autonomic activation (Nguyen et al. 2009).

6 Theoretical Consideration

Whether IEDs can be associated with actual clinical manifestations such as fear or autonomic responses is unknown, but it is a very important question for a number of reasons. Firstly, an IED lasts only a fraction of a second, whereas a typical panic attack can last several minutes. The current requirements to diagnose a panic attack as an actual seizure are difficult to meet. The American Academy of Neurology requires that for any behaviour to be labelled a seizure, evidence for ongoing ictal activity during the behaviour must be documented on a simultaneous EEG/video monitoring. If a single or a short run of IEDs can indeed induce or spark a panic attack, more research is needed to understand how a very short-duration event can induce a (usually) much longer experience. In fact, the relationship between the presence of epileptic discharges in the amygdala or insular regions and the emergence of panic symptoms is likely to be much more complex than the simple notion of an electric stimulus activating the function of a particular brain region. This is exemplified by the reports of Mintzer and Lopez (2002) showing panic symptoms getting worse after epileptic lesions were surgically removed (Mintzer and Lopez 2002). Some evidence is beginning to accumulate that abnormal discharges may be causing disturbances in the workings of the fear neural circuitry (Bartolomei et al. 2005). The group provided evidence that during ictal activity associated with intense emotional alterations, a significant decrease in synchrony between signals recorded from the neural networks known to be involved in emotional processing was present. In particular, loss of synchrony between the orbitofrontal cortex and the amygdala was noted (Bartolomei et al. 2005). The authors suggested that the disconnection releases the amygdala from the inhibitory control of the frontal lobe, allowing usually suppressed fear to emerge. Furthermore, Kellett and Kokkinidis (2004) suggested conditioning the epileptic discharges to fear experiences to be a prerequisite for the development of panic symptoms. They showed that animals failed to demonstrate fear-potentiated startle responses when electrical stimulation of the amygdala was unconditioned (Kellett and Kokkinidis 2004). This would explain why not all individuals with epileptic spikes in this location develop PA.

The proper neurochemical state for developing conditioned fear may also be a prerequisite. The dopamine-containing ventral tegmental area (VTA) is known to be activated by threatening environmental stimuli and has been shown to be an important variable related to neural excitability of the amygdala (Gelowitz and Kokkinidis 1999).

7 Conclusion

According to the EEG literature, patients with PD or PA present with a higher rate of unspecific EEG abnormalities, including paroxysmal EEG changes or epileptic discharges. QEEG studies showed differences between groups of PD patients and healthy controls related to the fear-associated networks, and PD patients may show increased beta power which could be in line with a state of hyperarousal. Regarding the association of PD and epilepsy, there are reports that ictal EEG patterns and epileptic discharges can be associated with symptoms of fear and panic. Thus, if pharmacological treatment approaches with SSRI, TCA, MAO inhibitors and psychotherapy do not lead to a clinical amelioration in patients suffering from PD, the use of AEDs, especially carbamazepine and valproic acid, should be considered (Adamaszek et al. 2011). Boutros et al. (2014) recently gave a comprehensive review of the current literature on the predictive value of IEDs for a favourable therapeutic response to AEDs in non-epileptic patients (Boutros et al. 2014). They found that the number of controlled studies in this field was extremely small and the majority of citations found were single case reports or case series. Another problem is that the clinical significance of IEDs is mainly discussed in neurology (Gorji and Speckmann 2009), but often unrecognised and rarely studied in psychiatry (Boutros 2009). Therefore, clinicians caring for psychiatric patients need to develop the necessary skills to be able to interpret results from advancing neurodiagnostic technologies (Boutros et al. 2013a, b). Psychiatric clinicians need to be more familiar with electroencephalography, its advantages as well as its limitations. More research is needed in this field to better define indications, adequate EEG workups, best AEDs to be used and optimal durations of treatment (Boutros et al. 2014). Multimodal EEG approaches, including resting and long-term recordings, and simultaneous video-EEG-recordings should be considered during the diagnostic workup of PD patients (Adamaszek et al. 2011). With recent developments in the investigation of altered synchronous neuronal activities (such as the combination of EEG/fMRI or MEG and coherence or microstate analysis), there are a couple of promising techniques emerging which may be helpful to better understand the underlying neurophysiologic mechanisms of PD in the future (Ray et al. 2007; Jacobs et al. 2009; Adamaszek et al. 2011).

References

Adamaszek M, Olbrich S et al (2011) The diagnostic value of clinical EEG in detecting abnormal synchronicity in panic disorder. Clin EEG Neurosci 42(3):166–174

Alvarez-Silva S, Alvarez-Rodriguez J et al (2006) Panic and epilepsy. J Anxiety Disord 20 (3):353–362

Asnaani A, Richey JA et al (2010) A cross-ethnic comparison of lifetime prevalence rates of anxiety disorders. J Nerv Ment Dis 198(8):551–555

Avoli M, Louvel J et al (2005) Cellular and molecular mechanisms of epilepsy in the human brain. Prog Neurobiol 77(3):166–200

Bartolomei F, Trebuchon A et al (2005) Acute alteration of emotional behaviour in epileptic seizures is related to transient desynchrony in emotion-regulation networks. Clin Neurophysiol: Official J Int Fed Clin Neurophysiol 116(10):2473–2479

Beyenburg S, Mitchell AJ et al (2005) Anxiety in patients with epilepsy: systematic review and suggestions for clinical management. Epilepsy Behav: E&B 7(2):161–171

Boutros N (2009) Epileptiform discharges in psychiatric patients: a controversy in need of resurrection. Clin EEG Neurosci 40(4):239–244

Boutros NN (2010) Interictal spikes in psychiatric patients: a controversy in need for resurrection. J Clin Neurophysiol: Official Publ Am Electroencephalogr Soc 27(6):484–485 (Author reply 485)

Boutros NN, Galloway MP et al (2013a) Abnormal coherence imaging in panic disorder: a magnetoencephalography investigation. NeuroReport 24(9):487–491

Boutros NN, Gjini K et al (2013b) Panic versus epilepsy: a challenging differential diagnosis. Clin EEG Neurosci 44(4):313–318

Boutros NN, Kirollos SB et al (2014) Predictive value of isolated epileptiform discharges for a favorable therapeutic response to antiepileptic drugs in nonepileptic psychiatric patients. J Clin Neurophysiol: Official Publ Am Electroencephalogr Soc 31(1):21–30

Britz J, Van De Ville D et al (2010) BOLD correlates of EEG topography reveal rapid resting-state network dynamics. Neuroimage 52(4):1162–1170

Clark CR, Galletly CA et al (2009) Evidence-based medicine evaluation of electrophysiological studies of the anxiety disorders. Clin EEG Neurosci 40(2):84–112

Coburn KL, Lauterbach EC et al (2006) The value of quantitative electroencephalography in clinical psychiatry: a report by the Committee on Research of the American Neuropsychiatric Association. J Neuropsychiatry Clin Neurosci 18(4):460–500

Deutsch SI, Rosse RB et al (2009) Temporal lobe epilepsy confused with panic disorder: implications for treatment. Clin Neuropharmacol 32(3):160–162

Dieler AC, Samann PG et al (2008) Independent component analysis applied to pharmacological magnetic resonance imaging (phMRI): new insights into the functional networks underlying panic attacks as induced by CCK-4. Curr Pharm Des 14(33):3492–3507

Doyle A, Pollack MH (2004) Long-term management of panic disorder. J Clin Psychiatry 65 (Suppl 5):24–28

Elisevich K, Shukla N et al (2011) An assessment of MEG coherence imaging in the study of temporal lobe epilepsy. Epilepsia 52(6):1110–1119

Engelbregt HJKD, Promes VHL, Verhagen-Schouten S, Deijen JB (2012) In vivo EEG changes during a panic attack in a patient with specific phobia. J Med Cases 3(1):34–38

Gallinat J, Hegerl U (1999) Limbic ictus as a condition for anxiety attacks. Der Nervenarzt 70 (3):206–215

Gallinat J, Stotz-Ingenlath G et al (2003) Panic attacks, spike-wave activity, and limbic dysfunction. A case report. Pharmacopsychiatry 36(3):123–126

Gelowitz DL, Kokkinidis L (1999) Enhanced amygdala kindling after electrical stimulation of the ventral tegmental area: implications for fear and anxiety. J Neurosci: Official J Soc Neurosci 19 (22): RC41

Gerez M, Sada A et al (2011) Amygdalar hyperactivity, a fear-related link between panic disorder and mesiotemporal epilepsy. Clin EEG Neurosci 42(1):29–39

Gloor P, Olivier A et al (1982) The role of the limbic system in experiential phenomena of temporal lobe epilepsy. Ann Neurol 12(2):129–144

Gordon E, Palmer DM et al (2010) EEG alpha asymmetry in schizophrenia, depression, PTSD, panic disorder, ADHD and conduct disorder. Clin EEG Neurosci 41(4):178–183

Gorji A, Speckmann EJ (2009) Epileptiform EEG spikes and their functional significance. Clin EEG Neurosci 40(4):230–233

Ham P, Waters DB et al (2005) Treatment of panic disorder. Am Fam Physician 71(4):733–739

Hirschfeld RM (1996) Panic disorder: diagnosis, epidemiology, and clinical course. J Clin Psychiatry 57(Suppl 10):3–8 (Discussion 9–10)

Hurley RA, Fisher R et al (2006) Sudden onset panic: epileptic aura or panic disorder? J Neuropsychiatry Clin Neurosci 18(4):436–443

Jabourian AP, Erlich M et al (1992) Panic attacks and 24-hour ambulatory EEG monitoring. Ann Med Psychol 150(2–3):240–244 (Discussion 245)

Jacobi F, Wittchen HU et al (2004) Prevalence, co-morbidity and correlates of mental disorders in the general population: results from the German Health Interview and Examination Survey (GHS). Psychol Med 34(4):597–611

Jacobs J, Levan P et al (2009) Hemodynamic changes preceding the interictal EEG spike in patients with focal epilepsy investigated using simultaneous EEG-fMRI. Neuroimage 45 (4):1220–1231

Kalynchuk LE (2000) Long-term amygdala kindling in rats as a model for the study of interictal emotionality in temporal lobe epilepsy. Neurosci Biobehav Rev 24(7):691–704

Kanner AM (2011) Ictal panic and interictal panic attacks: diagnostic and therapeutic principles. Neurol Clin 29(1):163–175 (ix)

Keele NB (2005) The role of serotonin in impulsive and aggressive behaviors associated with epilepsy-like neuronal hyperexcitability in the amygdala. Epilepsy Behav: E&B 7(3):325–335

Kellett J, Kokkinidis L (2004) Extinction deficit and fear reinstatement after electrical stimulation of the amygdala: implications for kindling-associated fear and anxiety. Neuroscience 127 (2):277–287

Kikuchi M, Koenig T et al (2011) EEG microstate analysis in drug-naive patients with panic disorder. PLoS ONE 6(7):e22912

Lanteaume L, Khalfa S et al (2007) Emotion induction after direct intracerebral stimulations of human amygdala. Cereb Cortex 17(6):1307–1313

Lewine JD, Andrews R et al (1999) Magnetoencephalographic patterns of epileptiform activity in children with regressive autism spectrum disorders. Pediatrics 104(3 Pt 1):405–418

Lin YY, Shih YH et al (2003) Magnetoencephalographic yield of interictal spikes in temporal lobe epilepsy. Comparison with scalp EEG recordings. Neuroimage 19(3):1115–1126

Meletti S, Tassi L et al (2006) Emotions induced by intracerebral electrical stimulation of the temporal lobe. Epilepsia 47(Suppl 5):47–51

Mintzer S, Lopez F (2002) Comorbidity of ictal fear and panic disorder. Epilepsy Behav: E&B 3 (4):330–337

Nguyen DK, Nguyen DB et al (2009) Insular cortex epilepsy: an overview. Can J Neurol Sci (Le journal canadien des sciences neurologiques) 36(Suppl 2):S58–S562

Ray A, Tao JX et al (2007) Localizing value of scalp EEG spikes: a simultaneous scalp and intracranial study. Clin Neurophysiol: Official J Int Fed Clin Neurophysiol 118(1):69–79

Sazgar M, Carlen PL et al (2003) Panic attack semiology in right temporal lobe epilepsy. Epileptic Dis: Int Epilepsy J Videotape 5(2):93–100

Skapinakis P, Lewis G et al (2011) Panic disorder and subthreshold panic in the UK general population: epidemiology, comorbidity and functional limitation. Eur Psychiatry: J Assoc Eur Psychiatrists 26(6):354–362

Toni C, Cassano GB et al (1996) Psychosensorial and related phenomena in panic disorder and in temporal lobe epilepsy. Compr Psychiatry 37(2):125–133

Towle VL, Carder RK et al (1999) Electrocorticographic coherence patterns. J Clin Neurophysiol: Official Publ Am Electroencephalogr Soc 16(6):528–547

Vaaler AE, Morken G et al (2009) Symptoms of epilepsy and organic brain dysfunctions in patients with acute, brief depression combined with other fluctuating psychiatric symptoms: a controlled study from an acute psychiatric department. BMC Psychiatry 9:63

Van de Ville D, Britz J et al (2010) EEG microstate sequences in healthy humans at rest reveal scale-free dynamics. Proc Natl Acad Sci USA 107(42):18179–18184

Weilburg JB, Schachter S et al (1993) Focal paroxysmal EEG changes during atypical panic attacks. J Neuropsychiatry Clin Neurosci 5(1):50–55

Weilburg JB, Schachter S et al (1995) EEG abnormalities in patients with atypical panic attacks. J Clin Psychiatry 56(8):358–362

Wieser HG (1983) Depth recorded limbic seizures and psychopathology. Neurosci Biobehav Rev 7(3):427–440

Young GB, Chandarana PC et al (1995) Mesial temporal lobe seizures presenting as anxiety disorders. J Neuropsychiatry Clin Neurosci 7(3):352–357

Physiological Correlates of Psychopathy, Antisocial Personality Disorder, Habitual Aggression, and Violence

Christopher J. Patrick

Abstract This chapter reviews the existing literature on physiological correlates of psychopathy, antisocial personality disorder, and persistent violence/aggression. Coverage is provided of findings from studies utilizing peripheral, electrocortical, and neuroimaging measures. The review begins with a discussion of how psychopathy and antisocial personality are defined, and how these conditions relate to one another and to violent behavior. A case is made that the relationships psychopathy and ASPD show with violent and aggressive behavior, and similarities and differences in associations of each with physiological measures of various types can be understood in terms of symptomatic features these conditions have in common versus features that distinguish them. Following this, an overview is provided of major lines of evidence emerging from psychophysiological and neuroimaging studies conducted to date on these conditions. The final section of the chapter summarizes what has been learned from these existing studies and discusses implications and directions for future research.

Keywords Psychopathy · Antisocial personality disorder · Aggression · Violence · Autonomic response · EEG/ERP · Neuroimaging

Contents

C.J. Patrick (✉)
Florida State University, Tallahassee, Florida, USA
e-mail: cpatrick@psy.fsu.edu

Curr Topics Behav Neurosci (2015) 21: 197–227
DOI: 10.1007/7854_2014_345

Published Online: 17 August 2014

1 Introduction

This chapter reviews what is known about physiological correlates of psychopathy, antisocial personality disorder (ASPD), and aggression/violence based on findings from studies employing peripheral, electrocortical, and neuroimaging measures. A key theme of the review is that the relationships psychopathy and ASPD show with violence and aggression can be understood in terms of diagnostic features these conditions share and those that distinguish them. In turn, divergences in observed physiological correlates of psychopathy as compared to antisocial personality and aggression can be understood in terms of common and distinctive features.

The chapter is organized into three sections. The first discusses conceptions of psychopathy and antisocial personality and their relations with one another and with violent behavior. The second section provides an overview of major lines of evidence emerging from psychophysiological and neuroimaging investigations of these conditions that have been published to date. The third section summarizes existing findings and discusses implications and directions for future research.

2 Phenotype Descriptions and Interrelations

2.1 Psychopathy and ASPD: Conceptions, Measures, and Distinguishable Facets

Historic conceptions of psychopathy have emphasized reckless, unrestrained behavior in conjunction with distinct affective–interpersonal symptoms including shallow affect, lack of close relationships, and an appearance of psychological stability ("a convincing mask of sanity"; Cleckley 1941/1976; see also Hare 1980, 2003; Lykken 1957). By contrast, the diagnosis of antisocial personality disorder (ASPD) in the third and fourth editions of the Diagnostic and Statistical Manual of Mental Disorders (DSM-III and IV; American Psychiatric Association (APA) 1980, 2000) focused predominantly on impulsive–antisocial tendencies—beginning in childhood, and continuing on into adulthood—with limited representation of affective–interpersonal features aside from deceptiveness and lack of remorse. The DSM-IV conception of ASPD was maintained without revision in the main diagnostic codes part (Section II) of the latest, fifth edition of the DSM (APA 2013). However, DSM-5 also contains a new dimensional system for characterizing personality pathology (in Section III, "Emerging Measures and Models") that includes alternative trait-based definitions of certain personality disorders, including ASPD. This trait-based definition provides more balanced coverage of affective–interpersonal and impulsive–antisocial features (Strickland et al. 2013; Anderson et al. in press), and includes a trait-based specifier for designating a classically low-anxious, socially efficacious (i.e., "primary psychopathic"; Karpman 1941; Skeem et al. 2007) variant of ASPD.

Alternative conceptions of psychopathy are embodied in differing contemporary assessment instruments. The dominant inventory used with adults in clinical and forensic settings is the interview-based Psychopathy Checklist-Revised (PCL-R; Hare 2003). Adaptations have been developed for children and adolescents, including an interview-based youth version (PCL-YV; Forth et al. 2003) and the child-oriented Antisocial Process Screening Device (APSD; Frick and Hare 2001) and Child Psychopathy Scale (CPS; Lynam 1997), which rely on informant ratings. Various self-report instruments also exist for assessing psychopathy. Some are patterned after the PCL-R, including Hare's Self-Report Psychopathy Scale (SRP; Williams et al. 2007), the Levenson Self-Report Psychopathy Scale (LSRP; Levenson et al. 1995), and the Youth Psychopathic Traits Inventory (YPI; Andershed et al. 2002). Others have been developed separately from the PCL-R. The most widely used of these in recent years has been the Psychopathic Personality Inventory (PPInv; Lilienfeld and Andrews 1996; Lilienfeld and Widows 2005).[1]

Subdimensions of psychopathy: A shift has occurred over the past several years from the idea of psychopathy as a unitary entity to a view of psychopathy as multifaceted—that is, as composed of distinguishable symptomatic subdimensions, or factors. The PCL-R, for example, contains distinct affective–interpersonal and impulsive–antisocial factors (labeled "1" and "2", respectively) even though its items were selected to index psychopathy as a unitary construct (Hare 1980). While intercorrelated, these factors show differing relationships with various criterion variables (Hare 2003; Patrick and Bernat 2009). PCL-R Factor 1 shows selective associations with narcissism, instrumental aggression, and adaptive qualities such as lack of anxiety or depression, whereas Factor 2 shows preferential positive relations with reactive aggression, substance use problems, and suicidal behavior. Factor 2 also accounts for the moderate-level relationship between the PCL-R and ASPD diagnoses or symptoms; controlling for overlap with Factor 1, scores on PCL-R Factor 2 are unrelated to ASPD (Verona et al. 2001).

Factor analyses of the PCL-R's main childhood counterpart, the APSD, have also revealed distinct callous–unemotional (CU) and impulsive/conduct problems (I/CP) subdimensions. Children who score high on both appear average or above average in intellect and show reduced reactivity to stressors, failure to learn from punishment, and high levels of both reactive and proactive aggression, whereas those high on the I/CP factor alone tend to be below average in intellect and show heightened stress reactivity and emotional liability along with increased reactive

---

[1] PPInv is used here in place of the more standard abbreviation PPI to avoid confusion with the psychophysiological phenomenon of prepulse inhibition, also abbreviated PPI.

(but not instrumental) aggression (Frick and Marsee 2006; Frick and White 2008). These findings served as the impetus for inclusion of a "low prosocial emotions" specifier for the diagnosis of conduct disorder in DSM-5—allowing for designation of a callous–unemotional (i.e., "psychopathic") variant of this child behavior disorder.

Distinct subdimensions are also evident in all contemporary self-report inventories for psychopathy. Like the PCL-R itself, inventories patterned after it have correlated factors (e.g., Andershed et al. 2002; Levenson et al. 1995; Williams et al. 2007). By contrast, the PPInv—which was developed to index basic trait dispositions associated with psychopathy without specific requirements for convergence—has two higher-order factors that are largely uncorrelated. These factors, labeled fearless dominance and impulsive antisociality by Benning (2005a), show differential relations with criterion variables in domains of self-report, interview based assessment, and physiology (for a review, see Patrick and Bernat 2009). Notably, the PPInv contains one subscale, Coldheartedness, which fails to load appreciably on either of these factors—instead emerging as a separate subdimension in structural analyses (Benning et al. 2003). As discussed further below, this subscale appears to index callous–unemotionality or meanness more exclusively than the other subscales of the PPInv.

Subdimensions of ASPD: The childhood criteria for ASPD in Section II of DSM-5 mirror those for conduct disorder (CD) and include aggressive and destructive behaviors along with theft/deceptiveness and non-aggressive rule-breaking acts. Factor analyses of the CD diagnostic criteria (e.g., Frick et al., 1991; Tackett et al. 2003) have shown that the aggressive and rule-breaking symptoms define separate, albeit correlated factors. Follow-up twin studies have demonstrated differing sources of genetic and environmental influence for these factors (Tackett et al. 2005; Kendler et al. (2013), with the proportion of symptom variance attributable to genes higher for the aggressive than the rule-breaking factor (see review by Burt 2009). Evidence for two distinct factors underlying the adult symptoms of ASPD—a disinhibition factor encompassing tendencies toward impulsivity, irresponsibility, and deceitfulness and an aggressive-disregard factor reflecting irritability/aggressiveness, reckless behavior, and lack of concern for self or others—has also been reported Kendler et al. (2012). Paralleling findings for the subdimensions of CD, these adult ASPD factors appear to reflect differing sources of genetic influence.

Clarifying relations among differing psychopathy measures and ASPD: The Triarchic model. The Triarchic model of psychopathy (Patrick et al. 2009) was advanced as a framework for integrating alternative conceptions, organizing findings pertaining to psychopathy subdimensions, and guiding research on neurobiological correlates and etiologic influences. The model characterizes psychopathy as encompassing three distinct but intersecting symptomatic (phenotypic) constructs: disinhibition, boldness, and meanness. *Disinhibition* entails impulsiveness, weak restraint, hostility and mistrust, and difficulties in regulating emotion; *meanness* entails deficient empathy, lack of affiliative capacity, contempt toward others, predatory exploitativeness, and empowerment through cruelty or destructiveness; and *boldness* encompasses tendencies toward confidence and social assertiveness, emotional resiliency, and venturesomeness. The Triarchic model provides a frame of reference for relating subdimensions of psychopathy to those of ASPD. In addition, the constructs of the model have biobehavioral referents and show replicable associations with physiological variables and thus can be helpful for relating psychopathy and ASPD to neurobiology (cf. Patrick et al. 2012).

The Triarchic Psychopathy Measure (TriPM; Patrick 2010) was developed to index the three constructs of this model. Its disinhibition and meanness subscales index disinhibitory externalizing and callous aggression factors of the externalizing psychopathology spectrum (Krueger et al. 2007); the TriPM's Boldness scale indexes fearless–dominant tendencies associated with the common factor among scale measures of fear and fearlessness (Kramer et al. 2012). The TriPM has been used as a referent to evaluate coverage of the Triarchic model facets in differing psychopathy inventories. The PPInv provides balanced coverage of boldness, meanness, and disinhibition as indexed by the TriPM (Drislane et al. 2014a; Sellbom and Phillips 2013). Subscales that demarcate the PPInv's fearless dominance factor relate very strongly to TriPM Boldness, scales demarcating PPInv impulsive antisociality relate very strongly to TriPM Disinhibition (particularly carefree non-planfulness and blame externalization/alienation), and moderately to meanness (mainly due to Machiavellianism Egocentricity), and the PPInv Coldheartedness scale relates specifically to TriPM Meanness. By contrast, other psychopathy inventories index meanness and disinhibition either more so than boldness (e.g., SRP, YPI) or to the exclusion of boldness (e.g., LSRP; Drislane et al. 2014a; Sellbom and Phillips 2013; Hall et al. 2014).

The TriPM has also been used to clarify similarities and differences between PCL-R psychopathy and ASPD in terms of the Triarchic model (Venables et al. 2014; Wall et al. in press). This work shows that (a) overall scores on the PCL-R contain variance associated with all three Triarchic model constructs, whereas ASPD indexes the meanness and disinhibition constructs only; (b) PCL-R Factor 1 is associated with boldness and meanness but not disinhibition; and (c) Factor 2 is associated with disinhibition and meanness but not boldness. These findings serve to clarify the relationship between PCL-R psychopathy and ASPD: The two overlap in terms of Factor 2, which includes common elements of disinhibition and meanness, but differ in elements of boldness and meanness that are represented exclusively in PCL-R Factor 1 (cf. Patrick et al. 2007).

2.2 ASPD and Psychopathy: Associations with Aggression and Violence

Disorders such as ASPD and substance abuse/dependence co-occur at high rates (Krueger 1999) and show relationships in common with disinhibitory personality traits (i.e., impulsivity, sensation seeking, nonconformity; Krueger et al. 2002). The variance in common among disorders within this externalizing spectrum has been shown to reflect a highly heritable liability factor (Krueger et al. 2002; Young et al.

2000), labeled externalizing proneness (Krueger et al. 2002, 2007), or disinhibition (Patrick et al. 2009, 2013a). By contrast, the variance specific to each disorder appears to be attributable more to non-shared environmental influences. From this perspective, aggressive behavior associated with ASPD in part reflects high levels of externalizing liability, shaped toward violent criminal expression by adverse physical and social experiences encountered by individuals across time.

However, research directed at modeling externalizing problems and traits more comprehensively (Krueger et al. 2007) demonstrates factors distinct from the general disinhibitory-externalizing factor, reflecting callous–aggressive tendencies and proneness to abuse various substances. The finding of distinct disinhibitory and callous–aggression factors coincides with aforementioned evidence for distinct etiologic influences contributing to aggressive versus rule-breaking subdimensions of CD and adult ASPD, and with evidence for separable callous–unemotional and impulsive–disruptive subdimensions to child psychopathy (Frick and Marsee 2006). Other above-noted work that has directly evaluated relations between PCL-R psychopathy and ASPD from the perspective of the Triarchic model (Venables et al. 2014; Wall et al. in press; see also Patrick et al. 2005, 2007; Venables and Patrick 2012) shows that PCL-R Factor 2 reflects externalizing proneness (disinhibition) and elements of callous–aggression (meanness) in common with ASPD, whereas PCL-R Factor 1 reflects boldness and other elements of callous–aggression separate from ASPD.

Considering these findings, a key question is whether documented predictive relations for PCL-R psychopathy with violent offending and criminal recidivism are accounted for by the features it shares in common with ASPD, or by tendencies that distinguish it from ASPD. As reviewed by Kennealy et al. (2010), the answer appears to be that PCL-R psychopathy is predictive of violent behavior largely as a function of features encompassed by Factor 2 (i.e., disinhibition and affiliated aspects of meanness). Using a meta-analytic, regression-based approach in which scores on the two PCL-R factors were evaluated as concurrent predictors, these authors found that the antisocial deviance features associated with PCL-R Factor 2 were substantially predictive of violence (effect size $d = 0.40$), whereas the affective–interpersonal features associated with Factor 1 were only mildly predictive ($d = 0.11$). Further analyses were undertaken to examine whether affective–interpersonal (Factor 1) features might interact with impulsive–antisocial (Factor 2) features to predict elevated risk for violence in a non-additive fashion (cf. Hare and Neumann 2009). Results indicated that these two components of PCL-R psychopathy did not contribute interactively to violence prediction.

Similarities and distinctions between psychopathy and ASPD, and relations of each with violent behavior, are important to consider in reviewing the literature on physiological correlates of these conditions. In particular, it can be expected that physiological correlates will be more similar among impulsive violence, ASPD, and Factor 2 of psychopathy than between impulsive violence or ASPD and Factor 1 of psychopathy or psychopathy as a whole.

3 Physiological Correlates of Psychopathy, Antisocial Personality, and Aggression

3.1 Peripheral Measures: Cardiovascular, Electrodermal, and Startle Blink Responses

Studies of children and adolescents exhibiting antisocial behavior have yielded consistent evidence of lower resting levels of autonomic activity—most notably heart rate (HR), but also to some extent skin conductance (SC)—in comparison with control youth (Lorber 2004; Ortiz and Raine 2004). The finding of low resting HR in particular is especially robust among children with aggressive behavioral tendencies (Scarpa and Raine 1997). Findings for autonomic *reactivity* to noxious or threatening stimuli have been more mixed, but as a whole the available evidence points to *enhanced* HR and SC response to stressors in children exhibiting aggressive conduct problems specifically (Lorber 2004). This is particularly the case for children exhibiting *reactive* aggression; proactively aggressive children if anything tend to show attenuated reactivity to stressors compared with control children (Hubbard et al. 2002).

Other studies focusing on parasympathetic versus sympathetic mediation of cardiovascular activity have yielded evidence of weaker vagal–parasympathetic regulation in children and adolescents with aggressive conduct problems—reflected in enhanced HR variability under circumstances involving stressors or challenges (Beauchaine et al. 2001; Mezzacappa et al. 1997). It has been hypothesized that this lack of vagal control combines with chronic underarousal and weak inhibitory capacity (reflected in lower resting HR and reduced spontaneous SC responses, respectively) to lower the threshold for impulsive aggressive behavior (Beauchaine et al. 2001). Other research has identified weak vagal control as a variable associated with the development of both internalizing (emotional dysregulation) and externalizing problems in at-risk children (e.g., El-Sheikh et al. 2001). Taken together, these findings are consistent with the idea that aggression in children entails difficulties regulating anger and other emotional reactions—with consequent enhancement of defensive reactivity under conditions of threat.

In studies of adults, one prominent focus has been on autonomic (particularly cardiac) reactivity differences in individuals high on aggression-related traits such as hostility, anger expression, and Type A personality. A meta-analysis of relations between trait hostility and cardiovascular reactivity by Suls and Wan (1993) reported that although effects in studies of this kind were generally small, positive results were especially evident in studies that examined relations between dispositional hostility (particularly when defined by overt expressions of anger such as verbal and physical aggression) and cardiac (particularly blood pressure) reactivity in situations involving interpersonal stress or provocation as opposed to physical stressors. Studies published since this meta-analysis have not yielded positive findings in all cases (see, e.g., Gallo et al. 2000), but significant effects when obtained have generally been in the direction of heightened autonomic reactivity for high trait-aggressive individuals during interpersonal stress (e.g., Smith and Gallo 1999; Peters et al. 2003).

Studies comparing autonomic reactivity in adults with and without a history of violent behavior have yielded less consistent results. For example, physiological studies of men who have assaulted their romantic partners have not revealed consistent differences in relation to non-assaultive men. Gottman et al. (1995) suggested a possible explanation for this in terms of two distinct subgroups of male batterers: one exhibiting decreases in HR activity during a marital interaction (Type 1) and the other exhibiting increases in HR (Type 2). Type 1 batterers scored higher on antisocial traits and were more hostile and contemptuous toward their spouses and more assaultive toward other people in general, whereas Type 2 batterers scored higher in social dependency. However, this pattern of results has not been replicated in subsequent studies (Babcock et al. 2004; Meehan et al. 2001).

Most published studies examining autonomic response have focused on detecting simple reactivity differences between aggressive and non-aggressive individuals. Only a few studies have sought to assess underlying psychological processes contributing to such differences. One of these was a study by Verona et al. (2002) that examined the mediating role of negative emotional activation in enhancing punitive behavior among aggression-prone individuals. These investigators used increased magnitude of reflexive startle responding to index unpleasant activation associated with threat versus absence of threat in a laboratory aggression paradigm. Individuals high on traits of anxiousness, alienation, and aggressiveness showed enhanced unpleasant activation during shock-threat periods (as evidenced by heightened startle reactivity to unwarned noise probes), and in conjunction with this, enhanced aggressive behavior (i.e., delivery of stronger shocks to a putative co-participant). This was interpreted as supporting the perspective that negative emotional activation operates to prime aggressive behavior (Berkowitz's 1990). Subsequent work (e.g., Verona and Curtin 2006) has shown this facilitative effect of negative emotion on aggression to be stronger in men than women, in line with prior research findings (cf. Hokanson 1970).

In sum, research to date has generally revealed lower baseline levels of autonomic arousal, but increased autonomic reactivity to stressful events, in aggressive children and adolescents. Findings for adults have been less consistent, but in general have indicated enhanced autonomic reactivity to stressors (interpersonal stressors in particular) in hostile–aggressive individuals. Notably, the general finding that aggression-prone individuals show enhanced autonomic reactivity to stressful events fits with the hypothesis that violent behavior entails a breakdown in normal affective regulatory capacity (Davidson et al. 2000).

In contrast, markedly different results are evident in the psychophysiological literature on adult psychopathy. Most of these studies have relied on diagnoses based on Cleckley's criteria or Hare's PCL-R. Adult psychopathic offenders, relative to non-psychopathic offenders, show *reduced* electrodermal response to aversive cues and during anticipation of stressful events (Arnett 1997; Hare 1978; Lorber 2004). However, psychopathy is not consistently associated with differential HR reactivity to aversive or stressful stimuli (Lorber 2004), or with differential baseline levels of either HR or electrodermal arousal (Arnett 1997; Hare 1978; but see Hansen et al. 2007). These contrasting results are noteworthy because higher levels of PCL-R psychopathy in offender samples are reliably associated with increased violence and violent recidivism (Porter and Woodworth 2006). However, as discussed earlier, it is the impulsive–antisocial (Factor 2) component of the PCL-R that is most predictive of violent offending. This factor reflects disinhibitory–externalizing tendencies along with elements of callous aggression (meanness), and a subset of offenders who score very high on the PCL-R exhibit personality profiles and behavioral tendencies characteristic of extreme externalizing individuals (i.e., high negative affectivity along with low behavioral restraint; Hicks et al. 2004). The other, affective–interpersonal factor of the PCL-R reflects boldness along with elements of meanness that cohere more with boldness than with disinhibition (Patrick et al. 2009; Venables et al. 2014; Wall et al. in press), and another subset of individuals who score as psychopathic on the PCL-R shows personality profiles characteristic of high fearless dominance or boldness (i.e., low anxiety, high social potency, low harm avoidance; Hicks et al. 2004). This factor of the PCL-R tends to be associated more with instrumental/proactive aggression than impulsive/reactive aggression, both in adult offenders (Porter and Woodworth 2006) and clinic-referred youth (Frick and Marsee 2006).

While at odds with results for impulsive–aggressive individuals, the finding of reduced autonomic (in particular electrodermal) reactivity to stressors in psychopathic individuals is consistent with theories that have emphasized insensitivity to punishment or diminished fear capacity in psychopathy (e.g., Lykken 1995)—especially in relation to its affective–interpersonal features (Patrick 1994; Patrick and Bernat 2009). Direct evidence for reduced electrodermal reactivity to stress in relation to affective–interpersonal features in offenders was provided by Patrick (1995), who reported opposing correlations (positive and negative, respectively) for scores on PCL-R Factors 1 and 2 with amplitude of skin conductance response during anticipation of aversive noise. In more recent work with non-offenders, Dindo and Fowles (2011) reported reduced electrodermal activation during stressor anticipation as a function of high scores on PPInv fearless dominance (akin to boldness), but not PPInv impulsive antisociality, which reflects disinhibition to a substantial degree along with lesser representation of meanness. Additionally, evidence for a selective association of PPInv fearless dominance with electrodermal reactivity to aversive picture stimuli was reported by Benning et al. (2005b).

Complementing these findings for electrodermal reactivity deficits is another body of literature demonstrating a lack of startle reflex potentiation during viewing of aversive picture stimuli in offenders rated high on PCL-R Factor 1 (Patrick et al. 1993; Vaidyanathan et al. 2011; Vanman et al. 2003; see also Patrick 1994) or community participants scoring high on PPInv fearless dominance (Benning et al. 2005b). This deficit in startle potentiation is not seen for individuals scoring high on PCL-R Factor 2 or PPInv impulsive antisociality alone. The absence of startle potentiation during aversive cuing provides strong evidence for deficient fear because startle potentiation is directly indicative of defensive motivational priming (Lang et al. 1990) and as such covaries with individual differences in dispositional fear (Kramer et al. 2012; Vaidyanathan et al. 2009). Extending this literature on startle modulation during picture viewing, Newman and colleagues have tested for psychopathy-related differences in startle reactivity to noise probes during exposure to shock threat versus safety cues, under conditions of concurrent distraction or no distraction (Baskin-Sommers et al. 2011; Dvorak-Bertsch et al. 2009; Newman et al. 2010). Findings from these shock threat studies indicate that deficits in startle potentiation for offenders high on the PCL-R or non-offenders high on PPInv fearless dominance occur mainly under conditions of concurrent distraction. This result has been interpreted as indicating that reduced reactivity to fear cues in individuals with affective–interpersonal features of psychopathy reflects "idiosyncrasies in attention that limit their processing of peripheral information" (Newman et al. 2010, p. 66). However, the finding can be also interpreted as evidence for deficient fear reactivity, insofar as threat cues normally exert an automatic "pull" on attentional resources (Bradley 2009; LeDoux 1995).

Regardless of interpretation, these findings for electrodermal reactivity and startle modulation indicate that individuals exhibiting core-affective features of psychopathy need to be considered separately from other types of violent offenders in attempting to understand physiological processes in aggressive behavior. A similar conclusion has emerged in the literature on antisocial behavior in youth, where it has been shown that children or adolescents who exhibit callous–-unemotional tendencies in conjunction with conduct problems show distinctly different behavioral responses to laboratory stressors and (as discussed further below) differential brain reactivity to fear-relevant cues.

3.2 Electrocortical Measures: EEG and ERP

Early investigations of brain differences in violent criminal offenders focused on abnormalities in electroencephalographic (EEG) activity. A relatively consistent finding in these early studies was enhanced cortical slow-wave activity, particularly in the delta (<4 Hz) frequency range (cf. Volavka 1990). While much of this early literature suffered from notable methodological weaknesses, subsequent research using improved designs and procedures has successfully replicated this finding, with some work demonstrating prediction of antisocial behavior later in life (i.e., official criminal convictions) from increased slow-wave EEG activity in adolescence (Raine et al. 1990). Theoretical interpretations of the association between slow-wave EEG and violent offending have focused on cortical immaturity resulting in impaired inhibitory control (Volavka 1990), and cortical underarousal that predisposes toward compensatory stimulation seeking (Raine et al. 1990).

Associations with aggression have also been reported for various components of the cortical event-related potential (ERP)—reflecting average changes in voltage at scalp recording sites across time following the presentation of a stimulus or the emission of a response. The most consistent finding has been reduced amplitude of the P300 response component in oddball tasks where participants respond to intermittent target stimuli interspersed with more frequently occurring nontargets. Diminished P300 has been reported especially among individuals exhibiting aggression of the impulsive variety (e.g., Barratt et al. 1997; Branchey et al. 1988; Gerstle et al. 1998). In view of theoretic models that interpret P300 response as reflecting brain activity associated with post-perceptual processing of salient stimuli within a task (Donchin and Coles 1988; Polich 2007), reduced P300 amplitude in impulsively aggressive individuals implies some impairment in higher cognitive–elaborative processing of stimulus events.

Reduced P300 response has also been reported in individuals with ASPD (Bauer et al. 1994) and other impulse control disorders—most notably alcohol dependence (cf. Polich et al. 1994), but also drug dependence, nicotine dependence, child conduct disorder, and attention-deficit hyperactivity disorder (Iacono et al. 2002). Given aforementioned evidence for a common liability factor underlying these various conditions (Krueger et al. 2002), Patrick et al. (2006) tested the hypothesis that reduced P300 amplitude reflects this shared liability factor and found clear supportive evidence. Subsequent research has corroborated this finding and demonstrated that the association between externalizing proneness (disinhibition) and reduced P300 reflects common genetic influences (Hicks et al. 2007; Yancey et al. 2013).

In contrast with findings from studies of impulsive aggressive individuals, Stanford et al. (2003) reported no difference in P300 amplitude to auditory target stimuli in psychiatric outpatients characterized as "premeditated aggressors" compared with controls. Similarly, Barratt et al. (1997) found no evidence of a relationship between premeditated aggression and P300. Results from these studies indicate that the association between reduced P300 and aggression may be specific to individuals who manifest aggression of an impulsive nature. Studies examining the relationship between psychopathy and P300 amplitude have yielded mixed results, with some showing a negative association, others a positive association, and still others no association (Gao and Raine 2009). As noted earlier, these inconsistent findings could reflect the fact that a diagnosis of psychopathy includes affective–interpersonal features in addition to impulsive–antisocial symptoms; the affective–interpersonal features, which tend to be associated more with proactive rather than impulsive aggression (Porter and Woodworth 2006), may moderate the relationship between psychopathy and P300 response in some samples. In support of this, Venables and Patrick (2014) examined effects for the two PCL-R factors separately and found the relationship with P300 amplitude to be specific to PCL-R Factor 2. Parallel results were reported by Carlson et al. (2009) for the two factors of the PPInv: Reduced P300 amplitude was associated significantly with scores on the Impulsive Antisocial factor, whereas no relationship was evident for the Fearless Dominance factor.

Although P300 response amplitude is the most widely studied ERP correlate of antisocial–externalizing conditions, some other brain potential correlates have been reported in the literature. One is the error-related negativity (ERN), a negative-polarity response occurring approximately 50 ms after the commission of errors in speeded performance tasks, believed to reflect early "endogenous" error processing associated with the neural signaling function of the anterior cingulate cortex. Reduced ERN has been reported for individuals high in impulsive–antisocial tendencies (Dikman and Allen 2000; Pailing and Segalowitz 2004) and in relation to externalizing proneness or disinhibition (Hall et al. 2007; Patrick et al. 2013b).

However, as with P300, findings for ERN and psychopathy have been mixed. Munro et al. (2007) examined relations between psychopathy as indexed by overall scores on the SRP (Williams et al. 2007) in two variants of a "flanker" task (Eriksen and Eriksen 1974): one involving discrimination of letter strings and the other discrimination of fearful versus angry faces. Task performance and ERN amplitude in the letter discrimination version of the task were comparable between high and low psychopathic participants, but high-psychopathy participants were less accurate and exhibited reduced ERN amplitude in the emotional face flanker task. In a subsequent study, Brazil et al. (2009) reported relatively intact amplitude of the ERN within a letter discrimination flanker task in high-PCL-R psychopathy forensic patients as compared to controls. However, a reduction in the psychopathic group was evident for amplitude of the post-error positivity (Pe), an ERP component considered similar to P3 and thought to reflect later evaluative stages of performance monitoring. The psychopathic group also evidenced decreased behavioral recognition of errors (i.e., reduced ability to signal via a button press when they noticed an error had occurred). A subsequent study by von Borries et al. (2010) reported reduced amplitude of ERN response along with increased error rates and impaired learning of task contingencies in psychopathic forensic patients during a probabilistic learning task that included feedback (either a monetary gain or loss) regarding performance accuracy on each trial.

Notably, studies of ERN response in psychopathy have not systematically evaluated effects for distinguishable factors or facets known to exhibit differential relations with measures of various types—including physiological response measures (e.g., aversive startle potentiation, P300). The more consistent finding of reduced ERN as well as P300 in relation to impulsive–externalizing tendencies, together with evidence for a selective association of P300 with Factor 2 of psychopathy (whether assessed via the PCL-R or the PPInv), suggests that more consistent evidence of reduced ERN is likely to be found in relation to this symptomatic component of psychopathy. To the extent this component of psychopathy is more related to impulsive–aggressive tendencies, it would be expected that reductions in ERN (indicative of impairments in online monitoring of behavior and recognition of incorrect or inappropriate responses) would also be associated with proclivities toward angry/reactive aggression. A study by Kramer et al. (2011) found no difference in ERN between groups scoring high versus low (upper/low quartile) on an aggression questionnaire, but the participants were college students rather than offenders or clinic patients, and the questionnaire measure focused on aggression of various types rather than angry/reactive aggression specifically.

3.3 Neuroimaging Studies of Psychopathy, ASPD, and Aggressive Behavior

Different neuroimaging methods have been used in studies of psychopathy, ASPD, and aggression. Many recent studies have used magnetic resonance imaging (MRI), which quantifies variations in the alignment of endogenous subatomic particles within a magnetic field to index anatomic details of the brain (structural MRI) or variations in blood flow and blood oxygenation (i.e., hemodynamic, or blood-oxygen-level-dependent [BOLD] response) associated with neuronal activity in specific brain regions (functional MRI, or fMRI). Computerized tomography (CT), a structural imaging method that measures regional density of neural tissue using X-ray beams passed through the brain, was used in some older studies of individuals identified as violent or antisocial. Other functional imaging techniques that have been used in studies of psychopathic and antisocial–aggressive individuals are single-photon emission computerized tomography (SPECT) and positron emission tomography (PET). Both rely on the injection of radioactive tracer isotopes into the blood in small amounts, with particles emitted by the isotope from brain regions of interest (photons in the case of SPECT, positrons in the case of PET) used to index either neuronal activity or neurotransmitter function in those regions.

Structural imaging studies: Two older studies by Tonkonogy (1991) and Wong et al. (1994) that tested for brain anatomic differences in psychiatric patients with violent behavior using CT reported evidence for abnormalities in temporal lobe regions. A third study by Blake et al. (1995) found evidence of abnormalities in frontal as well as temporal brain regions in a sample of 31 homicide offenders. More recent studies have used structural MRI to investigate neuroanatomic differences associated with impulsive–aggressive behavior and ASPD. Tiihonen et al. (2008) reported gray matter volume reductions in bilateral regions of frontal cortex (frontopolar, orbitofrontal) in persistently violent offenders. Two other studies—one involving temporal lobe epilepsy patients with aggressive–assaultive behavior (Woermann et al. 2000) and the other female patients diagnosed with borderline personality disorder (van Elst et al. 2003)—reported evidence of reduced gray matter volume in regions of prefrontal cortex, and the latter of these also reported volume reductions in anterior cingulate cortex (ACC), hippocampus, and amygdala. However, another study by Dolan et al. (2002) that compared impulsive–aggressive patients with controls reported a significant reduction in temporal but not frontal lobe volume, and two other studies that examined subcortical structures (Laakso et al. 2000; van Elst et al. 2000) found no difference between violent and nonviolent patient groups in hippocampal or amygdala volume. Studies reporting reduced gray matter volume in prefrontal regions in individuals diagnosed with ASPD include Raine et al. (2000), Laakso et al. (2002), and Narayan et al. (2007). The latter two of these studies focused on ASPD individuals exhibiting salient violent behavior.

Evidence for neuroanatomic abnormalities in individuals scoring high in psychopathy as defined by the PCL-R has emerged from more recent studies utilizing structural MRI. Reported findings include reduced volume of gray matter in frontal and temporal regions of cortex (Müller et al. 2008a; Yang et al. 2005), reduced volume bilaterally of the amygdala (Yang et al. 2009) and posterior hippocampus (particularly in relation to scores on PCL-R Factor 1; Laakso et al. 2001), other hippocampal abnormalities in the form of left/right volume asymmetry (Raine et al. 2004) or deviations in shape (Boccardi et al. 2010), increased volume of white matter in the corpus callosum (Raine et al. 2003), and increased volume of the striatum—with enhanced size of the lenticular nucleus in particular predicted by overall PCL-R scores, and increases in caudate body and caudate head volumes associated, respectively, with scores on PCL-R Factors 1 and 2 (Glenn et al. 2010a). In addition, a study by Craig et al. (2009) that used the MRI-based method of diffusion tensor imaging reported evidence for reduced structural integrity of the uncinate fasciculus, a neural pathway connecting the orbitofrontal cortex and the amygdala, in a sample of nine forensic patients scoring high (>25) on the PCL-R compared with a non-forensic control group. Notably, the one study to date that tested specifically for differences in the ACC and its dorsal and ventral subregions (Glenn et al. 2010b) found no associations with PCL-R psychopathy, either in comparisons of high- versus low-PCL-R total score groups or in correlational analyses utilizing continuous PCL-R total and factor scores.

In sum, the foregoing structural MRI studies of psychopathic individuals have yielded evidence for reduced volume of limbic structures (hippocampus, amygdala) and impaired structural connectivity between amygdala and orbitofrontal cortex, along with isolated indications of increased volume of corpus callosum and striatal structures and mixed evidence for volume reductions in prefrontal and temporal brain regions. Thus, some overlap is evident in findings for psychopathic and violent/antisocial samples (i.e., volume reductions in frontal and temporal regions), along with some divergence (i.e., subcortical–limbic volume reductions mainly in psychopathic samples). The question of how psychopathic participants compare neuroanatomically with violent–antisocial individuals was directly addressed in a recent structural MRI study by Gregory et al. (2012), who tested for gray matter volume differences in predefined brain regions across the following groups: (1) violent offenders meeting diagnostic criteria for both ASPD and PCL-R psychopathy (n = 17), (2) violent offenders meeting criteria for ASPD but not PCL-R psychopathy (n = 27), and (3) healthy non-offenders from the community at large (n = 22). The psychopathic/ASPD offender group showed significant gray matter reductions bilaterally in anterior rostral prefrontal cortex and temporal poles relative to both violent offenders with ASPD (group 2) and healthy controls, along with reductions in bilateral insula compared to violent/ASPD-only offenders.

These findings help to clarify the nature of brain structural anomalies in psychopathic offenders, while also raising questions about the pervasiveness of brain differences in individuals exhibiting violent behavior or ASPD per se. For psychopathic offenders, reduced volumes were evident in brain structures important for emotional processing as related to aversive learning, moral reasoning, and social interchange. While consistent with the notion of psychopathy as entailing deficits in affective sensitivity and social relatedness that give rise to behavioral deviance, this study did not report group comparisons for anterior brain regions involved more in cognitive control than affective processing (e.g., dorsal–lateral regions of prefrontal cortex), and thus, it remains unclear whether frontal anomalies in this psychopathic offender group were limited to affective processing regions. The fact that no differences were found between violent/ASPD-only offenders and healthy controls could also reflect the choice of brain regions for analyses. It may be the case that regions important for cognitive control and performance monitoring would have shown up as anomalous in this offender subgroup. However, it is important to bear in mind that cognitive and affective impairments can reflect deviations in brain function not necessarily reflected in structural–anatomic anomalies.

Functional imaging studies: Three SPECT imaging studies that assessed neuronal activity at rest in aggressive psychiatric patients (Amen et al. 1996; Hirono et al. 2000) and impulsive violent offenders (Soderstrom et al. 2000) found evidence of reduced blood flow in both the prefrontal cortex and the temporal lobes (the left temporal lobe, specifically, in two of the three studies). In conjunction with reductions in these brain regions, Amen et al. (1996) reported evidence of *increased* activity in basal ganglia and subcortical (limbic) regions in their aggressive patient sample. One other SPECT study (Kuruoglu et al. 1996) reported reduced blood flow in frontal brain regions in alcoholic individuals with comorbid ASPD relative to non-alcoholic controls. SPECT was also used in some earlier studies to examine differences in neurotransmitter function in impulsively violent offenders. Studies of this kind revealed evidence of abnormal dopaminergic neurotransmission in the striatum and diminished serotonin transporter density in the midbrain (Tiihonen et al. 1995, 1997).

A number of studies have used PET imaging to test for functional brain abnormalities in violent individuals compared with controls. The majority have reported evidence of prefrontal dysfunction. Some of these studies focused on brain activity at rest (e.g., Volkow and Tancredi 1987), others on activity during tasks designed to activate the prefrontal cortex (e.g., Raine et al. 1994, 1997). Raine et al. (1998) subdivided violent participants from the Raine et al. (1997) study, consisting of 41 convicted murderers, into predatory (proactive) and affective (impulsive) subgroups based on the nature of their crimes, and found that prefrontal dysfunction was specific to the affective subgroup. Other studies have used PET imaging to investigate brain reactivity to drugs that activate the serotonergic system in aggressive and non-aggressive individuals (e.g., New et al. 2002; Siever et al. 1999). These studies have reported blunted reactivity to serotonin agonists (as evidenced by lower levels of glucose metabolism) among impulsive–aggressive patients compared with controls in regions of prefrontal cortex, particularly orbitofrontal and ventromedial regions. Other brain regions implicated with some consistency in these and other PET imaging studies include temporal cortex, ACC, and, to a lesser degree, hippocampus and amygdala.

Findings of abnormal ACC activity in conjunction with prefrontal anomalies are of interest in view of earlier described work demonstrating reduced ERN in impulsive–aggressive individuals and a recent study by Aharoni et al. (2013) reporting significant prediction of post-release recidivism among criminal offenders from degree of ACC activation in a laboratory inhibitory control task. With regard to the amygdala and hippocampus, Raine et al. (1997) found evidence of abnormal asymmetry (i.e., decreased functioning on the left side and increased functioning on the right) in both these structures in murderers compared to controls. In a PET study of serotonin binding potential, Parsey et al. (2002) reported a significant negative relationship between reported lifetime aggression and binding in brain regions including the amygdala (but not hippocampus). George et al. (2004), in a study of domestic abusers with comorbid alcoholism, reported decreased correlations between glucose activity in the amygdala and glucose activity in various cortical structures compared with nonviolent controls. The authors postulated that these decreased associations reflected a lack of cortical input to the amygdala associated with increased sensitivity to environmental stressors among impulsively violent individuals.

Some studies of violent individuals have been also conducted using functional MRI. Raine et al. (2001) examined brain activation during a working memory task in small groups of community participants with histories of serious violent behavior and/or early abuse (*n*s = 4–5) relative to a healthy control group (*n* = 9). Compared to controls, violent individuals who had been abused as children showed reduced right hemisphere activation (particularly in right temporal regions), whereas abused individuals without violence showed lower left, but higher right activation of the superior temporal gyrus. In addition, both of these groups showed generally reduced cortical activation during task processing, particularly in the left hemisphere. The authors interpreted these findings as indicating a unique role of right hemisphere dysfunction, when combined with exposure to early abuse, in violent behavior. However, the findings of this study were quite tentative given the small sample sizes.

Several studies have used SPECT or fMRI to examine brain activation differences in individuals diagnosed as psychopathic using Hare's (2003) PCL-R, with a smaller number focusing on psychopathy as defined by self-report in adult samples or the informant-rated APSD in younger samples. Of studies that have focused on PCL-R psychopathy, most have examined brain reactivity in emotional processing paradigms entailing viewing of affective and neutral visual stimuli, aversive conditioning, anticipation of punishment to oneself or another person, processing of moral dilemmas entailing more or less emotion provocation, or performance of a cognitive task following manipulation of mood. Although no two of these PCL-R studies have used the same experimental task, some have used similar procedures. Both Intrator et al. (1997) and Kiehl et al. (2001) examined reactivity to emotional versus neutral words, within discrimination (word vs. nonword) and memory (encoding, rehearsal, recall) contexts, respectively. Using SPECT, Intrator et al. found *increased* bilateral activation for emotional versus neutral words in high-PCL-R participants within frontal–temporal cortex and adjacent subcortical regions. Kiehl et al. (2001) reported *decreased* activation in multiple a priori-defined limbic–subcortical regions, along with (in post hoc analyses) increased activation in right and left inferior lateral–frontal regions of cortex. Both Schneider et al. (2000) and Birbaumer et al. (2005) examined brain reactivity to CS+ and CS− stimuli in a differential aversive condition procedure, using foul odor and painful tactile pressure stimuli as USs, respectively. The first of these studies reported *increased* activation in amygdala and dorsolateral prefrontal cortex regions to the CS+ versus the CS− for psychopathic participants during the latter part of acquisition, whereas the second reported *decreased* differential activation for high-PCL-R participants in left amygdala and ventromedial prefrontal cortex regions, as well as in right insula, rostral anterior cingulate, and secondary somatosensory cortex.

Two other studies by Müller et al. (2003, 2008b) used emotional and neutral picture stimuli, but in quite different ways. Müller et al. (2003) examined reactivity to pictures as primary stimuli and reported a complex pattern of differences for psychopathic as compared to non-psychopathic participants (i.e., decreased activation in some cortical and subcortical brain regions, but increased activation in others, for both pleasant and unpleasant pictures relative to neutral—with specific regions of decrease and increase for unpleasant pictures overlapping only partly with regions of decrease/increase for pleasant pictures). Müller et al. (2008b) used unpleasant picture viewing as a mood induction and found that high-PCL-R offenders, in contrast to low-PCL-R controls, exhibited no impact of this induction on responding in a subsequent "cognitive" reaction-time task, either behaviorally or in terms of activity in distinct brain regions (R medial and L inferior frontal gyri, R superior temporal gyrus) during this task.

Converging results across these differing emotion-processing studies include increased activation in regions of frontal/prefrontal cortex (Intrator et al. 1997; Schneider et al. 2000; Kiehl et al. 2001; Müller et al. 2003), increased activation in temporal–subcortical regions including the amygdala in some studies (Intrator et al. 1997; Müller et al. 2003; Schneider et al. 2000) along with decreased amygdala activation in others (Kiehl et al. 2001; Birbaumer et al. 2005), decreased activation in anterior cingulate (Kiehl et al. 2001; Müller et al. 2003; Birbaumer et al. 2005) and posterior cingulate, hippocampal, and frontal gyrus regions (Kiehl et al. 2001; Müller et al. 2003), and decreased activation in inferior frontal and superior temporal gyri (Kiehl et al. 2001; Müller et al. 2008b). An additional four fMRI studies of high-PCL participants used emotional processing tasks of other types. Findings from these studies that converge with results from the six above-mentioned studies include the following: (1) increased activation in regions of prefrontal cortex (dorsolateral, evaluated post hoc (Glenn et al. 2009b); dorsal and ventral medial, selectively in relation to higher PCL-R Factor 2 (Veit et al. 2010), and (2) decreased activation in anterior cingulate (Veit et al. 2010), posterior cingulate (Glenn et al. 2009a), amygdala (Glenn et al. 2009a; Veit et al. 2010), and right fusiform gyrus (Deeley et al. 2006; also reported by Müller et al. 2003). The two studies from among this overall group that included conditions entailing receipt of physical punishment (Birbaumer et al. 2005; Veit et al. 2010) also converged in finding decreased activation of the insula—a region implicated in pain perception.

However, some clear opposing findings are also evident across these different emotion-processing studies, including the following: (1) *decreased* activation of frontal/prefrontal cortex in some studies (i.e., ventromedial orbitofrontal cortex in Birbaumer et al. 2005; post-central gyrus in Deeley et al. (2005); right medial and left inferior frontal gyri in Müller et al. (2008b); medial frontal cortex, selectively in relation to higher PCL-R Factor 1, in Glenn et al. (2009a) versus *increased* frontal/prefrontal activation in others (i.e., bilateral frontal/temporal cortex in Intrator et al. 1997; bilateral inferior lateral frontal cortex in Kiehl et al. 2001; bilateral precentral, bilateral inferior frontal, and right medial frontal gyri in Müller et al. 2003; right dorsolateral prefrontal cortex in Glenn et al. 2009b; dorsal and ventral medial prefrontal cortex, selectively in relation to PCL-R Factor 2, in Veit et al. 2010); and (2) *decreased* activation of the amygdala specifically in some studies (Kiehl et al. 2001; Birbaumer et al. 2005; Glenn et al. 2009a; Veit et al. 2010) versus *increased* amygdala activation in others (Müller et al. 2003; Schneider et al. 2000).

Other fMRI studies have investigated college or community adults varying in levels of psychopathy as assessed by self-report inventories (PPInv, four studies; TriPM, one study). All of the PPInv studies examined reactivity in affective processing or provocation tasks (i.e., affective picture viewing; affective face discrimination; anticipation of monetary reward; Prisoner's Dilemma). One study by Harenski et al. (2009) examined brain reactivity to unpleasant pictures, including depictions of moral dilemmas, under conditions of simple viewing and instructed emotion suppression in relation to scores on the PPInv as a whole and its Coldheartedness subscale. During simple viewing of moral-violation scenes, participants with high overall PPInv scores showed *decreased* activation in medial prefrontal cortex (Birbaumer et al. 2005; Müller et al. 2008b; Glenn et al. 2009a), and those specifically high in PPInv Coldheartedness showed decreased activation of the amygdala. Additionally, in the instructed suppression condition, high overall PPInv scorers showed *increased* activation in specific subdivisions of prefrontal cortex reported to be hypoactive in a number of PCL-R/emotion-processing studies (Intrator et al. 1997; Glenn et al. 2009b; Kiehl et al. 2001; Müller et al. 2003; Veit et al. 2010).

The other three PPInv studies evaluated effects for the inventory's distinctive fearless dominance and impulsive antisociality factors. Two of these studies (Gordon et al. 2004; Rilling et al. 2007) found higher scores on PPInv fearless dominance to be associated with differential activation in certain brain regions. However, no overlap was evident between effects observed by Gordon et al. (i.e., decreased activation in right amygdala, medial prefrontal cortex, right inferior temporal cortex, and increased activation in visual cortex and right dorsolateral prefrontal cortex) and the single effect reported by Rilling et al. (i.e., decreased activation in rostral anterior cingulate cortex). Additionally, Gordon et al. (but not Rilling et al.) reported evidence of *increased* activation in the right amygdala for participants classified as high versus low on PPInv impulsive antisociality. The other study that presented results separately for the two PPInv factors (Buckholtz et al. 2010) focused primarily on reactivity in the nucleus accumbens and found effects exclusively for PPInv impulsive antisociality—with higher scorers showing increased dopamine release in the accumbens both during anticipation of monetary reward and following administration of a dopamine agonist (amphetamine).

One other fMRI study tested for brain reactivity differences in an economic decision-making task as a function of overall scores on the TriPM (Vieira et al. in press). The major finding was that high TriPM scorers exhibited a different pattern of brain response when rejecting unfair offers, entailing enhanced activation of ventromedial prefrontal cortex relative to dorsolateral prefrontal cortex, compared to low TriPM scorers. The authors' interpretation was that economic decision making may be more strongly driven by frustration than perceived fairness in high psychopathic individuals. These findings are interesting in light of other recent work by Drislane et al. (2014b) demonstrating distinct subgroups among high overall scorers on the TriPM—namely, a classically low-neurotic, high-bold ("primary") subtype and a high-neurotic, high-disinhibited ("secondary") subtype. This work raises intriguing questions about the representation of these distinct variants in the Vieira et al. study and the contribution of one versus the other to reported differences in brain activation.

A further set of fMRI studies has focused on psychopathy in children or adolescents as indexed by the APSD (Frick and Hare 2001). Two of these studies used affective face processing procedures, and two examined brain reactivity in reward/punishment learning paradigms. The first of the two face processing studies (Marsh et al. 2008) compared young adolescent participants meeting criteria for psychopathy on both the APSD and the youth version of the PCL-R with two other age-matched groups: (1) participants who met criteria for attention-deficit hyperactivity disorder (ADHD) but scored low on APSD callous–unemotional features and (2) a non-disorder ("healthy comparison") group. Relative to these comparison groups, psychopathic participants showed decreased right amygdala activation for fearful versus neutral faces, along with decreased covariation of activity between the right amygdala and interconnected structures including ventromedial prefrontal cortex, anterior and posterior cingulate gyrus, insula, and inferior temporal/fusiform gyrus. Using a very similar task with younger participants, and employing somewhat different selection criteria for psychopathy (i.e., ASPD ratings in conjunction with ratings on a separate measure of conduct problems) and a single non-clinical control group, Jones et al. (2009) replicated Marsh et al.'s finding of decreased amygdala activation during processing of fearful versus neutral faces and also reported a concomitant reduction in activity of the anterior cingulate cortex. The latter of these findings coincides with results from a number of PCL-R/imaging studies (Kiehl et al. 2001; Müller et al. 2003; Birbaumer et al. 2005; Veit et al. 2010) and one of four PPInv/imaging studies (Rilling et al. 2007).

The other two studies that focused on psychopathy in young participants used the same dual-diagnostic criterion (ASPD + PCL:YV) employed by Marsh et al. (2008), but examined brain reactivity in reward/punishment learning tasks. Finger et al. (2008) used a probabilistic reversal-learning task and reported increased activation in relation to punished reversal errors in bilateral medial frontal gyrus and right caudate regions in high-psychopathy participants as compared to ADHD and healthy comparison groups. Within the high-psychopathy group, scores on the callous–unemotional factor of the APSD selectively predicted degree of enhanced activation for punished errors. Finger et al. (2011) compared brain reactivity during a passive avoidance learning task in psychopathic (APSD + PCL:YV) youth and health controls (no ADHD comparison group was included). Relative to controls, psychopathic youth showed decreased reactivity in right orbitofrontal cortex and caudate regions to earlier (as compared to later) occurrences of reinforced outcomes in the task, along with decreased reactivity in orbitofrontal cortex for correct rewarded response trials overall. A main effect of group was also evident for particular brain regions across the task as a whole, reflecting generally decreased activation for the psychopathic group in regions including the amygdala, caudate, and insula, and regions characterized by the authors as components of an "attention network" (i.e., prefrontal and parietal cortex).

4 Summary, Implications, and Future Directions

A number of consistent findings have emerged from psychophysiological studies of aggression and aggressive individuals. One is the finding of low resting HR, which has been interpreted as reflecting low dispositional arousal associated with tendencies toward impulsive stimulation seeking (Raine 1993, 2002; Ortiz and Raine 2004). However, this interpretation remains speculative, as no research to date has directly assessed the functional role of low cardiac arousal in the disinhibited behavior of antisocial–aggressive individuals. Two findings of related interest are enhanced EEG slow-wave activity in antisocial–aggressive individuals and reduced P300 brain response in individuals with externalizing problems more broadly. Reduced P300 response has also been reported specifically in relation to Factor 2 of psychopathy, whether indexed by the PCL-R (Venables and Patrick 2014) or the PPInv (Carlson et al. 2009). Enhanced EEG slow wave, like low resting HR, has been theorized to reflect low dispositional arousal that motivates stimulation seeking (Eysenck 1967; Zuckerman 1979). Differing explanations have been proposed for the finding of reduced P300 response. One that fits with findings of low resting HR and enhanced EEG slow wave is that anticipatory and preparatory activities are reduced in such individuals, resulting in a more stimulus-driven processing style (Malone et al. 2002; Taylor et al. 1999).

In contrast with these findings, other research has demonstrated *enhanced* phasic reactivity to stressful or aversive stimuli in hostile, aggressive, and abusive individuals—including enhanced cardiac and skin conductance reactivity to stressors, poor regulation of autonomic activity during anticipation of aversive events, and reduced cardiac vagal tone. Furthermore, some evidence exists to indicate that this pattern of heightened reactivity to aversive cues or events, like reduced P300 brain response, may be generally characteristic of individuals with impulse control problems, rather than specific to impulsive aggressive individuals (Taylor et al. 1999). Although the finding of enhanced reactivity to phasic stressors might seem inconsistent with data indicating low resting activation levels, the hypothesis that externalizing proneness (including proclivities toward impulsive aggression) entails a reactive, stimulus-driven processing style provides a framework for interpreting this overall configuration of results. From this perspective, high externalizing individuals are more reactive to immediate stressors or challenges because they anticipate and prepare for them less effectively (cf. Davidson et al. 2000).

These findings for impulsive aggression, and externalizing conditions more broadly, are clearly at odds with findings for psychopathy as defined by differing inventories. Adult psychopathic offenders do not show reliable differences in resting autonomic activity levels or P300 brain response (Raine 1993), but do show consistent reductions in phasic reactivity to aversive cues, including diminished SC response (cf. Hare 1978; Arnett 1997) and startle reflex potentiation (cf. Patrick 1994, 2007). The explanation for this divergence in findings almost certainly lies in the distinction between the affective–interpersonal versus the antisocial deviance features of psychopathy: It is the latter features that reflect heightened externalizing tendencies, including aggression and impulsiveness (Patrick 2007; Patrick et al. 2005). However, EEG/ERP and brain imaging studies have only recently begun to investigate effects for these two components of psychopathy separately. This is a key issue that should continue to be systematically addressed in future research.

However, it is important to keep in mind when considering findings from studies of these types that differing instruments for psychopathy index affective–interpersonal and impulsive–antisocial components of psychopathy differently. For example, whereas the two factors of the PCL-R are correlated and overlap in coverage of callous–aggressive tendencies (Patrick et al. 2009; Venables and Patrick 2012), the PPInv's two factors are uncorrelated, with PPInv fearless dominance indexing bold–fearless tendencies and PPInv impulsive–antisociality indexing disinhibition and to a secondary degree callous aggressiveness or meanness—and the PPInv's Coldheartedness scale indexing elements of meanness not captured by PPInv impulsive antisociality (Drislane et al. 2014a; Hall et al. 2014; Sellbom and Phillips 2013). Additionally, child and adult symptoms of ASPD reflect separable aggressive and non-normative/rule-breaking factors (Kendler et al. 2012, 2013; Tackett et al. 2003, 2005) that differentially reflect meanness versus disinhibition (Venables and Patrick 2012)—consistent with evidence from research demonstrating separable callous–unemotional and impulsive/conduct problem factors to child psychopathy. In light of this growing body of evidence, it seems likely that greater precision can be obtained in identifying reliable physiological correlates of psychopathy and related diagnostic conditions by routinely assessing distinct boldness, meanness, and disinhibition facets of these conditions in research studies. As an example of this, considerable progress has been made in indexing robust, replicable brain correlates of the disinhibition facet of psychopathy that exhibit correlations with one another, and thus can be combined to form composite brain-based indices of disinhibition or externalizing proneness (Nelson et al. 2011; Patrick et al. 2012). Through work of this kind, it will be possible in future brain electrophysiology and neuroimaging studies to characterize individuals along distinct dimensions of psychopathy or antisociality through combined use of physiological and clinical or psychometric measures (Patrick et al. 2013b). Phenotypes operationalized in this way would be more likely to exhibit consistent, meaningful biological correlates than phenotypes operationalized exclusively through diagnostic ratings or self-report (Patrick et al. 2012, 2013b).

Related to this, it will be important in future research to systematically examine alternative forms of aggression associated with differing underlying motives (e.g., proactive–instrumental versus reactive–impulsive) in relation to these two psychopathy factors in order to clarify relations with neurobiological measures. In particular, it is the impulsive–reactive subtype that appears to be most related to externalizing proneness and to impairments in brain systems that govern emotion regulation. It will also be valuable in future studies to include multiple measures of physiological response (peripheral–autonomic along with electrocortical; EEG together with structural or functional neuroimaging) so that findings for different measures can be directly compared within the same task procedures (cf. Nelson et al. 2011; Patrick et al. 2012, 2013b).

As a final point, it is important to note that most published psychophysiological studies of aggressive individuals to date (including neuroimaging studies) have focused either on differences in brain structure or differences in physiological activity at rest or in simple stimulus tasks. A pressing need exists for studies aimed at elucidating differences in online cognitive and affective processing with functional relevance to aggression—including cortical psychophysiology studies that capitalize on the fine-grained temporal and frequency information afforded by EEG/ERP, and functional neuroimaging studies that capitalize on the fine-grained spatial information provided by MRI. Along these lines, key questions for future research include the following: (1) What are the distinctive functional roles of brain regions that have been implicated in electrocortical and neuroimaging studies of aggression and how do these regions interact to achieve regulatory control over emotional states? Basic cognitive and affective neuroscience research is needed to elucidate this issue. (2) What specific impairments in the functioning of these brain systems predispose individuals toward aggressive behavior? To address this question, more EEG/ERP and functional neuroimaging studies are needed that examine online processing and brain reactivity within aggression-relevant task procedures, such as interpersonal provocation paradigms. (3) Do different types of brain dysfunction underlie impulsive–reactive and callous–proactive manifestations of aggressive behavior? The work of Raine et al. (1997), Marsh et al. (2008), and others suggests that these manifestations of aggression may reflect separate neuropathologies. Thus, an important challenge for future research will be to delineate the nature of processing impairments or deviations that underlie impulsive aggression associated with externalizing conditions compared with more callous–instrumental forms of aggression associated with psychopathic personality.

Acknowledgments Supported by grants MH089727 from the National Institute of Mental Health and W911NF-14-1-0027 from the US Army. The content of this chapter is solely the responsibility of the author and does not necessarily represent the official views of the US Government, Department of Defense, Department of the Army, Department of Veterans Affairs, or US Recruiting Command.

References

Aharoni E, Vincent GM, Harenski CL, Calhoun VD, Sinnott-Armstrong W Gazzaniga MS, Kiehl KA (2013) Neuroprediction of future arrest. In: Proceedings of the New York Academy of Sciences

Amen DG, Stubblefield M, Carmichael B, Thisted R (1996) Brain SPECT findings and aggressiveness. Ann Clin Psych 8:129–137

American Psychiatric Association (1980) Diagnostic and statistical manual of mental disorders, 3rd edn. Author, Washington, DC

American Psychiatric Association (2000) Diagnostic and statistical manual of mental disorders, 4th edn. Author, Washington, DC

American Psychiatric Association (2013) Diagnostic and statistical manual of mental disorders, 5th edn. Author, Washington, DC

Andershed H, Kerr M, Stattin H, Levander S (2002) Psychopathic traits in non-referred youths: initial test of a new assessment tool. In: Blaauw E, Philippa JM, Ferenschild KCMP, van Lodensteijn B (eds) Psychopaths: current international perspectives. Elsevier, The Hague, pp 131–158

Anderson JL, Sellbom M, Wygant DB, Salekin RJ, Krueger RF (in press) Examining the associations between DSM-5 Section III antisocial personality disorder traits and psychopathy in community and university samples. J Pers Disord

Arnett PA (1997) Autonomic responsivity in psychopaths: a critical review and theoretical proposal. Clin Psychol Rev 17:903–936

Babcock JC, Green CE, Webb SA, Graham KH (2004) A second failure to replicate the Gottman et al. (1995) typology of men who abuse intimate partners...and possible reasons why. J Family Psychol 18:396–400

Barratt ES, Stanford MS, Kent TA, Felthous AR (1997) Neuropsychological and cognitive psychophysiological substrates of impulsive aggression. Biol Psychiatry 41:1045–1061

Baskin-Sommers AR, Newman JP, Sathasivam N, Curtin JJ (2011) Evaluating the generalizability of a fear deficit in psychopathic African American offenders. J Abnorm Psychol 120:71–78

Bauer LO, O'Connor S, Hesselbrock VM (1994) Frontal P300 decrements in antisocial personality disorder. Alcohol Clin Exp Res 18:1300–1305

Beauchaine TP, Katkin ES, Strassberg Z, Snarr J (2001) Disinhibitory psychopathology in male adolescents: discriminating conduct disorder from attention-deficit/hyperactivity disorder through concurrent assessment of multiple autonomic states. J Abnorm Psychol 110:610–624

Benning SD, Patrick CJ, Hicks BM, Blonigen DM, Krueger RF (2003) Factor structure of the psychopathic personality inventory: validity and implications for clinical assessment. Psychol Assess 15:340–350

Benning SD, Patrick CJ, Blonigen DM, Hicks BM, Iacono WG (2005a) Estimating facets of psychopathy from normal personality traits: a step toward community-epidemiological investigations. Assessment 12:3–18

Benning SD, Patrick CJ, Iacono WG (2005b) Psychopathy, startle blink modulation, and electrodermal reactivity in twin men. Psychophysiology 42:753–762

Berkowitz L (1990) On the formation and regulation of anger and aggression: a cognitive-neoassociationistic analysis. Am Psychol 45:494–503

Birbaumer N, Veit R, Lotze M, Erb M, Hermann C, Grodd W et al (2005) Deficient fear conditioning in psychopathy: a functional magnetic resonance imaging study. Arch Gen Psychiatry 62:799–805

Blake PY, Pincus JH, Buckner C (1995) Neurological abnormalities in murderers. Neurology 45:1641–1647

Boccardi M, Ganzola R, Rossi R, Sabattoli F, Laakso MP, Repo-Tiihonen E et al (2010) Abnormal hippocampal shape in offenders with psychopathy. Hum Brain Mapp 31:438–447

Bradley MM (2009) Natural selective attention: orienting and emotion. Psychophysiology 46:1–11

Branchey MH, Buydens-Branchey L, Lieber CS (1988) P3 in alcoholics with disordered regulation of aggression. Psychiatry Res 25:49–58

Brazil IA, de Bruijn ER, Bulten BH, von Borries AK, van Lankveld JJ, Buitelaar JK et al (2009) Early and late components of error monitoring in violent offenders with psychopathy. Biol Psychiatry 65:137–143

Buckholtz JW, Treadway MT, Cowan RL, Woodward ND, Benning SD, Li R et al (2010) Mesolimbic dopamine reward system hypersensitivity in individuals with psychopathic traits. Nat Neurosci 13:419–421

Burt SA (2009) Are there meaningful etiological differences within antisocial behavior? Results of a meta-analysis. Clin Psychol Rev 29:163–178

Carlson SR, Thái S, McLaron ME (2009) Visual P3 amplitude and self-reported psychopathic personality traits: frontal reduction is associated with self-centered impulsivity. Psychophysiology 46:100–113

Cleckley H (1976) The mask of sanity, 5th edn. Mosby, St. Louis, Mo. (Original edition published in 1941)

Craig MC, Catani M, Deeley Q, Latham R, Daly E, Kanaan R et al (2009) Altered connections on the road to psychopathy. Mol Psychiatry 14:946–953

Davidson RJ, Putnam KM, Larson CL (2000) Dysfunction in the neural circuitry of emotion regulation—a possible prelude to violence. Science 289:591–594

Deeley Q, Daly E, Surguladze S, Tunstall N, Mezey G, Beer D et al (2006) Facial emotion processing in criminal psychopathy. Preliminary functional magnetic resonance imaging study. Br J Psychiatry 189:533–539

Dikman ZV, Allen JJ (2000) Error monitoring during reward and avoidance learning in high- and low-socialized individuals. Psychophysiology 37:43–54

Dindo L, Fowles DC (2011) Dual temperamental risk factors for psychopathic personality: evidence from self-report and skin conductance. J Pers Soc Psychol 100:556–566

Dolan MC, Deakin JFW, Roberts N, Anderson IM (2002) Quantitative frontal and temporal structural MRI studies in personality-disordered offenders and control subjects. Psychiatry Res Neuroimaging 116:133–149

Donchin E, Coles MGH (1988) Is the P300 component a manifestation of context updating? Behav Brain Sci 11:355–372

Drislane LE, Patrick CJ, Arsal G (2014a) Clarifying the content coverage of differing psychopathy inventories through reference to the Triarchic Psychopathy Measure. Psychol Assess 26:350–362

Drislane LE, Patrick CJ, Sourander A, Sillanmäki L, Aggen SH, Elonheimo H, Parkkola K, Multimäki P, Kendler KS (2014b) Distinct variants of extreme psychopathic individuals in society at large: evidence from a population-based sample. Pers Disord Theory Res Treat 5:154–163

Dvorak-Bertsch JD, Curtin JJ, Rubinstein TJ, Newman JP (2009) Psychopathic traits moderate the interaction between cognitive and affective processing. Psychophysiology 46:913–921

El-Sheikh M, Harger J, Whitson SM (2001) Exposure to interparental conflict and children's adjustment and physical health: the moderating role of vagal tone. Child Dev 72:1617–1636

Eriksen B, Eriksen C (1974) Effects of noise letters upon the identification of a target letter in a non-search task. Percept Psychophysics 16:143–149

Eysenck HJ (1967) The biological basis of personality. Charles C. Thomas, Springfield

Finger EC, Marsh AA, Mitchell DGV, Reid ME, Sims C, Budhani S et al (2008) Abnormal ventromedial prefrontal cortex function in children with psychopathic traits during reversal learning. Arch Gen Psychiatry 65:586–594

Finger EC, Marsh AA, Blair KS, Reid ME, Sims C, Ng P et al (2011) Disrupted reinforcement signaling in the orbitofrontal cortex and caudate in youths with conduct disorder or oppositional defiant disorder and a high level of psychopathic traits. Am J Psychiatry 168:152–162

Forth AE, Kosson DS, Hare RD (2003) The psychopathy checklist: youth version manual, 2nd edn. Multi-Health Systems, Toronto (1st edition published in 1996)

Frick PJ, Hare RD (2001) Antisocial process screening device. Multi Health Systems, Toronto
Frick PJ, Marsee MA (2006) Psychopathy and developmental pathways to antisocial behavior in youth. In: Patrick CJ (ed) Handbook of psychopathy. Guilford Press, New York, pp 353–374
Frick PJ, White S (2008) The importance of callous-unemotional traits for developmental models of aggressive and antisocial behavior. J Child Psychol Psychiatry 49:359–375
Frick PJ, Lahey BB, Loeber R, Stouthamer-Loeber M, Green S, Hart EL, Christ MAG (1991) Oppositional defiant disorder and conduct disorder in boys: patterns of behavioral covariation. J Clin Child Psychol 20:202–208
Gallo LC, Smith TW, Kircher JC (2000) Cardiovascular and electrodermal responses to support and provocation: Interpersonal methods in the study of psychophysiological reactivity. Psychophysiology 37:289–301
Gao Y, Raine A (2009) P3 event-related potential impairments in antisocial and psychopathic individuals: A meta-analysis. Biol Psychol 82:199–210
George DT, Rawlings RR, Williams WA, Phillips MJ, Fong G, Kerich M et al (2004) A select group of perpetrators of domestic violence: evidence of decreased metabolism in the right hypothalamus and reduced relationships between cortical/subcortical brain structures in position emission tomography. Psychiatry Res Neuroimaging 130:11–25
Gerstle JE, Mathias CW, Stanford MS (1998) Auditory P300 and self-reported impulsive aggression. Prog Neuropsychopharmacol Biol Psychiatry 22:575–583
Glenn AL, Raine A, Schug RA (2009a) The neural correlates of moral decision-making in psychopathy. Mol Psychiatry 14:5–6
Glenn AL, Raine A, Schug RA, Young L, Hauser M (2009b) Increased DLPFC activity during moral decision-making in psychopathy. Mol Psychiatry 14:909–911
Glenn AL, Raine A, Yaralian PS, Yang Y (2010a) Increased volume of the striatum in psychopathic individuals. Biol Psychiatry 67:52–58
Glenn A, Yang Y, Raine A, Colletti P (2010b) No volumetric differences in the anterior cingulate cortex of psychopathic individuals. Psychiatry Res Neuroimaging 183:140–143
Gordon HL, Baird AA, End A (2004) Functional differences among those high and low on a trait measure of psychopathy. Biol Psychiatry 56:516–521
Gottman JM, Jacobson NS, Rushe RH, Shortt JW, Babcock J et al (1995) The relationship between heart rate reactivity, emotionally aggressive behavior, and generaln violence in batterers. J Fam Psychol 9:227–248
Gregory S, Ffytche D, Simmons A, Kumari V, Howard M, Hodgins S, Blackwood N (2012) The antisocial brain: psychopathy matters. Arch Gen Psychiatry 69:962–972
Hall JR, Bernat EM, Patrick CJ (2007) Externalizing psychopathology and the error-related negativity. Psychol Sci 18:326–333
Hall JR, Drislane LE, Murano M, Patrick CJ, Lilienfeld SO, Poythress NG (2014) Development and validation of Triarchic construct scales from the psychopathic personality inventory. Psychol Assess 26:447–461
Hansen AL, Johnsen BH, Thornton D, Waage L, Thayer JF (2007) Facets of psychopathy, heart rate variability and cognitive function. J Pers Disord 21:568–582
Hare RD (1978) Electrodermal and cardiovascular correlates of psychopathy. In: Hare RD, Schalling D (eds) Psychopathic behavior: approaches to research. Wiley, Chichester, pp 107–143
Hare RD (1980) A research scale for the assessment of psychopathy in criminal populations. Pers Individ Differ 1:111–119
Hare RD (2003) Manual for the hare psychopathy checklist-revised, 2nd edn. Multi-Health Systems, Toronto
Hare RD, Neumann CS (2009) Psychopathy: assessment and forensic implications. Can J Psychiatry 54:791–802
Harenski CL, Kim S, Hamann S (2009) Neuroticism and psychopathic traits predict brain activity during moral and non-moral emotion regulation. Cogn Affect Behav Neurosci 9:1–15
Hicks BM, Markon KE, Patrick CJ, Krueger RF, Newman JP (2004) Identifying psychopathy subtypes based on personality structure. Psychol Assess 16:276–288

Hicks BM, Bernat EM, Malone SM, Iacono WG, Patrick CJ, Krueger RF, McGue M (2007) Genes mediate the association between P300 amplitude and externalizing disorders. Psychophysiology 44:98–105

Hirono N, Mega MS, Dinov ID, Mishkin F, Cummings JL (2000) Left fronto-temporal hypoperfusion in associated with aggression in patients with dementia. Arch Neurol 57:861–866

Hokanson JE (1970) Psychophysiological evaluation of the catharsis hypothesis. In: Megargee EI, Hokanson JE (eds) The dynamics of aggression. Harper-Collins, New York, pp 74–86

Hubbard JA, Smithmyer CM, Ramsden SR, Parker EH, Flanagan KD, Dearing KF, Relyea N, Simons RF (2002) Observational, physiological, and self-report measures of children's anger: Relations to reactive versus proactive aggression. Child Dev 73:1101–1118

Iacono WG, Carlson SR, Malone SM, McGue M (2002) P3 event-related potential amplitude and risk for disinhibitory disorders in adolescent boys. Arch Gen Psychiatry 59:750–757

Intrator J, Hare R, Stritzke P, Brichtswein K (1997) A brain imaging (single photon emission computerized tomography) study of semantic and affective processing in psychopaths. Biol Psychiatry 42:96–103

Jones AP, Laurens KR, Herba CM, Barker GJ, Viding E (2009) Amygdala hypoactivity to fearful faces in boys with conduct problems and callous-unemotional traits. Am J Psychiatry 166:95–102

Karpman B (1941) On the need of separating psychopathy into two distinct clinical types: the symptomatic and the idiopathic. J Crim Psychopathol 3:112–137

Kendler KS, Aggen SH, Patrick CJ (2012) A multivariate twin study of the DSM-IV criteria for antisocial personality disorder. Biol Psychiatry 71:247–253

Kendler KS, Aggen SH, Patrick CJ (2013) Familial influences on conduct disorder criteria in males reflect two genetic factors and one shared environmental factor: a population-based twin study. JAMA Psychiatry 70:78–86

Kennealy P, Skeem J, Walters G, Camp J (2010) Do core interpersonal and affective traits of PCL-R psychopathy interact with antisocial behavior and disinhibition to predict violence? Psychol Assess 22:569–580

Kiehl KA, Smith AM, Hare RD, Mendrek A, Forster BB, Brink J et al (2001) Limbic abnormalities in affective processing by criminal psychopaths as revealed by functional magnetic resonance imaging. Biol Psychiatry 50:677–684

Kramer UM, Kopyciok RPJ, Richter S, Rodriguez-Fornells A, Münte TF (2011) The role of executive functions in the control of aggressive behavior. Front Psychol 2:152. doi:10.3389/fpsyg.2011.00152

Kramer MD, Patrick CJ, Krueger RF, Gasperi M (2012) Delineating physiologic defensive reactivity in the domain of self-report: phenotypic and etiologic structure of dispositional fear. Psychol Med 42:1305–1320

Krueger RF (1999) The structure of common mental disorders. Arch Gen Psychiatry 56:921–926

Krueger RF, Hicks B, Patrick CJ, Carlson S, Iacono WG, McGue M (2002) Etiologic connections among substance dependence, antisocial behavior, and personality: modeling the externalizing spectrum. J Abnorm Psychol 111:411–424

Krueger RF, Markon KE, Patrick CJ, Benning SD, Kramer M (2007) Linking antisocial behavior, substance use, and personality: an integrative quantitative model of the adult externalizing spectrum. J Abnorm Psychol 116:645–666

Kuruoglu AC, Arikan Z, Vural G, Karatas M (1996) Single photon emission computerised tomography in chronic alcoholism: antisocial personality disorder may be associated with decreased frontal perfusion. Br J Psychiatry 169:348–354

Laakso MP, Vaurio O, Savolainen L, Repo E, Soininen H, Aronen HJ, Tiihonen J (2000) A volumetric MRI study of the hippocampus in type 1 and 2 alcoholism. Behav Brain Res 109:177–186

Laakso MP, Vaurio O, Koivisto E, Savolainen L, Eronen M, Aronen HJ (2001) Psychopathy and the posterior hippocampus. Behav Brain Res 118:187–193

Laakso MP, Gunning-Dixon F, Vaurio O, Repo-Tiihonen E, Soininen H, Tiihonen J (2002) Prefrontal volumes in habitually violent subjects with antisocial personality disorder and type 2 alcoholism. Psychiatry Res Neuroimaging 114:95–102
Lang PJ, Bradley MM, Cuthbert BN (1990) Emotion, attention, and the startle reflex. Psychol Rev 97:377–398
LeDoux JE (1995) Emotion: clues from the brain. Ann Rev Psychol 46:209–215
Levenson MR, Kiehl KA, Fitzpatrick CM (1995) Assessing psychopathic attributes in a noninstitutional population. J Pers Soc Psychol 68:151–158
Lilienfeld SO, Andrews BP (1996) Development and preliminary validation of a self report measure of psychopathic personality traits in noncriminal populations. J Pers Assess 66:488–524
Lilienfeld SO, Widows MR (2005) Psychopathic personality inventory—revised professional manual. Psychological Assessment Resources, Odessa
Lorber MF (2004) Psychophysiology of aggression, psychopathy, and conduct problems: a meta-analysis. Psychol Bull 130:531–552
Lykken DT (1957) A study of anxiety in the sociopathic personality. J Abnorm Clin Psychol 55:6–10
Lykken DT (1995) The antisocial personalities. Erlbaum, Hillsdale
Lynam DR (1997) Pursuing the psychopath: capturing the fledgling psychopath in a nomological net. J Abnorm Psychol 106:425–438
Malone SM, Bernat E, Patrick CJ, Iacono WG (2002) P300 and prestimulus EEG power: relationship to externalizing psychopathology in adolescent males. Psychophysiology 39:S54
Marsh AA, Finger EC, Mitchell DG, Reid ME, Sims C, Kosson DS et al (2008) Reduced amygdala response to fearful expressions in children and adolescents with callous-unemotional traits and disruptive behavior disorders. Am J Psychiatry 165:712–720
Meehan JC, Holtzworth-Munroe A, Herron K (2001) Maritally violent men's heart rate reactivity to martial interactions: a failure to replicate the Gottman et al. (1995) typology. J Family Psychol 15:394–408
Mezzacappa E, Tremblay RE, Kindlon D, Saul JP, Arseneault L, Seguin J et al (1997) Anxiety, antisocial behavior and heart rate regulation in adolescent males. J Child Psychol Psychiatry 38:457–469
Müller JL, Sommer M, Wagner V, Lange K, Taschler H, Roder CH et al (2003) Abnormalities in emotion processing within cortical and subcortical regions in criminal psychopaths: evidence from a functional magnetic resonance imaging study using pictures with emotional content. Biol Psychiatry 54:152–162
Müller JL, Ganssbauer S, Sommer M, Dohnel K, Weber T, Schmidt-Wilcke T et al (2008a) Gray matter changes in right superior temporal gyrus in criminal psychopaths. Evidence from voxel-based morphometry. Psychiatry Res 163:213–222
Müller JL, Sommer M, Dohnel K, Weber T, Schmidt-Wilcke T, Hajak G (2008b) Disturbed prefrontal and temporal brain function during emotion and cognition interaction in criminal psychopathy. Behav Sci Law 26:131–150
Munro GE, Dywan J, Harris GT, McKee S, Unsal A, Segalowitz SJ (2007) ERN varies with degree of psychopathy in an emotion discrimination task. Biol Psychol 76:31–42
Narayan VM, Narr KL, Kumari V, Woods RP, Thompson PM, Toga AW et al (2007) Regional cortical thinning in subjects with violent antisocial personality disorder or schizophrenia. Am J Psychiatry 164:1418–1427
Nelson LD, Patrick CJ, Bernat EM (2011) Operationalizing proneness to externalizing psychopathology as a multivariate psychophysiological phenotype. Psychophysiology 48:64–72
New AS, Hazlett EA, Buchsbaum MS, Goodman M, Reynolds et al (2002) Blunted prefrontal cortical superscript 1-sup-8 fluorodeoxyglucose positron emission tomography response to meta-chlorophenylpiperazine in impulsive aggression. Arch Gen Psychiatry 59:621–629
Newman JP, Curtin JJ, Bertsch JD, Baskin-Sommers A (2010) Attention moderates the fearlessness of psychopathic offenders. Biol Psychiatry 67:66–70

Ortiz J, Raine A (2004) Heart rate level and antisocial behavior in children and adolescents: a meta-analysis. J Am Acad Child Adolesc Psychiatry 43:154–162
Pailing PE, Segalowitz SJ (2004) The error-related negativity as a state and trait measure: motivation, personality, and ERPs in response to errors. Psychophysiology 41:84–95
Parsey RV, Oquendo MA, Simpson NR, Ogden RT, Van Heertum R, Arango V, Mann JJ (2002) Effects of sex, age, and aggressive traits in man on brain serotonin 5-HT-sub(1A) receptor binding potential measured by PET using [C-11]WAY-100635. Brain Res 954:173–182
Patrick CJ (1994) Emotion and psychopathy: startling new insights. Psychophysiology 31:319–330
Patrick CJ (1995) Emotion and temperament in psychopathy. Clin Sci 5–8
Patrick CJ (2007) Getting to the heart of psychopathy. In: Hervé HF, Yuille JC (eds) Psychopathy: theory, research, and social implications. Erlbaum, Hillsdale, pp 207–252
Patrick CJ (2010) Operationalizing the Triarchic conceptualization of psychopathy: Preliminary description of brief scales for assessment of boldness, meanness, and disinhibition. Unpublished test manual, Florida State University, Tallahassee, FL. Test is available on-line at: https://www.phenxtoolkit.org/index.php?pageLink=browse.protocoldetails&id=121601
Patrick CJ, Bernat EM (2009) Neurobiology of psychopathy: a two-process theory. In: Berntson GG, Cacioppo JT (eds) Handbook of neuroscience for the behavioral sciences. Wiley, New York, pp 1110–1131
Patrick CJ, Bradley MM, Lang PJ (1993) Emotion in the criminal psychopath: startle reflex modulation. J Abnorm Psychol 102:82–92
Patrick CJ, Hicks BM, Krueger RF, Lang AR (2005) Relations between psychopathy facets and externalizing in a criminal offender sample. J Pers Disord 19:339–356
Patrick CJ, Bernat EM, Malone SM, Iacono WG, Krueger RF, McGue M (2006) P300 amplitude as an indicator of externalizing in adolescent males. Psychophysiology 43:84–92
Patrick CJ, Hicks BM, Nichol PE, Krueger RF (2007) A bifactor approach to modeling the structure of the psychopathy checklist-revised. J Pers Disord 21:118–141
Patrick CJ, Fowles DC, Krueger RF (2009) Triarchic conceptualization of psychopathy: developmental origins of disinhibition, boldness, and meanness. Dev Psychopathol 21:913–938
Patrick CJ, Durbin CE, Moser JS (2012) Conceptualizing proneness to antisocial deviance in neurobehavioral terms. Dev Psychopathol 24:1047–1071
Patrick CJ, Kramer MD, Krueger RF, Markon KE (2013a) Optimizing efficiency of psychopathology assessment through quantitative modeling: development of a brief form of the externalizing spectrum inventory. Psychol Assess 25:1332–1348
Patrick CJ, Venables NC, Yancey JR, Hicks BM, Nelson LD, Kramer MD (2013b) A construct-network approach to bridging diagnostic and physiological domains: application to assessment of externalizing psychopathology. J Abnorm Psychol 122:902–916
Peters ML, Godaert GLR, Ballieux RE, Heijnen CJ (2003) Moderation of physiological stress responses by personality traits and daily hassles: less flexibility of immune system responses. Biol Psychol 65:21–48
Polich J (2007) Updating the P300: an integrative theory of P3a and P3b. Clin Neurophysiol 118:2128–2148
Polich J, Pollock VE, Bloom FE (1994) Meta-analysis of P300 amplitude from males at risk for alcoholism. Psychol Bull 115:55–73
Porter S, Woodworth M (2006) Psychopathy and aggression. In: Patrick CJ (ed) Handbook of psychopathy. Guilford Press, New York, pp 481–494
Raine A (1993) The psychopathology of crime. Academic Press, San Diego
Raine A (2002) Biosocial bases of antisocial and violent behavior: a review. J Abnorm Child Psychol 30:311–326
Raine A, Venables PH, Williams M (1990) Relationships between N1, P300, and CNV recorded at age 15 and criminal behavior at age 24. Psychophysiology 27:567–575
Raine A, Buchsbaum MS, Stanley J, Lottenberg S, Abel L, Stoddard J (1994) Selective reductions in pre-frontal glucose metabolism in murderers. Biol Psychiatry 36:365–373

Raine A, Buchsbaum M, LaCasse L (1997) Brain abnormalities in murderers indicated by positron emission tomography. Biol Psychiatry 42:495–508

Raine A, Meloy JR, Bihrle S, Stoddard J, LaCasse L, Buchsbaum MS (1998) Reduced prefrontal and increased subcortical brain functioning assessed using positron emission tomography in predatory and affective murderers. Behav Sci Law 16:319–332

Raine A, Lencz T, Bihrle S, LaCasse L, Colletti P (2000) Reduced prefrontal gray matter volume and reduced autonomic activity in antisocial personality disorder. Arch Gen Psychiatry 57:119–127

Raine A, Park S, Lencz T, Bihrle S, LaCasse L, Widom CS et al (2001) Reduced right hemisphere activation in severely abused violent offenders during a working memory task: an fMRI study. Aggressive Behav 27:111–129

Raine A, Lencz T, Taylor K, Hellige JB, Bihrle S, Lacasse L et al (2003) Corpus callosum abnormalities in psychopathic antisocial individuals. Arch Gen Psychiatry 60:1134–1142

Raine A, Ishikawa SS, Arce E, Lencz T, Knuth KH, Bihrle S, LaCasse L, Colletti P (2004) Hippocampal structural asymmetry in unsuccessful psychopaths. Biological Psychiatry 55: 185–191

Rilling JK, Glenn AL, Jairam MR, Pagnoni G, Goldsmith DR, Elfenbein HA, Lilienfeld SO (2007) Neural correlates of social cooperation and non-cooperation as a function of psychopathy. Biol Psychiatry 61:1260–1271

Scarpa A, Raine A (1997) Psychophysiology of anger and violent behavior. Psychiatric Clin North Am 20:375–394

Schneider F, Habel U, Kessler C, Posse S, Grodd W, Muller-Gartner H (2000) Functional imaging of conditioned aversive emotional responses in antisocial personality disorder. Neuropsychobiology 42:192–201

Sellbom M, Phillips TR (2013) An examination of the triarchic conceptualization of psychopathy in incarcerated and non-incarcerated samples. J Abnorm Psychol 122:208–214

Siever LJ, Buchsbaum MS, New AS, Spiegel-Cohen J, Wei T, Hazlett EA et al (1999) d, l-Fenfluramine response in impulsive personality disorder assessed with [18F]flurodeoxyglucose positron emission tomography. Neuropsychopharmacology 20:413–423

Skeem JL, Johansson P, Andershed H, Kerr M, Eno Louden J (2007) Two subtypes of psychopathic violent offenders that parallel primary and secondary variants. J Abnorm Psychol 116:395–409

Smith TW, Gallo LC (1999) Hostility and cardiovascular reactivity during marital interaction. Psychosom Med 61:436–445

Soderstrom H, Tullberg M, Wikkelsoe C, Ekholm S, Forsman A (2000) Reduced regional cerebral blood flow in non-psychotic violent offenders. Psychiatry Res Neuroimaging 98:29–41

Stanford MS, Houston RJ, Villemarette-Pittman NR, Greve KW (2003) Premeditated aggression: clinical assessment and cognitive psychophysiology. Pers Individ Differ 34:773–781

Strickland CM, Drislane LE, Lucy MD, Krueger RF, Patrick CJ (2013) Characterizing psychopathy using DSM-5 personality traits. Assessment 20:327–338

Suls J, Wan CK (1993) The relationship between trait hostility and cardiovascular reactivity: a quantitative review and analysis. Psychophysiology 30:615–626

Tackett JL, Krueger RF, Sawyer MG, Graetz BW (2003) Subfactors of DSM-IV conduct disorder: Evidence and connections with syndromes from the child behavior checklist. J Abnorm Child Psychol 31:647–654

Tackett JL, Krueger RF, Iacono WG, McGue M (2005) Symptom-based subfactors of DSM-defined conduct disorder: evidence for etiologic distinctions. J Abnorm Psychol 114:483–487

Taylor J, Carlson SR, Iacono WG, Lykken DT, McGue M (1999) Individual differences in electrodermal responsivity to predictable aversive stimuli and substance dependence. Psychophysiology 36:193–198

Tiihonen J, Kuikka J, Bergstrom K, Hakola P, Karhu J, Ryynanen OP, Fohr J (1995) Altered striatal dopamine re-uptake site densities in habitually violent and non-violent alcoholics. Nat Med 1:654–657

Tiihonen J, Kuikka JT, Bergstrom KA, Karhu J, Viinamaki H, Lehtonen J et al (1997) Single-photon emission tomography imaging of monoamine transporters in impulsive violent behaviour. Eur J Nucl Med Mol Imaging 24:1253–1260

Tiihonen J, Rossi R, Laakso MP, Hodgins S, Testa C, Perez J et al (2008) Brain anatomy of persistent violent offenders: more rather than less. Psychiatry Res Neuroimaging 163:201–212

Tonkonogy JM (1991) Violence and temporal lobe lesion: head CT and MRI data. J Neuropsychiatry Clin Neurosci 3:189–196

Vaidyanathan U, Patrick CJ, Bernat EM (2009) Startle reflex potentiation during aversive picture viewing as an index of trait fear. Psychophysiology 46:75–85

Vaidyanathan U, Hall JR, Patrick CJ, Bernat EM (2011) Clarifying the role of defensive reactivity deficits in psychopathy and antisocial personality using startle reflex methodology. J Abnorm Psychol 120:253–258

van Elst LT, Woermann FG, Lemieux L, Thompson PJ, Trimble MR (2000) Affective aggression in patients with temporal lobe epilepsy: a quantitative MRI study of the amygdala. Brain 123:234–243

van Elst LT, Hesslinger B, Thiel T, Geiger E, Haegele K, Lemieux L et al (2003) Frontolimbic brain abnormalities in patients with borderline personality disorder: a volumetric magnetic resonance imaging study. Biol Psychiatry 54:163–171

Vanman EJ, Mejia VY, Dawson ME, Schell AM, Raine A (2003) Modification of the startle reflex in a community sample: do one or two dimensions of psychopathy underlie emotional processing? Pers Individ Differ 35:2007–2021

Veit R, Lotze M, Sewing S, Missenhardt H, Gaber T, Birbaumer N (2010) Aberrant social and cerebral responding in a competitive reaction time paradigm in criminal psychopaths. NeuroImage 49:3365–3372

Venables NC, Patrick CJ (2012) Validating factors of the externalizing spectrum inventory in a criminal offender sample: relations with disinhibitory psychopathology, personality, and psychopathic features. Psychol Assess 24:88–100

Venables NC, Patrick CJ (2014) P3 brain response amplitude in criminal psychopathy: distinct relations with impulsive-antisocial versus affective-interpersonal features. Psychophysiology 51:427–436

Venables NC, Hall JR, Patrick CJ (2014) Differentiating psychopathy from antisocial personality disorder: a Triarchic model perspective. Psychol Med 44:1005–1013

Verona E, Curtin JJ (2006) Gender differences in the negative affective priming of aggression. Emotion 6:115–124

Verona E, Patrick CJ, Joiner TE (2001) Psychopathy, antisocial personality, and suicide risk. J Abnorm Psychol 110:462–470

Verona E, Patrick CJ, Lang AR (2002) A direct assessment of the role of state and trait negative emotion in aggressive behavior. J Abnorm Psychol 111:249–258

Vieira JB, Almeida PR, Ferreira-Santos F, Barbosa F, Marques-Teixeira J, Marsh AA (in press) Distinct neural activation patterns underlie economic decisions in high and low psychopathy scorers. Soc Cogn Affect Neurosci

Volavka J (1990) Aggression, electroencephalography, and evoked potentials: a critical review. Neuropsychiatry Neuropsychology Behav Neurol 3:249–259

Volkow ND, Tancredi L (1987) Neural substrates of violent behaviour: a preliminary study with positron emission tomography. Br J Psychiatry 151:668–673

von Borries AK, Brazil IA, Bulten BH, Buitelaar JK, Verkes RJ, de Bruijn ER (2010) Neural correlates of error-related learning deficits in individuals with psychopathy. Psychol Med 40:1559–1568

Wall TD, Wygant DB, Sellbom M (in press) Boldness explains a key difference between psychopathy and antisocial personality disorder. Psychiatry Psychol Law

Williams KM, Paulhus DL, Hare RD (2007) Capturing the four-factor structure of psychopathy in college students via self-report. J Pers Assess 88:205–219

Woermann FG, Van Elst LT, Koepp MJ, Free SL, Thompson PJ, Trimble MR, Duncan JS (2000) Reduction of frontal neocortical grey matter associated with affective aggression in patients with temporal lobe epilepsy: an objective voxel by voxel analysis of automatically segmented MRI. J Neurol Neurosurg Psychiatry 68:162–169
Wong MTH, Lumsden J, Fenton GW, Fenwick PBC (1994) Electroencephalography, computed tomography and violence ratings of male patients in a maximum-security mental hospital. Acta Psychiatr Scand 90:97–101
Yancey JR, Venables NC, Hicks BM, Patrick CJ (2013) Evidence for a heritable brain basis to deviance-promoting deficits in self-control. J Crim Justice 41:309–317
Yang Y, Raine A, Lencz T, Bihrle S, LaCasse L, Colletti P (2005) Volume reduction in prefrontal gray matter in unsuccessful criminal psychopaths. Biol Psychiatry 57:1103–1108
Yang Y, Raine A, Narr KL, Colletti P, Toga AW (2009) Localization of deformations within the amygdala in individuals with psychopathy. Arch Gen Psychiatry 66:986–994
Young SE, Stallings MC, Corley RP, Krauter KS, Hewitt JK (2000) Genetic and environmental influences on behavioral disinhibition. Am J Med Gen (Neuropsychiatric Genetics) 96:684–695
Zuckerman M (1979) Sensation seeking: beyond the optimal level of arousal. Erlbaum, Hillsdale

Physiological Correlates of Premenstrual Dysphoric Disorder (PMDD)

Inger Sundström Poromaa

Abstract Premenstrual dysphoric disorder (PMDD) is a mood disorder with onset of functionally impairing or distressing mood symptoms in the late luteal phase of the menstrual cycle. Psychophysiologic findings in PMDD broadly fall into two categories: vulnerability trait findings, thus categorized because they are present in the asymptomatic phases of the menstrual cycle, and state findings, which are only present in the symptomatic late luteal phase and which are potentially representative of the hormonal events and biological mechanisms that lead to PMDD. Trait vulnerability markers in PMDD include diminished cardiovascular stress responses, lower heart rate variability (reflecting increased vagal tone), and lower P300 amplitude, eventually suggesting that women with PMDD share a number of physiological correlates with related anxiety and mood disorders. State findings in PMDD include lower luteal phase prepulse inhibition and altered luteal phase emotion processing. Lower prepulse inhibition in the late luteal phase may be an important ovarian steroid-influenced indicative of altered serotonergic neurotransmission, of relevance for women with PMDD. Attempts to determine the neural correlates of emotion processing in the late luteal phase are thus far inconsistent, but promising.

Keywords Premenstrual dysphoric disorder · Estradiol · Progesterone · Prepulse inhibition · Emotion processing · Functional magnetic resonance imaging

Contents

I. S. Poromaa (✉)
Uppsala University, Uppsala, Sweden
e-mail: inger.sundstrom@kbh.uu.se

Curr Topics Behav Neurosci (2015) 21: 229–243
DOI: 10.1007/7854_2014_296

Published Online: 4 March 2014

1 Introduction

Premenstrual dysphoric disorder (PMDD) is categorized as a mood disorder with onset of functionally impairing or distressing mood and physical symptoms in the late luteal phase of the menstrual cycle, a decline in symptom severity after onset of menstruation, and an absence of symptoms in the postmenstrual week (O'Brien et al. 2011). Hallmark mood symptoms include mood lability, irritability, anxiety, tension, and depression (O'Brien et al. 2011). The disorder affects roughly 5 % of women of reproductive age (Wittchen et al. 2002), and has a moderate heritability (Kendler et al. 1998).

PMDD is defined by the relation to the late luteal phase of the menstrual cycle. As progesterone is only present in the luteal phase, PMDD is commonly regarded as a disorder caused by the variation in (or mere presence) of progesterone levels. Research in support of this include findings of symptom relief during anovulatory cycles (Wyatt et al. 2004), the reinstatement of symptoms when add-back hormone therapy is administered together with gonadotropin-releasing hormone (GnRH) agonists (Segebladh et al. 2009), and findings of progestagen-induced mood symptoms in postmenopausal women (Andreen et al. 2003, 2005, 2006). Notably, no consistent hormonal differences between women with PMDD and healthy controls have been reported (Backstrom et al. 2003). Hence, the exact mechanism by which progesterone precipitates the symptoms of PMDD is unknown, although interactions with the serotonin (Jovanovic et al. 2006; Brown et al. 2009) and the GABAergic systems (Epperson et al. 2002; Sundstrom Poromaa et al. 2003) are plausible.

Estradiol and progesterone are both highly lipophilic and easily pass through the blood–brain barrier. In fact, animal studies and post-mortem studies in reproductive and postmenopausal women indicate that estradiol and progesterone are accumulated in the brain (Bixo et al. 1986, 1995, 1997), with the highest concentration of progesterone found in the amygdala (Bixo et al. 1997). The estradiol receptors (ERα and ERβ) and the progesterone receptors (PRA and PRB) are highly expressed in brain areas associated with reproductive function, such as the hypothalamus and the limbic system (for review, see (Gruber et al. 2002; Brinton et al. 2008). For example, the expression of the estradiol receptors has been demonstrated in the human amygdala, hippocampus, claustrum, hypothalamus, and the cerebral cortex. Within the cerebral cortex, the most distinct expression of estradiol receptors is found in the temporal cortex (Osterlund et al. 2000a, b). The progesterone receptors are also distributed throughout the amygdala, hippocampus, hypothalamus, thalamus, and the frontal cortex (Kato et al. 1994; Bixo et al. 1997; Guerra-Araiza et al. 2000, 2002, 2003). The receptor distribution suggests that the ovarian hormones have the capability to modulate emotional processing, cognitive function, sensory input, attention, decision making, and motor function. Progesterone can also be metabolized into neuroactive steroids, among which allopregnanolone and pregnanolone are the two neurosteroids most studied in PMDD. Neurosteroids potentiate the $GABA_A$ receptor, where they increase hyperpolarization and act in a similar manner to barbiturates and benzodiazepines (Melcangi et al. 2011). As GABA is the major inhibitory transmitter in the central nervous system, acute administration of allopregnanolone has sedative, anxiolytic, and anticonvulsant properties (Melcangi et al. 2011). A functionally relevant amount of allopregnanolone is synthesized in the brain, but the main source of brain and serum allopregnanolone in nonpregnant women is the corpus luteum (Ottander et al. 2005).

Over the past years menstrual cycle studies have greatly improved in quality by use of counterbalanced, longitudinal designs, and by correct classification of menstrual cycle stage by use of hormonal levels and assessment of the luteinizing hormone surge rather than just the onset of menses. For this review, only studies where these design standards have been met have been included. The most relevant menstrual cycle contrast in PMDD women is obviously the late follicular phase (when patients are symptom-free) and the late luteal phase (when symptoms culminate). In addition, this review has only considered studies where PMDD diagnosis is based on criteria devised by the Diagnostic and Statistical Manual of Mental Disorders (DSM). Besides the typical symptom profile and a clear-cut impact on daily life, the diagnosis must also be confirmed by use of prospective daily symptom ratings. Unfortunately, many PMDD studies typically include relatively small sample sizes. While this is a problem in terms of statistical power or high probability of chance findings, it is also understandable as diagnostic procedures are time consuming and a large number of women need to be screened before the target population is reached.

2 Trait Vulnerability Markers

Life-time comorbidity with depressive and anxiety disorders is common in women with PMDD (Wittchen et al. 2002, 2003), and many PMDD women display typical vulnerability markers for psychiatric morbidity such as neurotic personality traits (Freeman et al. 1995; Gingnell et al. 2010) and history of violence or abuse (Girdler et al. 2003). Furthermore, several biochemical and neurophysiological abnormalities encountered in depressed or anxious patients are also found in women with PMDD. If such findings are present in the asymptomatic phases of the menstrual cycle, they are commonly regarded as vulnerability traits for PMDD, or alternatively, vulnerability traits for the depressive and anxiety disorders that are commonly associated with PMDD.

2.1 Cardiovascular Stress Reactivity

Many women with PMDD state that their symptoms worsen during time periods of intense workload and stress (Sadler et al. 2010), and hypothetically, women with PMDD could have an altered stress response in comparison with healthy women (Perkonigg et al. 2004; Sadler et al. 2010). In a series of studies, Girdler and colleagues have reported on cardiovascular stress responses in healthy women and PMDD women across the menstrual cycle. Besides a lack of menstrual cycle influence on stress-induced hemodynamic responses such as cardiac output and peripheral resistance, women with PMDD also showed diminished diastolic blood pressure and heart rate responses and a tendency to blunted cardiac output to the stress tests, irrespective of cycle phase (Girdler et al. 1993). These findings were later partly replicated in a study where women with PMDD in both cycle phases had lower cardiac output and stroke volume and higher peripheral resistance during mental stress than healthy controls (Girdler et al. 1998). Further attempts to replicate these findings have either failed (van den Akker and Steptoe 1989; Girdler et al. 2003), or only been confirmed in women with PMDD who also had a history of abuse (Girdler et al. 2007). Although the positive physiological findings that have been reported are suggestive of trait vulnerability markers, it should also be noted that endocrine measures of stress, such as allopregnanolone levels, are elevated in PMDD women in the luteal phase (Girdler et al. 2001).

Heart rate variability (HRV) measures provide a sensitive noninvasive measure of cardiac autonomic regulation via the parasympathetic (vagal) and sympathetic nervous systems. Time domain variables of beat-to-beat variability of the heart rhythm are assumed to reflect mainly vagal tone while frequency domain variables, derived from power spectral analysis, can be used to distinguish between sympathetic and vagal predominance. While two studies have failed to demonstrate any difference between PMDD patients and controls (although within group differences in responsiveness to the menstrual cycle changes were reported) (Baker et al. 2008; de Zambotti et al. 2013), two studies have suggested lower heart rate variability, reflecting increased vagal tone, in PMDD women (Landen et al. 2004; Matsumoto et al. 2007). Again, these findings were not confined to the symptomatic late luteal phase. Instead, women with PMDD appear to have lower HRV indices across both cycle phases (Matsumoto et al. 2007) or only in the follicular phase (Landen et al. 2004). These findings support the notion that reduced HRV is a feature shared by a number of related psychiatric disorders that are characterized by symptoms such as depressed mood and anxiety, including PMDD. In addition, these findings imply that women with lower autonomic function regardless of the menstrual cycle are vulnerable to more severe premenstrual disorders.

2.2 Attention and Alertness

Event-related potentials (ERPs) are electroencephalogram (EEG) changes that are time locked to a stimulus event. One of the most prominent components of the ERP is the positive P300 waveform that occurs between 300 and 500 ms after stimulus onset. The P300 is thought to reflect higher processing of the psychological meaning of stimuli and the P300 amplitude is affected by numerous factors, including task or stimulus complexity, expectancy, vigilance, and attention. Two studies have investigated the P300 in women with PMDD, one reporting lower P300 amplitude to both auditory and visual stimuli in both menstrual cycle phases (Baker et al. 2010), and the other reporting longer P300 latency to auditory stimuli across both cycle phases in PMDD women (Ehlers et al. 1996). As PMDD women in the study by Baker and colleagues also performed poorly on a psychomotor vigilance task in the luteal phase, the authors suggested that women with PMDD allocated less attentional resources to the tasks than did controls at both symptomatic and symptom-free menstrual cycle phases, although response output processing (reaction times on the PVT) was only affected in the symptomatic phase (Ehlers et al. 1996).

Despite common reports of fatigue during the luteal phase, no difference in waking EEG power density in the theta/alpha range, as possible indicators of alertness (Cajochen et al. 1997; Baker et al. 2010), has been noted in PMDD women. Women with PMDD, however, are reported to have a lower saccadic eye velocity across both cycle phases (Sundstrom and Backstrom 1998).

2.3 Startle Response

The acoustic startle reflex is a withdrawal reflex to sudden or noxious auditory stimuli which can be measured as an eyeblink in humans or as a whole-body response in laboratory animals. This paradigm is a useful bridge between preclinical and human data, since it has a similar circuitry and pharmacology in humans as it does in animals. While healthy female controls do not show cyclic changes in this measure of physiologic arousal (Epperson et al. 2007), PMDD is associated with an increase in baseline startle magnitude, although it is unclear if this present in both cycle phases (Kask et al. 2008) or confined to the symptomatic late luteal phase (Epperson et al. 2007). Again, increased startle response is a feature shared by a range of anxiety disorders, but the relevance in PMDD is strengthened by the influence of ovarian steroids on the acoustic startle reflex (Toufexis et al 1999; Van den Buuse and Eikelis 2001; Vaillancourt et al. 2002; Byrnes et al. 2007) and by the fact that the startle response is increased in an animal model of PMDD (the progesterone withdrawal model) (Gulinello et al. 2003; Gulinello and Smith 2003). In addition, the acoustic startle response is also regulated by the agents thought to be critical to the etiology PMDD symptoms, notably progesterone fluctuations and underlying alterations in inhibitory neurotransmission via the $GABA_A$ receptor (Gulinello et al. 2003; Gulinello and Smith 2003; Toufexis et al. 2004).

3 State Findings

As understood by the previous section, relatively few psychophysiology findings in PMDD women have been confined to the symptomatic late luteal phase. While trait findings are important for the overall understanding of PMDD, psychophysiology findings during the symptomatic phase would aid in our understanding of the biological mechanisms that lead to the symptom surfacing in the premenstrual phase.

3.1 Prepulse Inhibition

Prepulse inhibition of the startle magnitude is a sexually dimorphic measure which also varies across the menstrual cycle. In fact, it is one of the most reliable and consistent menstrual cycle findings that PPI is lower at times when estradiol and progesterone levels are high, such as during the mid-luteal phase of the menstrual cycle and in pregnancy (reviewed in (Kumari 2011)). In addition, women of childbearing ages have lower PPI than postmenopausal women (Bannbers et al. 2011), and the sex difference also disappears following menopause (Kumari et al. 2008).

Women with PMDD patients exhibit lower levels of PPI compared to control subjects in the luteal but not in the follicular phase (Kask et al. 2008). Furthermore, PMDD patients with pronounced anxiety and depression symptoms during the cycle in which they were tested had even more impaired PPI than less symptomatic patients (Kask et al. 2008). As variable hormone levels from menstrual cycle to menstrual cycle within individual subjects may result in variable symptom expressions (Wang et al. 1996), this finding underlies the assumption that the hormonal events that trigger PMDD symptoms in a specific menstrual cycle could also affect the circuits modulating PPI. The relevance of this measure for PMDD symptom expression is further strengthened by similar findings of low PPI in women suffering from depression and irritability while using combined oral contraceptives (Borgstrom et al. 2008).

While progesterone appears to be the most relevant hormone for PMDD, estradiol has been of greater interest in the field of PPI, presumably due to its role in schizophrenia (Gogos and Van den Buuse 2004; Gogos et al. 2006a, b, 2009; Guille et al. 2011; Hill et al. 2013; Thwaites et al. 2014; Wu et al. 2013). Menstrual phase-related variability in PPI has been suggested to be mediated by fluctuating estrogen level, based on the observations of more PPI in women during the follicular, relative to the luteal, phase. Both estrogen receptors are found in the nucleus accumbens and amygdala (Gruber et al. 2002; Brinton et al. 2008) and other areas of the PPI circuit (Charitidi et al. 2012), and ERα has been suggested to play a key role (Charitidi et al. 2012). Estrogen induces a dose-dependent increase in PPI in ovariectomized rats (Van den Buuse and Eikelis 2001; Charitidi et al. 2012) and estrogen (or combined estrogen-progestagen) treatment in ovariectomized female rats may also prevent 5-HT_{1A}–, dopamine-, and NMDA receptor-induced disruptions of PPI (Gogos and Van den Buuse 2004; Gogos et al. 2010, 2012; Thwaites et al. 2014). Similar findings have also been obtained in females where estrogen treatment prevented buspirone-induced PPI deficits (Gogos et al. 2006a). However, treatment with 2 mg estradiol during the early follicular phase did not affect PPI in healthy women (Gogos et al. 2006b), and no direct correlations between estradiol levels and PPI have been reported (Kask et al. 2008b; Kumari et al. 2008,2010; Talledo et al. 2009; Bannbers et al. 2011). Hence, an alternative explanation to the menstrual cycle effects and the PMDD findings be equally well be that of progesterone-induced inhibition of estradiol effects in the luteal phase. Recently, a role for progesterone in PPI was also suggested (Kumari et al. 2010) as a larger increase in progesterone was associated with a smaller decrease in PPI from the follicular to the luteal phase, which could be of relevance for PMDD women. This effect is presumably mediated by the progesterone receptors, as GABA-active progesterone metabolites have no influence on PPI (Kask et al. 2009).

In terms of PMDD pathophysiology, the finding of reduced PPI in the symptomatic late luteal phase may suggest altered serotonergic or dopaminergic neurotransmission. Of these two, the serotonin system is by far the most researched in PMDD. A role for serotonin in PMDD is predominantly suggested by the fact that serotonin reuptake inhibitors (SSRI) can be used for treatment (Marjoribanks et al. 2013). Not only so, the SSRIs appear to have a distinctly different route of action in PMDD, as opposed to when it is used for depressive or anxiety disorders. For instance, it is equally effective when used intermittently, i.e., only during the luteal phase, or continuously, and the onset of action is reportedly as short as 14 hours after first drug intake (Landen et al. 2009). These temporal relationships suggest that SSRIs may facilitate serotonin transmission shortly after the onset of treatment in anger-modulating pathways, by increasing synaptic levels of serotonin (Landen et al. 2009). Further pharmacological challenges in PMDD women are, however, needed to establish the role of the serotonin system for the impaired late luteal phase PPI in PMDD women.

3.2 Emotion Processing: Startle Reactivity

Startle reactivity may also be used as an measure of emotional processing (Lang et al. 2000). Animal studies as well as human studies show the acoustic startle response to be enhanced during arousal and fearful situations, such as during threat of shock, noxious noise, or aversive pictures, while it is reduced when presented with rewarding stimuli such as pictures of food or erotica (Lang and Davis 2006). Emotion-induced modulation of the startle response, as far as it has been studied, appears not to be different in women with PMDD, either in comparison with healthy controls or across menstrual cycle phases (Epperson et al. 2007; Bannbers et al. 2011). However, PMDD patients display an increased startle modulation by positive and negative anticipation stimuli in the late luteal phase. Possibly one reason why anticipation, as opposed to image viewing, results in an increased startle response in women with PMDD, could be that startle reactivity in this case is not dependent upon the image itself; thus differences in how subjects respond to a particular image based on different life experiences do not confound the interpretation of startle effects. The latter may be of specific relevance in PMDD women, as women with a history of trauma, or PTSD, are more likely to experience PMDD (Wittchen et al. 2003; Pilver et al. 2011), and as PMDD women with a history of trauma have an abnormal neuroendocrine stress response (Girdler et al. 1998, 2003, 2007; Segebladh et al. 2011). Anticipation of negative events may be adaptive and promote behavior that increases the possibility for survival in response to threat, but could also be dysfunctional and initiate anxious responses also to nonthreatening stimuli (Grupe and Nitschke 2013). In the context of PMDD, the anticipation of the luteal phase symptoms has been shown to influence the severity of symptoms (Segebladh et al. 2009). Further, the increased startle modulation by positive and negative anticipation stimuli have later been validated by functional magnetic resonance imaging (fMRI), where women with PMDD had higher luteal phase reactivity in the anterior medial prefrontal cortex and dorsolateral prefrontal cortex during negative anticipation than healthy controls, while they did not differ from healthy controls in their response to the emotional images (Gingnell et al. 2013).

3.3 Emotion Processing: Functional Magnetic Resonance Imaging

A growing number of studies have attempted to pinpoint the neural correlates of the late luteal phase emotion processing in PMDD women. Most attention has been given to the hypothesized corticolimbic emotion processing network (reviewed by (Shin and Liberzon 2010)), where the amygdala and insula are activated by bottom-up emotional processes and the anterior cingulate cortex involved in top-down regulation. While increased amygdala reactivity characterizes negative affective states like anxiety and depression (Etkin and Wager 2007; Shin and Liberzon 2010), studies on amygdala reactivity in PMDD have thus far been inconsistent (Protopopescu et al. 2008; Gingnell et al. 2012, 2013), possibly because none of the paradigms used have tapped into the PMDD-specific symptomatology. Protopopescu et al. (2008) reported increased late luteal amygdala reactivity in response to emotional words, but the results appeared to reflect alterations in reactivity over the menstrual cycle in healthy controls rather than in women with PMDD. Gingnell et al. (2012) reported a follicular phase increase in amygdala reactivity to emotional faces, whereas the expected luteal phase increase was only noted in a subgroup of PMDD patients who also had high trait anxiety. Furthermore, no differences in the amygdala and insula or ACC reactivity between patients and controls and no menstrual phase modulation were observed to negative emotional stimuli (Gingnell et al. 2013). Although these findings appear inconsistent and not specific to the late luteal phase, some details may be of importance for the understanding of PMDD. First, the increased bilateral amygdala reactivity in PMDD women during the follicular phase noted by Gingnell and colleagues was highly and positively correlated with progesterone levels, which was not the case in the healthy controls. While progesterone levels at this point of the menstrual cycle are low, and also not associated with mood worsening, women with PMDD are in general more sensitive to progesterone than controls (Schmidt et al. 1998). Hence, this finding may suggest that the amygdalae in PMDD women respond already at very low levels of progesterone. Indeed, the progesterone levels needed to induce amygdala reactivity in healthy women are approximately 20-fold higher (van Wingen et al. 2008). As progesterone receptors are present throughout the limbic system (Kato et al. 1994; Bixo et al. 1997; Guerra-Araiza et al. 2000, 2002, 2003) and progesterone concentrations are high in the amygdala (Bixo et al. 1997), the follicular phase increase in amygdala reactivity might reflect a trait-like sensitivity in the amygdala to low levels of progesterone, which is abolished in the luteal phase when ovarian steroid levels are increased. It has been suggested that tolerance to progesterone may develop during the luteal phase and that women with PMDD have an aberrant tolerance development (Backstrom et al. 2014). Although this hypothesis mainly relies on findings of GABAergic progesterone metabolites [GABA plays a major role in amygdala functioning (Roberto et al. 2012)], it is equally possible that progesterone receptors in the CNS may be saturated during the luteal phase and direct associative couplings between amygdala reactivity and small stress-induced changes in progesterone may not be detectable.

Yet another important issue is that the emotional stimuli used in startle and fMRI experiments may not have been specific enough to discriminate the PMDD women from healthy controls in the luteal phase, especially since amygdala reactivity is increased in the luteal phase already in healthy women (Andreano and Cahill 2010; Goldstein et al. 2010; Gingnell et al. 2012). Using a paradigm that contrasted brain reactivity in response to social as opposed to nonsocial aversive stimuli, women with PMDD displayed increased amygdala and insular cortex reactivity during the luteal phase which was paralleled by a reduced reactivity in areas previously reported to be involved in emotion regulation, i.e., the anterior cingulate cortex (Gingnell in press). Among women with PMDD reactivity in the amygdala increased between the follicular and luteal phase and was positively correlated to the progesterone increase across cycle phases (Gingnell in manuscript). Reactivity alterations in PMDD to social stimuli might thus be similar to anxiety disorders with increased reactivity in the amygdala and insula but decreased reactivity in the regulatory parts of ACC which is restricted to symptom provoking situations (Shin and Liberzon 2010).

Attempts to use cognitive paradigms have thus far yielded trait-like findings across both cycle phases. For instance, Baller and colleagues (Baller et al. 2013) reported increased activity in the prefrontal cortex of women with PMDD as compared to healthy controls during a short-term memory task, but this effect was unrelated to hormonal levels. Using a nonemotional Go/NoGo-task, women with PMDD have also been reported to have lower parietal reactivity than healthy controls regardless of menstrual cycle phase (Bannbers et al. 2012).

4 Conclusion

Although PMDD is common, and although it is an important model for our understanding of how ovarian steroids influence mood and anxiety in women, the number of studies attempting to elucidate its pathophysiology is strikingly low. Most physiologic findings in PMDD women, such as altered cardiovascular responses to stress or decreased heart rate variability, have been reported in both cycle phases, suggesting that they represent vulnerability traits for PMDD, or alternatively, vulnerability traits for the depressive and anxiety disorders that are commonly associated with PMDD. Relatively few, and yet for the most part, unconfirmed findings have pinpointed the late luteal phase as the crucial determinant for differences between women with PMDD and healthy controls. Among these, lower prepulse inhibition in the late luteal phase may be an important ovarian steroid-influenced indicative of altered serotonergic neurotransmission, of relevance not only for women with PMDD but also for women suffering from mood and anxiety disorders. Attempts to determine the neural correlates of emotion processing in the late luteal phase are thus far inconsistent, but promising.

References

Andreano JM, Cahill L (2010) Menstrual cycle modulation of medial temporal activity evoked by negative emotion. Neuroimage 53:1286–1293

Andreen L, Bixo M, Nyberg S et al (2003) Progesterone effects during sequential hormone replacement therapy. Eur J Endocrinol 148:571–577

Andreen L, Sundstrom-Poromaa I, Bixo M et al (2005) Relationship between allopregnanolone and negative mood in postmenopausal women taking sequential hormone replacement therapy with vaginal progesterone. Psychoneuroendocrino 30:212–224

Andreen L, Sundstrom-Poromaa I, Bixo M et al (2006) Allopregnanolone concentration and mood–a bimodal association in postmenopausal women treated with oral progesterone. Psychopharmacology (Berl) 187:209–221

Backstrom T, Andersson A, Andree L et al (2003) Pathogenesis in menstrual cycle-linked CNS disorders. Ann N Y Acad Sci 1007:42–53

Backstrom T, Bixo M, Johansson M et al (2014) Allopregnanolone and mood disorders. Prog Neurobiol 113:88–94

Baker FC, Colrain IM (2010) Daytime sleepiness, psychomotor performance, waking EEG spectra and evoked potentials in women with severe premenstrual syndrome. J Sleep Res 19:214–227

Baker FC, Colrain IM, Trinder J (2008) Reduced parasympathetic activity during sleep in the symptomatic phase of severe premenstrual syndrome. J Psychosom Res 65:13–22

Baller EB, Wei SM, Kohn PD et al (2013) Abnormalities of dorsolateral prefrontal function in women with premenstrual dysphoric disorder: a multimodal neuroimaging study. Am J Psychiatry 170:305–14

Bannbers E, Gingnell M, Engman J et al (2012) The effect of premenstrual dysphoric disorder and menstrual cycle phase on brain activity during response inhibition. J Affect Disord 142:347–350

Bannbers E, Kask K, Wikstrom J et al (2011) Patients with premenstrual dysphoric disorder have increased startle modulation during anticipation in the late luteal phase period in comparison to control subjects. Psychoneuroendocrino 36:1184–1192

Bixo M, Andersson A, Winblad B et al (1997) Progesterone, 5alpha-pregnane-3,20-dione and 3alpha-hydroxy-5alpha-pregnane-20-one in specific regions of the human female brain in different endocrine states. Brain Res 764:173–178

Bixo M, Backstrom T, Winblad B et al (1995) Estradiol and testosterone in specific regions of the human female brain in different endocrine states. J Steroid Biochem Mol Biol 55:297–303

Bixo M, Backstrom T, Winblad B et al (1986) Comparison between pre- and postovulatory distributions of oestradiol and progesterone in the brain of the PMSG-treated rat. Acta Physiol Scand 128:241–246

Borgstrom A, Kask K, Gulinello M et al (2008) Patients with adverse mood effects from combined oral contraceptives have lower levels of prepulse inhibition than healthy controls. Psychoneuroendocrino 33:487–496

Brinton RD, Thompson RF, Foy MR et al (2008) Progesterone receptors: form and function in brain. Front Neuroendocrinol 29:313–339

Brown J, O'Brien PM, Marjoribanks J et al (2009) Selective serotonin reuptake inhibitors for premenstrual syndrome. Cochrane Database Syst Rev CD001396

Byrnes EM, Bridges RS, Scanlan VF et al (2007) Sensorimotor gating and dopamine function in postpartum rats. Neuropsychopharmacolog 32:1021–1031

Cajochen C, Krauchi K, Wirz-Justice A (1997) The acute soporific action of daytime melatonin administration: effects on the EEG during wakefulness and subjective alertness. J Biol Rhythms 12:636–643

Charitidi K, Meltser I, Canlon B (2012) Estradiol treatment and hormonal fluctuations during the estrous cycle modulate the expression of estrogen receptors in the auditory system and the prepulse inhibition of acoustic startle response. Endocrinology 153:4412–4421

de Zambotti M, Nicholas CL, Colrain IM et al (2013) Autonomic regulation across phases of the menstrual cycle and sleep stages in women with premenstrual syndrome and healthy controls. Psychoneuroendocrinology 38:2618–2627

Ehlers CL, Phillips E, Parry BL (1996) Electrophysiological findings during the menstrual cycle in women with and without late luteal phase dysphoric disorder: relationship to risk for alcoholism? Biol Psychiatry 39:720–732

Epperson CN, Haga K, Mason GF et al (2002) Cortical gamma-aminobutyric acid levels across the menstrual cycle in healthy women and those with premenstrual dysphoric disorder: a proton magnetic resonance spectroscopy study. Arch Gen Psychiatry 59:851–858

Epperson CN, Pittman B, Czarkowski KA et al (2007) Luteal-phase accentuation of acoustic startle response in women with premenstrual dysphoric disorder. Neuropsychopharmacology 32:2190–2198

Etkin A, Wager TD (2007) Functional neuroimaging of anxiety: a meta-analysis of emotional processing in PTSD, social anxiety disorder, and specific phobia. Am J Psychiatry 164: 1476–1488

Freeman EW, Schweizer E, Rickels K (1995) Personality factors in women with premenstrual syndrome. Psychosom Med 57:453–459
Gingnell M, Ahlstedt V, Bannbers E et al (2014) Social stimulation and corticolimbic reactivity in premenstrual dysphoric disorder: a preliminary study. Biol Mood Anxiety Disord 4:3
Gingnell M, Bannbers E, Wikstrom J et al (2013) Premenstrual dysphoric disorder and prefrontal reactivity during anticipation of emotional stimuli. Eur Neuropsychopharmacol 23:1474–1483
Gingnell M, Comasco E, Oreland L et al (2010) Neuroticism-related personality traits are related to symptom severity in patients with premenstrual dysphoric disorder and to the serotonin transporter gene-linked polymorphism 5-HTTPLPR. Arch Womens Ment Health 13:417-423
Gingnell M, Morell A, Bannbers E et al (2012) Menstrual cycle effects on amygdala reactivity to emotional stimulation in premenstrual dysphoric disorder. Horm Behav 62:400–406
Girdler SS, Leserman J, Bunevicius R et al (2007) Persistent alterations in biological profiles in women with abuse histories: influence of premenstrual dysphoric disorder. Health Psychol 26:201–213
Girdler SS, Pedersen CA, Stern RA et al (1993) Menstrual cycle and premenstrual syndrome: modifiers of cardiovascular reactivity in women. Health Psychol 12:180–192
Girdler SS, Pedersen CA, Straneva PA et al (1998) Dysregulation of cardiovascular and neuroendocrine responses to stress in premenstrual dysphoric disorder. Psychiatry Res 81:163–178
Girdler SS, Sherwood A, Hinderliter AL et al (2003) Biological correlates of abuse in women with premenstrual dysphoric disorder and healthy controls. Psychosom Med 65:849–856
Girdler SS, Straneva PA, Light KC et al (2001) Allopregnanolone levels and reactivity to mental stress in premenstrual dysphoric disorder. Biol Psychiatry 49:788–797
Gogos A, Kwek P, Chavez C et al (2010) Estrogen treatment blocks 8-hydroxy-2-dipropylaminotetralin- and apomorphine-induced disruptions of prepulse inhibition: involvement of dopamine D1 or D2 or serotonin 5-HT1A, 5-HT2A, or 5-HT7 receptors. J Pharmacol Exp Ther 333:218–227
Gogos A, Kwek P, den Buuse M (2012) The role of estrogen and testosterone in female rats in behavioral models of relevance to schizophrenia. Psychopharmacology (Berl) 219:213–224
Gogos A, Martin S, Jones ME et al (2006a) Oestrogen modulation of the effect of 8-OH-DPAT on prepulse inhibition: effects of aromatase deficiency and castration in mice. Psychopharmacology (Berl) 188:100–110
Gogos A, Nathan PJ, Guille V et al (2006b) Estrogen prevents 5-HT1A receptor-induced disruptions of prepulse inhibition in healthy women. Neuropsychopharmacology 31:885–889
Gogos A, Van den Buuse M (2004) Estrogen and progesterone prevent disruption of prepulse inhibition by the serotonin-1A receptor agonist 8-hydroxy-2-dipropylaminotetralin. J Pharmacol Exp Ther 309:267–274
Gogos A, van den Buuse M, Rossell S (2009) Gender differences in prepulse inhibition (PPI) in bipolar disorder: men have reduced PPI, women have increased PPI. Int J Neuropsychopharmacol 12:1249–1259
Goldstein JM, Jerram M, Abbs B et al (2010) Sex differences in stress response circuitry activation dependent on female hormonal cycle. J Neurosci 30:431–438
Gruber CJ, Tschugguel W, Schneeberger C et al (2002) Production and actions of estrogens. N Engl J Med 346:340–352
Grupe DW, Nitschke JB (2013) Uncertainty and anticipation in anxiety: an integrated neurobiological and psychological perspective. Nat Rev Neurosci 14:488–501
Guerra-Araiza C, Cerbon MA, Morimoto S et al (2000) Progesterone receptor isoforms expression pattern in the rat brain during the estrous cycle. Life Sci 66:1743–1752
Guerra-Araiza C, Coyoy-Salgado A, Camacho-Arroyo I (2002) Sex differences in the regulation of progesterone receptor isoforms expression in the rat brain. Brain Res Bull 59:105–109
Guerra-Araiza C, Villamar-Cruz O, Gonzalez-Arenas A et al (2003) Changes in progesterone receptor isoforms content in the rat brain during the oestrous cycle and after oestradiol and progesterone treatments. J Neuroendocrinol 15:984–990

Guille V, Gogos A, Nathan PJ et al (2011) Interaction of estrogen with central serotonergic mechanisms in human sensory processing: loudness dependence of the auditory evoked potential and mismatch negativity. J Psychopharmacol 25:1614–1622
Gulinello M, Orman R, Smith SS (2003) Sex differences in anxiety, sensorimotor gating and expression of the alpha4 subunit of the GABAA receptor in the amygdala after progesterone withdrawal. Eur J Neurosci 17:641–648
Gulinello M, Smith SS (2003) Anxiogenic effects of neurosteroid exposure: sex differences and altered GABAA receptor pharmacology in adult rats. J Pharmacol Exp Ther 305:541-548
Hill RA, Wu YW, Gogos A et al (2013) Sex-dependent alterations in BDNF-TrkB signaling in the hippocampus of reelin heterozygous mice: a role for sex steroid hormones. J Neurochem 126:389–399
Jovanovic H, Cerin A, Karlsson P et al (2006) A PET study of 5-HT1A receptors at different phases of the menstrual cycle in women with premenstrual dysphoria. Psychiatry Res 148:185–193
Kask K, Backstrom T, Gulinello M et al (2008a) Lower levels of prepulse inhibition of startle response in pregnant women compared to postpartum women. Psychoneuroendocrinology 33:100–107
Kask K, Backstrom T, Lundgren P et al (2009) Allopregnanolone has no effect on startle response and prepulse inhibition of startle response in patients with premenstrual dysphoric disorder or healthy controls. Pharmacol Biochem Behav 92:608–613
Kask K, Gulinello M, Backstrom T et al (2008b) Patients with premenstrual dysphoric disorder have increased startle response across both cycle phases and lower levels of prepulse inhibition during the late luteal phase of the menstrual cycle. Neuropsychopharmacology 33:2283–2290
Kato J, Hirata S, Nozawa A et al (1994) Gene expression of progesterone receptor isoforms in the rat brain. Horm Behav 28:454–463
Kendler KS, Karkowski LM, Corey LA et al (1998) Longitudinal population-based twin study of retrospectively reported premenstrual symptoms and lifetime major depression. Am J Psychiatry 155:1234–1240
Kumari V (2011) Sex differences and hormonal influences in human sensorimotor gating: implications for schizophrenia. Curr Top Behav Neurosci 8:141–154
Kumari V, Aasen I, Papadopoulos A et al (2008) A comparison of prepulse inhibition in pre- and postmenopausal women and age-matched men. Neuropsychopharmacology 33:2610–2618
Kumari V, Konstantinou J, Papadopoulos A et al (2010) Evidence for a role of progesterone in menstrual cycle-related variability in prepulse inhibition in healthy young women. Neuropsychopharmacology 35:929–937
Landen M, Erlandsson H, Bengtsson F et al (2009) Short onset of action of a serotonin reuptake inhibitor when used to reduce premenstrual irritability. Neuropsychopharmacology 34:585–592
Landen M, Wennerblom B, Tygesen H et al (2004) Heart rate variability in premenstrual dysphoric disorder. Neuropsychopharmacology 29:733–740
Lang PJ, Davis M (2006) Emotion, motivation, and the brain: reflex foundations in animal and human research. Prog Brain Res 156:3–29
Lang PJ, Davis M, Ohman A (2000) Fear and anxiety: animal models and human cognitive psychophysiology. J Affect Disord 61:137–159
Marjoribanks J, Brown J, O'Brien PM et al (2013) Selective serotonin reuptake inhibitors for premenstrual syndrome. Cochrane Database Syst Rev 6:CD001396
Matsumoto T, Ushiroyama T, Kimura T et al (2007) Altered autonomic nervous system activity as a potential etiological factor of premenstrual syndrome and premenstrual dysphoric disorder. Biopsychosoc Med 1:24
Melcangi RC, Panzica G, Garcia-Segura LM (2011) Neuroactive steroids: focus on human brain. Neuroscience 191:1–5
O'Brien PM, Backstrom T, Brown C et al (2011) Towards a consensus on diagnostic criteria, measurement and trial design of the premenstrual disorders: the ISPMD Montreal consensus. Arch Womens Ment Health 14:13–21

Osterlund MK, Gustafsson JA, Keller E et al (2000a) Estrogen receptor beta (ERbeta) messenger ribonucleic acid (mRNA) expression within the human forebrain: distinct distribution pattern to ERalpha mRNA. J Clin Endocrinol Metab 85:3840–3846

Osterlund MK, Keller E, Hurd YL (2000b) The human forebrain has discrete estrogen receptor alpha messenger RNA expression: high levels in the amygdaloid complex. Neuroscience 95:333–342

Ottander U, Poromaa IS, Bjurulf E et al (2005) Allopregnanolone and pregnanolone are produced by the human corpus luteum. Mol Cell Endocrinol 239:37–44

Perkonigg A, Yonkers KA, Pfister H et al (2004) Risk factors for premenstrual dysphoric disorder in a community sample of young women: the role of traumatic events and posttraumatic stress disorder. J Clin Psychiatry 65:1314–1322

Pilver CE, Levy BR, Libby DJ et al (2011) Posttraumatic stress disorder and trauma characteristics are correlates of premenstrual dysphoric disorder. Arch Womens Ment Health 14:383–93

Protopopescu X, Tuescher O, Pan H et al (2008) Toward a functional neuroanatomy of premenstrual dysphoric disorder. J Affect Disord 108:87–94

Roberto M, Gilpin NW, Siggins GR (2012) The central amygdala and alcohol: role of gamma-aminobutyric acid, glutamate, and neuropeptides. Cold Spring Harb Perspect Med 2:a012195

Sadler C, Smith H, Hammond J et al (2010) Lifestyle factors, hormonal contraception, and premenstrual symptoms: the United Kingdom Southampton women's survey. J Womens Health (Larchmt) 19:391–396

Schmidt PJ, Nieman LK, Danaceau MA et al (1998) Differential behavioral effects of gonadal steroids in women with and in those without premenstrual syndrome. N Engl J Med 338:209–216

Segebladh B, Bannbers E, Kask K et al (2011) Prevalence of violence exposure in women with premenstrual dysphoric disorder in comparison with other gynecological patients and asymptomatic controls. Acta Obstet Gynecol Scand 90:746–752

Segebladh B, Borgstrom A, Nyberg S et al (2009) Evaluation of different add-back estradiol and progesterone treatments to gonadotropin-releasing hormone agonist treatment in patients with premenstrual dysphoric disorder. Am J Obstet Gynecol 201(139): e1–8

Shin LM, Liberzon I (2010) The neurocircuitry of fear, stress, and anxiety disorders. Neuropsychopharmacology 35:169–191

Sundstrom I, Backstrom T (1998) Patients with premenstrual syndrome have decreased saccadic eye velocity compared to control subjects. Biol Psychiatry 44:755–764

Sundstrom Poromaa I, Smith S, Gulinello M (2003) GABA receptors, progesterone and premenstrual dysphoric disorder. Arch Women Ment Health 6:23–41

Talledo JA, Sutherland Owens AN, Schortinghuis T et al (2009) Amphetamine effects on startle gating in normal women and female rats. Psychopharmacology (Berl) 204:165–175

Thwaites SJ, van den Buuse MGogos A (2014) Differential effects of estrogen and testosterone on auditory sensory gating in rats. Psychopharmacology (Berl) 231:243–56

Toufexis DJ, Davis C, Hammond A et al (2004) Progesterone attenuates corticotropin-releasing factor-enhanced but not fear-potentiated startle via the activity of its neuroactive metabolite, allopregnanolone. J Neurosci 24:10280–10287

Toufexis DJ, Rochford J, Walker CD et al (1999) Lactation-induced reduction in rats' acoustic startle is associated with changes in noradrenergic neurotransmission. Behav Neurosci 113:176–184

Vaillancourt C, Cyr M, Rochford J et al (2002) Effects of ovariectomy and estradiol on acoustic startle responses in rats. Pharmacol Biochem Behav 74:103–109

van den Akker O, Steptoe A (1989) Psychophysiological responses in women reporting severe premenstrual symptoms. Psychosom Med 51:319–328

van den Buuse M, Eikelis N (2001) Estrogen increases prepulse inhibition of acoustic startle in rats. Eur J Pharmacol 425:33–41

van Wingen GA, van Broekhoven F, Verkes RJ et al (2008) Progesterone selectively increases amygdala reactivity in women. Mol Psychiatry 13:325–333

Wang M, Seippel L, Purdy RH et al (1996) Relationship between symptom severity and steroid variation in women with premenstrual syndrome: study on serum pregnenolone, pregnenolone sulfate, 5 alpha-pregnane-3,20-dione and 3 alpha-hydroxy-5 alpha-pregnan-20-one. J Clin Endocrinol Metab 81:1076–1082
Wittchen HU, Becker E, Lieb R et al (2002) Prevalence, incidence and stability of premenstrual dysphoric disorder in the community. Psychol Med 32:119–132
Wittchen HU, Perkonigg A, Pfister H (2003) Trauma and PTSD—an overlooked pathogenic pathway for premenstrual dysphoric disorder? Arch Women Ment Health 6:293–297
Wu YC, Hill RA, Gogos A et al (2013) Sex differences and the role of estrogen in animal models of schizophrenia: interaction with BDNF. Neuroscience 239:67–83
Wyatt KM, Dimmock PW, Ismail KM et al (2004) The effectiveness of GnRHa with and without 'add-back' therapy in treating premenstrual syndrome: a meta analysis. BJOG 111:585–593

Postmenopausal Physiological Changes

Robert R. Freedman

Abstract The hallmark of menopause is the marked reduction of estradiol levels due to ovarian failure. This, among other factors result in hot flashes, the most common menopausal symptom. Hot flashes (HFs) can be measured objectively, both inside and outside the laboratory, using sternal skin conductance, an electrical measure of sweating. We have found that HFs are triggered by small elevations in core body temperature (T_C), acting within a greatly reduced thermoneutral zone. This reduction is caused by elevated central sympathetic activation, among other factors. There is a circadian rhythm of HFs peaking at 1825 h. Imaging studies have shown that hot flash activation begins in the brainstem, followed by the insula and by the prefrontal cortex. HFs in the first, but not the second half of the night can produce awakenings and arousals. This is because rapid eye movement (REM) sleep suppresses thermoregulatory effector responses, which include hot flashes.

Keywords Hot flash · Menopause · Sleep · Thermoregulation

Contents

R. R. Freedman (✉)
Departments of Psychiatry and Behavioral Neurosciences and Obstetrics and Gynecology, Wayne State University School of Medicine, 275 E. Hancock, Detroit, MI 48201, USA
e-mail: aa2613@wayne.edu

Curr Topics Behav Neurosci (2014) 21: 245–256
DOI: 10.1007/7854_2014_325

Published Online: 15 June 2014

1 Introduction

Menopause, defined as the final menstrual period, represents the permanent cessation of menses due to loss of ovarian follicular function, generally caused by aging. In practice, 1 year after the final menstrual period is usually employed to mark the onset of menopause. In Western societies, menopause occurs at an average of 51.4 ± 5 years, ranging from approximately 40–60 years (North American Menopause Society 2007).

Physiologically, the hallmark of menopause is the marked reduction of estradiol levels to 10 % or less of those found during reproductive years (North American Menopause Society 2007). Additionally, levels of FSH (follicular stimulating hormone) and LH (luteinizing hormone) increase following menopause. The most common symptom of menopause is the hot flash (HF) which forms the basis of the present chapter.

Hot flashes (HFs) are reported as feelings of intense warmth along with sweating, flushing, and chills. Sweating is generally reported in the face, neck, and chest. HFs usually last for 1–5 min, with some lasting as long as an hour (Kronenberg 1990). The median duration of symptoms is about 4 years, with some lasting as long as 20 years (Feldman et al. 1985). In one U.S. study, 87 % of the women reported daily HFs and about a third of those reported more than 10 per day (Kronenberg 1990). There is some racial and ethnic variation of HFs with Caucasian women reporting the highest prevalence and Japanese and Chinese women reporting the lowest (Gold et al. 2006).

2 Physiologic Events of the Hot Flash

Peripheral vasodilation, demonstrated by elevated skin temperature and blood flow, occurs during HFs in all areas that have been studied (Fig. 1). Skin temperature increases in the digits, face, arms, chest, abdomen, back, and legs (Freedman 1998; Molnar 1975; Kronenberg et al. 1984; Tataryn et al. 1980; Ginsburg et al. 1981) and blood flow in these areas is elevated, as well (Kronenberg et al. 1984; Tataryn et al. 1980; Ginsburg et al. 1981).

Sweating and skin conductance, an electrical measure of this, also increases during HFs (Fig. 1). Molnar (1975) determined the whole body sweat rate to be about 1.3 g/min in one subject. We measured sweating and skin conductance from the sternum at the same time in 14 women (Freedman 1998). We found a close temporal correspondence between both measures, which were significantly elevated. Measurable sweating occurred in 90 % of the HFs.

Core body temperature (T_C) also increases prior to HFs. We measured T_C and sternal skin conductance during 77 HFs in 10 menopausal women who reported frequent symptoms (Freedman et al. 1995). We found small but significant T_C elevations before the majority of HFs and replicated these findings in two subsequent studies (Freedman 1998; Freedman and Woodward 1996).

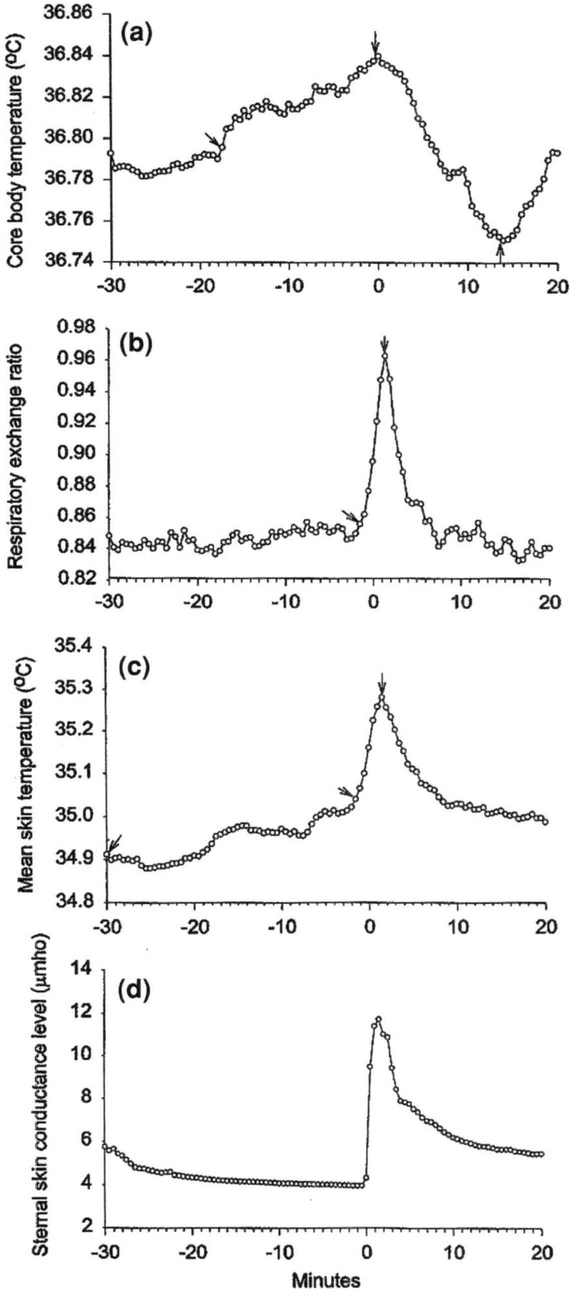

Fig. 1 **a** Core body temperature (means) during menopausal hot flashes. **b** Respiratory exchange ratio (means) during hot flashes. **c** Mean skin temperature (means) during hot flashes. **d** Sternal skin conductance (means) during hot flashes. Time 0 is the beginning of the sternal skin conductance response. Intervals between arrows are significantly different from each other at $P < 0.05$, Duncan's test

The T_C elevations could be caused by increased metabolic rate (heat production) and/or peripheral vasoconstriction (decreased heat loss). We did find significant increases in metabolic rate (Fig. 1), but they occurred at the same time as the peripheral vasodilation and sweating; peripheral vasoconstriction did not occur (Freedman 1998). Therefore, the T_C elevations are not caused by metabolic rate elevations. Small increases in heart rate, about 7–15 beats/min do occur along with the metabolic rate increases (Molnar 1975; Kronenberg et al. 1984) and palpitations are sometimes reported.

3 Objective Measurement of Hot Flashes

Typically, diaries are used to assess treatment outcome in HF studies. However, there are several problems with these measures. Errors in compliance are major sources of bias (Takarangi et al. 2006). Also, HFs occurring during sleep are not accurately reported because recall of these events is usually poor and many HFs do not produce awakenings (Freedman 2010). Finally, placebo effects as large as 40–50 %, occur with self-reports (Stone et al. 2003). Therefore, objective measures of HFs have been developed.

Increased skin conductance recorded from the sternum is presently the best objective marker of HFs. An increase in this measure ≥ 2 μmho (electrical unit of conductance) within 30 s corresponded with 95 (Freedman 1989), 90 (Freedman et al. 1992), and 80 % of reported HFs (de Bakker and Everaerd 1996). These results have been independently replicated (de Bakker and Everaerd 1996). Moreover, these results have been extended to men with HFs due to androgen therapy for prostate cancer (Hanisch et al. 2007).

The skin conductance measure is also useful because it can be recorded outside the laboratory in daily life. Using the same recording methods with ambulatory monitors, we found an agreement of 85 % between the skin conductance criterion and patient event marks (Freedman 1989). A second study found an agreement of 77 % (Freedman et al. 1992).

However, the major drawback of skin conductance recording is that it requires the use of electrodes and gel, which must be changed every 24 h. Therefore, the author has invented a miniature, hygrometric HF recorder which requires neither electrodes nor gel (Freedman and Wasson 2007). This device will record all HFs for 1 month using a single hearing aid battery. It attaches to the skin with a double-sided sticky collar. A simple computer scoring program has been developed which will score 1 week of 24 h data in <5 min.

In a recent study of patient satisfaction with the recorder, the author obtained positive responses from patients regarding the ease of use and appearance of the recorder (Freedman and Wasson 2007; Freedman 2009).

4 Endocrinology of Hot Flashes

Since HFs occur in the vast majority of women having natural or surgical menopause, estrogens are clearly involved in their etiology. This is consistent with the fact that estrogen therapy virtually eliminates HFs. However, estrogen reduction alone does not explain the occurrence of HFs because there are no relationships between these symptoms and plasma, urinary, or vaginal (Askel et al. 1976) levels of estrogens, nor are there differences in plasma levels between women with and without HFs (Freedman et al. 1995; Askel et al. 1976). Additionally, clonidine reduces HF frequency but does not change estrogen levels (Schindler et al. 1979), and prepubertal girls have low estrogen levels but no HFs. Therefore, estrogen withdrawal is necessary but not sufficient to explain the occurrence of HFs.

A temporal relationship was observed between HFs and LH pulses (Casper et al. 1979; Tataryn et al. 1979). However, further work demonstrated that women with isolated gonadotropin deficiency had HFs but no LH pulses (Gambone et al. 1984), and those with hypothalamic amenorrhea had LH pulses but no HFs. Also, HFs occurs in women with LH suppression from GnRH compounds (Casper and Yen 1981; DeFazio et al. 1983), in women with pituitary insufficiency and hypoestrogenism (Meldrum et al. 1981), and in hypophysectomized women, who have no LH pulses (Mulley et al. 1977).

Subsequently, an opiate system was hypothesized in the etiology of HFs. Jeffcoate (1981) showed that an opiate antagonist reduced HF and LH pulse frequencies, although other research failed to replicate these results (DeFazio et al. 1984). Thus, the evidence for opiate involvement in HFs is inconsistent.

Norepinephrine (NE) plays an important role in thermoregulation acting, in part, through α_2-adrenergic receptors. Injected into the preoptic hypothalamus, NE causes heat dissipation responses followed by a decline in T_C (Brück and Zeisberger 1990). Additionally, gonadal steroids modulate central NE activity (Insel and Motulsky 1987). Although plasma NE levels do not change during HFs (Kronenberg et al. 1984; Casper et al. 1979), these do not represent levels in the brain (Kopin et al. 1984).

We addressed these issues using pharmacologic probes. In a controlled, laboratory investigation (Freedman et al. 1990), we showed that yohimbine, a α_2-adrenergic antagonist that elevates brain NE (Goldberg and Robertson 1983), triggered HFs in symptomatic but not asymptomatic menopausal women, while clonidine, a α_2 agonist, ameliorated them. Postmortem studies have shown that most α_2 receptors in the human brain are inhibitory (Sastre and Garcia-Sevilla 1994). Blockade of these receptors with yohimbine would increase NE release, whereas clonidine would reduce it (Starke et al. 1989; Charney et al. 1982).

Furthermore, estrogens modulate brain adrenergic receptors (Etgen et al. 2001; Ansonoff and Etgen 2001). Taken together, these data formed the basis of our theory that elevated brain NE, in conjunction with estrogen withdrawal, are part of the etiology of HFs.

5 Thermoregulation and Hot Flashes

T_C in homeotherms is regulated between an upper threshold for sweating and a lower threshold for shivering. Between these thresholds is a neutral zone within which major thermoregulatory responses (sweating, shivering) do not occur (Savage and Brengelmann 1996). Small adjustments within the neutral zone are performed by changes in peripheral blood flow.

According to this theory, the heat dissipation responses of the hot flash (sweating, peripheral vasodilation) would be provoked if T_C crossed the upper threshold. We had already demonstrated T_C increases before most HFs (Freedman 1998; Freedman et al. 1995; Freedman and Woodward 1996). We therefore chose to study the width of the thermoneutral zone in women with and without HFs.

Previous research showed that warm, ambient temperatures and peripheral body heating could provoke HFs (Freedman 1989; Freedman et al. 1992) suggesting that the upper threshold is lowered in women with HFs. We then demonstrated that the lower threshold is elevated in these women by inducing shivering while measuring T_C (Freedman and Woodward 1995). We then measured the upper and lower thresholds using ambient heating and cooling in women with and without HFs. We measured the thermoneutral zone to be 0.0 °C in the symptomatic women and 0.4 °C in the asymptomatic women (Freedman and Krell 1999). We then replicated the T_C sweating threshold findings using exercise. When sweating thresholds were reached, all symptomatic but no asymptomatic women demonstrated objective and subjective HFs. Sweat rates in the former group were twice those of the later group.

Thus, we believe that HFs are triggered by T_C elevations acting within a greatly narrowed thermoneutral zone in postmenopausal women with HFs. A HF, consisting of sweating and peripheral vasodilation, is provoked when T_C reaches the upper threshold. T_C then declines, and when the lower threshold is crossed, shivering occurs. What biochemical mechanisms account for this?

Basic science investigations have found that increased brain NE narrows the width of the thermoneutral zone (Brück and Zeisberger 1990). Conversely, clonidine lowers NE release, raises the sweating threshold, and reduces the shivering threshold. We therefore hypothesize that increased brain NE narrows the thermoneutral zone in menopausal women with HFs.

We determined the T_C sweating threshold in women with and without HFs during I.V. clonidine and placebo (Freedman and Dinsay 2000). We showed that clonidine significantly elevated the sweating threshold compared to placebo in the women with HFs, whereas the opposite occurred in the women without HFs. We therefore believe that clonidine reduces HFs by raising the T_C sweating threshold.

We then conducted a similar study to examine the mechanism through which estrogen ameliorates HFs (Freedman and Blacker 2002). Symptomatic menopausal women were randomly assigned to receive 1 mg/day 17 β-estradiol P.O. or placebo for 90 days. We found that the T_C sweating threshold was significantly elevated and HF frequency significantly ameliorated in the E_2 but not the placebo group. Thus, estrogen ameliorates HFs by raising the T_C sweating threshold, but we do not know the precise mechanisms of this.

6 Circadian Rhythm of Hot Flashes

Given this mechanism, we sought to determine if HF occurrence was related to the T_C circadian rhythm. Using 24 h ambulatory monitoring, we recorded sternal SCL to detect HFs, ambient temperature, skin temperature, and T_C (ingested radio telemetry pill) (Freedman et al. 1995). Cosinor analysis revealed an HF circadian rhythm with a peak at 1825 h (Fig. 2). The majority of HFs were preceded by T_C elevations ($P < 0.05$). HFs began at significantly higher T_C levels (36.82 ± 0.04 °C) compared with all nonflash periods (36.70 ± 0.005 °C). We then replicated these findings in symptomatic women with breast cancer using a whole-room calorimeter (Carpenter et al. 2004).

7 Imaging Studies

We were interested to determine the brain areas associated with the physiologic and phenomenological aspects of the HF and employed functional magnetic resonance imaging (fMRI) to do this. In the first study, we used symptomatic menopausal women and asymptomatic amenorrheic women and induced HFs and sweating (measured with sternal SCL) in the scanner (Freedman et al. 2006). Significant areas of activation in the symptomatic women included the insular and the anterior cingulate cortex. Sweating in the amenorrheic women was associated with activation in the anterior cingulate and superior frontal gyrus. We believe the insular activation is associated with the "rush of heat" described during menopausal HFs.

In a second investigation (Diwadkar et al. 2013), we sought to determine the temporal sequencing of the neuronal events underlying the HF. Methods were similar to those described above. We performed fMRI in a group of postmenopausal women to measure neuronal activity in the brainstem, insula, and prefrontal cortex around the onset of an HF (detected using synchronously acquired skin conductance responses). Rise in brainstem activity occurred before the detectable onset of an HF. Insula and prefrontal activity trailed activity in the brainstem, appearing following HF onset (Fig. 3). Pre-HF brainstem responses may reflect the functional origins of internal thermoregulatory events such as HFs. By comparison, insula, and prefrontal activity may be associated with the phenomenological correlates of HFs.

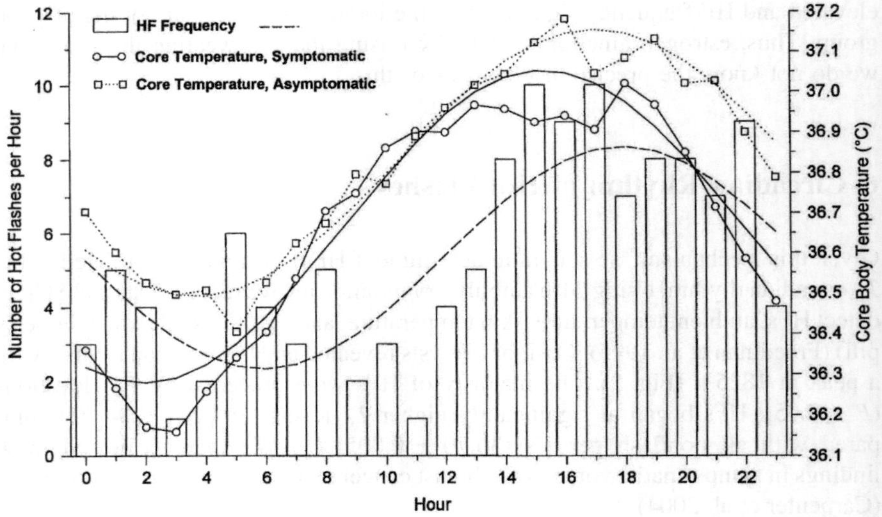

Fig. 2 Hot flash frequency and T_C during 24 h. Hot flash frequency in 10 symptomatic women (*bars*); best-fit cosine *curve* for hot flash frequency (*dashed line*); 24 h T_C data for 10 symptomatic women (O) with best-fit cosine *curve* (*solid line*); 24 h T_C data in 6 asymptomatic women (□) with best-fit cosine *curve* (*dotted line*)

8 Hot Flashes (HFs) and Sleep

Although most epidemiologic studies have found increased reports of sleep disturbance at menopause (Kravitz et al. 2003), this has not been found in most laboratory studies (Young et al. 2003). A study in our laboratory (Freedman and Roehrs 2004) found no differences among age-matched premenopausal women, postmenopausal symptomatic women, and postmenopausal asymptomatic women on any sleep measure, performance test, or questionnaire measure. Additionally, HFs did not appear to trigger awakenings or arousals based on analysis of whole-night data.

A subsequent study analyzed this last issue in greater depth by analyzing data by halves of the night (Freedman and Roehrs 2006). This was done because there is more rapid eye movement (REM) sleep in the second half of the night. It has been shown that REM sleep suppresses thermoregulatory effector responses, such as sweating and peripheral vasodilation, which constitute HFs. Indeed, it was found that HFs in the second half of the night occurred after the awakenings and arousals, whereas those in the first half of the night preceded them and could therefore trigger them.

This temporal relationship was replicated in a recent laboratory study of 102 women, 44–56 years of age, who complained of poor sleep (Freedman and Roehrs 2007). Fifty-three percent (53 %) of the women had apnea, restless legs, or both. The best predictors of objective sleep quality (laboratory sleep efficiency) were

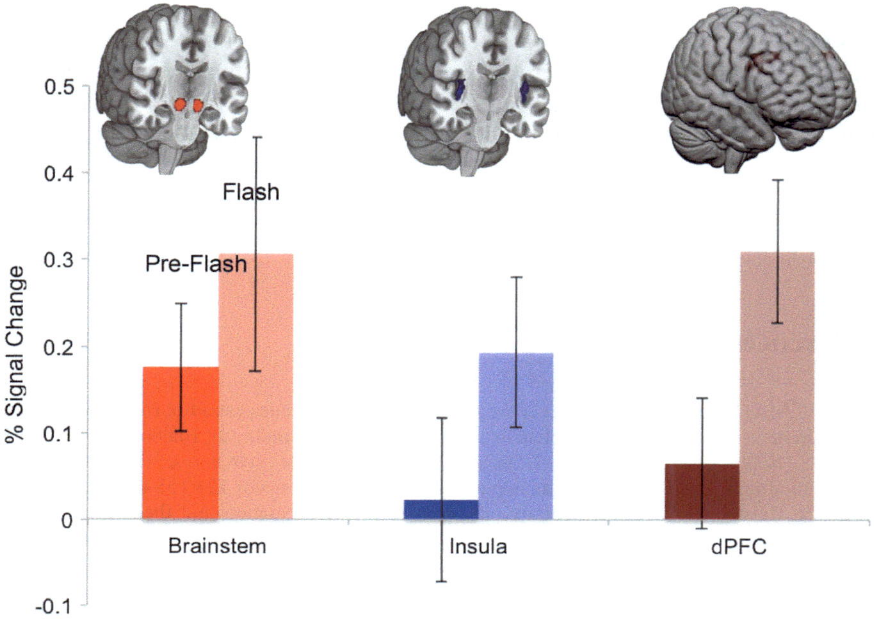

Fig. 3 Averaged activity in the pre-flash and flash windows across the three regions of interest. Images depict volume-rendered activations in the brainstem, the insula, and the dorsal prefrontal cortex successively. Data are shown collapsed across image within the pre-flash and flash windows for each subject. Relative to the pre-flash window, significantly increased activity in the flash window is seen in the insula and prefrontal cortex, but not in the brainstem. Error bars are ± SEM

apneas, periodic limb movements, and arousals ($R^2 = 0.44$, $P < 0.0001$). The best predictors of subjective sleep quality (Pittsburgh Sleep Quality Index global score) were the Hamilton anxiety score and the number of HFs in the first half of the night ($R^2 = 0.19$, $P < 0.001$). It is, therefore, possible that anxiety mediates some reports of poor sleep.

These results may explain the difference between our first laboratory study which did not analyze data by halves of the night (Young et al. 2003), and self-report studies of increased sleep disturbance at menopause (Kravitz et al. 2003). Our findings also emphasize the importance of detecting primary sleep disorders, such as apnea and periodic limb movements which are highly disruptive of sleep and can have serious medical consequences.

As described throughout this chapter, we have been interested in the role of sympathetic activation in HFs, many of which occur during sleep. It has been shown that measures of heart rate variability (HRV) in the frequency domain reflect the different components of sympathetic and parasympathetic activation. Components computed using the spectral analysis, power in HRV frequencies greater than 0.15 Hz reflects parasympathetic activation, whereas power in frequencies below 0.15 Hz reflects, in part, sympathetic activation. We performed these measurements during laboratory sleep in 16 postmenopausal women demonstrating at least four HFs per night (Freedman et al. 2011). For the frequency bin of 0–0.15 Hz, we found that, during stage 1 sleep, power was greater during HFs compared with preceding and subsequent periods. For waking HFs, power in this band was higher before HFs than during or after them. These data are consistent with our theory of elevated sympathetic activation as a trigger for menopausal HFs.

Acknowledgments This work was supported by NIH Merit Award, R37-AG05233 and by R01-MH63089.

References

Ansonoff MA, Etgen AM (2001) Receptor phosphorylation mediates estradiol reduction of alpha 2-adrenoceptor coupling to G protein in the hypothalamus of female rats. Endocrine 14:165–174
Askel S, Schomberg DW, Tyrey L et al (1976) Vasomotor symptoms, serum estrogens, gonadotropin levels in surgical menopause. Am J Obstet Gynecol 126:165–169
Brück K, Zeisberger E (1990) Adaptive changes in thermoregulation and their neuropharmacological basis. In: Schönbaum E, Lomax P (eds) Thermoregulation: physiology and biochemistry. Pergamon, New York, pp 255–307
Carpenter JS, Gilchrist JM, Chen K et al (2004) Hot flashes, core body temperature, and metabolic parameters in breast cancer survivors. Menopause 11:375–381
Casper RF, Yen SSC (1981) Menopausal flushes: effect of pituitary gonadotropin desensitization by a potent luteinizing hormone releasing factor agonist. J Clin Endocrinol Metab 53:1056–1058
Casper RF, Yen SSC, Wilkes MM (1979) Menopausal flushes: a neuroendocrine link with pulsatile luteinizing hormone secretion. Science 204:823–825
Charney DS, Heninger GR, Sternberg DE (1982) Assessment of α2-adrenergic autoreceptor function in humans: effects of oral yohimbine. Life Sci 30:2033–2041
de Bakker IPM, Everaerd W (1996) Measurement of menopausal hot flushes: validation and cross-validation. Maturitas 24:87–98
DeFazio J, Meldrum DR, Laufer L et al (1983) Induction of hot flashes in premenopausal women treated with a long-acting GnRH agonist. J Clin Endocrinol Metab 56:445–448
DeFazio J, Vorheugen C, Chetkowski R et al (1984) The effects of naloxone on hot flashes and gonadotropin secretion in postmenopausal women. J Clin Endocrinol Metab 58:578–581
Diwadkar VA, Murphy ER, Freedman RR (2013) Temporal sequencing of brain activations during naturally occurring thermoregulatory events. Cereb cortex Jun 19 [Epub ahead of print]
Etgen AM, Ansonoff MA, Quesada A (2001) Mechanisms of ovarian steroid regulation of norepinephrine receptor-mediated signal transduction in the hypothalamus: implications for female reproductive physiology. Horm Behav 40:169–177
Feldman BM, Voda A, Groseth E (1985) The prevalence of hot flash and associated variables among perimenopausal women. Res Nurs Health 8:261–268
Freedman RR (1989) Laboratory and ambulatory monitoring of menopausal hot flashes. Psychophysiology 26:573–579
Freedman RR (1998) Biochemical, metabolic, and vascular mechanisms in menopausal hot flushes. Fertil Steril 70:1–6
Freedman RR (2009) Patient satisfaction with miniature, ambulatory, postmenopausal hot flash recorder. Open Med Devices J 1:1–2
Freedman RR (2010) Objective or subjective measurement of hot flashes in clinical trials: Quo Vadis. Maturitas 67:99–100
Freedman RR, Blacker CM (2002) Estrogen raises the sweating threshold in postmenopausal women with hot flashes. Fertil Steril 77:487–490

Freedman RR, Dinsay R (2000) Clonidine raises the sweating threshold in symptomatic but not in asymptomatic postmenopausal women. Fertil Steril 74:20–23
Freedman RR, Krell W (1999) Reduced thermoregulatory null zone in postmenopausal women with hot flashes. Am J Obstet Gynecol 181:66–70
Freedman RR, Roehrs TA (2004) Lack of sleep disturbance from menopausal hot flashes. Fertil Steril 82:138–144
Freedman RR, Roehrs TA (2006) Effects of REM sleep and ambient temperature on hot flash-induced sleep disturbance. Menopause 13:576–583
Freedman RR, Roehrs TA (2007) Sleep disturbance in menopause. Menopause 14:826–829
Freedman RR, Wasson S (2007) Miniature, hygrometric hot flash recorder. Fertil Steril 88:494–496
Freedman RR, Woodward S (1995) Altered shivering threshold in postmenopausal women with hot flashes. Menopause 2:163–168
Freedman RR, Woodward S (1996) Core body temperature during menopausal hot flushes. Fertil Steril 65:1141–1144
Freedman RR, Woodward S, Sabharwal SC (1990) Adrenergic mechanism in menopausal hot flushes. Obstet Gynecol 76:573–578
Freedman RR, Woodward S, Norton D (1992) Laboratory and ambulatory monitoring of menopausal hot flushes: comparison of symptomatic and asymptomatic women. J Psychophysiol 6:162–166
Freedman RR, Norton D, Woodward S et al (1995) Core body temperature and circadian rhythm of hot flashes in menopausal women. J Clin Endocrinol Metab 80:2354–2358
Freedman RR, Benton MD, Genik RJ II et al (2006) Cortical activation during menopausal hot flashes. Fertil Steril 85:674–678
Freedman RR, Kruger ML, Wasson SL (2011) Heart rate variability in menopausal hot flashes during sleep. Menopause 18:1–4
Gambone J, Meldrum DR, Laufer L et al (1984) Further delineation of hypothalamic dysfunction responsible for menopausal hot flashes. J Clin Endocrinol Metab 59:1097–1102
Ginsburg J, Swinhoe J, O'Reilly B (1981) Cardiovascular responses during the menopausal hot flush. Br J Obstet Gynaecol 88:925–930
Gold EB, Colvin A, Avis N et al (2006) Longitudinal analysis of the association between vasomotor symptoms and race/ethnicity across the menopausal transition: study of women's health across the nation. Am J Pub Health 96:1–10
Goldberg M, Robertson D (1983) Yohimbine: a pharmacological probe for study of α_2-adrenoceptor. Pharmacol Rev 35:143–180
Hanisch LJ, Palmer SC, Donahue A et al (2007) Validation of sternal skin conductance for detection of hot flashes in prostate cancer survivors. Psychophysiol 44:189–193
Insel PA, Motulsky HJ (1987) Physiologic and pharmacologic regulation of adrenergic receptors. In: Insel PA (ed) Adrenergic receptors in man. Marcel Dekker, New York, pp 201–236
Jeffcoate SL (1981) Climacteric flushing: clinical and endocrine response to infusion of naloxone. Br J Obstet Gynaecol 88:919–924
Kopin IJ, Blombery P, Ebert MH et al (1984) Disposition and metabolism of MHPG-CD3 in humans: plasma MHPG as the principal pathway of norepinephrine metabolism and as an important determinant in CSF levels of MHPG. In: Usdin E et al (eds) Frontiers in biochemical and pharmacological research in depression. Raven, New York, pp 57–68
Kravitz HM, Ganz PA, Bromberger J et al (2003) Sleep difficulty in women at midlife: a community survey of sleep and the menopausal transition. Menopause 10:19–28
Kronenberg F (1990) Hot flashes: epidemiology and physiology. Ann N Y Acad Sci 592:52–86
Kronenberg F, Cote LJ, Linkie DM et al (1984) Menopausal hot flashes: thermoregulatory, cardiovascular, and circulating catecholamine and LH changes. Maturitas 6:31–43
Meldrum DR, Erlik Y, Lu JKH et al (1981) Objectively recorded hot flushes in patients with pituitary insufficiency. J Clin Endocrinol Metab 52:684–687
Molnar GW (1975) Body temperature during menopausal hot flashes. J Appl Physiol: Respir Environ Exerc Physiol 38:499–503

Mulley G, Mitchell RA, Tattersall RB (1977) Hot flushes after hypophysectomy. BMJ 2:1062
North American Menopause Society (2007) Menopause practice: a clinician's guide. Normal physiology, pp 19–27
Sastre M, Garcia-Sevilla JA (1994) Density of alpha-2A adrenoceptors and Gi proteins in the human brain: ratio of high-affinity agonist sites to antagonist sites and effect of age. J Pharmacol Exp Ther 269:1062–1072
Savage MV, Brengelmann GL (1996) Control of skin blood flow in the neutral zone of human body temperature regulation. J Appl Physiol 80:1249–1257
Schindler AE, Muller D, Keller E et al (1979) Studies with clonidine (Dixarit) in menopausal women. Arch Gynecol 227:341–347
Starke K, Gothert M, Kilbringer H (1989) Modulation of neurotransmitter release of presynaptic autoreceptors. Physiol Rev 69:864–989
Stone AA, Shiffman S, Schwartz JE et al (2003) Patient compliance with paper and electronic diaries. Control Clin Trials 24:182–199
Takarangi MK, Garry M, Loftus EF (2006) Dear diary, is plastic better than paper? I can't remember. Comment on Green, Bolger, Shrout, and Reis. Psychol Methods 11:119–122
Tataryn IV, Meldrum DR, Lu KH et al (1979) FSH, and skin temperature during menopausal hot flush. J Clin Endocrinol Metab 49:152–154
Tataryn IV, Lomax P, Bajorek JG et al (1980) Postmenopausal hot flushes: a disorder of thermoregulation. Maturitas 2:101–107
Young T, Rabago D, Zgierska A et al (2003) Objective and subjective sleep quality in premenopausal, perimenopausal, and postmenopausal women in the Wisconsin Sleep Cohort Study. Sleep 26:667–672

ASD: Psychopharmacologic Treatments and Neurophysiologic Underpinnings

Ian Kodish, Carol M. Rockhill and Sara J. Webb

Abstract Autism Spectrum Disorder encompasses a range of neurodevelopmental disorders characterized by early deficits in social communication in addition to restricted and repetitive behaviors. Symptoms are increasingly understood to be associated with abnormalities in the coordination of neuronal assemblies responsible for processing information essential for early adaptive behaviors. Pharmacologic treatments carry evidence for clinically significant benefit of multiple impairing symptoms of ASD, yet these benefits are limited and range across a broad spectrum of medication classes, making it difficult to characterize associated neurochemical impairments. Increasing prevalence of both ASD and its pharmacologic management calls for greater understanding of the neurophysiologic basis of the disorder. This paper reviews underlying alterations in local brain regions and coordination of brain activation patterns during both resting state and task-related processes. We propose that new pharmacologic treatments may focus on realigning trajectories of network specialization across development by working in combination with behavioral treatments to enhance social and emotional learning by bolstering the impact of experience-induced plasticity on neuronal network connectivity.

Keywords ASD · Autism · Neuroimaging · Pharmacology · Physiology · Review

---

I. Kodish · C. M. Rockhill
University of Washington Department of Psychiatry and Behavioral Sciences, Seattle, United States of America

S. J. Webb (✉)
University of Washington Department of Psychiatry and Behavioral Sciences, Seattle Children's Research Institute, Po Box 5371,M/S CW8-6, Seattle 98145, United States of America
e-mail: sjwebb@u.washington.edu

Curr Topics Behav Neurosci (2014) 21: 257–275
DOI: 10.1007/7854_2014_298

Published Online: 30 March 2014

Contents

1 Introduction

Children with ASD exhibit characteristic behavioral and functional abnormalities, including core deficits in interpersonal functioning (e.g., social-emotional reciprocity, nonverbal communication, adjustment of behavior to suit context) and stereotypic repetitive behaviors (e.g., abnormal repetitive movements, inflexible adherence to routine, abnormal sensory reactivity)(APA 2013). Impairments manifest early in development, by age 3, yet can range significantly across the lifespan within the individual and across individuals, highlighting the multifaceted nature of the disorder. ASD is also associated with several comorbid conditions including sleep disorders, seizure disorders, inflammatory disorders, anxiety disorders, and attention-deficit hyperactivity disorder (ADHD). Prevalence rates of autism spectrum disorder are estimated at 1/88, with a male prevalence reaching 1/54 (CDC 2012).

ASD is now widely accepted as a disorder of brain development. Recent utilization of neurophysiological and neuroimaging methods have begun to elucidate the neural mechanisms that may underlie the course and presentation of autism behaviors. Electroencephalography (EEG) measures characteristic brain waveform patterns, and analyses of event-related potentials (ERP) reveal changes in EEG wave patterns as a function of cognitive or motor operations, or states of alertness. These neurophysiological methods provide information about neural pathways at multiple levels of the neuroaxis and within selected aspects of sensory, motor, cognitive, and social function. Neural oscillations reflect the synchronous firing of large populations of neurons mediated by excitatory and inhibitory interactions. Fluctuations in various EEG frequency bands are thought to represent abnormalities in network organization, and can further characterize the timing of processing abnormalities. Neuroimaging techniques, including single-photon emission computed tomography (SPECT), diffusion tensor imaging (DTI), magnetic resonance imaging (MRI), and functional MRI (fMRI) further enable detection of anatomical changes and alterations in the functional utilization of brain regions during resting states and under task demands. Recent efforts and advanced statistical methods have fostered cross-utilization of these techniques to detect patterns of regional synchrony and coactivation, allowing further characterization of functional connectivity of brain networks.

This paper aims to review both the medication treatments for ASD and the emerging patterns of neurophysiologic and neuroanatomic alterations in networks associated with ASD. The goal is to integrate these disparate literatures, highlighting important new targets of treatment that can be derived from and assessed by neurophysiological measures.

2 Pharmacologic Treatment

The clinical impairments associated with ASD are often difficult to alleviate, and are increasingly managed using pharmacologic interventions. While core symptoms of communication deficits and circumscribed interests are difficult to address with medication, other clinical impairments are often targets of treatment, including comorbid anxiety, difficulty with sustained attention, aggressive behaviors, sleep disturbances, and stereotypic movements.

Despite the lack of extensive evidence base, examination of prescribing patterns for youth with ASD reveals that pharmacotherapy is very common (e.g., Hsia et al. 2013; Mire et al. 2013; Schubart et al. 2013). The Mire et al. study examined over 1,600 North American youth with Autism, and found that 41.7 % of parents reported that their child or adolescent had used psychotropic medications, with ADHD medications most commonly used. Correlational analyses indicated that the likelihood for medication use in this sample was higher for children who had social impairment and for those with low cognitive function. Similarly, a large longitudinal study of U.S. Medicaid claims for patients with ASD showed that over a 4-year period, 65 % were prescribed psychiatric medication. In contrast to the primary use of stimulants in the Mire study, antipsychotics were the most frequently prescribed in this sample, and the use of more than one medication was very common (Schubart et al. 2013). An international study of prescribing practices showed that North American youth with ASD received the highest percentage of prescriptions for psychotropic medication compared to European, South American, and Asian countries. Risperidone was the most commonly prescribed medication in North America for ASD, in contrast to methylphenidate being most commonly prescribed in the UK, and haloperidol being most commonly prescribed in Japan (Hsia et al. 2013).

2.1 Antipsychotic Medications

Risperidone and aripiprazole are now approved by the United States Food and Drug Administration (USDA) to address irritability associated with autism. Their effectiveness is supported by seven randomized controlled trials (RCTs) showing significant differences between risperidone and placebo (Hellings et al. 2006; Nagaraj 2006; Pandina et al. 2007; McCracken et al. 2002; Aman et al. 2005; Troost et al. 2005; Shea et al. 2004) and two for aripiprazole (Owen et al. 2009; Marcus et al. 2009), with most studies measuring improvement on the irritability subscale of the Aberrant Behavior Checklist. Although two trials of risperidone did not achieve statistical significance in comparison with placebo (Luby et al. 2006; Miral et al. 2008), several articles reviewing this literature all conclude that the data supporting effectiveness is strong, while cautioning that behavioral intervention should be tried first and that side effects including metabolic abnormalities, weight gain, and potential for extrapyramidal side effects warrant caution in their use (Elbe and Lalani 2012; Parikh et al. 2008; Pringsheim and Gorman 2012; Sharma and Shaw 2012).

2.2 Stimulants

A review of relevant studies concluded that 41–78 % of youth with autism meet criteria for attention deficit hyperactivity disorder (ADHD) (Murray 2010). A recent review of randomized and nonrandomized trials concluded that, after careful symptom assessment, treatment of comorbid ADHD symptoms with stimulant medication is indicated for youth with ASD (Mahajan et al. 2012). In addition, there is some evidence for effectiveness of non-stimulant ADHD medications in youth with ASD, with one randomized controlled trial each for atomoxetine and guanfacine having shown superiority over placebo (Arnold et al. 2006; Handen et al. 2008), and further studies underway.

2.3 SSRIs

Providers also often consider the use of selective serotonin reuptake inhibitors (SSRIs), targeting impairing symptoms of anxiety, including obsessive thoughts, and compulsive behaviors frequently associated with ASD. However, data to support their effectiveness in ASD populations is mixed, and reviews highlight a positive publication bias, making the literature difficult to accurately interpret (Williams et al. 2013; Carrasco et al. 2012; Kolevzon et al. 2006). RCTs examining the impact of SSRIs on compulsive behaviors failed to show improvements beyond placebo, although several open-label studies have demonstrated effectiveness of SSRIs for anxiety. A subsequent Cochrane Review concluded that there was no systematic evidence in support of the use of SSRIs to treat ASD (Williams et al. 2013). Despite this, prescribing practices in populations of youth with ASD indicates that the use of SSRIs is relatively common (e.g., Lopata et al. 2013). Older medication classes such as tricyclic antidepressants are not recommended due to the lack of evidence supporting their use, as well as significant side effects (Hurwitz et al. 2012).

2.4 Other Agents

Because of its relevance to social behaviors, oxytocin has also received attention as a potential treatment for ASD. Although no RCTs have been completed, an open label study of oxytocin treatment showed improved performance on tasks of emotion recognition compared to placebo (Guastella et al. 2010). Furthermore, initial findings from oxytocin treatment in ASD reveal enhanced regional activation during social tasks in several distributed brain areas relevant to social processing (Gordon et al. 2013). Despite these positive preliminary findings, a recent editorial cautioned against premature clinical use as treatment modality until research focused on long-term implications and potential side effects or problems can be completed (Harris and Carter 2013).

Disrupted sleep is also a frequent problem for youth with autism, and often impacts sleep quality of the entire family. A meta-analysis (Rossignol and Frye 2013) and a controlled trial (Wright et al. 2011) of melatonin in autism found consistent positive effects. In addition, one open-label study supported the use of clonidine to address insomnia in youth with autism (Ming et al. 2008). However, given the limited data available, a recent review suggested that the most prudent initial course is to formally evaluate patients for sleep disorders, without clear support for any one particular sleep medication (Malow et al. 2012).

Other novel agents have also been used in the treatment of ASD symptoms, with some evidence for positive effectiveness. Amantadine, which impacts the N-methyl-D-aspartate (NMDA) receptor, may act by limiting excitotoxicity of the glutamatergic neurotransimitter system. This receptor class is thought to be essential for modulating synaptic plasticity and represents a new class of pharmacologic targets with the potential to impact neurophysiologic and cognitive functioning. One RCT of amantadine treatment in youth with ASD found improved control of irritability and hyperactivity (King et al. 2001), and another reported beneficial effects for ADHD (Aman and Langworthy 2000). Although amantadine is not commonly used, it may be considered when other treatments do not provide adequate symptom control, particularly for distractibility, and hyperactivity.

3 Neuroanatomic Alterations Through Development

Macrocephaly is common in infants with ASD, and neuroanatomic studies have demonstrated larger brain sizes in children with ASD during early development (e.g., Aylward et al. 2002; Sparks et al. 2002). MRI studies indicate that brain enlargement is partially due to an increase in gray matter. For example, toddlers subsequently diagnosed with autism exhibit significant early increases in overall gray matter, distributed across frontal and temporal regions (Schumann et al. 2010), with evidence for the greatest enlargement in the frontal cortex (Carper and Courchesne 2005). Postmortem studies further suggest that these changes represent significant increases in neuronal number (Courchesne et al. 2011) and dendritic spine density (Hutsler and Zhang 2010).

Increased synaptic connectivity in ASD is also paralleled by an increase in the number of cortical minicolumns (Casanova et al. 2006), vertical clusters of interconnected neurons thought to represent fundamental processing units of cortical architecture (Silberberg et al. 2002). These units are each devoted to processing a specific type of information, such as a specific orientation of lines in visual space,or in animals a specific input from an individual whisker. Each is composed of a core excitatory assembly wrapped by an inhibitory network to provide intercolumnar dampening and heightened specificity of response. More synapses, neurons, and minicolumn assemblies suggest an overelaboration of sensory units in ASD, creating greater processing demands. As brain regions coordinate information at progressive levels of integration, particularly in prefrontal regions,these heightened processing demands become even more magnified.

The normative trajectory of gray matter development involves dramatic synaptic overproduction during prenatal and early neurodevelopment, with subsequent elimination of less adaptive neuronal circuits and their connections through synaptic pruning (Huttenlocher and Dabholkar 1997). While autism is associated with excessive early over proliferation of synaptic connectivity and gray matter in multiple brain regions, this overgrowth trajectory isthen thought to reverse during later childhood, with more advanced synaptic atrophy and neuronal loss leading to total brain volumes similar to typical development (Redcay and Courchesne 2005). Consistent with this notion of abnormal pruning trajectories, youth 8–12-years old with ASD exhibited faster rates of gray matter loss in several cortical regions over a 30-month interval compared to age-matched typically developing youth (Hardan et al. 2009), suggesting more dramatic rates of both synaptic overproduction and their subsequent elimination across development in ASD.

Functional connectivity further relies on the anatomical integrity of axonal tracts within neural networks, and several lines of evidence point to white matter (WM) abnormalities in ASD. While typical WM development linearly increases over time, MRI studies of individuals with ASD reveal accelerated early overgrowth of frontal WM followed by reductions in adolescence and adulthood (Herbert et al. 2004; Courchesne et al. 2001, 2004).Compromised interhemispheric WM connectivity is also revealed by modest reductions in overall corpus callosum size (e.g., Vidal et al. 2006), a finding also associated with underconnectivity in the prefrontal cortex(Lo et al. 2011). Callosal fiber reductions may further contribute to laterality differences seen in fMRI studies, which show that individuals with ASD tend to excessively utilize networks within the right hemisphere, including those underlying executive functioning (Cardinale et al. 2013; Gilbert et al. 2008). Even during sleep, rightward asymmetry is exhibited in 1-year-old infants affected by ASD, indicating abnormal lateralization is an early feature of neurodevelopment that predates language acquisition (Eyler et al. 2012).

Differences in axonal patterning are also seen early in development, highlighting altered early connectivity driven by both white and gray matter changes. In studies of WM development, high-risk infants who were eventually diagnosed with ASD exhibited greater fractional anisotropy, a measure of unidirectional patterning of axonal tracts, in comparison with high risk infants who did not develop ASD. This advanced spatial organization was suggestive of precocious development and heightened early network patterning (Wolff et al. 2012). Despite this heightened early specialization of WM tracts in ASD, infants subsequently exhibited slower changes in WM patterning, so that by 24 months of age, infants with ASD exhibited lower FA values than unaffected infants.

Local trajectories of synaptic overelaboration and subsequent pruning vary in step with developmental expectancies and regional specialization (Greenough et al. 1987). While primary sensory regions exhibit pruning refinements in early childhood, more highly integrative areas have a more protracted course (Gogtay and Thompson 2010). Synaptic pruning serves to drive specialization of circuitry, with a developmental sequence toward increasingly integrative regions culminating with those involved in complex thought, self-awareness, and cognitive flexibility. In contrast, ASD has been described by some authors as a disorder of mistimed critical periods driven by early imbalances in excitatory and inhibitory inputs, resulting in abnormal unfolding of developmental processes (Leblanc and Fagiolini 2011). Each iteration of regional specialization may therefore be increasingly affected at successive levels of cortical elaboration and network complexity. Aberrant early development may also contribute to downstream effects, impairing connectivity normally supported by coactivation of distributed neuronal assemblies, thus preventing the formation of stable networks, leading to further alterations in functional specialization and integration.

Similar to neocortical regions, amygdala sizes are also abnormally large in ASD during early development (Nordahl et al. 2012), yet are thought to gradually normalize into adolescence, with variable findings in older subjects (Ecker et al. 2012). Perceptual and attention networks that coordinate with amygdala nuclei may therefore receive heightened input during early periods of network patterning, and result in emotional networks more highly tuned to features of lower order sensory input that become specialized early in development. However, these highly connected early systems may restrict subsequent integration with higher order networks that typically foster specialized attunement to social and emotional functions.

4 Neurophysiology of Resting State and Social Processing

ASD is associated with neurophysiological differences in regional activation, such that examination of patterns of fMRI or EEG activation reveals different responses to task-related cognitive demands and differences during resting state compared to control subjects. Atypicalities in the "default mode network," neuronal circuits engaged during resting state conditions, have been found in studies of ASD. This resting state network is thought to involve distributed regions, including those also involved in complex tasks, suggesting a baseline level of impaired executive functioning even during rest (Uddin et al. 2013).

Resting state EEG and magnetoencephalography (MEG), a technique for detecting magnetic fields produced by electrical brain activity, have been used to quantify the (absolute or relative) amount of power at a given oscillatory frequency. A number of studies suggest an overall pattern of differential power profiles in ASD, with excess power at low frequency (delta, theta) and high frequency (beta, gamma) bands, but reduced power in middle frequency (alpha)bands (for review, see Webb et al. 2009).This pattern appears to be relatively consistent through development and is exhibited across multiple brain regions during resting state conditions. ASD has also been associated with reduced long-range coherence patterns during the resting condition, most commonly with reductions between frontal regions and more posterior primary sensory regions (Barttfeld et al. 2011; Ghanbari et al. 2013; Murias et al. 2007; Duffy and Als 2012).

Some resting state fMRI studies of adolescents and adults with ASD also reveal decreased functional connectivity(Kennedy and Courchesne 2008; Monk et al. 2009; Weng et al. 2011), with some findings of disconnect specific to integrative regions of prefrontal cortex (Assaf et al. 2010). However, in contrast to distributed underconnectivity, Keown et al. (2013) found increased local connectivity in teens with ASD during resting state conditions. This overconnectivity was primarily found in visual and extrastriate cortex, as well as temporal lobe; and functional hyperconnectivity was more marked in those with higher symptom severity. Interestingly, underconnectivity was seen in anterior regions, which tend to exhibit specialized pruning refinements later in development. This pattern of hyperconnectivity in posterior regions and hypoconnectivity in anterior may reflect impaired anterior progression of regional synaptic refinements that typically drives functional specialization seen across normative development(Gogtay and Thompson 2010). Moreover, in a younger sample of children with ASD (aged 7–13 years), hyperconnectivity was found at both long and short ranges (Supekar et al. 2013), with increased overconnectivity associated with increased social deficits. The authors suggest that hyperconnectivity may be a feature of early neurodevelopment in ASD which limits flexibility in the allocation of coordinated activity required to enable functioning of adaptive distributed networks. Neurodevelopmental shifts from early overelaboration of synaptic connectivity to later under elaboration may be responsible for developmental shifts in network activation and specialization (Uddin et al. 2013).

Neurophysiological alterations associated with ASD are also revealed during tasks designed to activate networks thought to support social functioning. The superior temporal region specifically recruited during attention tasks involving face processing is known as the Face Fusiform Area (FFA). Social processing requires attention to facial expressions and identification of subtle contextual visual cues to appreciate others' intentions and emotions. Highlighting the relationship between social and emotional processing, the FFA is further regulated by inputs from emotional networks, including amygdala nuclei (e.g., Geschwind et al. 2012).

When assessing basic face perception and identification, a number of studies show FFA hypoactivation in high functioning teens and adults with ASD comparison to controls (e.g., for review see Schultz 2005), although other studies have found no differences (e.g., Hadjikhani et al. 2004; Kleinhans et al. 2008). Variations in FFA activation may be due to individual differences in attention to eye regions (Dalton et al. 2005) or to additional task requirements (Koshino et al. 2008). It has also been suggested that attention modulation in face specific regions in ASD is impaired for social but not nonsocial information, which could account for task-related difference in performance (Bird et al. 2006; also see Dichter and Belger 2007). Perception and identification of emotional facial expressions is also altered in ASD, with decreased amygdala responses in some studies (e.g., Ashwin et al. 2007; see Baron-Cohen et al. 2000). Consistent with impaired attunement to emotional cues, highly emotional expressions failed to modulate FFA activation among subjects with ASD (Lauvin et al. 2012). During emotion processing tasks, SPECT imaging also revealed hypoactivity in frontal regions extending to the amygdala bodies, and the degree of hypoactvity correlated with symptom severity (Ohnishi et al. 2000). These findings together suggest that ASD is associated with a deficit in FFA activation and connectivity between frontal networks, evident in situations requiring social processing.

EEG and ERP studies also reveal an association between ASD and alterations in activation patterns triggered by aspects of face processing. Children with ASD show larger ERP responses to direct eye gaze, perhaps accounting for behaviors that modulate sensory input through eye contact aversion instead of more typical modulation by prefrontal dampening (Grice et al. 2005; Kyllianinen et al. 2006). Atypical or delayed temporal processing of social compared to nonsocial information has been found early in the development of autism (e.g., Webb et al. 2006, 2011) and this pattern extends through childhood and into adulthood (see Webb et al. 2009 for review). Emotional cues of face stimuli also elicit altered lower order cortical processing during early development (Dawson et al. 2004), but this pattern may normalize in late childhood (e.g., Wong et al. 2008).

EEG studies also show alterations in neuronal responses during observation of manual motor actions in ASD, with less desynchronization of neuronal assemblies (i.e., attenuation of the mu rhythm) during conditions of observation compared to imitation or execution. Unlike typically developing subjects, who exhibited attenuation of the mu rhythm during observation, imitation, and execution conditions, ASD subject only demonstrated attenuation during conditions when they executed or immediately imitated a manual motor action (Bernier et al. 2007; Oberman et al. 2005); and the degree of attenuation was correlated with imitative behaviors (Bernier et al. 2007). Because "decreased" power in the mu band (attenuation) reflects increased neural activity, the failure to increase neural resources during observation may reflect compromised integrative neurophysiological processes.

Deficits in the ability to imitate other people's actions are also commonly seen in studies of ASD, revealing another bias toward object-oriented tasks and away from direct action imitation (Williams et al. 2004). A number of neuroimaging studies have suggested that alterations inregional activation during observation and imitation of manual motor movements in individuals with ASD are indicative of impairments in mirror neuron systems (Williams et al. 2006; Marsh and Hamilton 2011); however, the extent to which patterns of hypoactivation are directly related to alterations in inferior frontal gyrus and the mirror neuron system is debated (for review, Hamilton 2013). Instead, like face processing, the pattern of responses is more complicated, and results include features of hyperactivity (Martineau et al. 2010), hypoactivity (Dapretto et al. 2005), and equivalent activity (Grezes et al. 2009; Schulte-Rüther et al. 2007).

Overall, these neurophysiologic findings indicate that alterations in cortical organization in ASD contribute to alterations inactivation responses to diverse processing demands. Findings also highlight early changes in connectivity, contributing to developmental shifts in activation patterns both at rest and within regions typically associated with specialized tasks essential for social learning. These connectivity differences are also observed within both local and distributed networks in ASD, and are likely impacted by alterations in developmental trajectories of synaptic refinements which affect regional specialization.

5 Alterations in Neuronal Network Specialization

Clinically, children with autism demonstrate an extremely high rate of idiosyncratic sensory responsivity. Heightened reactivity to aversive sensory stimuli in ASD is thought to be driven by hyperconnectivity of local sensory networks in conjunction with decreased modulation from integrative frontal networks. This systems connectivity deficit provides an alternate perspective on local sensory hyperactivity, which has been proposed as a model accounting for alternate sensory processing strategies in individuals with autism. Specifically, quicker neuronal responsivity based on EEG is seen in some visual paradigms, highlighting earlier activation patterns in ASD (Boeschoten et al. 2007). However, the P1 response, an event-related EEG component thought to be generated by extrastriate activity, was smaller in a study of PDD subjects compared to controls; as well, inferior medial sources were also found to be weaker. Increased activity was instead observed in the superior lateral visual area, suggesting alterations in anatomic separation and specialization, specifically in early visual processing networks.

Alterations in the specialization of visual networks in ASD is also demonstrated in heightened maintenance of sensitivity in peripheral visual fields with less enhancement of foveal regions. Clinical signs of aberrant eye contact and lateral glance behavior may therefore be related to behavioral modulation of sensory input due to differential activation patterns at early sensory levels, which may otherwise overwhelm processing capabilities (e.g., Mottron et al. 2007). Perceptual modulation is normally afforded by neurodevelopmental dampening of less relevant sensory input. Reduced developmental refinement of peripheral visual field sensitivity further highlights impaired perceptual modulation in ASD.

ASD is also marked by stereotypic behaviors, which are hypothesized to represent an effort to enhance proprioceptive processing by conveying greater contextual proprioceptive and visual cues. This stereotypic behavior may serve to strengthen proprioceptive circuits, resulting in greater allocation of attention resources to seeking and maintaining high levels of sensory stimulation, particularly those related to visual and proprioceptive cues. Heightened reliance on local sensory cortical regions is also demonstrated in behavioral enhancements, including improved processing of elemental visual features and better auditory tone discrimination (see Mottron et al. 2006; Dakin and Frith 2005). However, as perceptual processing tasks begin to involve more complex visual features or noise with high variability, performance on these tasks may start to show impairments (Simmons et al. 2009).

Clinically, ASD is also associated with rigid behavioral routines and difficulties adapting to shifting environmental demands. Anterior cingulate cortex (ACC), a region involved in monitoring and set-shifting to adapt to new conditions and demands, is thought to assist in cognitive flexibility. Subjects with ASD exhibited deficits in performance of tasks requiring cognitive flexibility, and the tasks elicited less activation than controls in ACC and other frontal regions (Shafritz et al. 2008). Hypoactivity was also seen in basal ganglia and parietal regions, but decreased activation in ACC was specifically associated with clinical severity of repetitive behavior.

Further, ASD has been associated with impairments in top-down regulation, affecting how cognitions and expectancies influence even basic perceptions. Top-down modulation is also required for shaping perceptual processes to accord with beliefs and appreciation of contextual cues. Consistent with impaired top-down modulation, individuals with ASD exhibit differences in illusory perceptions and their impact on modulating behavior. For example, the rubber-hand illusion, when individuals are exposed to visual information of a rubber hand being touched while their true hand is also touched simultaneously, typically elicits a sense that touch is indeed felt on the rubber hand, and a heightened sense of ownership of the rubber hand. This requires cross modal integration of visual, tactile, and proprioceptive information, which can be influenced by perceptual beliefs relevant to the context. Even when individuals with ASD cognitively appreciate the illusory effect, they subsequently maintain more accurate proprioception than individuals without ASD traits, as their motor responses are less impacted by these higher order perceptual shifts (Palmer et al. 2013). This increased reliance on lower level sensory estimates may serve to strengthen these local systems over time and may further perpetuate deficits in higher order contextual processing (Mottron et al. 2006; Dakin and Frith 2005; Simmons et al. 2009).

Other perceptual tasks that require more abstract social and emotional reasoning have also highlighted a lack of higher order integration in ASD. For example, among individuals with ASD, regional fMRI activation was not appropriately affected by modulating videos to depict good versus bad intentions of actors (Pinkham et al. 2008), nor by ironic content (Wang et al. 2007). Appreciation of these more nuanced aspects of others' behavior requires integration of diverse networks, including those thought to underlie empathy and one's sense of self. Altered utilization of these networks may thus contribute to the conceptualization of ASD as a deficiency in Theory of Mind and contribute to idiosyncratic alterations in social repertoires.

6 Conclusions and Treatment Implications

Studies examining the neurophysiology of ASD describe distributed alterations in activation patterns of regional networks, highlighting impacts on perceptual processing in basic sensory tasks as well as modulation from higher order regions typically involved in complex social and emotional processing. Despite evidence for underutilization and hypoconnectivity of regions important for social processing in ASD, earlier in neurodevelopment ASD appears to be associated with heightened anatomic and functional connectivity. This developmental trajectory may place greater demands on primary sensory systems, enhancing hyperconnectivity of localized networks and shifting developmental specialization away from more integrative networks. This shift in network specialization may contribute to alterations in processing strategies, behaviorally expressed as idiosyncratic patterns that favor proprioceptive and other lower order sensory cues. Behavioral stereotypies manipulating sounds, smells or movement often seen in children with ASD likely serve to enrich and bind this sensory input, similar to how slight head movement allows visual perception to become richer and more seamless, even when looking through a mesh screen. Emotional networks may also shift to become integrated into sensory experiences apart from interpersonal relatedness, contributing to alterations in communication, and social reactivity.

This conceptualization suggests that treatments should be focused on reestablishing specialization of networks underlying higher order contextual and social processing. This type of behavioral intervention often initially requires increasing environmental structure to provide a more restricted context for learning, thereby allowing more attentional resources to be paid to other cognitive demands, such as social interactions or academic performance. Evidence also points to enhanced neurophysiologic functioning when attention is restricted to cues that are highly relevant for social perception, such as by improving eye contact and actively attending to the perspective of others. These experiences may enable temporal coordination of networks to become increasingly specialized and attuned to effectively process relevant information and enhance behavioral functioning.

The refinement of networks underlying clinical benefit in ASD may further be fostered by pharmacologic approaches, particularly those that can drive adaptive specialization. Some EEG studies of pharmacotherapy for ADHD suggest enhanced coherence of distributed cortical regions following treatment (Dupuy et al. 2010), suggesting medication-mediated changes in neurophysiologic patterns may cause or contribute to clinical improvement. Increasing evidence for plasticity induced changes associated with SSRIs suggests they may alter normative developmental timing, which highlights the role that medications may play in enhancing the influence of experience on neuroanatomy and neurophysiology (Castren 2013).

One caveat to neurophysiological findings in ASD is that high medication rates may confound studies, as many include subjects already on psychotropic medications, of which several commonly prescribed are known to affect neural oscillations (e.g., Blume 2006; Dumont et al. 2005; Loo et al. 1999). Additionally, while there is a scarcity of knowledge about the effects of medication on EEG and fMRI measures, even less is known about short and long-term effects of medication on neural anatomy.

Medication treatments for ASD have nevertheless increasingly been focused on targeting alterations in synaptic connectivity (Won et al. 2013). Little is known about the direct effects of atypical antipsychotics on brain activity; however in adult clinical trials, risperidone was found to normalize EEG findings in autistic subjects by increasing EEG theta power (Liem-Moolenaar et al. 2011). Aripiprazole treatment initially increased delta frequency power (Kim et al. 2006), but resulted in decreased delta after 8 weeks (Canive et al. 1998). The Early Start Denver comprehensive behavioral treatment for ASD also reveal evidence of clinical improvements and normalization of EEG profiles, suggesting both pharmacologic and behavioral treatments may address features of neurophysiological alterations (Dawson et al. 2012).

Individuals with ASD may benefit from medication interventions that enable more adaptive functioning of social and emotional processing networks through enhancements in neuronal plasticity and regional specialization. These treatments should be maximized through coordination with ongoing behavioral supports, and integrative research approaches to study early intervention strategies are important to determine mediators of clinical response. Studies of ASD must also be viewed within a shifting neurodevelopmental landscape, and longitudinal characterization of neurophysiological functioning and patterns of regional specialization are needed. These markers of network connectivity can offer insights into the clinical impairments associated with ASD, and more importantly, reveal the impact of important treatment interventions on brain functioning.

Acknowledgment Support for this project was provided to Dr. Webb by the National Institutes of Health (R01-HD064820, R01-MH10028, HHSN271201200005I).

References

Aman MG, Langworthy KS (2000) Pharmacotherapy for hyperactivity in children with autism and other pervasive developmental disorders. J Autism Dev Disord 30(5):451–459

Aman MG, Arnold LE, Lindsay R, Nash P, Hollway J, McCracken JT, Shah B, McMahon D (2005) Risperidone treatment of autistic disorder: longer-term benefits and blinded discontinuation after 6 months. Am J Psychiatry 162(7):1361–1369

American Psychological Association (2013) Diagnostic and statistical manual of mental disorders (DSM-5). American Psychiatric Publishing, Arlington

Arnold LE, Aman MG, Cook AM, Witwer AN, Hall KL, Thompson S, Ramadan Y (2006) Atomoxetine for hyperactivity in autism spectrum disorders: placebo-controlled crossover pilot trial. J Am Acad Child Adolesc Psychiatry 45(10):1196–1205

Ashwin C, Baron-Cohen S, Wheelwright S, O'Riordan M, Bullmore ET (2007) Differential activation of the amygdala and the 'social brain'during fearful face-processing in Asperger Syndrome. Neuropsychologia 45(1):2–14

Assaf M, Jagannathan K, Calhoun VD, Miller L, Stevens MC, Sahl R, O'Boyle J, Schultz R, Pearlson GD (2010) Abnormal functional connectivity of default mode sub-networks in autism spectrum disorder patients. Neuroimage 53(1):247–256

Aylward EH, Minshew NJ, Field K, Sparks BF, Singh N (2002) Effects of age on brain volume and head circumference in autism. Neurology 59(2):175–183

Baron-Cohen S, Ring HA, Bullmore ET, Wheelwright S, Ashwin C, Williams SCR (2000) The amygdala theory of autism. Neurosci Biobehav Rev 24(3):355–364

Barttfeld P, Wicker B, Cukier S, Navarta S, Lew S, Sigman M (2011) A big-world network in ASD: dynamical connectivity analysis reflects a deficit in long-range connections and an excess of short-range connections. Neuropsychologia 49(2):254–263

Bernier R, Dawson G, Webb S, Murias M (2007) EEG mu rhythm and imitation impairments in individuals with autism spectrum disorder. Brain Cogn 64(3):228–237

Bird G, Catmur C, Silani G, Frith C, Frith U (2006) Attention does not modulate neural responses to social stimuli in autism spectrum disorders. Neuroimage 31(4):1614–1624

Blume WT (2006) Drug effects on EEG. J Clin Neurophysiol 23(4):306–311

Boeschoten MA, Kenemans JL, Engeland HV, Kemner C (2007) Abnormal spatial frequency processing in high-functioning children with pervasive developmental disorder (PDD). Clin Neurophysiol 118(9):2076–2088

Canive JM, Lewine JD, Edgar JC, Davis JT, Miller GA, Torres F, Tuason VB (1998) Spontaneous brain magnetic activity in schizophrenia patients treated with aripiprazole. Psychopharmacol Bull 34(1):101

Cardinale RC, Shih P, Fishman I, Ford LM, Müller RA (2013) Pervasive rightward asymmetry shifts of functional networks in Autism Spectrum Disorder. JAMA Psychiatry 70(9):975–982

Carper RA, Courchesne E (2005) Localized enlargement of the frontal cortex in early autism. Biol Psychiatry 57(2):126–133

Carrasco M, Volkmar FR, Bloch MH (2012) Pharmacologic treatment of repetitive behaviors in autism spectrum disorders: evidence of publication bias. Pediatrics 129(5):e1301–e1310

Casanova MF, van Kooten IA, Switala AE, van Engeland H, Heinsen H, Steinbusch HW, Hof P, Trippe J, Stone J, Schmitz C (2006) Minicolumnar abnormalities in autism. Acta Neuropathol 112(3):287–303

Castrén E (2013) Neuronal network plasticity and recovery from depression. JAMA Psychiatry 70(9):983–989

Center for Disease Control (2012) Prevalence of autism spectrum disorders–autism and developmental disabilities monitoring network, 14 Sites, United States, 2008

Courchesne E, Karns CM, Davis HR, Ziccardi R, Carper RA, Tigue ZD, Chisum H, Courchesne RY (2001) Unusual brain growth patterns in early life in patients with autistic disorder an MRI study. Neurology 57(2):245–254

Courchesne E, Mouton PR, Calhoun ME, Semendeferi K, Ahrens-Barbeau C, Hallet MJ, Barnes C, Pierce K (2011) Neuron number and size in prefrontal cortex of children with autism. J Am Med Assoc 306(18):2001–2010

Dakin S, Frith U (2005) Vagaries of visual perception in autism. Neuron 48(3):497–507

Dalton KM, Nacewicz BM, Johnstone T, Schaefer HS, Gernsbacher MA, Goldsmith HH, Alexander A, Davidson RJ (2005) Gaze fixation and the neural circuitry of face processing in autism. Nat Neurosci 8(4):519–526

Dapretto M, Davies MS, Pfeifer JH, Scott AA, Sigman M, Bookheimer SY, Iacoboni M (2005) Understanding emotions in others: mirror neuron dysfunction in children with autism spectrum disorders. Nat Neurosci 9(1):28–30

Dawson G, Jones EJ, Merkle K, Venema K, Lowy R, Faja S, Kamara D, Murias M, Greenson J, Winter J, Smith M, Rogers S, Webb SJ (2012) Early behavioral intervention is associated with normalized brain activity in young children with autism. J Am Acad Child Adolesc Psychiatry 51(11):1150–1159

Dawson G, Webb SJ, Carver L, Panagiotides H, McPartland J (2004) Young children with autism show atypical brain responses to fearful versus neutral facial expressions of emotion. Dev Sci 7(3):340–359

Dichter GS, Belger A (2007) Social stimuli interfere with cognitive control in autism. Neuroimage 35(3):1219–1230

Duffy FH, Als H (2012) A stable pattern of EEG spectral coherence distinguishes children with autism from neuro-typical controls-a large case control study. BMC Med 10(1):64

Dumont GJH, De Visser SJ, Cohen AF, Van Gerven JMA (2005) Biomarkers for the effects of selective serotonin reuptake inhibitors (SSRIs) in healthy subjects. Br J Clin Pharmacol 59(5):495–510

Dupuy FE, Clarke AR, Barry RJ, McCarthy R, Selikowitz M (2010) EEG coherence in children with attention-deficit/hyperactivity disorder: differences between good and poor responders to methylphenidate. Psychiatry Res 180(2–3):114–119

Ecker C, Suckling J, Deoni S.C, Lombardo M.V, Bullmore E.T, Baron-Cohen S, Williams C (2012) Brain anatomy and its relationship to behavior in adults with autism spectrum disorder: a multicenter magnetic resonance imaging study. Arch Gen Psychiatry 69(2):195–209

Elbe D, Lalani Z (2012) Review of the pharmacotherapy of irritability in autism. J Can Acad Child Adolesc Psychiatry 21(2):130–146

Eyler LT, Pierce K, Courchesne E (2012) A failure of left temporal cortex to specialize for language is an early emerging and fundamental property of autism. Brain 135(Pt 3):949–960

Geschwind M, Pourtois G, Schwartz S, Van De Ville D, Vuilleumier P (2012) White-matter connectivity between face-responsive regions in the human brain. Cereb Cortex 22(7):1564–1576

Ghanbari Y, Bloy L, Edgar JC., Blaskey L, Verma R, Roberts TP (2013) Joint analysis of band-specific functional connectivity and signal complexity in autism. J. Autism Dev Disord 2(1):17

Gilbert SJ, Bird G, Brindley R, Frith CD, Burgess PW (2008) Atypical recruitment of medial prefrontal cortex in autism spectrum disorders: an fMRI study of two executive function tasks. Neuropsychologia 46(9):2281–2291

Gogtay N, Thompson PM (2010) Mapping gray matter development: implications for typical development and vulnerability to psychopathology. Brain Cogn 72(1):6–15

Gordon I, Vander Wyk BC, Bennett RH, Cordeaux C, Lucas MV, Eilbott JA, Pelphrey KA (2013) Oxytocin enhances brain function in children with autism. Proc Nat Acad Sci U.S.A 110(52):20953–20958

Greenough WT, Black JE, Wallace CS (1987) Experience and brain development. Child Dev 58(3), 539–559

Grèzes J, Wicker B, Berthoz S, De Gelder B (2009) A failure to grasp the affective meaning of actions in autism spectrum disorder subjects. Neuropsychologia 47(8):1816–1825

Grice SJ, Halit H, Farroni T, Baron-Cohen S, Bolton P, Johnson MH (2005) Neural correlates of eye-gaze detection in young children with autism. Cortex 41(3):342–353

Guastella AJ, Einfeld SL, Gray KM, Rinehart NJ, Tonge BJ, Lambert TJ, Hickie IB (2010) Intranasal oxytocin improves emotion recognition for youth with autism spectrum disorders. Biol Psychiatry 67(7):692–694

Hadjikhani N, Joseph RM, Snyder J, Chabris CF, Clark J, Steele S, McGrath L, Tager-Flusberg H (2004) Activation of the fusiform gyrus when individuals with autism spectrum disorder view faces. Neuroimage 22(3):1141–1150

Hamilton AF (2013) The mirror neuron system contributes to social responding. Cortex 49(10):2957–2959

Handen BL, Sahl R, Hardan AY (2008) Guanfacine in children with autism and/or intellectual disabilities. J Dev Behav Pediatr 29(4):303–308

Hardan AY, Libove RA, Keshavan MS, Melhem NM, Minshew NJ (2009) A preliminary longitudinal magnetic resonance imaging study of brain volume and cortical thickness in autism. Biol Psychiatry 66(4):320–326

Harris JC, Carter CS (2013) Therapeutic interventions with oxytocin: current status and concerns. J Am Acad Child Adolesc Psychiatry 52(10):998–1000

Hellings JA, Zarcone JR, Reese RM, Valdovinos MG, Marquis JG, Fleming KK, Schroeder SR (2006) A crossover study of risperidone in children, adolescents and adults with mental retardation. J Autism Dev Disord 36(3):401–411

Herbert MR, Ziegler DA, Makris N, Filipek PA, Kemper TL, Normandin JJ, Caviness, VS (2004). Localization of white matter volume increase in autism and developmental language disorder. Ann Neurol 55(4):530–540

Hsia Y, Wong AY, Murphy DG, Simonoff E, Buitelaar JK, Wong, IC (2013) Psychopharmacological prescriptions for people with autism spectrum disorder (ASD): a multinational study. Psychopharmacology 13:1–11

Hurwitz R, Blackmore R, Hazell P, Williams K, Woolfenden S (2012) Tricyclic antidepressants for autism spectrum disorders (ASD) in children and adolescents. Cochrane Database Syst Rev (3)

Hutsler JJ, Zhang H (2010) Increased dendritic spine densities on cortical projection neurons in autism spectrum disorders. Brain Res 1309:83–94

Huttenlocher PR, Dabholkar AS (1997) Regional differences in synaptogenesis in human cerebral cortex. J Comp Neurol 387(2):167–178

Kennedy DP, Courchesne E (2008) The intrinsic functional organization of the brain is altered in autism. Neuroimage 39(4):1877–1885

Keown CL, Shih P, Nair A, Peterson N, Mulvey ME, Müller RA (2013) Local functional over connectivity in posterior brain regions is associated with symptom severity in autism spectrum disorders. Cell Rep 5(3):567–572

Kim E, Yu KS, Cho JY, Shin YW, Yoo SY, Kim YY, Jang I-J, Shin S-G, Kwon JS (2006) Effects of DRD2 and CYP2D6 genotypes on delta EEG power response to aripiprazole in healthy male volunteers: a preliminary study. Hum Psychopharmacol: Clin Exp 21(8):519–528

King BH, Wright D, Handen BL, Sikich L, Zimmerman AW, McMahon W, Cantwell E, Cook EH Jr (2001) Double-blind, placebo-controlled study of amantadine hydrochloride in the treatment of children with autistic disorder. J Am Acad Child Adolesc Psychiatry 40(6): 658–665

Kleinhans NM, Richards T, Sterling L, Stegbauer KC, Mahurin R, Johnson LC, Greenson J, Dawson G, Aylward E (2008) Abnormal functional connectivity in autism spectrum disorders during face processing. Brain 131(4):1000–1012

Kolevzon A, Mathewson KA, Hollander E (2006) Selective serotonin reuptake inhibitors in autism: a review of efficacy and tolerability. J Clin Psychiatry 67(3):407–414

Koshino H, Kana RK, Keller TA, Cherkassky VL, Minshew NJ, Just MA (2008) fMRI investigation of working memory for faces in autism: visual coding and underconnectivity with frontal areas. Cereb Cortex 18(2):289–300

Kylliäinen A, Braeutigam S, Hietanen JK, Swithenby SJ, Bailey AJ (2006) Face- and gaze-sensitive neural responses in children with autism: a magnetoencephalographic study. Eur J Neurosci 24(9):2679–2690

Lauvin MA, Martineau J, Destrieux C, Andersson F, Bonnet-Brilhault F, Gomot M, El-Hage W, Cottier JP (2012) Functional morphological imaging of autism spectrum disorders: current position and theories proposed. Diagn Interv Imaging 93(3):139–147

LeBlanc JJ, Fagiolini M (2011) Autism: a "critical period" disorder? Neural Plasticity 18(3):264–276

Liem-Moolenaar M, Rad M, Zamuner S, Cohen AF, Lemme F, Franson KL, van Gerven J, Pich EM (2011) Central nervous system effects of the interaction between risperidone (single dose) and the 5-HT6 antagonist SB742457 (repeated doses) in healthy men. Br J Clin Pharmacol 71(6):907–916

Lo YC, Soong WT, Gau SSF, Wu YY, Lai MC, Yeh FC, Chiang W-Y, Kuo L-W, Jaw F-S, Tseng WYI (2011) The loss of asymmetry and reduced interhemispheric connectivity in adolescents with autism: a study using diffusion spectrum imaging tractography. Psychiatry Res: Neuroimaging 192(1):60–66

Loo SK, Teale PD, Reite ML (1999) EEG correlates of methylphenidate response among children with ADHD: a preliminary report. Biol Psychiatry 45(12):1657–1660

Lopata C, Toomey JA, Fox JD, Thomeer ML, Volker MA, Lee GK (2013) Prevalence and predictors of psychotropic use in children with high-functioning ASDs. Autism Res Treat 103(5):881–888

Luby J, Mrakotsky C, Stalets MM ABelden A, Heffelfinger A, Williams M, Spitsnagel E (2006) Risperidone in preschool children with autistic spectrum disorders: An investigation of safety and efficacy. J Child Adolesc Psychopharmacol 16(5):575–587

Mahajan R, Bernal MP, Panzer R, Whitaker A, Roberts W, Handen B, Veenstra-VanderWeele J (2012) Clinical practice pathways for evaluation and medication choice for attention-deficit/hyperactivity disorder symptoms in autism spectrum Disorders. Pediatrics 130(Supplement 2):S125–S138

Malow B, Adkins KW, McGrew SG, Wang L, Goldman SE, Fawkes D, Burnette C (2012) Melatonin for sleep in children with autism: a controlled trial examining dose, tolerability, and outcomes. J Autism Dev Disord 42(8):1729–1737

Marcus RN, Owen R, Kamen L, Manos G, McQuade R, Carson WH, Aman MG (2009) A placebo-controlled fixed dose study of aripiprazole in children and adolescents with irritability associated with autistic disorder. J Am Acad Child Adolesc Psychiatry 48(11):1110–1119

Marsh LE, Hamilton AFDC (2011) Dissociation of mirroring and mentalising systems in autism. Neuroimage 56(3):1511–1519

Martineau J, Andersson F, Barthélémy C, Cottier JP, Destrieux C (2010) Atypical activation of the mirror neuron system during perception of hand motion in autism. Brain Res 12(1320):168–175

McCracken JT, McGough J, Shah B, Cronin P, Hong D, Aman MG, Arnold LE, McMahon D (2002) Risperidone in children with autism and serious behavioral problems. N Engl J Med 347(5):314–321

Ming X, Gordon E, Kang N, Wagner GC (2008) Use of clonidine in children with autism spectrum disorders. Brain Dev 30(7):454–460

Miral S, Gencer O, Baykara B, Baykara A, Dirik E (2008) Risperidone versus haloperidol in children and adolescents with Autistic Disorder. Eur Child Adolesc Psychiatry 17(1):1–8

Mire SS, Nowell KP, Kubiszyn T, Goin-Kochel RP (2013) Psychotropic medication use among children with autism spectrum disorders within the Simons Simplex Collection: are core features of autism spectrum disorder related? Autism

Monk CS, Peltier SJ, Wiggins JL, Weng SJ, Carrasco M, Risi S, Lord C (2009) Abnormalities of intrinsic functional connectivity in autism spectrum disorders. Neuroimage 47(2):764–772

Mottron L, Dawson M, Soulieres I, Hubert B, Burack J (2006) Enhanced perceptual functioning in autism: an update, and eight principles of autistic perception. J Autism Dev Disord 36(1): 27–43

Mottron L, Mineau S, Martel G, Bernier CS, Berthiaume C, Dawson M, Faubert, J (2007) Lateral glances toward moving stimuli among young children with autism: early regulation of locally oriented perception? Dev Psychopathol 19(1):23–36

Murias M, Webb SJ, Greenson J, Dawson G (2007) Resting state cortical connectivity reflected in EEG coherence in individuals with autism. Biol Psychiatry 62(3):270–273

Murray M (2010) Attention-deficit/hyperactivity disorder in the context of autism spectrum disorders. Curr Psychiatry Rep 123(5):382–388

Nagaraj R, Singhi P, Malhi P (2006) Risperidone in children with autism: randomized, placebo-controlled, double-blind study. J Child Neurol 21(6):450–455

Nordahl CW, Scholz R, Yang X, Buonocore MH, Simon T, Rogers S, Amaral DG (2012) Increased rate of amygdala growth in children aged 2 to 4 years with autism spectrum disorders: a longitudinal study. Arch Gen Psychiatry 69(1):53–61

Oberman LM, Hubbard EM, McCleery JP, Altschuler EL, Ramachandran VS, Pineda JA (2005) EEG evidence for mirror neuron dysfunction in autism spectrum disorders. Cogn Brain Res 24(2):190–198

Ohnishi T, Matsuda H, Hashimoto T, Kunihiro T, Nishikawa M, Uema T, Sasaki M (2000) Abnormal regional cerebral blood flow in childhood autism. Brain 123(Pt 9):1838–1844

Owen R, Sikich L, Marcus RN, Corey-Lisle P, Manos G, McQuade RD, Findling RL (2009) Aripiprazole in the treatment of irritability in children and adolescents with autistic disorder. Pediatrics 124(6):1533–1540

Palmer CJ, Paton B, Hohwy J, Enticott PG (2013) Movement under uncertainty: the effects of the rubber-hand illusion vary along the nonclinical autism spectrum. Neuropsychologia 51(10):1942–1951

Pandina GJ, Bossie CA, Youssef E, Zhu Y, Dunbar F (2007) Risperidone improves behavioral symptoms in children with autism in a randomized, double-blind, placebo-controlled trial. J Autism Dev Disord 37(2):367–373

Parikh MS, Kolevzon A, Hollander E (2008) Psychopharmacology of aggression in children and adolescents with autism: A critical review of efficacy and tolerability. J Child Adolesc Psychopharmacol 18(2):157–178

Pinkham AE, Hopfinger JB, Pelphrey KA, Piven J, Penn DL (2008) Neural bases for impaired social cognition in schizophrenia and autism spectrum disorders. Schizophr Res 99(1):164–175

Pringsheim T, Gorman D (2012) Second-generation antipsychotics for the treatment of disruptive behaviour disorders in children: a systematic review. Can J Psychiatry 57(12):722–727

Redcay E, Courchesne E (2005) When is the brain enlarged in autism? A meta-analysis of all brain size reports. Biol Psychiatry 58(1):1–9

Rossignol DA, Frye RE (2013) Melatonin in Autism Spectrum Disorders. Curr Clin Pharmacol. [Epub ahead of print]

Schubart JR, Camacho F, Leslie D (2013) Psychotropic medication trends among children and adolescents with autism spectrum disorder in the Medicaid program. Autism. [Epub ahead of print]

Schulte-Rüther M, Markowitsch HJ, Fink GR, Piefke M (2007) Mirror neuron and theory of mind mechanisms involved in face-to-face interactions: a functional magnetic resonance imaging approach to empathy. J Cogn Neurosci 19(8):1354–1372

Schultz RT (2005) Developmental deficits in social perception in autism: the role of the amygdala and fusiform face area. Int J Dev Neurosci 23(2):125–141

Schumann CM, Bloss CS, Barnes CC, Wideman GM, Carper RA, Akshoomoff N, Courchesne E (2010) Longitudinal magnetic resonance imaging study of cortical development through early childhood in autism. J Neurosci 30(12):4419–4427

Shafritz KM, Dichter GS, Baranek GT, Belger A (2008) The neural circuitry mediating shifts in behavioral response and cognitive set in autism. Biol Psychiatry 63(10):974–980

Sharma A, Shaw SR (2012) Efficacy of risperidone in managing maladaptive behaviors for children with autistic spectrum disorder: a meta-analysis. J Pediatric Health Care 26(4): 291–299

Shea S, Turgay A, Carroll A, Schulz M, Orlik H, Smith I, Dunbar F (2004) Risperidone in the treatment of disruptive behavioral symptoms in children with autistic and other pervasive developmental disorders. Pediatrics 114(5):e634–e641

Silberberg G, Gupta A, Markram H (2002) Stereotypy in neocortical microcircuits. Trends Neurosci 25(5):227–230

Simmons DR, Robertson AE, McKay LS, Toal E, McAleer P, Pollick FE (2009) Vision in autism spectrum disorders. Vision Res 49(22):2705–2739

Sparks BF, Friedman SD, Shaw DW, Aylward EH, Echelard D, Artru AA, Maravilla K, Dager SR (2002) Brain structural abnormalities in young children with autism spectrum disorder. Neurology 59(2):184–192

Supekar K, Uddin LQ, Khouzam A, Phillips J, Gaillard WD, Kenworthy LE, Yerys B, Vaidya C, Menon V (2013) Brain hyperconnectivity in children with autism and its links to social deficits. Cell Reports 5(3):738–747

Troost PW, Lahuis BE, Steenhuis MP, Ketelaars CE, Buitelaar JK, van Engeland H, Hoekstra PJ (2005) Long-term effects of risperidone in children with autism spectrum disorders: a placebo discontinuation study. J Am Acad Child Adolesc Psychiatry 44(11):1137–1144

Uddin LQ, Superkar K, Lynch CJ, Khouzam A, Phillips J, Feinstein C et al (2013) Salience network based classification and prediction of symptom severity in children with autism. JAMA Psychiatry 70(8):869–879

Uddin LQ, Supekar K, Menon V (2010) Typical and atypical development of functional human brain networks: insights from resting-state FMRI. Frontiers Syst Neurosci 4(21)

Vidal CN, Nicolson R, DeVito TJ, Hayashi KM, Geaga JA, Drost DJ, Williamson P, Thompson PM (2006) Mapping corpus callosum deficits in autism: an index of aberrant cortical connectivity. Biol Psychiatry 60(3):218–225

Wang AT, Lee SS, Sigman M, Dapretto M (2007) Reading affect in the face and voice: neural correlates of interpreting communicative intent in children and adolescents with autism spectrum disorders. Arch Gen Psychiatry 64(6):698

Webb SJ, Dawson G, Bernier R, Panagiotides H (2006) ERP evidence of atypical face processing in young children with autism. J Autism Dev Disord 36(7):881–890

Webb S, Bernier R, Burner K, Murias M (2009) Electrophysiology and evoked potentials in autism research. In: Amaral D, Dawson G, Geschwind D (eds) Autism spectrum disorders. Oxford Library of Psychology, New York

Webb SJ, Jones EJ, Merkle K, Venema K, Greenson J, Murias M, Dawson G (2011) Developmental change in the ERP responses to familiar faces in toddlers with autism spectrum disorders versus typical development. Child Dev 82(6):1868–1886

Weng SJ, Carrasco M, Swartz JR, Wiggins JL, Kurapati N, Liberzon I, Monk CS (2011) Neural activation to emotional faces in adolescents with autism spectrum disorders. J Child Psychol Psychiatry 52(3):296–305

Williams JH, Whiten A, Singh T (2004) A systematic review of action imitation in autistic spectrum disorder. J Autism Dev Disord 34(3):285–299

Williams K, Wheeler DM, Silove N, Hazell P (2013) Selective serotonin reuptake inhibitors (SSRIs) for autism spectrum disorders (ASD). Cochrane Database of Syst Rev (8)

Williams JH, Waiter GD, Gilchrist A, Perrett DI, Murray AD, Whiten A (2006) Neural mechanisms of imitation and 'mirror neuron'functioning in autistic spectrum disorder. Neuropsychologia 44(4):610–621

Wolff JJ, Gu H, Gerig G, Elison JT, Styner M, Gouttard S, Botteron K, Piven J (2012) Differences in white matter fiber tract development present from 6 to 24 months in infants with autism. Am J Psychiatry 169(6):589–600

Won H, Mah W, Kim E (2013) Autism spectrum disorder causes, mechanisms, and treatments: focus on neuronal synapses. Front Mol Neurosci 6(19)

Wong TK, Fung PC, Chua SE, McAlonan GM (2008) Abnormal spatiotemporal processing of emotional facial expressions in childhood autism: dipole source analysis of event-related potentials. Eur J Neurosci 28(2):407–416

Wright B, Sims D, Smart S, Alwazeer A, Alderson-Day B, Allgar V, Whitton C, Miles J (2011) Melatonin versus placebo in children with autism spectrum conditions and severe sleep problems not amenable to behaviour management strategies: a randomised controlled crossover trial. J Autism Dev Disord 41(2):175–184

Physiological Correlates of Insomnia

Timothy Roehrs, Valentina Gumenyuk, Christopher Drake and Thomas Roth

Abstract Insomnia is a prevalent sleep disorder that is typically comorbid with medical, psychiatric, and other sleep disorders. Yet, it is a disorder with its own course and morbidity that can persist if untreated. This chapter describes the physiological correlates of insomnia expressed during sleep and during the daytime. Together, the data from nighttime and daytime electrophysiology, event-related brain potential recording, neuroimaging studies, sympathetic nervous system, and HPA axis monitoring all suggest that insomnia is a 24 h disorder of hyperarousal.

Keywords Polysomnography · Event related potentials · Multiple Sleep Latency Test · Sympathetic nervous system · HPA

Contents

T. Roehrs (✉) · V. Gumenyuk · C. Drake · T. Roth
Sleep Disorders & Research Center, Henry Ford Hospital, Detroit, MI, USA
e-mail: TROEHRS1@hfhs.org

T. Roehrs · C. Drake · T. Roth
Department of Psychiatry and Behavioral Neuroscience, Wayne State University, School of Medicine, Detroit, MI, USA

V. Gumenyuk
MEG Imaging Center, Meadowlands Hospital, Secaucus, NJ, USA

Curr Topics Behav Neurosci (2014) 21: 277–290
DOI: 10.1007/7854_2014_324

Published Online: 12 June 2014

1 Introduction

Insomnia is the most prevalent sleep disorder in the population with approximately 30 % of respondents reporting insomnia symptoms and 20 % meeting diagnostic criteria (Roth et al. 2007). Over the past decade there has been a major paradigm shift in understanding the pathophysiology, diagnosis, and treatment of insomnia. It is now recognized that insomnia is a disorder with significant morbidity and its own course that can persist for years if untreated (NIH 2005). It typically coexists with any number of medical and psychiatric disorders and other primary sleep disorders. Of particular relevance to the focus of this book is the fact that persistent insomnia is a risk factor for the development of major depressive disorder (MDD), is an important symptom predictive of MDD relapse, and can remain after successful treatment of the MDD (Krystal 2006). Unlike MDD, insomnia coexisting with anxiety disorders typically follows the onset of the specific anxiety disorder and follows relapse of the anxiety disorder (Johnson et al. 2006). This chapter will describe the physiological correlates of *insomnia,* expressed both during sleep and during the daytime. As this chapter will indicate, insomnia can be considered a 24 h disorder of arousal.

2 Insomnia Diagnosis

The *Diagnostic and Statistical Manual of Mental Disorders,* 5th edition (DSM-V) defines *insomnia* (see Table 1) as difficulty initiating sleep, maintaining sleep, and/or early awakening with inability to return to sleep. These complaints are present despite there being an adequate opportunity and circumstances to sleep (American Psychiatric Association 2013). The sleep disturbance must have a frequency of at least three nights per week, a duration of at least 3 months, and cause clinically significant impairment in daytime functioning. The sleep disturbance cannot occur *exclusively* during the course of another sleep disorder and is not attributable to drugs of abuse or medications. When coexisting with a psychiatric or medical disorder the insomnia symptom is prominent and can persist after resolution of the psychiatric or medical disorder.

The DSM-V reflects a number of significant changes from the DSM-IV-R. First, *insomnia* is no longer dichotomized into *primary insomnia* and *comorbid insomnia* (i.e., insomnia related to another disorder). This change in nomenclature moves from the causal attribution that is inherent in a primary versus secondary distinction. DSM-V then directs that any comorbid psychiatric or medical disorder

Table 1 DSM-V diagnostic criteria for insomnia 780.52

(A) A predominant complaint of dissatisfaction with sleep quantity or quality associated with one (or more) of the following symptoms
1. Difficulty initiating sleep
2 Difficulty maintaining sleep
3. Early-morning awakening awakening with inability to return to sleep
(B) The sleep disturbance causes clinically significant distress or impairment in social, occupational, educational, academic, behavioral, or other important areas of functioning
(C) The sleep difficulty occurs at least 3 nights per week
(D) The sleep difficulty is present at least 3 months
(E) The sleep difficulty occurs despite adequate opportunity for sleep
(F) The insomnia is not better explained by and does not occur exclusively during the course of another sleep-wake disorder (e.g., narcolepsy, a breathing-related sleep disorder, a circadian rhythm sleep-wake disorder, a parasomnia)
(G) The insomnia is not attributable to the direct physiological effects of a substance (e.g., a drug of abuse, a medication)
(H) Coexisting mental disorders and medical conditions do not adequately explain the predominant complaint of insomnia

be listed. A minimal frequency criteria of insomnia symptoms, three times per week, has been added and the necessary duration of symptoms has been increased from 1 to 3 months. Both of these changes better establish the severity and clinical significance of an insomnia complaint. The DSM-IV-R symptom of "non-restorative" sleep has been removed as it is poorly defined and nonspecific.

A critical feature of *insomnia* carried over from the DSM-IV-R is the inclusion of daytime symptoms in addition to any nighttime sleep symptoms. These daytime symptoms may vary from patient to patient and the diagnostic criteria lists impairment in social, occupational, educational, academic, behavioral, or other important areas of functioning as examples. Importantly, these daytime symptoms cause the patient significant distress and impairment and often are the motivation for a patient to seek treatment. Recognition of the clinical importance of daytime consequences associated with sleep disturbance and the emerging understanding of the pathophysiology of insomnia, which is expressed during the daytime as well as at night (see discussion) has lead to the view that insomnia is a 24 h disorder of arousal (NIH 2005).

3 Electrophysiological Assessment of Insomnia

3.1 Polysomnography

Polysomnography (PSG) is the term used to refer to the continuous recording of multiple physiological parameters during sleep including orbital electrooculograms, at least a centrally derived electroencephalogram, and submental electromyogram, which are the basic parameters used to differentiate sleep and wake and to score the standard NREM and REM sleep stages (Carskadon and Dement 2011). Clinically, to identify other sleep disorders additional physiologic measures are recorded, such as tibialis electromyograms to assess restless legs and periodic leg movements during sleep and respiratory airflow, effort, and oximetry for sleep-related breathing disorder.

Numerous PSG studies of people with insomnia, defined using various diagnostic criteria and recruited from clinics or the general population, have compared them to volunteers without sleep complaints variously described as "good sleepers" or "normal sleepers". The results have been inconsistent, but some studies have shown increased wake after sleep onset and reduced overall sleep efficiency (e.g., min of wake /min of recording time) in people with insomnia (Bonnet and Arand 2006). One reason for the inconsistency of results is revealed in a frequently referenced early study. Carskadon et al. showed considerable overlap in the sleep efficiency distribution of 122 drug-free people with insomnia to the sleep efficiency distribution of those without insomnia (Carskadon et al. 1976). Insomnia is a symptom-based diagnosis and as currently done PSG is not needed for a diagnosis of insomnia. Consequently, by expert consensus PSG is not performed in the routine evaluation of insomnia unless there is suspicion of sleep-related breathing disorder or restless legs/periodic leg movement disorder (Schutte-Rodin et al. 2008).

The PSG also yields information regarding what is termed "sleep architecture," which refers to the amount of the various sleep stages and the progression through NREM to REM sleep cycles across the night. Some studies have reported reduced amounts of REM and NREM stage 3–4 sleep in insomniacs compared to age-matched controls (Mendelson et al. 1986; Gillin et al. 1979; Roth et al. 2014; Baglioni et al. 2014). Parenthetically, age-matching in such studies is critically important as the amount of stage 3–4 sleep normally declines as a function of age (Carskadon and Dement 2011). Some studies have also found elevated amounts of stage 1 NREM sleep, particularly when multiple nights of sleep are sampled (Salin-Pascual et al. 1992).

Several other analytic approaches to the PSG have assessed what is referred to as the "microstructure" of sleep or "sleep continuity". Identification of primary sleep disorders such as sleep apnea and periodic leg movements raised attention to the clinical significance of the brief (3–15 s) EEG arousal during sleep that follows a breathing disturbance or leg movement event. Brief EEG arousal scoring rules have been developed, which have produced reliable and valid results (Bonnet et al. 2007). Some studies of people with insomnia have found elevated arousal frequencies compared to people without insomnia (Terzano et al. 2003), while other studies have not (Stepanski et al. 1984). Brief arousal frequency relates well to the level of excessive daytime sleepiness (Bonnet et al. 2007), and as will be seen in the next section, excessive daytime sleepiness is not the problem of a majority of people with insomnia.

The issue in insomnia may not be the frequency of arousal, but rather the inability to return to sleep after the arousal. A recently developed analytic approach assesses sleep and wake bout frequency and duration using the standard 30 s PSG epoch-by-epoch sleep versus wake scoring. The number and duration of each bout of consecutive epochs scored during sleep or wake is tabulated. A recent paper compared fibromyalgia and rheumatoid arthritis patients with comorbid insomnia to age-matched controls and found comparable sleep efficiency among the patients, which was lower than that of the controls (Roehrs et al. 2013). The wake bout duration of both patient groups was twice as long as that of controls. A much larger study ($n = 293$) comparing people with primary insomnia and people with fibromyalgia to healthy controls showed those with primary insomnia compared to the controls had more frequent and longer wake bouts, as in the previous study of comorbid insomnia (Roth et al. 2014). The larger study also showed those with insomnia had more frequent and shorter sleep bouts than controls. Compared to those with insomnia, the people with fibromyalgia had more sleep bouts and more and longer wake bouts. As in the smaller study, the sleep efficiency of both patient groups was comparable and lower than that of the controls.

3.2 EEG Spectral Analyses

The EEG signal during sleep has been submitted to EEG spectral frequency analyses. Spectral frequency analysis separates the EEG signal into frequency bands and then quantifies signal amplitude over a given frequency band, which is defined as signal power. Some studies have failed to find differences in EEG power in the delta (0.5–3.5 Hz) and theta (3–7 Hz) bands (Buysse et al. 2008), whereas others have reported declines among people with insomnia (Mercia et al. 1998). One of the more consistent EEG spectral findings is an increase in beta (14–35 Hz) and gamma (>35 Hz) EEG activity in people with insomnia compared to age-matched controls (Mercia et al. 1998; Perlis et al. 2001). The International Classification of Sleep Disorders, 2nd edition (ICSD-II) includes the diagnostic entity "paradoxical insomnia" to account for people who report disturbed sleep, but show normal PSGs. Spectral analyses of the sleep EEG of patients with "paradoxical" insomnia have found lower delta and greater alpha (8–12 Hz), sigma (12–14 Hz), and beta activity than controls (Krystal et al. 2002). In such studies the sleep efficiency and sleep architecture of the patients with "paradoxical" insomnia appears normal, despite a patient complaint of disturbed sleep. Since beta and gamma frequencies are a waking EEG component associated with the processing of sensory information or with attentional focus, observing such frequencies during sleep suggests people with insomnia continue to process information while asleep, suggesting an elevated state of arousal (i.e., are "hyperaroused").

3.3 Multiple Sleep Latency Test

As noted earlier, a critical feature of insomnia is the inclusion of daytime symptoms in addition to any nighttime sleep symptoms. The Multiple Sleep Latency

Test (MSLT) uses standard PSG technology and scoring to assess level of sleepiness/alertness, calculated as the average time to fall asleep on 4–5 tests conducted at 2 h intervals across the day. Data are now emerging to indicate that people with insomnia have unusually high MSLT latencies relative to control subjects (Bonnet and Arand 1995; Stepanski et al. 1988; Roehrs et al. 2011). Difficulty falling asleep during the daytime after a night of disturbed sleep, despite feeling fatigued, is a frequent daytime symptom reported by people with insomnia. A large study (n = 95) of people with insomnia, diagnosed by DSM-IV-R criteria, and the additional criterion of a sleep efficiency ≤85 %, found that elevated MSLT scores were stable for over 8 months (Roehrs et al. 2011). Importantly, those with the highest MSLTs had the shortest total sleep times, which is opposite to what is seen in control subjects (e.g., short sleep times produce short MSLTs). Parenthetically, the study also showed that there is a wide distribution of MSLT latencies among people with insomnia and that there is a small subset of people with insomnia who are "sleepy" by MSLT (e.g., average latency <8 min) criteria.

Given that among healthy controls, reduced nocturnal sleep duration is associated with greater sleepiness (e.g., shorter sleep latencies) on the MSLT (Drake et al. 2001), it was hypothesized that homeostatic sleep mechanisms are weakened in people with insomnia (Stepanski et al. 2000). To test the homeostatic mechanisms in insomnia, the effects of total sleep deprivation on the MSLT and on recovery nocturnal sleep were compared to age-matched healthy controls (Stepanski et al. 2000). Although elevated at baseline relative to controls, the average daily sleep latency on the MSLT after total deprivation was reduced in those with insomnia to similar levels as controls. The total sleep time of people with insomnia was less than that of controls at baseline (6.1 versus 7.6 h), and it increased to that of the controls (7.5 versus 7.8 h) on the recovery night, suggesting normally responsive homeostatic sleep mechanisms. These data then support the hypothesis that most people with insomnia show a reliable disorder of "hyperarousal" (see discussion) with increased wake drive both at night and during the day.

3.4 Auditory Evoked Potentials

Another method to assess hypothesized "hyperarousal" in insomnia is analysis of event-related brain potentials (ERPs), which is a classic electrophysiological method employed in a variety of psychiatric disorders. The electrophysiological response to any stimuli presented repeatedly and then averaged, yields a wave form with consistent negative and positive polarity components at different temporal points (ms) after stimulus onset. ERPs to auditory tasks have been measured in people with insomnia during the daytime, at sleep onset, and during sleep. Using an oddball paradigm (i.e., stimuli presented during a reading task) in people with insomnia, a self-rated bad night of sleep was followed during the day by an enlarged amplitude P300 (e.g., a positive wave at 300 ms) and a lower P300 after a good night (Devoto et al. 2003, 2005). In comparing people with insomnia to self-described "good sleepers" heightened N1 (most negative peak between 60 and 110 ms) amplitudes were seen in those with insomnia on both morning and evening testing (Bastien et al. 2008). Recently, the ERPs of night workers with insomnia were compared to night workers with sleepiness and asymptomatic night workers (Guymenyuk et al. 2014). Tested during their night shift, the night workers with only insomnia showed an enhanced N100 amplitude and a decreased P200 amplitude relative to asymptomatic night workers and to those with sleepiness in addition to insomnia (see Fig. 1). During sleep onset people with insomnia only showed increased P2 (most positive peak between 120 and 200 ms) amplitudes and decreased N350 (most negative peak between 250 and 350 ms) amplitudes, interpreted by the investigators as an inability to "disengage from waking sensory processing" during the sleep onset process (Bastien et al. 2008). In another study the oddball paradigm was applied during sleep and the study found increased N1 (most negative peak between 76 and 150 ms) and decreased P2 (most positive peak between 150 and 260 ms) amplitudes during the first 5 min of stage 2 NREM sleep, but thereafter during the remainder of the night *no* ERP differences from the control subjects were seen (Yang and Lo 2007). These studies of ERPs in people with insomnia, while differing in methodology and specific ERP findings, do suggest increased sensitivity to auditory stimulation during waking and reduced sensory inhibition during the sleep onset process. These ERP data are consistent with the MSLT data of people with insomnia, both assessments suggesting hyperarousal.

4 Sympathetic Nervous System (SNS) and Hypothalamic-Pituitary-Adrenal (HPA) Correlates

Insomnia is hypothesized to reflect a 24 h state of hyperarousal. As discussed earlier, this hyperarousal is evident in the prolonged sleep latencies (i.e., about one standard deviation above the mean of a representative population sample) on the MSLT during the day, despite disrupted and shortened nocturnal sleep the previous night seen in people with insomnia (Bonnet and Arand 1995; Stepanski et al. 1988; Roehrs et al. 2011). Evidence also suggests that this physiologic hyperarousal is associated with activation of the sympathetic nervous system (SNS) and hypothalamic-pituitary-adrenal (HPA) axis. Insomniacs show elevated levels of circulating daytime and nighttime catecholamines (Vgontzas et al. 1998), increased metabolic rates (Bonnet and Arand 1995), increased body temperature (Lushington et al. 2000), decreased high frequency heart rate variability (Lichstein et al. 1994), and altered pupillometry patterns (Lichstein and Johnson 1994). HPA augmentation in insomnia is indicated by elevated levels of nighttime urinary free cortisol proportional to the amount of wakefulness during the night (Vgontzas et al. 1998). An activated SNS and HPA axis suggests there may be an underlying central mechanism for the hyperarousal of insomnia, possibly involving corticotropin releasing factor (CRF) neurons.

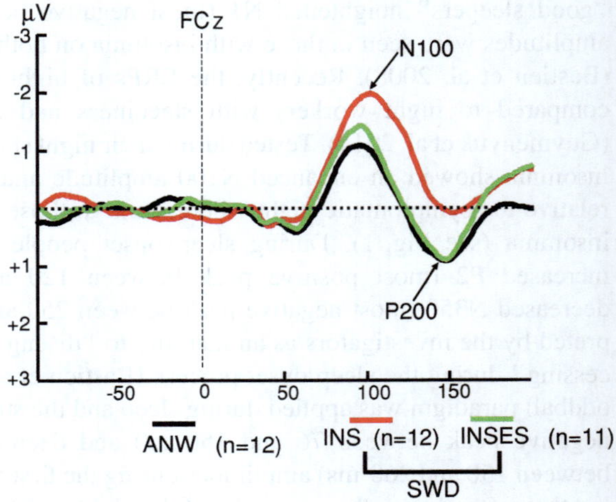

Fig. 1 ERPs at FCz electrode, elicited by tones (1200 Hz, 100 ms duration) in three groups of participants: asymptomatic night workers (ANW) and shift work disorder (SWD) patients with insomnia only (INS) and with insomnia/excessive sleepiness (INSES). Enlarged N100 and decreased P200 are indicators of cortical hyperarousal in insomnia without excessive sleepiness

All these studies have compared people with insomnia to people without insomnia, rather than assessing individual differences among those with insomnia as a function of level of hyperarousal. As noted earlier, the presence of hyperarousal is seen in some people with insomnia, about two-thirds, as defined by the MSLT (Roehrs et al. 2011). In a large sample of people with insomnia it was found that MSLT elevation in insomnia is a stable "trait-like" finding (Roehrs et al. 2011). Recently, it was reported that elevated MSLTs were associated with elevated levels of daytime urinary norepinephrine (see Fig. 2) (Roehrs et al. 2012). This is a first validation of the construct of hyperarousal in insomnia using two concurrent "independent" physiological measures.

5 Neuroimaging

Neuroimaging findings in people with insomnia also support a hyperarousal hypothesis. In PET studies primary insomniacs, relative to healthy controls, showed greater cerebral glucose metabolism during sleep, while awake, and at the transition from wake to sleep and particularly in the ascending reticular activation system (ARAS) (Nofzinger et al. 2004). The hypothalamus, thalamus, amygdala, hippocampus, and prefrontal cortices were also activated during sleep. In a follow-up study also using PET, it was shown that there was a positive correlation between the amount of wake time after sleep onset and the level of cerebral glucose metabolism in the pontine tegmentum and thalamocortical networks in a frontal, anterior temporal, and anterior cingulate distribution (Nofzinger et al. 2006).

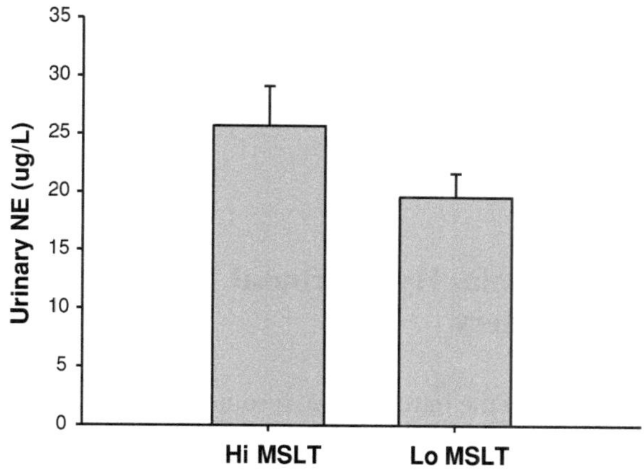

Fig. 2 Daytime (0700–1500 h) urinary NE (ug/L) level in people with insomnia and high MSLT latencies versus those with low MSLT latencies ($p < 0.01$)

6 Pathophysiology of Insomnia

It is hypothesized that the pathophysiology underlying the hyperarousal and sleep disturbance of insomnia is at least in part a HPA axis dysfunction, specifically increased corticotrophin releasing factor (CRF) activity (Richardson and Roth 2001; Drak et al. 2003). First, studies in normal volunteers without insomnia have shown IV administration of cortisol or adrenocorticotropic hormone (ACTH) reduces rapid eye movement sleep (Born et al. 1989) and IV administration of CRF reduces slow wave sleep (Holsboer et al. 1988), both of which are consistent with PSG findings in insomnia (Baglioni et al. 2014). In other words, HPA axis activation including CRF elevation is disruptive of sleep in healthy volunteers. In addition, a study comparing people with chronic primary insomnia to control subjects showed that those with insomnia had increased evening and nocturnal plasma ACTH and cortisol concentrations (Vgontzas et al. 2001; Vgontzas et al. 2013). Moreover, among those with insomnia the level of cortisol was proportional to the level of the degree of objective PSG-defined sleep disturbance (Vgontzas et al. 2013). In a study of severe chronic insomnia, the area under the curve of cortisol levels was strongly correlated ($r = -0.91$) with the level of sleep efficiency (Rodenbeck et al. 2002). Some of the studies that have failed to find cortisol elevations in people with insomnia relative to controls, also do not show PSG-defined sleep disturbances. For example, in one negative study the sleep efficiency of subjects with insomnia was 88.2 % and that of the controls was 88.6 % (Riemann et al. 2002). This suggests that cortisol elevation may be a marker of severe insomnia.

Data from animal studies suggest that CRF functions centrally as a neurotransmitter in the locus coeruleus. Microinjection of CRF into the locus coeruleus of the rat elicits fear-specific behaviors and a general behavioral activation (Butler et al. 1990) and local CRF infusion increases locus coeruleus cell firing and the release of norepinephrine (Page and Abercombie 1999). In addition, CRF positive cells and fibers have been localized to the locus coeruleus through immunohistochemical labeling (Swanson et al. 1983). That brain CRF has a sleep-disruptive effect has been shown in several animal models. Stress normally reduces sleep time and an antagonist of CRF blocks the stress-induced reduction of sleep time; that is, sleep time normalizes (Matsumoto et al. 1997). Importantly, the antagonist itself has no effect on baseline sleep.

7 Insomnia, Hyperarousal, and Mood and Anxiety Disorders

As noted in the introduction, insomnia is a risk factor for the development of major depressive disorder (MDD), is an important symptom predictive of MDD relapse, and can remain after successful treatment of the MDD (Krystal 2006). This raises questions as to how insomnia and MDD are related (e.g., whether insomnia causes MDD, MDD causes insomnia, or either condition can cause the other) or rather than a cause-effect relation, whether insomnia and MDD share components of a common pathophysiology. It should be noted that not all insomnia leads to MDD and that MDD is not highly predictive of the development of new insomnia. An epidemiological study in adolescents explored the strength of the directionality of the insomnia and MDD association (Johnson et al. 2006). Prior insomnia carried a 3.8 times greater risk of MDD, adjusting for gender, race/ethnicity, and anxiety, while prior depression was not associated with onset of insomnia.

In attempting to understand the relation of hyperarousal, CRF/HPA axis dysregulation and insomnia comorbid with MDD, it also must be noted that MDD is not a homogeneous entity. MDD has been clinically characterized as two distinct subtypes of contrasting psychological and neurovegetative symptoms (Gold and Chrousos 2002). Melancholic patients are anxious, anorectic, unresponsive to psychosocial stimuli, more depressed in the morning and exhibit insomnia. Patients with atypical depression are lethargic, fatigued, hyperphagic, reactive to the environment, more depressed in the evening and exhibit hypersomnia. These two subtypes exhibit two distinct CRF/NE states (Gold and Chrousos 2002). Melancholic patients show an overactive CRF/HPA axis, while atypical patients show a down-regulated CRF/HPA axis.

It was also noted in the introduction that insomnia coexisting with anxiety disorders (AD) follows, rather than precedes, the onset of the specific AD. The epidemiological study cited earlier also assessed the directionality of the insomnia AD association (Johnson et al. 2006). Prior AD carried a 3.5 risk for subsequent insomnia adjusting for gender, race/ethnicity, and depression, while prior insomnia was not associated with the development of AD. As to hypothesized pathophysiology, regardless of the specific AD, it is critical to note that this literature has not distinguished AD with and without insomnia. The focus of the hypothesized pathophysiology of the various ADs has been on central NE dysfunction (Gold and Chrousos 2002). In addition to neglecting the possible role of comorbid insomnia, peripheral measures of resting NE (i.e., plasma, urine) are assessed to reflect central NE. But, such measures predominantly reflect peripheral, not, central NE. The literature has not consistently shown that resting NE measures in the various ADs are different from those of controls (Gold and Chrousos 2002). NE challenge studies, in which drugs (e.g., yohimbine) are administered producing NE dysfunction, have been more successful. Relative to controls, patients with panic disorder have shown increased panic attacks, increased ratings of anxiety or nervousness, and increases in various physiological measures reflecting arousal (i.e., blood pressure, pulse, and cortisol) (Kalk et al. 2011). The extent to which the daytime NE challenge that also disrupts nocturnal sleep has not been documented. As noted, we also are unaware of studies that have assessed patients with ADs and comorbid insomnia versus those without comorbid insomnia.

8 Summary

There has been a major paradigm shift over the last decade in understanding the pathophysiology, diagnosis, and treatment of insomnia. That shift is reflected in the way insomnia is defined and diagnosed according to DSM-V. Insomnia is viewed as a disorder with its own course and morbidity that is typically coexistent with other medical, psychiatric, and sleep disorders. Electrophysiological study of the sleep of people with insomnia using standard methods termed polysomnography (PSG) in which sleep/wake and sleep stages are scored in 30 s epochs have not consistently found elevated wake time or disrupted sleep stage progression relative to age-matched controls. Studies using analyses of the microstructure of the PSG have found increased fragmentation of sleep with brief arousals (<15 s) and spectral analyses of the sleep EEG has shown increased fast frequency [e.g., beta (15–35 Hz) and gamma (>35 Hz)] EEG activity. Testing speed of sleep onset during the daytime using PSG methods [Multiple Sleep Latency Test (MSLT)] have shown usual daytime alertness, not sleepiness, particularly given short nocturnal sleep times and daytime sleepiness/fatigue symptoms. Auditory evoked potential (ERP) assessment during sleep and during the day have suggested increased sensitivity to auditory stimulation during waking and reduced sensory inhibition during sleep. This increased CNS activation is also reflected in neuroimaging studies. Sympathetic nervous system and hypothalamic-pituitary-adrenal assessments also support the hypothesis that insomnia reflects a 24 h state of hyperarousal. Given that insomnia is a common symptom in both mood and anxiety disorders, a future focus on understanding the relation of insomnia and its hyperarousal and CRF/HPA axis dysregulation in mood and anxiety disorders will be important. It may be that insomnia shares components of a common pathophysiology with depression or anxiety?

References

American Psychiatric Association (2013) Diagnostic and statistical manual of mental disorders (DSM5). American Psychiatric Association, Washington, DC

Baglioni C, Regen W, Teghen A, Spiegelhalder K, Feige B, Nissen C, Riemann D (2014) Sleep changes in the disorder of insomnia: a meta-analysis of polysomnographic studies. Sleep Med Rev 18:195–213

Bastien C, St-Jean G, Morin C (2008) Chronic psychophysiological insomnia: hyperarousal and/or inhibition deficits? an ERPs investigation. Sleep 31:887–898

Bonnet MH, Arand DL (1995) 24-hour metabolic rate in insomniacs and matched normal sleepers. Sleep 18:581–588

Bonnet MH, Arand DL (2006) Consequences of insomnia. Sleep Med Clin 1:351–358

Bonnet MH, Doghramji K, Roehrs T et al (2007) The scoring of arousal in sleep: reliability, validity, and alternatives. J Clin Sleep Med 3:133–145

Born J, Spath-Schwalbe E, Schwakenhofer H, Kern W, Fehm HL (1989) Influences of corticotrophin-releasing hormone, adrenocorticotropin, and cortisol on sleep in normal man. J Clin Endocrinol Metab 68:904–911

Butler PD, Weiss JM, Stout JC, Nemeroff CB (1990) Corticotropin-releasing factor produces fear-enhancing and behavioral activation effects following infusion into the locus coeruleus. J Neurosci 10:176–183

Buysse DJ, Germain A, Hall M et al (2008) EEG spectral analysis in primary insomnia: NREM period effects and sex differences. Sleep 31:1673–1682

Carskadon MA, Dement WC, Mitler MM (1976) Self reports versus sleep laboratory findings in 122 drug-free subjects with complaints of primary insomnia. Am J Psychiatry 133:1382–1387

Carskadon MA, Dement WC (2011) Normal human sleep: an overview. In: Kryger MH, Roth T, Dement WC (eds) Principles and practice of sleep medicine, 5th edn. Elsevier Saunders, St Louis, pp 16–26

Devoto A, Violani C, Lucidi F, Lombardo C (2003) P300 amplitude in subjects with primary insomnia is modulated by their sleep quality. J Psychosom Res 54:3–10

Devoto A, Manganelli S, Lucidi F, Lombardo C, Russo M, Violani C (2005) Quality of sleep and P300 amplitude in primary insomnia: a preliminary study. Sleep 28:859–863

Drake C, Roehrs T, Burduvali E, Bonahoom A, Rosekind M, Roth T (2001) Effects of rapid versus slow accumulation of eight hours of sleep loss. Psychophysiology 38:979–987

Drake CL, Roehrs T, Roth T (2003) Insomnia causes, consequences, and therapeutics: an overview. Dep Anx 18:163–176

Gold PW, Chrousos GP (2002) Organization of the stress system and its dysregulation in melancholic and atypical depression: high versus low CRH/NE states. Mol Psychiatry 7:254–275

Gillin JC, Duncan W, Pettigrew KD et al (1979) Successful separation of depressed, normal, and insomniac subjects by EEG sleep data. Arch Gen Psychiatry 36:85–90

Guymenyuk V, Belcher R, Drake CL, Spear L, Roth T (2014) Cortical arousal is present in alert insomniacs, but absent in sleepy insomniacs with shift work disorders. Presented at Sleep 2014, Minneapolis, June 2014

Holsboer F, von Bardeleben U, Steiger A (1988) Effects of intravenous corticotrophin-releasing hormone upon sleep-related growth hormone surge and sleep EEG in man. Neuroendocrinology 48:32–38

Johnson EO, Roth T, Breslau N (2006) The association of insomnia with anxiety disorders and depression: exploration of the direction of risk. J Psychiatric Res 40:700–708

Kalk NJ, Nutt DJ, Lingford-Hughes AR (2011) The role of central noradrenergic dysregulation in anxiety disorders: evidence from clinical studies. J Psychopharm 25:3–16

Krystal AD, Edinger JD, Wohlgemuth WK et al (2002) NREM sleep EEG frequency correlates of sleep complaints in primary insomnia subtypes. Sleep 25:626–636

Krystal AD (2006) Psychiatric comorbidity: the case for treating insomnia. Sleep Med Clin 1:359–365

Lichstein KL, Wilson NM, Noe SL, Aguillard RN, Bellur SN (1994) Daytime sleepiness in insomnia: behavioral, biological and subjective indices. Sleep 17:693–702

Lichstein KL, Johnson RS (1994) Pupillometric discrimination in insomniacs. Behav Res Ther 32:123–129

Lushington K, Dawson D, Lack L (2000) Core body temperature is elevated during constant wakefulness in elderly poor sleepers. Sleep 23:504–510

Matsumoto K, Ojima K, Watanabe H (1997) Central corticotrophin-releasing factor and benzodiazepine receptor systems are involved in the social isolation stress-induced decrease in ethanol sleep in mice. Brain Res 753:318–321

Mendelson WB, James SP, Garnett D et al (1986) A psychophysiological study of insomnia. Psychiatry Res 16:267–284

Mercia H, Blois R, Gaillard JM (1998) Spectral characteristics of sleep EEG in chronic insomnia. Eur J Neurosci 10:1826–1834

NIH: National Institutes of Health State of the Science Conference Statement (2005) Manifestations and management of chronic insomnia in adults. Sleep 28:1049–1057 June 13–15

Nofzinger EA, Buysee DJ, Germain A, Price JC, Miewald JM, Kupfer DJ (2004) Functional neuroimaging evidence for hyperarousal in insomnia. Am J Psychiatry 161:2126–2129

Nofzinger EA, Nissen C, Germain A, Moul D, Hall M, Price J (2006) Reginal cerebral metabolic correlates of WASO during NREM sleep in insomnia. J Clin Sleep Med 2:316–322

Page ME, Abercrombie ED (1999) Discrete local application of corticotrophin-releasing factor increases locus coeruleus discharge and extracellular norepinephrine. Synapse 33:304–313

Perlis ML, Smith MT, Andrews PJ et al (2001) Beta/gamma EEG activity in patients with primary and secondary insomnia and good sleeper controls. Sleep 24:110–117

Richardson GS, Roth T (2001) Future directions in the management of insomnia. J Clin Psychiatry 62(suppl 10):39–45

Riemann D, Klein T, Rodenbeck A, Feige B, Horney A, Hummel R (2002) Nocturnal cortisol and melatonin secretion in primary insomnia. Psychiatry Res 113:17–27

Roehrs TA, Randall S, Harris E et al (2011) MSLT in primary insomnia: stability and relation to nocturnal sleep. Sleep 13:1647–1652

Roehrs T, Diederichs C, Gillis M et al (2013) Nocturnal sleep, daytime sleepiness and fatigue in fibromyalgia patients compared to rheumatoid arthritis patients and healthy controls: a preliminary study. Sleep Med 14:109–115

Roehrs T, Randall S, Roth T (2012) Daytime urinary norepinepherine levels in hyperaroused insomniacs. Presented at Sleep 2012, Boston, June 2012

Rodenbeck A, Huether G, Ruether E, Hajak G (2002) Interactions between evening and nocturnal cortisol secretion and sleep parameters in patients with severe chronic primary insomnia. Neurosci Lett 324:163–459

Roth T, Roehrs T, Pies R (2007) Insomnia: pathophysiology and implications for treatment. Sleep Med Rev 11:71–79

Roth T, Bhadra-Brown P, Pitman VW et al (2014) Characteristics of disturbed sleep in patients with fibromyalgia compared with insomnia or with healthy volunteers. J Rheumatol (Submitted paper in review)

Salin-Pascual RJ, Roehrs TA, Merlotti LA et al (1992) Long-term study of the sleep of insomnia patients with sleep-state misperception and other insomnia patients. Am J Psychiatry 149:904–908

Schutte-Rodin S, Broach L, Buysse D et al (2008) Clinical guideline for evaluation and management of chronic insomnia in adults. J Clin Sleep Med 4:487–504

Stepanski E, Lamphere J, Badia P et al (1984) Sleep fragmentation and daytime sleepiness. Sleep 7:18–26

Stepanski E, Zorick F, Roehrs T et al (1988) Daytime alertness in patients with chronic insomnia compared with asymptomatic control subjects. Sleep 11:54–60

Stepanski E, Zorick F, Roehrs T et al (2000) Effects of sleep deprivation on day sleepiness in primary insomnia. Sleep 23:1–5

Swanson LW, Sawchenko PE, River J, Vale WW (1983) Organization of ovine corticotrophin-releasing factor immunoreactive cells and fibers in the rat brain: an immunohistochemical study. Neuroendocrinology 36:165–186

Terzano MG, Parrino L, Spaggiari MC et al (2003) CAP variables and arousals as sleep electroencephalogram markers for primary insomnia. Clin Neurophysiol 114:1715–1723

Yang CM, Lo HS (2007) ERP evidence of enhanced excitatory and reduced inhibitory processes of auditory stimuli during sleep in patients with primary insomnia. Sleep 30:585–592

Vgontzas AN, Tsigos C, Bixler EO (1998) Chronic insomnia and activity of the stress system: a preliminary study. J Psychosom Res 45:21–31

Vgontzas AN, Bixler EO, Lin HM, Prolo P, Mastorakos G, Vela-Bueno A et al (2001) Chronic insomnia is associated with nycthemeral activation of the hypothalamic-pituitary-adrenal axis: clinical implications. J Clin Endocrinol Metab 86:3787–3794

Vgontzas AN, Fernandez-Mendoza F, Liao D (2013) Insomnia with objective short sleep duration: the most biologically severe phenotype of the disorder. Sleep Med Rev 24:241–254

Part II
Psychophysiology Measurements and Analytical Tools: New Perspectives

Neurophysiological Biomarkers Informing the Clinical Neuroscience of Schizophrenia: Mismatch Negativity and Prepulse Inhibition of Startle

Gregory A. Light and Neal R. Swerdlow

Abstract With the growing recognition of the heterogeneity of major brain disorders, and particularly the schizophrenias (SZ), biomarkers are being sought that parse patient groups in ways that can be used to predict treatment response, prognosis, and pathophysiology. A primary focus to date has been to identify biomarkers that predict damage or dysfunction within brain systems in SZ patients, that could then serve as targets for interventions designed to "undo" the causative pathology. After almost 50 years as the predominant strategy for developing SZ therapeutics, evidence supporting the value of this "find what's broke and fix it" approach is lacking. Here, we suggest an alternative strategy of using biomarkers to identify evidence of spared neural and cognitive function in SZ patients, and matching these residual neural assets with therapies toward which they can be applied. We describe ways to extract and interpret evidence of "spared function," using neurocognitive, and neurophysiological measures, and, suggest that further evidence of available neuroplasticity might be gleaned from studies in which the response to drug challenges and "practice effects" are measured. Finally, we discuss examples in which "better" (more normal) performance in specific neurophysiological measures predict a positive response to a neurocognitive task or therapeutic intervention. We believe that our field stands to gain tremendous therapeutic leverage by focusing less on what is "wrong" with our patients, and instead, focusing more on what is "right".

Keywords Biomarker · Cognitive remediation · Mismatch negativity · Neurocognition · Prepulse inhibition · Schizophrenia

G. A. Light · N. R. Swerdlow (✉)
Department of Psychiatry, School of Medicine, University of California, San Diego, 9500 Gilman Dr. La Jolla, CA 92093-0804, USA
e-mail: nswerdlow@ucsd.edu

Curr Topics Behav Neurosci (2014) 21: 293–314
DOI: 10.1007/7854_2014_316

Published Online: 22 May 2014

Contents

1 Overview: Biomarkers of Health

Our field's ability to understand and effectively treat mental illness has advanced in developmental "stages," each new stage building on the ideas and technologies that preceded it. The current stage of neuroscientific inquiry into brain disorders is most notable for the increasing acceptance by the scientific community of the notion—long held within clinical spheres—that many major psychiatric "disorders" are biologically heterogeneous syndromal endpoints of different etiological pathways. In embracing the heterogenous underpinnings of conditions like the schizophrenias (SZ), our field has both de-prioritized pursuits that seek to uncover unitary pathogenic processes (i.e., single causative gene), and has prioritized those designed to clarify the more complex biologies and treatments of these syndromes. One clear priority in this new "stage" is the development and application of biomarkers for mental illness.

Biomarkers are objective measures that can be informative about a variety of different clinical characteristics, such as an individual's normal biology, their pathology including the trajectory of illness, or their response to a therapeutic intervention. While it is clear that symptom-based diagnostic schema can distinguish patients in a manner that to some degree predicts their trajectory and therapeutic sensitivity (e.g., in the parsing of a primary anxiety disorder vs. a primary psychotic disorder), it is equally clear that these schema have reached their limits of resolution in terms of pathophysiology and the development of novel and individualized therapeutics. Biomarkers offer the hope that despite great heterogeneity and multivariate interactions in the pathogenesis of brain disorders, meaningful clusters of individuals can be associated with an objective measure, and then reliably stratified in terms of the cause, course and/or treatment sensitivity of a given disorder. Of course, this is not a new hope—the search for biomarkers for mental illness can be traced back decades, and perhaps centuries—nor is it a hope fulfilled, as we presently have no biomarkers that add in a meaningful way to our treatment of any major psychiatric syndrome.

One assumption driving the search for psychiatric biomarkers is that the biology of these biomarkers will be simpler, more easily understood and less heterogeneous than the biology of clinical psychiatric syndrome. But if the pathogenic pathways leading to the syndrome are highly heterogeneous, we might expect that the biomarkers for these pathways might also be highly heterogeneous. In essence, if we identify a biomarker of "Bill's schizophrenia," this information might be very important to Bill, but not generalizable to larger populations. For this reason, we have endorsed an approach in which biomarkers in psychiatric disorders are used to identify not pathological processes, but rather intact, healthy processes (e.g., brain circuitry). This approach has several advantages over the search for pathology biomarkers. For example, it is in many ways easier to interpret a biomarker of health than one of pathology. In a simple analogy, if you enter a room, flip on the light switch and no light goes on, there can be numerous explanations for this deficit. However, if you flip on the light switch and the light DOES go on, there can be only one parsimonious explanation: electrons are getting to where they need to be.

It is not simply that biomarkers of health are easier to understand than biomarkers of pathology, but rather, that they may be more "actionable," i.e., their presence may lead more directly to a predicted therapeutic intervention. To underscore this point, let's examine some putative biomarkers of pathology for schizophrenia. Hippocampus, amygdala, anterior cingulate cortex, and other structures are reduced in volume and/or functionally impaired in SZ patients *as well as their asymptomatic first-degree relatives* (cf. Swerdlow 2011a) One implication of these neuroanatomical "" of pathology is that while these circuit disturbances may be associated with a heritable vulnerability for SZ, they are insufficient to produce the disorder. Thus, while these biomarkers of impairment may inform us about various different etiologies and perhaps even preventative interventions, they do not by themselves provide "actionable" targets for corrective interventions: after all, most people with these abnormalities *do not have SZ*, so why would "correcting" this circuitry be of benefit to someone who does?

By contrast, biomarkers of healthy brain function in "system X" might provide more direct "actionable" evidence that a patient with SZ is likely to benefit therapeutically from "intervention Y." Several clinical models support this approach. For example, many interventions in stroke rehabilitation are designed not to re-grow brain circuitry that is lost or damaged, but rather to engage the normal physiological and anatomical properties of healthy brain circuits (e.g., in neighboring regions or parallel circuits) to restore or subsume the function of damaged ones (cf. Taub et al. 2002). In many forms of psychotherapy, the therapist's task is to identify an individual's psychological strengths (ego, intellectual, social, or otherwise) and then to engage them to overcome damaging thoughts or behaviors that are otherwise sustained by areas of psychological weakness. At a neural level, both stroke rehabilitation and psychotherapy engage viable and healthy systems to compensate for, or re-establish, functions lost to illness. Similarly, biomarkers of "health" can reveal a patient's neural "assets," which can then be leveraged in the service of therapy. There are several hurdles to clear in this process, e.g., (1) it requires biomarkers that identify these assets with sufficient sensitivity, specificity, and other limits of resolution discussed below, and (2) it requires therapies that can engage these assets to improve function. There is growing evidence that both of these hurdles can be cleared.

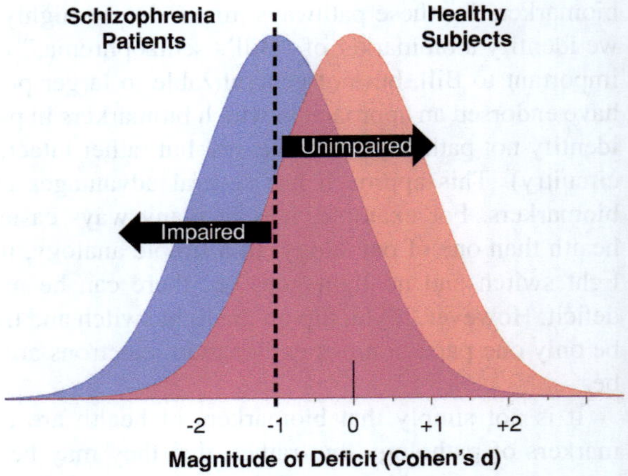

Fig. 1 Example of overlapping distributions in a robust ($d = 1$) effect size biomarker deficit in schizophrenia patients. In neuropsychological assessments, $d = 1$ standard deviations below the mean is commonly used for impairment classification. In this case, 50 % of patients exhibit unimpaired/normal range biomarker values

For example, as discussed below, reliable, repeatable, and robust measures can quantify working memory (WM) in SZ patients. Certain cognitive therapies place demands on SZ patients to engage WM to develop compensatory strategies for learning and applying information (Twamley et al. 2003). In doing so, these therapies specifically activate prefrontal regions subserving WM and attention (Haut et al. 2010). It is both parsimonious and testable that patients with the available "neural asset" of relatively intact WM—demonstrated by sensitive, specific and reliable laboratory measures—and hence frontal circuits that subserve WM, will benefit most from WM-targeted cognitive therapies.

What is the likelihood of identifying "healthy" biomarkers in patients who are suffering from obvious brain dysfunction associated with profound functional impairment? We view this likelihood to be substantial: even in the most "robust" biomarkers suggesting "pathology" in the most severe cohorts of chronic SZ patients, many and sometimes most patients "score" in the normal range. This is true in markers using volumetric or functional neuroimaging, or neurophysiology, or neurocognition. Biomarkers that identify differences in SZ patient versus healthy comparison subjects with a Cohen's standardized effect size of $d = 1.0$ are generally considered "robust"; *in fact, most of the highly replicable SZ biomarkers fail to reach this level of group separation*. Notably, falling 1 standard deviation below normative samples (i.e., effect size $d = 1.0$) is commonly used as a cutoff for impairment classification in neuropsychological assessments. This means that even in the case of a $d = 1.0$ biomarker impairment, 50 % of patients will *by definition* fall in the "normal" range (Fig. 1)—an often overlooked fact. Moreover, in this "best case" example of a pathology biomarker, "only" 54.5 % of the patients versus healthy group distributions are nonoverlapping. Whether the metric is hippocampal volume (Simm et al. 2006) or prepulse inhibition of startle (PPI) (Swerdlow et al. 2014) or WM (Horan et al. 2008) or mismatch negativity (MMN; (Umbricht and Krljes 2005; Rissling et al. 2012; Light and Braff 2005a, b; Kiang et al. 2009), some or even most SZ patients exhibit evidence of intact function: the "light switch works," and thus the neural assets can conceivably be applied toward a therapeutic response.

Does the search for biomarkers of health imply that we simply forego therapeutic options for patients whose biomarkers suggest a lack of health? Of course not. Given the heterogeneity of performance across measures, it is often the case that patients exhibiting deficits in one biomarker or neural domain will perform normally in others. Indeed, many of the common neurophysiological biomarkers and endophenotypes of SZ are uncorrelated with one another even when measuring similar operational constructs (e.g., sensory/sensorimotor gating: Schwarzkopf et al. 1993; Light and Braff 2001; Braff et al. 2007; sensory discrimination: Rissling et al. 2012; Horvath et al. 2008). The key to using this strategy in a heterogeneous population is to be able identify areas of neural strengths using a battery of well-validated and dissociable battery of laboratory-based biomarkers. And while cognitive therapies are generally benign and not prone to adverse events as traditionally measured in medicine, they are time-consuming, resource-intensive, and taxing; they require many hours of time, in addition to the logistical complexities involved in accessing treatment for a severely impaired individual, and the psychological implications of treatment failure if unsuccessful. Thus, a haphazard pairing of an individual with severe impairments in a biomarker of, say, WM, with a time- and resource-intensive cognitive intervention that places heavy demands on WM, is likely to be unsuccessful. *Unfortunately, such incidental couplings of individual patient characteristics with therapies represent the current state of the art.* Treatment "failures" are far too common and have the potential to cost the patient, family, therapist, and larger social system dearly. In contrast, biomarkers of health can guide patients toward viable therapies, and their absence can steer patients away from therapies that are not likely to be successful and whose failure carries significant real-life consequences.

There may be ways to "uncover" biomarkers of potential function in SZ, even among neural systems that appear by some biomarker evidence to be defective. The general principle behind this strategy is that a neural system at baseline may perform poorly, but may still respond to a "push" of a pharmacologic challenge. In this case, evidence for the requisite "spared" neural circuitry, and hence a target for therapeutic intervention, might be provided by specific neurophysiological or neurocognitive changes in response to a "push" produced by a drug challenge. This approach parallels the use of a "test dose" to predict clinical benefits from treatments ranging from hormones (Biller 2007) to anti-Parkinsonian drugs (Hughes et al. 1990) to bronchodilators (Fruchter and Yigla 2009). If a patient generates a specific neurobehavioral signal in response to a drug challenge—e.g., increased neurocognitive or neurophysiological performance, or enhanced performance of a computerized cognitive training task (discussed below)—this suggests that neural circuits spared by their SZ remain viable targets under the right conditions. Creating such conditions is the goal of "Pharmacologically Augmented Cognitive Therapy," as described previously (Swerdlow 2011a, b), and departs significantly from what has been a 50-year-old largely failed strategy of trying to use drugs to "undo" the neuropathology of SZ (Lieberman et al. 2005).

In the process of using a "drug challenge" to probe a biomarker of spared neural function, it is often the case that neurocognitive and/or neurophysiological measures are repeated under drug-free, placebo, and active drug conditions. One complexity of this experimental design is that brain mechanisms at many levels, and particularly at levels of higher order functions, exhibit "learning," as detected in practice /order effects. These changes in performance with repeated task experience are typically viewed as experimental confounds, as they can arithmetically complicate the interpretation of a drug effect (Chou et al. 2013b). However, it is possible that the brain's ability to "learn," particularly in specific neurocognitive domains, may—in and of itself—be a valuable biomarker of spared neural function. We have no data yet to support this notion, but it makes intuitive sense that the ability to increase one's performance with experience provides evidence of neuroplasticity that might be harnessed in the service of an appropriate therapy. Our data (Chou et al. 2013a) is consistent with previous reports (Nuechterlein et al. 2008) that different neurocognitive domains are differentially sensitive to such "learning" processes, and that the amount of learning exhibited by SZ patients varies substantially within any given domain. These data are typically collected as part of any "procognitive" drug trial involving more than one drug condition; that such personalized profiles of cognitive "neuroplasticity" are predictive of sensitivity to specific cognitive or medication therapies is a testable hypothesis.

Regardless of whether the intended use of a biomarker is to identify health or pathology in SZ, its utility will depend on its ability to meet a number of important criteria. The background for the development and application of such criteria for biomarkers relevant to SZ is discussed in the next section of this chapter.

2 SZ Biomarkers

What are the optimal characteristics of biomarkers for informing the clinical neuroscience and future treatments of SZ? Over the past decade, several expert consensus panels were convened to attempt to overcome some of the obstacles of developing treatments to improve cognition and psychosocial functioning in schizophrenia. The first initiative, the Measurement and Treatment Research to Improve Cognition in Schizophrenia (MATRICS), brought together academia, the pharmaceutical industry, and the Food and Drug Administration to identify cognitive targets in schizophrenia and develop a brief, repeatable, and standardized battery of tasks for use in clinical outcome studies (Green et al. 2004). In this context, a RAND panel carefully evaluated the desired measurement characteristics of individual tests for inclusion in the final FDA-approved battery and concluded that measures should exhibit: (1) high test–retest reliability; (2) utility as a repeated measure; (3) a relationship to functional outcome; (4) potential response to pharmacologic agents; and (5) practicality/tolerability. Clearly, both the *process* for evaluating measures and the *specific criteria* is also highly relevant for evaluating promising neurophysiologic biomarkers that can inform the development of next-generation personalized treatments.

The benefits of neurophysiologic biomarkers were also recognized in the MATRICS initiative since such measures can probe the earliest stages of sensory-perceptual information processing and the subsequent transitions to higher order cognitive operations with millisecond resolution. In many cases, responses can be automatically elicited in the absence of directed attention and do not require substantial effort or motivation on the part of the participant (Braff and Light 2004). Neuroscience-derived biomarkers are also well-suited for linking cognitive deficits to specific neural systems using source-imaging, pharmacology, and animal models. Thus, the Cognitive Neuroscience Treatment Research to Improve Cognition in Schizophrenia initiative (CNTRICS) was launched after MATRICS to identify the most promising brain-based tools for measuring cognition and testing new treatments in schizophrenia (Carter et al. 2008). This panel extended the five MATRICS criteria of cognitive tests described above by adding requirements that measures exhibit construct validity, clear links to neural circuits and cognitive mechanisms, and have an available an animal model (Barch et al. 2009). Out of this extensive process of evaluating the many promising measures in the existing literature, several tests were selected for further study and development. Critically, two neurophysiological measures were deemed already "mature," fulfilling all of the MATRICS/CNTRICS criteria and suitable for immediate incorporation into multi-site clinical studies: MMN and Prepulse Inhibition (Butler et al. 2012; Green et al. 2009)—the focus of this chapter. Below, we also provide examples of rational and deliberate matching of patients with intact biomarker functioning with appropriately targeted cognitive therapies that depend upon the engagement of the neural substrates of these measures.

3 Mismatch Negativity

Mismatch negativity is a preattentive event-related potential (ERP) component with tremendous promise as a biomarker for predicting and tracking response to novel therapeutic interventions (Light and Näätänen 2013; Nagai et al. 2013; Light et al. 2012; Perez et al. 2014a, b; Kawakubo et al. 2007). MMN is a negative-going deflection in the ERP that is evoked when a sequence of repetitive "standard" stimuli is occasionally interrupted by infrequent oddball or "deviant" stimuli that differ in some physical characteristic such as duration or pitch. The onset of MMN typically occurs within 50 ms of stimulus deviance, and peaks after an additional 100–150 ms. Since MMN requires no overt behavioral response and can be elicited even in the absence of directed attention (Näätänen 1992; Rinne et al. 2001; Sussman et al. 2003; Rissling et al. 2013), it is presumed to reflect a predominantly automatic, preconscious process of detecting a "mismatch" between the deviant stimulus and a sensory-memory trace (Näätänen et al. 1989).

MMN amplitude reduction in schizophrenia was first reported over 20 years ago (Shelley et al. 1991) with subsequent studies consistently identifying deficits in chronic ($d \cong 1.00$ Javitt et al. 1994; Shelley et al. 1991; Catts et al. 1995; Javitt et al. 2000; Michie 2001; Umbricht et al. 2003; Umbricht and Krljes 2005; Salisbury et al. 2002; Oknina et al. 2005; Oades et al. 2006; Light and Braff 2005a, b; Rissling et al. 2012, 2013), recent onset (Salisbury et al. 2002, 2007; Brockhaus-Dumke et al. 2005; Umbricht et al. 2006; Oknina et al. 2005; Oades et al. 2006; Hermens et al. 2010; Bodatsch et al. 2011; Jahshan et al. 2012; Atkinson et al. 2012; Perez et al. 2013), and even unmedicated schizophrenia patients (Rissling et al. 2012; Bodatsch et al. 2011; Brockhaus-Dumke et al. 2005; Kirino and Inoue 1999; Catts et al. 1995). MMN is supported by a distributed network of fronto-temporal sources with deficits in schizophrenia prominent in medial frontal brain regions (e.g., Takahashi et al. 2012) and a sensitive index of *N*-methyl D-aspartate (NMDA) receptor functioning (Javitt et al. 1996; Umbricht et al. 2000, 2002; Gil-da-Costa et al. 2013; Lavoie et al. 2007; Ehrlichman et al. 2008; Nakamura et al. 2011). The temporal window indexed by MMN may serve as a gateway to some higher order cognitive operations necessary for psychosocial functioning (e.g., Rissling et al. 2013). MMN accounts for substantial portions of variance in cognition (Baldeweg et al. 2004; Näätänen et al. 2011; Light et al. 2007b; Kawakubo et al. 2006), psychosocial functioning (Light and Braff 2005a, b; Kawakubo et al. 2007; Wynn et al. 2010; Rasser et al. 2011), and level of independence in community living (Light and Braff 2005a). MMN amplitude also exhibits utility as a repeated measure with high test–retest stability over short and long (e.g., 12-month) retest intervals in both healthy subjects and schizophrenia patients (Light et al. 2012). Indeed, MMN reliability coefficients are comparable to or even exceed those obtained from neuropsychological tests over 1 year (ICCs $\cong 0.90$; Light et al. 2012; Light and Braff 2005b). This collection of attributes has contributed to the view of MMN as a "breakthrough biomarker" (Light and Näätänen 2013) that is "translatable" (Nagai et al. 2013) and potentially "the one we have been waiting for" (Belger et al. 2012) in neuropsychiatry.

There is certainly ample evidence that MMN is an informative biomarker index and correlate of "what's wrong" in schizophrenia. In fact, we have previously argued for pharmacologic and nonpharmacologic treatments that target early auditory perceptual processing with the hope that an amelioration of MMN deficits might accompany or even precede improvements in highly associated cognitive and psychosocial functioning (Braff and Light 2004; Perez et al. 2014a, b). We now consider a "figure-ground" reversal: rather than focus on the 50 % of patients with deficient MMN, perhaps the remaining 50 % with normal range MMN will be most likely to benefit from therapies that are designed to target low-level auditory perceptual processes.

MMN may be particularly sensitive to one particular form of "bottom-up" CT—termed Targeted Cognitive Training (TCT; Fisher et al. 2009). TCT uses "neuroplasticity-based" computerized exercises designed to improve the accuracy and fidelity of auditory sensory information processing and auditory/verbal WM. TCT relies on intensive, attentionally engaging, adaptive, and reinforcing tasks to facilitate procedural learning (Adcock et al. 2009)—a mechanism that is largely intact in SZ (Perry et al. 2000). Conceptually, the goal of TCT is to capitalize on plastic changes within the neural substrates of low-level auditory information processing, which then feed forward to improve higher order cognitive operations such as attention, WM, and the encoding and retrieval of verbal information. Fisher and colleagues have shown (Vinogradov et al. 2012) that SZ patients exhibit large effect size ($d = 0.86$–0.89; Fisher et al. 2009) gains in auditory-dependent cognitive domains (verbal learning and memory), global cognition, and quality of life after 50 h of training. Importantly, these gains persist for at least 6 months after the cessation of training (Fisher et al. 2010). Although TCT is efficacious at the group level, individual patient responses to this resource and time-intensive intervention vary considerably; some patients exhibit little or no benefit after even an extended 100 h course of training (Fisher et al. 2014). Could MMN or other neurophysiological biomarkers of auditory sensory processing be used to predict whether an individual patient is likely to respond to this time- and resource-intensive intervention?

In addition to the emerging applications in neuropsychiatry, MMN is supported by a substantial cognitive neuroscience literature. Indeed, MMN is already regarded a dynamic index of central auditory system neuroplasticity that predicts cognitive enhancement in response to specific TCT-like auditory training interventions (Menning et al. 2000; Näätänen 2008). For example, Menning and colleagues (2000) demonstrated that 3 weeks of intensive ($\sim$1 h/day) auditory frequency discrimination training produced significant increases in MMN amplitude that persisted for several weeks after the cessation of training in healthy volunteers. Other studies have shown that MMN *both predicts and corresponds* to changes in language acquisition, musical training, and other auditory-dependent cognitive tasks in nonpsychiatric individuals (for review, Näätänen 2008). Likewise, MMN exhibits malleability after even a single 3 h session of auditory training in dyslexic children, which was associated with a significant amelioration of cognitive impairment in phonological processing, reading, and writing (Lovio et al. 2012). Thus, changes in MMN are detectable in the early stages of cognitive training, predict generalized improvements in nontrained higher order cognitive domains, and correspond to measurable changes of cortical plasticity in intact and impaired neuropsychiatric populations. In all instances, larger MMN (i.e., associated with healthy function) was associated with greater training gains.

Little is known about the neural mechanisms that underlie enhanced global cognition and inter-individual variation in TCT response in schizophrenia. Better characterization of biomarkers of TCT response will lead to more selective targeting of patients and neurobiological systems for preventive interventions. We have conducted a proof of concept validation study to begin to understand the potential relationship between MMN and immediate TCT effects (Perez et al. 2014b). MMN was assessed immediately before and after a 1 h TCT session (PositScience, Frequency Sweeps) in 31 chronic, medicated SZ patients. MMN

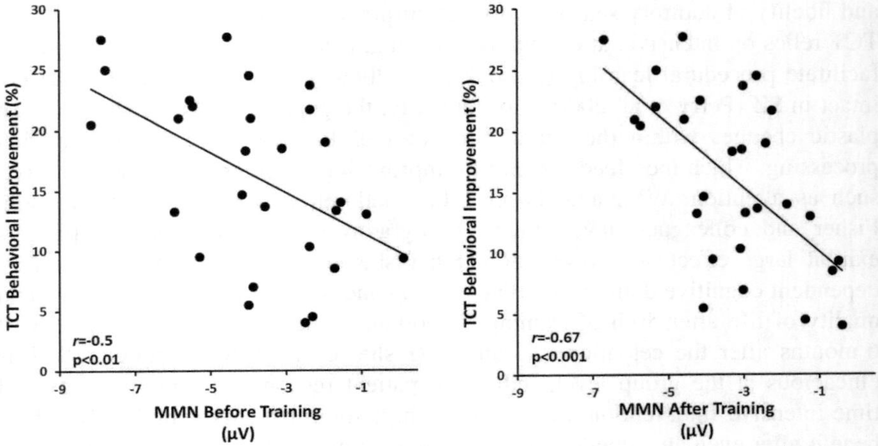

Fig. 2 MMN recorded before and after 1 h of training is associated with initial behavioral performance gains during TCT in schizophrenias patients. Larger pre-training MMN significantly predicted greater TCT improvements; post-training MMN was also significantly associated with performance gains (Perez et al. 2013)

amplitude exhibited significant change at frontocentral electrodes ($p < 0.02$) confirming our prediction that MMN is sensitive to early "target engagement" after just 1 h of training. In addition, patients with larger pretraining MMN amplitude exhibited greatest improvements across the single TCT session ($r = -0.5$, $p < 0.01$), confirming our hypothesis that baseline MMN predicted initial TCT performance gains (Fig. 2). In addition, post-training MMN accounted for 45 % of the variance in TCT performance gains. Thus, patients with larger (i.e., more normal) levels of MMN (i.e., "what's right") exhibited a larger initial response to training. While these results are encouraging, it is important to emphasize that behavioral response to a single TCT session is not known to predict longer term neurocognitive or functional gains in schizophrenia patients undergoing a full course of training. In support of this model of larger biomarker values predicting treatment response, Kawakubo and colleagues (2007) showed that larger baseline MMN predicted greater response to an intensive, 3 month social skills training program (Fig. 3). This approach may therefore serve as a useful platform for identifying patients who are likely to be "responders" to TCT (Light and Näätänen 2013; Perez et al. 2014b), social skills training (Kawakubo et al. 2007), or perhaps other forms of cognitive remediation. Such predictive biomarkers may also facilitate screening drugs to augment cognitive training.

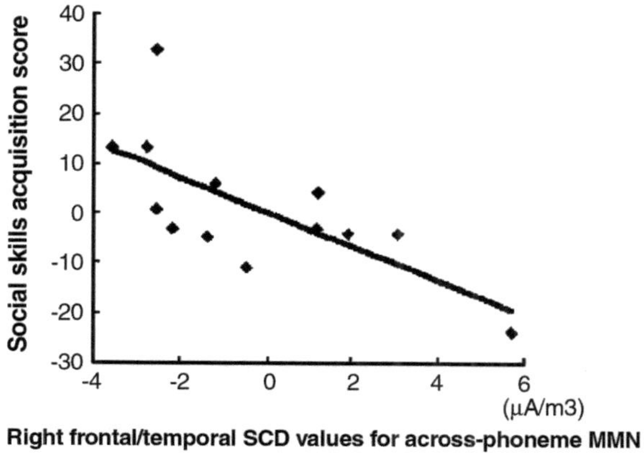

Fig. 3 Larger pre-training MMN amplitude predicts greater acquisition of social skills following an intensive 3-month training program (reprinted from Kawakubo et al. 2007)

4 Prepulse Inhibition of Startle

When a neuro- or psychophysiological biomarker can be studied across species, there is the potential that the measure can be used to elucidate neural and cellular substrates underlying its predictive properties. This concept has motivated studies of PPI as one potential neurophysiological biomarker predicting pro-cognitive drug effects. PPI is a laboratory-based operational measure of sensorimotor gating, in which a weak prepulse inhibits the magnitude of a startle response to an intense, abrupt "pulse" occurring 30–120 ms later. PPI is easily studied in animal models, including mice, rats, guinea pigs, pigs, and infrahuman primates, using stimulus parameters and equipment for stimulus delivery and response acquisition that are similar or identical to what are used in humans (cf. Swerdlow et al. 2008). While there appear to be differences in the neurochemical regulation of PPI across species, the basic parametric properties of PPI exhibit striking similarities from rodents to humans (e.g., Swerdlow et al. 1994). PPI is under significant genetic control in both rodents (Francis et al. 2003) and humans (Greenwood et al. 2011). While it has been advertised as a "simple" behavior, in reality PPI is a complex, heritable phenotype regulated by numerous different genes, as described in many reports (e.g., Greenwood et al. 2011, 2012, 2013). Reduced PPI is not specific to patients with SZ: in addition to SZ (Braff et al. 1978), PPI has been reported to be deficient in patients with Huntington's Disease (Swerdlow et al. 1995; Valls-Sole et al. 2004), Obsessive Compulsive Disorder (Swerdlow et al. 1993a, Hoenig et al. 2005; Ahmari et al. 2012), nocturnal enuresis (Ornitz et al. 1992), Asperger's Syndrome (McAlonan et al. 2002), 22q11 Syndrome (Sobin et al. 2005), Klein-felter Syndrome (Van Rijn et al. 2011), Fragile-X Syndrome (Frankland et al. 2004) and blepharospasm (Gomez-Wong et al. 1998) and Tourette Syndrome (Castellanos et al. 1996; Swerdlow et al. 2001b). However, PPI deficits in SZ patients have been perhaps the best studied: over 40 PubMed reports describe PPI deficits in SZ or "prodromal" patients (cf. Swerdlow et al. 2014).

There are a number of measures that show strong structural similarity to PPI, in that they all assess the amount of behavioral and/or neural inhibition generated by a lead stimulus, as determined by the amount to which the response to a second stimulus is suppressed. In measures of "recovery cycle" (also called "blink excitability," Smith and Lees 1989), "paired pulse inhibition" (Swerdlow et al. 2002), or "intracortical inhibition" (Ziemann et al. 1997), the dependent measure is the motor response to a target stimulus ("pulse" or "S2"), presented either alone or shortly after the presentation of a lead stimulus ("prepulse" or "S1"). A "healthy" response is generally indicated by a diminished motor response to S2 in the presence of S1, compared with the response to S2 alone. Thus, in its simplest view, PPI is a measure of the degree to which a motor response is inhibited by a sensory event, i.e., sensorimotor inhibition. With only 10–120 ms separating prepulses and startling pulses in the "uninstructed" PPI paradigm, PPI is generally viewed as a measure of largely automatic, preattentional inhibitory processes (Graham 1975; Filion et al. 1993). Nonetheless, the amount of PPI at relatively longer (60–120 ms) prepulse intervals correlates significantly with higher cognitive processes, including WM (Letter-Number Span (Greenwood et al. 2013; Light et al. 2007a, b)), strategy formation, measures of cognitive efficiency (Bitsios et al. 2006; Giakoumaki et al. 2006; Light et al. 2007a, b; van der Linden et al. 2006) and even global functioning (Swerdlow et al. 2006a).

PPI deficits in SZ patients might potentially reflect abnormalities at any one or more levels of PPI-regulatory circuitry that stretches from the prefrontal cortex to the pons. Thus, reduced PPI might be found under conditions of excessive dopamine (DA) neurotransmission in subcortical structures, deficient DA, or glutamate transmission in cortical structures, excessive serotonin, or deficient GABA transmission in pallidum (cf. Swerdlow et al. 2008), etc. In fact, PPI *deficits* in a particular patient might reflect an almost infinite number of deficits in isolation or combination. But for a SZ patient to exhibit *robust levels* of PPI requires functionality within some or all of PPI-regulatory circuitry, and perhaps more importantly the integrity of the process of sensorimotor gating. So, compared to PPI deficits, a biomarker of "normal" PPI might be more interpretable.

PPI levels in SZ are highly stable, with 1-year ICC's approaching 0.80 (Light et al. 2012). While some groups have reported medium-to-large effect size deficits in PPI in SZ versus healthy cohorts, our most recent large single-site reports have detected deficits with 60 ms prepulse intervals with effect sizes that ranged from 0.24 (Swerdlow et al. 2006a, b) to 0.58 (Light et al. 2012). Factors that may contribute to "artificially" small PPI differences between SZ and healthy cohorts include: (1) SZ-linked PPI deficits generally appear to be most robust under specific sets of stimulus parameters, i.e., the type of prepulse used [auditory vs. tactile; tone vs. noise; prepulse intensity over background and prepulse interval, etc. (Braff et al. 2001; Swerdlow et al. 2006a, b)]; (2) women have lower PPI than do men (Swerdlow et al. 1993a, b), and in most studies, healthy subjects are predominantly women, while SZ patients are predominantly men; (3) PPI is generally increased by nicotine (Hong et al. 2008; Kumari et al. 2001), and smoking is both more common and heavier among SZ patients versus healthy

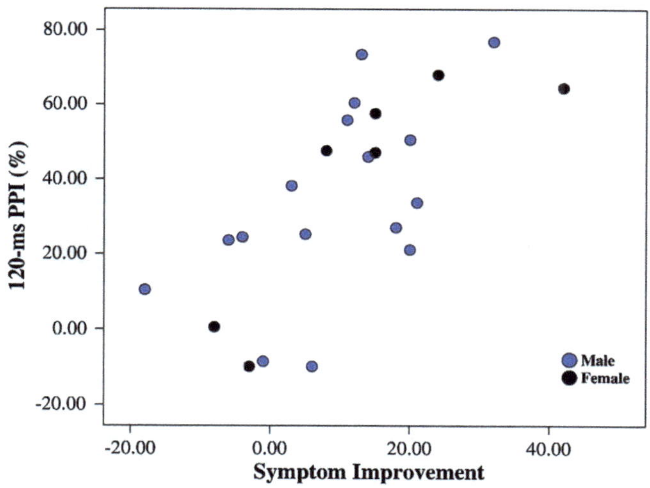

Fig. 4 Higher levels of baseline PPI predict positive response to cognitive-behavioral therapy in schizophrenia patients (reprinted from Kumari et al. 2012)

subjects; (4) PPI is higher in medicated versus unmedicated SZ patients, and especially in patients medicated with second-generation antipsychotics (SGAPs) (Csomor et al. 2009; Kumari et al. 1999; Swerdlow et al. 2006a, b; Weike et al. 2000); SGAPs are used by more than 80 % of SZ patients in most recent studies.

One implication of the relative modest effect sizes of PPI deficits in SZ cohorts is that many SZ patients exhibit PPI levels at, or above, HS "mean" values. Presumably, these "normal" PPI levels in SZ patients can serve as a biomarker of normal function in PPI neural circuitry; this does not mean that the entire PPI-regulatory apparatus is intact in these individuals, but simply that the overall circuit properties—with or without the influence of nicotine, SGAPs, and other moderating factors—remain adequately intact to perform its "function" of sensorimotor gating. Thus, "higher" neurocognitive processes that rely on intact sensorimotor gating would be not be expected to impaired in these individuals based solely on this reliance; certainly, these processes might nonetheless be impaired, based on deficits in other basic information processing mechanisms.

In keeping with our model of using biomarkers to identify residual "intact" neural mechanisms and function, it is reasonable to consider whether "high PPI" could be used as a biomarker of SZ patients who might be capable of marshaling adequate cognitive resources to meet the demands of, and reap the benefits of a particular therapeutic intervention. Consistent with such a model, Kumari et al. (2012) were able to demonstrate that baseline PPI levels positively predicted the therapeutic response to cognitive-behavioral therapy (CBT; Fig. 4). Schizophrenia patients who exhibited the highest pretherapy PPI levels were the ones who benefitted most from CBT, in terms of reductions in symptom severity. This finding supports the notion that evidence of intact, functioning neural mechanisms,

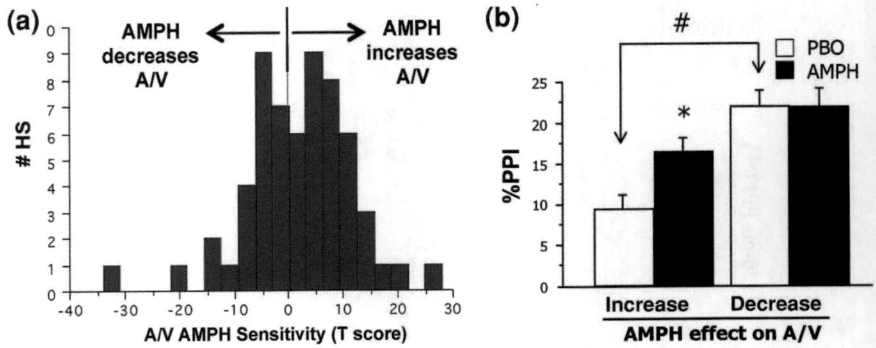

Fig. 5 Using a drug challenge to identify residual plasticity in sensorimotor gating and attentional capacity: a "proof of concept" in healthy subjects. **a** Distribution of the change in MCCB A/V T-scores after amphetamine (AMPH; 20 mg p.o.) versus placebo, corrected for order effects, in 60 healthy subjects (Swerdlow et al. 2013). **b** Baseline PPI was significantly lower (#) and more sensitive to AMPH-enhancement (*), among subjects in whom AMPH increased versus decreased A/V in "A"

provided by a psychophysiological biomarker, can positively predict the therapeutic response to a higher cognitive intervention.

The neural circuitry regulating PPI includes neurotransmitters and receptors that are targets of many of the major classes of psychotropic medications. Drugs acting at prominent nodes in this circuitry have potent effects on PPI, which have been studied extensively in rodents, and more recently in humans (cf. Swerdlow et al. 2008). Among our proposed applications of biomarkers for SZ therapeutics, we suggested a model in which a neural system at baseline may perform poorly, but still respond to a "push" of a pharmacologic challenge; in this case, evidence for the requisite "spared" neural circuitry, and hence a target for therapeutic intervention, is provided by changes in response to a drug challenge. In essence, the acute drug challenge is used to determine whether the impaired system retains sufficient plasticity to respond to therapeutic input.

To date, our "proof of concept" studies for the potential to detect residual plasticity in neurocognitive substrates via drug challenges in "low performers" have exclusively involved healthy subjects. We reported that in specific subgroups of HS—groups characterized by low basal PPI, low novelty-, or sensation-seeking traits—a single dose of the psychostimulant, amphetamine (AMPH, 20 mg p.o.) potently enhances PPI (Talledo et al. 2009). This suggests that among some individuals—even (though not exclusively) in the presence of low basal PPI—the neural circuitry regulating PPI retains significant plasticity, in that it can respond positively to a drug challenge. We also reported that this same dose of AMPH enhances MCCB performance, particularly in the domain of attention /vigilance (A/V), among 60 healthy individuals with low baseline A/V performance (Chou et al. 2013b; Fig. 5). When we stratified these 60 subjects according to baseline PPI, those with low baseline PPI were the ones most sensitive to both the PPI- and

A/V-enhancing effects of AMPH. Presumably, the neural circuit plasticity evident in low PPI healthy subjects predicts the likelihood of exhibiting a pro-attentional response to AMPH. We are currently testing these relationships in SZ patients. Ultimately, among patients exhibiting deficits in biomarkers, we might use a "challenge" paradigm to reveal those whose residual plasticity would predict benefits from the addition of a specific drug to a cognitive intervention. Clearly, we are several steps from fully testing this "drug challenge" biomarker model, but the path to such a test is clear.

5 Discussion

One of the challenges facing the use of biomarkers in SZ populations is that, for the most part, biomarkers are being applied "after the fact." In other words, if we acknowledge that SZ is a neurodevelopmental disorder (or set of disorders), likely reflecting perturbations of in utero neural development, then the events (genetic, environmental or otherwise) that lead to the late-adolescent/early adult manifestations of the disorder have come and gone, decades before biomarker data are measured. And the number of variations in the expression of these early events—e.g., variable neuronal migratory routes, the adjustments of the surrounding developing brain to them, the consequent alterations in premorbid behavior and the reflected impact of environmental responses onto brain development—from in utero causative event to adult manifestation are substantial if not limitless. And unlike disorders of adult onset in which an anatomically or neurochemically constrained "lesion" is superimposed on a normally developed brain, in SZ, the absent connections lost to cells that did not arrive, and the aberrant connections formed in their place, are infused throughout the matrix of a very complex forebrain circuitry. Making sense of "right" and "wrong" in this circuit context, as a basis for understanding the biology of SZ, its courses or treatments, may not be feasible, or even productive, in the foreseeable future.

We have proposed an alternative use of biomarkers in predicting treatment response in SZ patients, that is consistent both with the therapeutic goals of personalized medicine and the scientific strategies of experimental medicine (Insel 2014). Individuals are characterized via measures of brain activity that are associated with neurocognition and function, and areas of "healthy" or "normal range" performance are identified. In this process, drugs or other experimental manipulations and designs can be used as clinical probes to identify targets of residual neuroplasticity. Treatments are then identified that leverage the intact neural circuit or neurocognitive resources so that the individual patient can utilize their capacities to reap the gains of the therapeutic intervention. In truth, the basic principles of the "biomarkers of health" approach are simple ones, long espoused by disciplines ranging from childhood education to career counseling: a successful outcome is best achieved by matching residual strengths—areas of "resiliency"—with task demands. In the frenzied search for the genetic and molecular markers and mechanisms of what's wrong with individuals with SZ, our field and its treatments may not have fully appreciated and leveraged all that is "right".

One key to the successful use of biomarkers in this model is the ability to link a "healthy" biomarker with a positive response to a specific therapy. For example, as we allude to in our Introduction, some forms of cognitive training put demands on processes requiring "healthy" WM and attention (Haut et al. 2010), and thus would be best pursued in patients with biomarker evidence of relatively intact WM and attentional capacity. Alternatively, evidence that WM and attentional performance could be enhanced in that patient by a psychostimulant challenge (e.g. Barch and Carter 2005) might predict benefits of psychostimulant augmentation of cognitive training. Different biomarkers of neurocognitive and neural circuit strengths might predict optimal responses of SZ patients to CBT, (e.g. Kumari et al. 2009) computerized cognitive training, social skills training, or even medications like the pro-extinction drug, D-cycloserine (Gottlieb et al. 2011), or the "pro-social" drug, oxytocin (Davis et al. 2014). While there is substantial evidence that baseline cognitive deficits generally predict poor outcomes in cognitive interventions (Becker et al. 1998; Green 1996; McGurk and Meltzer 2000; McGurlk and Mueser 2004; Spaulding et al. 1999), we are not yet at a point where we can apply specific algorithms other than "clinical intuition" to match biomarkers of intact neural function in a SZ patient, with treatment response to different types of therapies. Developing such algorithms will be advanced by incorporating informative biomarkers, like MMN and PPI, and detailed neurocognitive assessments, into the designs of trials of cognitive interventions for SZ. Importantly, the fidelity and optimal methods for many potential biomarkers have already been established in multi-site studies, where deficits in these measures have been used as endophenotypes to identify risk genes for SZ (Turetsky et al. 2007). In the figure-ground reversal proposed here, these biomarkers are used not to predict a risk of illness, but rather, they are used to predict a likelihood of recovery.

Acknowledgments NRS is supported by NIMH awards MH59803, MH93453, MH42228 and MH094320. GL is supported by MH42228, MH065571, MH094151, MH093453, MH094320, UL1TR000100, MH081944, NARSAD, the Veterans Medical Research Foundation and the VISN-22 Mental Illness, Research, Education, and Clinical Center. GL has served as a consultant for Astellas Inc, Envivo Pharmaceuticals, and Neuroverse for work unrelated to this chapter. The authors have no conflicts of interest.

References

Adcock RA, Dale C, Fisher M et al (2009) When top-down meets bottom-up: auditory training enhances verbal memory in schizophrenia. Schizophr Bull 35:1132–1141

Ahmari SE, Risbrough VB, Geyer MA et al (2012) Impaired sensorimotor gating in unmedicated adults with obsessive-compulsive disorder. Neuropsychopharmacology 37:1216–1223

Atkinson RJ, Michie PT, Schall U (2012) Duration mismatch negativity and P3a in first-episode psychosis and individuals at ultra-high risk of psychosis. Biol Psychiatry 71:98–104

Baldeweg T, Klugman A, Gruzelier J et al (2004) Mismatch negativity potentials and cognitive impairment in schizophrenia. Schizophr Res 69:203–217

Barch DM, Carter CS (2005) Amphetamine improves cognitive function in medicated individuals with schizophrenia and in healthy volunteers. Schizophr Res 77:43–58

Barch DM, Carter CS, Arnsten A et al (2009) Selecting paradigms from cognitive neuroscience for translation into use in clinical trials: proceedings of the third CNTRICS meeting. Schizophr Bull 35:109–114

Becker DR, Drake RE, Bond GR et al (1998) Job terminations among persons with severe mental illness participating in supported employment. Community Ment Health J 34:71–82

Belger A, Yucel GH, Donkers FC (2012) In search of psychosis biomarkers in high-risk populations: is the mismatch negativity the one we've been waiting for? Biol Psychiatry 71:94–95

Biller BM (2007) Concepts in the diagnosis of adult growth hormone deficiency. Horm Res 68(Suppl 5):59–65

Bitsios P, Giakoumaki SG, Theou K et al (2006) Increased prepulse inhibition of the acoustic startle response is associated with better strategy formation and execution times in healthy males. Neuropsychologia 44:2494–2499

Bodatsch M, Ruhrmann S, Wagner M et al (2011) Prediction of psychosis by mismatch negativity. Biol Psychiatry 69:959–966

Braff DL, Light GA (2004) Preattentional and attentional cognitive deficits as targets for treating schizophrenia. Psychopharmacology 174:75–85

Braff D, Stone C, Callaway E et al (1978) Prestimulus effects on human startle reflex in normals and schizophrenics. Psychophysiology 15:339–343

Braff DL, Geyer MA, Light GA et al (2001) Impact of prepulse characteristics on the detection of sensorimotor gating deficits in schizophrenia. Schizophr Res 49:171–178

Braff DL, Light GA, Swerdlow NR (2007) Prepulse inhibition and P50 suppression are both deficient but not correlated in schizophrenia patients. Biol Psychiatry 61:1204–1207

Brockhaus-Dumke A, Tendolkar I, Pukrop R et al (2005) Impaired mismatch negativity generation in prodromal subjects and patients with schizophrenia. Schizophr Res 73:297–310

Butler PD, Chen Y, Ford JM et al (2012) Perceptual measurement in schizophrenia: promising electrophysiology and neuroimaging paradigms from CNTRICS. Schizophr Bull 38:81–91

Carter CS, Barch DM, Buchanan RW et al (2008) Identifying cognitive mechanisms targeted for treatment development in schizophrenia: an overview of the first meeting of the Cognitive Neuroscience Treatment Research to Improve Cognition in Schizophrenia Initiative. Biol Psychiatry 64:4–10

Castellanos FX, Fine EJ, Kaysen DL et al (1996) Sensorimotor gating in boys with Tourette's syndrome and ADHD: preliminary results. Biol Psychiatry 39:33–41

Catts SV, Shelley AM, Ward PB et al (1995) Brain potential evidence for an auditory sensory memory deficit in schizophrenia. Am J Psychiatry 152:213–219

Chou HH, Bhakta SG, Talledo JA et al (2013a) Memantine effects on MATRICS Consensus Cognitive Performance Battery in healthy adults and schizophrenia patients. Society Biol Psychiatry, San Francisco, 16–18 May

Chou HH, Talledo J, Lamb S et al (2013b) Amphetamine effects on MATRICS consensus cognitive battery performance in healthy adults. Psychopharmacology 227:165–176

Csomor PA, Yee BK, Feldon J et al (2009) Impaired prepulse inhibition and prepulse-elicited reactivity but intact reflex circuit excitability in unmedicated schizophrenia patients: a comparison with healthy subjects and medicated schizophrenia patients. Schizophr Bull 35:244–255

Davis MC, Green MF, Lee J et al (2014) Oxytocin-augmented social cognitive skills training in schizophrenia. Neuropsychopharmacology. doi:10.1038/npp.2014.68

Ehrlichman RS, Maxwell CR, Majumdar S et al (2008) Deviance-elicited changes in event-related potentials are attenuated by ketamine in mice. J Cogn Neurosci 20:1403–1414

Filion DL, Dawson ME, Schell AM (1993) Modification of the acoustic startle-reflex eyeblink: a tool for investigating early and late attentional processes. Biol Psychol 35:185–200

Fisher M, Holland C, Merzenich MM et al (2009) Using neuroplasticity-based auditory training to improve verbal memory in schizophrenia. Am J Psychiatry 166:805–811
Fisher M, Holland C, Subramaniam K et al (2010) Neuroplasticity-based cognitive training in schizophrenia: an interim report on the effects 6 months later. Schizophr Bull 36:869–879
Fisher M, Loewy R, Carter C et al (2014) Neuroplasticity-based auditory training via laptop computer improves cognition in young individuals with recent onset schizophrenia. Schizophr Bull. doi:10.1093/schbul/sbt232
Francis DD, Szegda K, Campbell G et al (2003) Epigenetic sources of behavioral differences in mice. Nat Neurosci 6:445–446
Frankland PW, Wang Y, Rosner B et al (2004) Sensorimotor gating abnormalities in young males with fragile X syndrome and Fmr1-knockout mice. Mol Psychiatry 9:417–425
Fruchter O, Yigla M (2009) Bronchodilator response after negative methacholine challenge test predicts future diagnosis of asthma. J Asthma 46:722–725
Giakoumaki SG, Bitsios P, Frangou S (2006) The level of prepulse inhibition in healthy individuals may index cortical modulation of early information processing. Brain Res 1078:168–170
Gil-da-Costa R, Stoner GR, Fung R et al (2013) Nonhuman primate model of schizophrenia using a noninvasive EEG method. Proc Natl Acad Sci USA 110:15425–15430
Gomez-Wong E, Marti MJ, Tolosa E et al (1998) Sensory modulation of the blink reflex in patients with blepharospasm. Arch Neurol 55:1233–1237
Gottlieb JD, Cather C, Shanahan M et al (2011) D-cycloserine facilitation of cognitive behavioral therapy for delusions in schizophrenia. Schizophr Res 131:69–74
Graham F (1975) The more or less startling effects of weak prestimuli. Psychophysiology 12:238–248
Green MF (1996) What are the functional consequences of neurocognitive deficits in schizophrenia? Am J Psychiatry 153:321–330
Green MF, Nuechterlein KH, Gold JM et al (2004) Approaching a consensus cognitive battery for clinical trials in schizophrenia: the NIMH-MATRICS conference to select cognitive domains and test criteria. Biol Psychiatry 56:301–307
Green MF, Butler PD, Chen Y et al (2009) Perception measurement in clinical trials of schizophrenia: promising paradigms from CNTRICS. Schizophr Bull 35:163–181
Greenwood TA, Lazzeroni LC, Murray SS et al (2011) Analysis of 94 candidate genes and twelve endophenotypes for schizophrenia from the consortium on the genetics of schizophrenia. Am J Psychiatry 168:930–946
Greenwood TA, Light GA, Swerdlow NR et al (2012) Association analysis of 94 candidate genes and schizophrenia-related endophenotypes. PLoS One 7:e29630
Greenwood TA, Swerdlow NR, Gur RE et al (2013) Genome-wide linkage analyses of 12 endophenotypes for schizophrenia from the consortium on the genetics of schizophrenia. Am J Psychiatry 170:521–532
Haut KM, Lim KO, MacDonald A (2010) Prefrontal cortical changes following cognitive training in patients with chronic schizophrenia: effects of practice, generalization and specificity. Neuropsychopharmacology 35:1850–1859
Hermens DF, Ward PB, Hodge MA et al (2010) Impaired MMN/P3a complex in first-episode psychosis: cognitive and psychosocial associations. Prog Neuropsychopharmacol Biol Psychiatry 34:822–829
Hoenig K, Hochrein A, Quednow BB et al (2005) Impaired prepulse inhibition of acoustic startle in obsessive-compulsive disorder. Biol Psychiatry 57:1153–1158
Hong LE, Wonodi I, Lewis J et al (2008) Nicotine effect on prepulse inhibition and prepulse facilitation in schizophrenia patients. Neuropsychopharmacology 33:2167–2174
Horan WP, Braff DL, Nuechterlein KH et al (2008) Verbal working memory impairments in individuals with schizophrenia and their first-degree relatives: findings from the consortium on the genetics of schizophrenia. Schizophr Res 103:218–228

Horvath J, Winkler I, Bendixen A (2008) Do N1/MMN, P3a, and RON form a strongly coupled chain reflecting the three stages of auditory distraction? Biol Psychol 79(2):139–147. doi:10.1016/j.biopsycho.2008.04.001
Hughes AJ, Lees AJ, Stern GM (1990) Apomorphine test to predict dopaminergic responsiveness in parkinsonian syndromes. Lancet 336:32–34
Insel T (2014) Director's blog: a new approach to clinical trials. http://www.nimh.nih.gov/about/director/2014/a-new-approach-to-clinical-trials.shtml
Jahshan C, Cadenhead KS, Rissling AJ et al (2012) Automatic sensory information processing abnormalities across the illness course of schizophrenia. Psychol Med 42:85–97
Javitt DC, Steinschneider M, Schroeder CE et al (1994) Detection of stimulus deviance within primate primary auditory cortex: intracortical mechanisms of mismatch negativity (MMN) generation. Brain Res 667:192–200
Javitt DC, Steinschneider M, Schroeder CE et al (1996) Role of cortical N-methyl-D-aspartate receptors in auditory sensory memory and mismatch negativity generation: implications for schizophrenia. Proc Natl Acad Sci USA 93:11962–11967
Javitt DC, Shelley AM, Silipo G et al (2000) Deficits in auditory and visual context-dependent processing in schizophrenia: defining the pattern. Arch Gen Psychiatry 57:1131–1137
Kawakubo Y, Kasai K, Kudo N et al (2006) Phonetic mismatch negativity predicts verbal memory deficits in schizophrenia. NeuroReport 17:1043–1046
Kawakubo Y, Kamio S, Nose T et al (2007) Phonetic mismatch negativity predicts social skills acquisition in schizophrenia. Psychiatry Res 152:261–265
Kiang M, Braff DL, Sprock J et al (2009) The relationship between preattentive sensory processing deficits and age in schizophrenia patients. Clin Neurophysiol: Off J Int Fed Clin Neurophysiol 120:1949–1957
Kirino E, Inoue R (1999) The relationship of mismatch negativity to quantitative EEG and morphological findings in schizophrenia. J Psychiatr Res 33:445–456
Kumari V, Soni W, Sharma T (1999) Normalization of information processing deficits in schizophrenia with clozapine. Am J Psychiatry 156:1046–1051
Kumari V, Soni W, Sharma T (2001) Influence of cigarette smoking on prepulse inhibition of the acoustic startle response in schizophrenia. Hum Psychopharmacol 16:321–326
Kumari V, Peters ER, Fannon D et al (2009) Dorsolateral prefrontal cortex activity predicts responsiveness to cognitive-behavioral therapy in schizophrenia. Biol Psychiatry 66:594–602
Kumari V, Premkumar P, Fannon D et al (2012) Sensorimotor gating and clinical outcome following cognitive behaviour therapy for psychosis. Schizophr Res 134:232–238
Lavoie S, Murray MM, Deppen P, Knyazeva MG, Berk M, Boulat O, Bovet P, Bush AI, Conus P, Copolov D (2007) Glutathione precursor, N-acetyl-cysteine, improves mismatch negativity in schizophrenia patients. Neuropsychopharmacology 33(9):2187–2199
Lieberman JA, Stroup TS, McEvoy JP et al (2005) Effectiveness of antipsychotic drugs in patients with chronic schizophrenia. N Engl J Med 353:1209–1223
Light GA, Braff DL (2001) Measuring P50 suppression and prepulse inhibition in a single recording session. Am J Psychiatry 158:2066–2068
Light GA, Braff DL (2005a) Mismatch negativity deficits are associated with poor functioning in schizophrenia patients. Arch Gen Psychiatry 62:127–136
Light GA, Braff DL (2005b) Stability of mismatch negativity deficits and their relationship to functional impairments in chronic schizophrenia. Am J Psychiatry 162:1741–1743
Light GA, Näätänen R (2013) Mismatch negativity is a breakthrough biomarker for understanding and treating psychotic disorders. Proc Natl Acad Sci USA 110:15175–15176
Light GA, Braff DL, Sprock J et al (2007a) Prepulse inhibition of startle is positively associated with higher order cognition in women. Soc Neurosci, San Diego, 3–7 Nov
Light GA, Swerdlow NR, Braff DL (2007b) Preattentive sensory processing as indexed by the MMN and P3a brain responses is associated with cognitive and psychosocial functioning in healthy adults. J Cogn Neurosci 19:1624–1632

Light GA, Swerdlow NR, Rissling AJ et al (2012) Characterization of neurophysiologic and neurocognitive biomarkers for use in genomic and clinical outcome studies of schizophrenia. PLoS One 7:e39434

Lovio R, Halttunen A, Lyytinen H et al (2012) Reading skill and neural processing accuracy improvement after a 3-hour intervention in preschoolers with difficulties in reading-related skills. Brain Res 1448:42–55

McAlonan GM, Daly E, Kumari V et al (2002) Brain anatomy and sensorimotor gating in Asperger's syndrome. Brain 125:1594–1606

McGurk SR, Meltzer HY (2000) The role of cognition in vocational functioning in schizophrenia. Schizophr Res 45:175–184

McGurk SR, Mueser KT (2004) Cognitive functioning, symptoms, and work in supported employment: a review and heuristic model. Schizophr Res 70:147–173

Menning H, Roberts LE, Pantev C (2000) Plastic changes in the auditory cortex induced by intensive frequency discrimination training. NeuroReport 11:817–822

Michie PT (2001) What has MMN revealed about the auditory system in schizophrenia? Int J Psychophysiol 42:177–194

Näätänen R (2008) Mismatch negativity (MMN) as an index of central auditory system plasticity. Int J Audiol 47(Suppl 2):S16–S20

Näätänen R (1992) Attention and brain function. Erlbaum, Hillsdale

Näätänen R, Kujala T, Kreegipuu K et al (2011) The mismatch negativity: an index of cognitive decline in neuropsychiatric and neurological diseases and in ageing. Brain: J Neurol 134(Pt 12):3435–3453

Näätänen R, Paavilainen P, Reinikainen K (1989) Do event-related potentials to infrequent decrements in duration of auditory stimuli demonstrate a memory trace in man? Neurosci Lett 107:347–352

Nagai T, Tada M, Kirihara K et al (2013) Mismatch negativity as a "translatable" brain marker toward early intervention for psychosis: a review. Front Psychiatry 4:115

Nakamura T, Michie PT, Fulham WR et al (2011) Epidural auditory event-related potentials in the rat to frequency and duration deviants: evidence of mismatch negativity? Front Psychology 2:367

Nuechterlein KH, Green MF, Kern RS et al (2008) The MATRICS consensus cognitive battery, part 1: test selection, reliability, and validity. Am J Psychiatry 165:203–213

Oades RD, Wild-Wall N, Juran SA et al (2006) Auditory change detection in schizophrenia: sources of activity, related neuropsychological function and symptoms in patients with a first episode in adolescence, and patients 14 years after an adolescent illness-onset. BMC Psychiatry 6:7

Oknina LB, Wild-Wall N, Oades RD et al (2005) Frontal and temporal sources of mismatch negativity in healthy controls, patients at onset of schizophrenia in adolescence and others at 15 years after onset. Schizophr Res 76:25–41

Ornitz EM, Hanna GL, de Traversay J (1992) Prestimulation-induced startle modulation in attention deficit hyperactivity disorder and nocturnal enuresis. Psychophysiology 29:437–451

Perez VB, Woods SW, Roach BJ, Ford JM, McGlashan TH, Srihari VH, Mathalon DH (2013) Automatic auditory processing deficits in schizophrenia and clinical high-risk patients: forecasting psychosis risk with mismatch negativity. Biol Psychiatry. doi:10.1016/j.biopsych.2013.07.038

Perez VB, Swerdlow NR, Braff DL, Näätänen R, Light GA (2014a) Using biomarkers to inform diagnosis, guide treatments and track response to interventions in psychotic illnesses. Biomark Med 8(1):9–14. doi:10.2217/bmm.13.133

Perez VB, Braff DL, Pianka S, Swerdlow NR, Light GA (2014b) Mismatch negativity predicts and corresponds to behavioral improvements following an initial dose of cognitive training in schizophrenia patients. Presented at Society of Biological Psychiatry, New York, NY

Perry W, Light GA, Davis H et al (2000) Schizophrenia patients demonstrate a dissociation on declarative and non-declarative memory tests. Schizophr Res 46:167–174

Quednow BB, Wagner M, Mössner R et al (2010) Sensorimotor gating of schizophrenia patients depends on catechol O-Methyltransferase Val158Met polymorphism. Schizophr Bull 36:341–346

Rasser PE, Schall U, Todd J et al (2011) Gray matter deficits, mismatch negativity, and outcomes in schizophrenia. Schizophr Bull 37:131–140

Rinne T, Antila S, Winkler I (2001) Mismatch negativity is unaffected by top-down predictive information. NeuroReport 12:2209–2213

Rissling AJ, Braff DL, Swerdlow NR et al (2012) Disentangling early sensory information processing deficits in schizophrenia. Clin Neurophysiol: Off J Int Fed Clin Neurophysiol 123:1942–1949

Rissling AJ, Park SH, Young JW et al (2013) Demand and modality of directed attention modulate "pre-attentive" sensory processes in schizophrenia patients and nonpsychiatric controls. Schizophr Res 146:326–335

Salisbury DF, Shenton ME, Griggs CB et al (2002) Mismatch negativity in chronic schizophrenia and first-episode schizophrenia. Arch Gen Psychiatry 59:686–694

Salisbury DF, Kuroki N, Kasai K et al (2007) Progressive and interrelated functional and structural evidence of post-onset brain reduction in schizophrenia. Arch Gen Psychiatry 64:521–529

Schwarzkopf SB, Lamberti JS, Smith DA (1993) Concurrent assessment of acoustic startle and auditory P50 evoked potential measures of sensory inhibition. Biol Psychiatry 33:815–828

Shelley AM, Ward PB, Catts SV et al (1991) Mismatch negativity: an index of a preattentive processing deficit in schizophrenia. Biol Psychiatry 30:1059–1062

Sim K, DeWitt I, Ditman T et al (2006) Hippocampal and parahippocampal volumes in schizophrenia: a structural MRI study. Schizophr Bull 32:332–340

Smith SJ, Lees AJ (1989) Abnormalities of the blink reflex in Gilles de la Tourette syndrome. J Neurol Neurosurg Psychiatry 52:895–898

Sobin C, Kiley-Brabeck K, Karayiorgou M (2005) Lower prepulse inhibition in children with the 22q11 deletion syndrome. Am J Psychiatry 162:1090–1099

Spaulding WD, Fleming SK, Reed D et al (1999) Cognitive functioning in schizophrenia: implications for psychiatric rehabilitation. Schizophr Bull 25:275–289

Sussman E, Winkler I, Wang W (2003) MMN and attention: competition for deviance detection. Psychophysiology 40:430–435

Swerdlow NR (2011a) Are we studying and treating schizophrenia correctly? Schizophr Res 130:1–10

Swerdlow NR (2011b) Beyond antipsychotics: pharmacologically-augmented cognitive therapies (PACTs) for schizophrenia. Neuropsychopharmacology 37:310–311

Swerdlow NR, Benbow CH, Zisook S et al (1993a) A preliminary assessment of sensorimotor gating in patients with obsessive compulsive disorder. Biol Psychiatry 33:298–301

Swerdlow NR, Monroe SM, Hartston HJ et al (1993b) Men are more inhibited than women by weak prepulses. Biol Psychiatry 34:253–260

Swerdlow NR, Braff DL, Taaid N et al (1994) Assessing the validity of an animal model of deficient sensorimotor gating in schizophrenic patients. Arch Gen Psychiatry 51:139–154

Swerdlow NR, Paulsen J, Braff DL et al (1995) Impaired prepulse inhibition of acoustic and tactile startle response in patients with Huntington's disease. J Neurol Neurosurg Psychiatry 58:192–200

Swerdlow NR, Geyer MA, Braff DL (2001a) Neural circuit regulation of prepulse inhibition of startle in the rat: current knowledge and future challenges. Psychopharmacology 156:194–215

Swerdlow NR, Karban B, Ploum Y et al (2001b) Tactile prepuff inhibition of startle in children with Tourette's syndrome: In search of an "fMRI-friendly" startle paradigm. Biol Psychiatry 50:578–585

Swerdlow NR, Shoemaker JM, Stephany N et al (2002) Prestimulus effects on startle magnitude: sensory or motor? Behav Neurosci 116:672–681

Swerdlow NR, Light GA, Cadenhead KS et al (2006a) Startle gating deficits in a large cohort of patients with schizophrenia: relationship to medications, symptoms, neurocognition, and level of function. Arch Gen Psychiatry 63:1325–1335

Swerdlow NR, Talledo J, Sutherland AN et al (2006b) Antipsychotic effects on prepulse inhibition in normal 'low gating' humans and rats. Neuropsychopharmacology 31:2011–2021

Swerdlow NR, Weber M, Qu Y et al (2008) Realistic expectations of prepulse inhibition in translational models for schizophrenia research. Psychopharmacology 199:331–388

Swerdlow NR, Bhakta SG, Talledo JA et al (2013) Sensorimotor gating predicts sensitivity to pro-attentional effects of amphetamine in healthy adults. Soc Neurosci, San Diego, 9–13 Nov

Swerdlow NR, Light GA, Sprock J et al (2014) Deficient prepulse inhibition in schizophrenia detected by the multi-site COGS. Schizophr Res 152:503–512

Takahashi H, Rissling AJ, Pascual-Marqui R et al (2012) Neural substrates of normal and impaired preattentive sensory discrimination in large cohorts of nonpsychiatric subjects and schizophrenia patients as indexed by MMN and P3a change detection responses. NeuroImage 66C:594–603

Talledo JA, Sutherland Owens AN, Schortinghuis T et al (2009) Amphetamine effects on startle gating in normal women and female rats. Psychopharmacology 204:165–175

Taub E, Uswatte G, Elbert T (2002) New treatments in neurorehabilitation founded in basic research. Nat Rev Neurosci 3:228–236

Turetsky BI, Calkins ME, Light GA et al (2007) Neurophysiological endophenotypes of schizophrenia: the viability of selected candidate measures. Schizophr Bull 33:69–94

Twamley EW, Jeste DV, Bellack AS (2003) A review of cognitive training in schizophrenia. Schizophr Bull 29:359–382

Umbricht D, Krljes S (2005) Mismatch negativity in schizophrenia: a meta-analysis. Schizophr Res 76:1–23

Umbricht D, Schmid L, Koller R, Vollenweider FX, Hell D, Javitt DC (2000) Ketamine-induced deficits in auditory and visual context-dependent processing in healthy volunteers: implications for models of cognitive deficits in schizophrenia. Arch Gen Psychiatry 57 (12):1139–1147. yoa9366b [pii]

Umbricht D, Koller R, Vollenweider FX, Schmid L (2002) Mismatch negativity predicts psychotic experiences induced by NMDA receptor antagonist in healthy volunteers. Biol Psychiatry 51(5):400–406. S0006322301012422 [pii]

Umbricht D, Koller R, Schmid L et al (2003) How specific are deficits in mismatch negativity generation to schizophrenia? Biol Psychiatry 53:1120–1131

Umbricht DS, Bates JA, Lieberman JA et al (2006) Electrophysiological indices of automatic and controlled auditory information processing in first-episode, recent-onset and chronic schizophrenia. Biol Psychiatry 59:762–772

Valls-Sole J, Munoz JE, Valldeoriola F (2004) Abnormalities of prepulse inhibition do not depend on blink reflex excitability: a study in Parkinson's disease and Huntington's disease. Clin Neurophysiol 115:1527–1536

van der Linden D, Massar SA, Schellekens AF et al (2006) Disrupted sensorimotor gating due to mental fatigue: preliminary evidence. Int J Psychophysiol 62:168–174

van Rijn S, Swaab H, Magnée M et al (2011) Psychophysiological markers of vulnerability to psychopathology in men with an extra X chromosome (XXY). PLoS One 6:e20292

Vinogradov S, Fisher M, de Villers-Sidani E (2012) Cognitive training for impaired neural systems in neuropsychiatric illness. Neuropsychopharmacology 37:43–76

Weike AI, Bauer U, Hamm AO (2000) Effective neuroleptic medication removes prepulse inhibition deficits in schizophrenia patients. Biol Psychiatry 47:61–70

Wynn JK, Sugar C, Horan WP et al (2010) Mismatch negativity, social cognition, and functioning in schizophrenia patients. Biol Psychiatry 67:940–947

Ziemann U, Paulus W, Rothenberger A (1997) Decreased motor inhibition in Tourette's disorder: evidence from transcranial magnetic stimulation. Am J Psychiatry 154:1277–1284

Connectivity Measurements for Network Imaging

Susan M. Bowyer

Abstract Communication across the brain networks is dependent on neuronal oscillations. Detection of the synchronous activation of neurons can be used to determine the well-being of the connectivity in the human brain networks. Well-connected highly synchronous activity can be measured by MEG, EEG, fMRI, and PET and then analyzed with several types of mathematical algorithms. Coherence is one mathematical method that can detect how well 2 or more sensors or brain regions have similar oscillatory activity with each other. Phase synchrony can be used to determine if these oscillatory activities are in sync or out of sync with each other. Correlation is used to determine the strength of interaction between two locations or signals. Granger causality can be used to determine the direction of the information flow in the neuronal brain networks. Statistical analysis can be performed on the connectivity results to verify evidence of normal or abnormal network activity in a patient.

Keywords Networks · Neuronal oscillations · Correlation · Coherence · Phase synchrony · Granger causality

Contents

S.M. Bowyer (✉)
Henry Ford Hospital, Detroit, USA
e-mail: SBOWYER1@hfhs.org

Curr Topics Behav Neurosci (2014) 21: 315–330

DOI: 10.1007/7854_2014_348

Published Online: 17 August 2014

1 Introduction

The interest in detecting and imaging the functional properties of the brain's networks is driving the development of advance mathematical imaging techniques and analysis. This in turn guides the need to understand the different techniques for measuring and for analyzing the location and strengths of these functional brain networks (Cabral et al. 2014). Brain connectivity networks can be subdivided into 3 main categories: neuroanatomical, functional and effective connectivity (Friston et al. 1993; Sakkalis 2011; Greenblatt et al. 2012).

Neuroanatomical connectivity or structural connectivity is based on detection of the fiber tracts that physically connect the regions of the brain. Figure 1 is a representative image of how the fiber tracts in the human brain appear (Kubicki et al. 2005). These fiber tracts are detected using Diffusion magnetic resonance imaging (MRI). Diffusion MRI measures the water anisotropy in the white matter of the brain. Diffusion tensor imaging (DTI) (Le Bihan et al. 2001) estimates the direction and strength of anisotropic diffusion in each voxel while diffusion spectrum imaging (DSI) explores the strength of anisotropy in all directions, allowing the detection of the crossing of multiple fibers in a single voxel (Wedeen et al. 2008). To date, several DTI studies support the notion that frontal–temporal connectivity in the brain is likely disrupted in schizophrenia along with reduced organization of the cortico-cortical connections (Kubicki et al. 2005; Rotarska-Jagiela et al. 2008; White et al. 2011). DTI is a method to determine local fiber tract orientation which can be used to identify and analyze fiber tract pathways. The Human Connectome Project is compiling a neural connectivity database of diffusion MRI studies (Van Essen et al. 2013). More information can be found at their website: www.humanconnectomeproject.org.

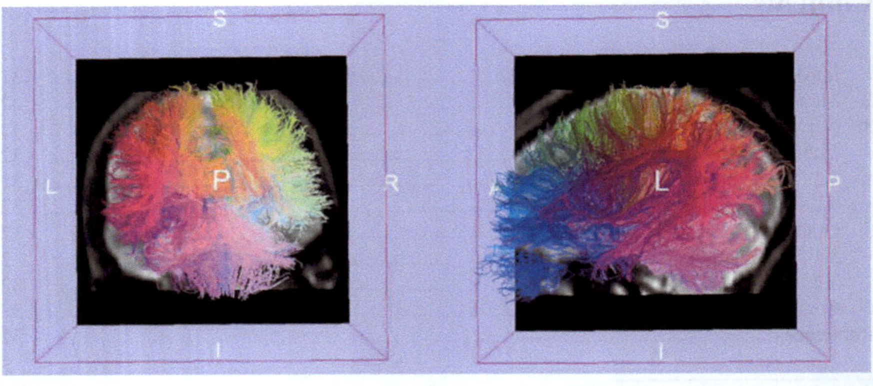

Fig. 1 Fiber traces from a human brain are colored such that fibers with similar endpoints are assigned similar colors. Slices from a T2-weighted volume add additional understanding of the anatomy. The view of the fiber traces on the left is coronal, while the view on the right is sagittal. (Kubicki et al. 2005)

Functional connectivity is established by identifying correlations of activity between multiple regions of the brain involved in basal brain/body function or higher order information processing that is required for sensory responses, motor responses, and intellectual or emotional processing. During cognitive and sensory processing, brain activity is characterized by bursts of information flow and correlated network activity. These bursts of regional brain activity are called nodes, and the links to other nodes are called edges (the fiber pathways). These regions (nodes) may only be active for a short period of time which emphasizes the dynamic fluctuation of information flowing around the brain during cognitive or sensory processing, or they may be active for minutes, hours, or even days as in the case of the epileptic network (Towle et al. 2007). Further, brain networks have frequency-dependent characteristics that differ with the scale of brain region that is measured. Recent studies of functional connectivity in patients with schizophrenia have shown beta- and gamma-band activity is abnormal (Hinkley et al. 2011). The dysfunctional oscillations in these frequencies may be due to abnormalities in the rhythm generating networks of GABA interneurons and cortico-cortical connections (Uhlhaas and Singer 2010). Coherence and phase synchrony are common mathematical methods for quantifying frequency-dependent coordination of brain activity. Figure 2 is an example of power distribution verses frequency graphs, frequency verses time as well as sensor space topography map of the phase synchrony differences between groups in the lower panel (Uhlhaas and Singer 2010). Functional connectivity does not determine the specific direction of information flow in the brain or an underlying structural model. It just shows that these regions are connected.

Effective connectivity takes functional connectivity one step further and determines the direct or indirect influence that one neural system may have over another, more specifically the direction of the dynamic information flow in the brain (Cabral et al. 2014; Horwitz 2003). Using mathematical techniques such as Granger causality, Hilbert transform, transfer entropy, and correlation, locations in the brain can be identified as a sender or receiver of the information flowing in the brain. Flow of information in the brain networks during a finger-tapping task is displayed in Fig. 3, where functional connectivity is seen in yellow and the effective connectivity is depicted by the arrows (Gross et al. 2002).

Analytical techniques for estimating functional or effective connectivity of the brain to determine if or how 2 or more sensors or locations are connected/coupled fall under these main categories linear and nonlinear, bivariate, and multivariate. Bivariate techniques are mathematical algorithms that determine how activity at 2 brain locations or electrodes is related to each other based on the evaluation of the frequencies and patterns of neuronal oscillations. After performing the connectivity analysis, then the results can be put into graph or matrix format for further analysis. Figure 4 represents modes of brain connectivity in the macaque cortex along with their corresponding matrices (Honey et al. 2007). Weighted undirected functional connectivity forms a full symmetric matrix, with each of the elements encoding statistical dependence or proximity between two nodes (recording sites, regions). A threshold may be applied to yield binary *undirected* graphs. Effective connectivity

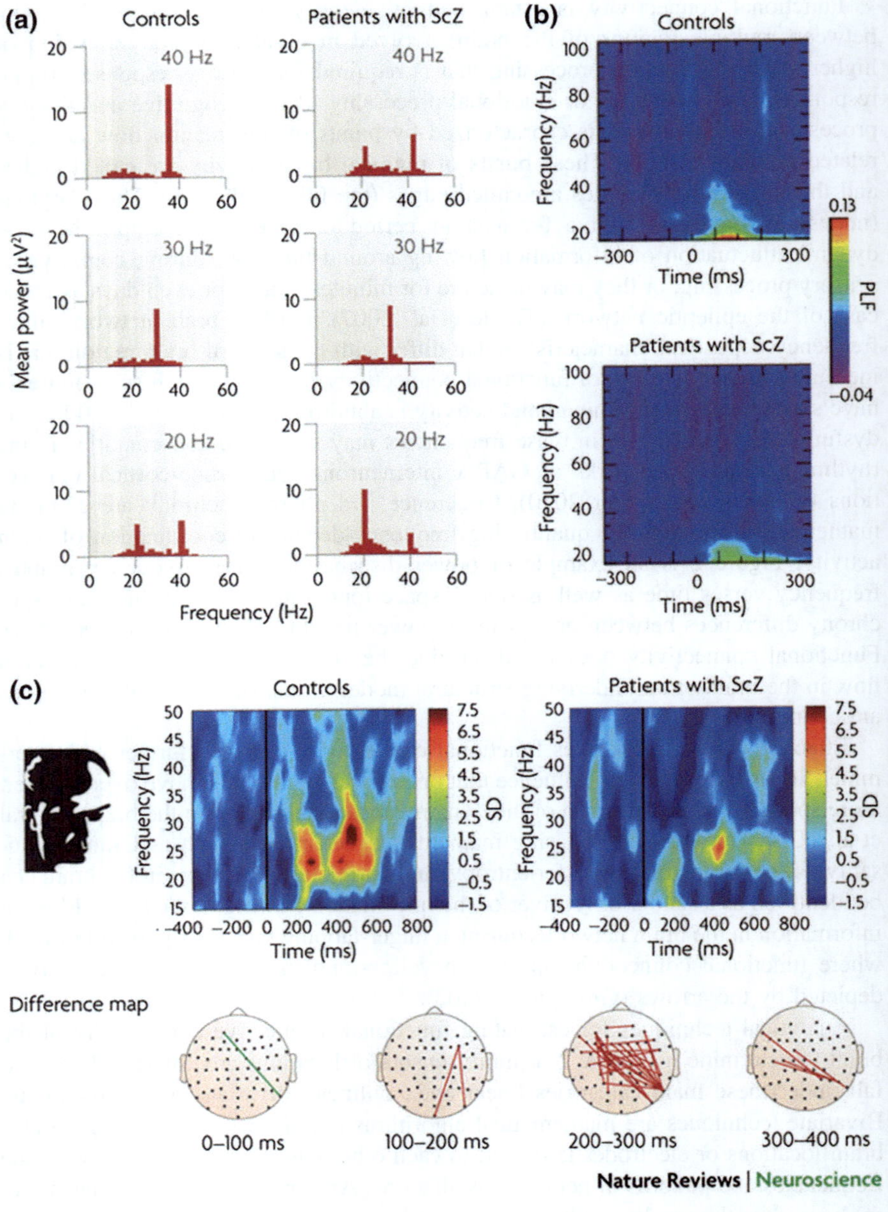
(a)
Controls
Patients with ScZ
40 Hz
30 Hz
20 Hz
Mean power (μV²)
Frequency (Hz)
(b)
Controls
Patients with ScZ
Frequency (Hz)
Time (ms)
PLF
0.13
−0.04
(c)
Controls
Patients with ScZ
Frequency (Hz)
Time (ms)
SD
Difference map
0–100 ms
100–200 ms
200–300 ms
300–400 ms
Nature Reviews | Neuroscience

◀ **Fig. 2** Neural oscillations and synchrony in schizophrenia. **a** Auditory steady state responses in patients with schizophrenia (*ScZ*) show lower power to stimulation at 40 Hz than control subjects. **b** Sensory evoked oscillations during a visual oddball task in patients with schizophrenia indicate the phase-locking factor of oscillations in the 20–100 Hz frequency range in the occipital cortex for healthy controls and patients with schizophrenia. **c** Dysfunctional phase synchrony during Gestalt perception in schizophrenia was significantly reduced relative to controls. In addition, patients with schizophrenia showed a desynchronization in the gamma band (30–55 Hz) in the 200–280 ms interval. The *bottom panel* shows differences in the topography of phase synchrony in the 20–30 Hz frequency range between groups. *Red lines* indicate less synchrony between two electrodes in patients with schizophrenia than in controls. *Green lines* indicate greater synchrony for patients with schizophrenia (Uhlhaas and Singer 2010)

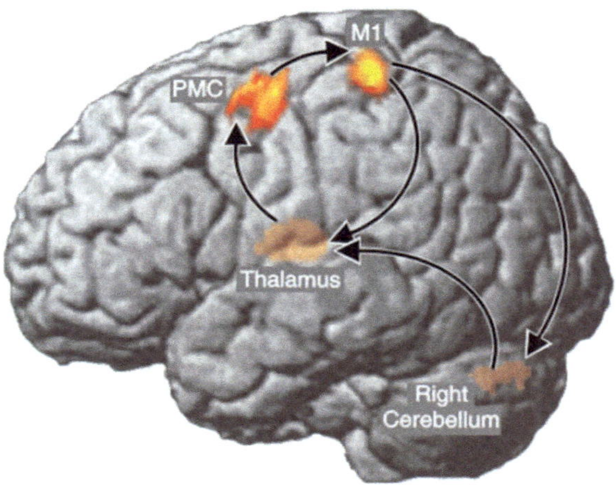

Fig. 3 The map represents spatial distribution of coherence (in the 6–9 Hz range). Dynamic imaging of coherent sources were applied to MEG data. The dominant coupling direction is indicated by arrows. Note that left thalamus and right cerebellum are projected to the left surface for easier visualization (Gross et al. 2002)

yields a full non-symmetric matrix. Applying a threshold to these matrices yields binary *directed* graphs.

Brain signals are usually recorded by electroencephalography (EEG), magnetoencephalography (MEG), functional magnetic resonance imaging (fMRI), and positron emission tomography (PET). Recent developments have advanced the ability for connectivity to image directly into the specific regions of the brain (called source space, Fig. 3); in past years, connectivity was seen by connecting lines between electrodes or coils that had similar frequency profiles on the brain surface (called sensor space, bottom of Fig. 2). Figures 2 and 3 depict these 2 different types of images. In this review, we will discuss how brain connectivity can be detected and recorded and the different measures of connectivity used to display the brain networks. We will also include various studies of connectivity where the results may be clinically applicable.

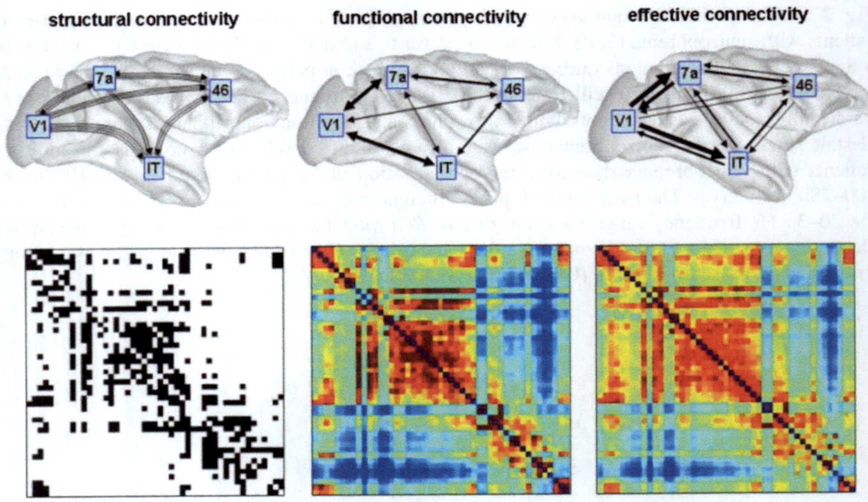

Fig. 4 Modes of brain connectivity. Sketches at the *top* illustrate structural connectivity (fiber pathways), functional connectivity (correlations), and effective connectivity (information flow) among four brain regions in macaque cortex. Matrices at the *bottom* show binary structural connections (*left*), symmetric mutual information (*middle*), and non-symmetric transfer entropy (*right*) (Sporns 2007). Data were obtained from a large-scale simulation of cortical dynamics (Honey et al. 2007)

2 Measures

When we investigate how the human brain functions, we tend to compare it to a computer's circuit board. There are locations in the brain that perform certain tasks such as feeling, tasting, smelling, hearing, and seeing. These last two are similar to the computer's ability to produce sounds and display images. Connectivity measures of the brain are performed to try to map out the communication networks (cortical circuits) needed for the brain to function. These networks are made up of neurons that function in unison to send signals to other parts of the brain. There are several properties of the neuron that play an important role in generating membrane potential oscillations that can be detected by neuroimaging devices. Neurons communicate with other neurons by releasing one of over 50 different types of neurotransmitters in the brain some of which are excitatory (stimulate the brain) and some are inhibitory (calm the brain) (Chana et al. 2013). Voltage-gated ion channels generate action potentials and periodic spiking membrane potentials which produces oscillatory activity and facilitates synchronous activity in neighboring neurons (Llinas 1988; Llinas et al. 1991). Coherent neuronal communications are based on neurotransmission dynamics dictated by major neurotransmitters such as the amino acids glutamate and GABA. Other important neurotransmitters include acetylcholine, dopamine, adrenaline, histamine, serotonin, and melatonin (Stephan et al. 2009; Haenschel and Linden 2011; Wang 2010). There is growing evidence that glutamatergic dysregulation may underlie schizophrenia and psychosis (Chana et al. 2013).

Synchronized activity of large numbers of neurons can give rise to large magnetic field and electric field oscillations, which are detected by EEG/MEG (Hamalainen et al. 1993) and the secondary metabolic responses are detected by fMRI/PET (Ogawa et al. 1990). Coherent activity within the whole brain is evidence for a network of dynamic links (edges) between different brain regions (nodes) that distribute information (Varela et al. 2001). Detection of normal or abnormal networks can provide information on the underlying developmental and/ or neurological disorder.

EEG uses electrodes pasted or glued to specific locations on the scalp that record the electric potential at the skin surface. MEG on the other hand uses coils suspended in a helmet placed around the head to detect the changing magnetic fields just outside the head. Both EEG and MEG signals come from the activation of neurons. EEG measures voltage potentials determined by electrical impedance boundaries of head structures that shape volume currents (return currents) where as MEG measures the magnetic field of primary or intracellular current flow. For a review of mathematical equations of connectivity measures used in EEG and MEG for neurologic disorders see Sakkalis (2011) or Greenblatt et al. (2012).

PET and FMRI measure the secondary or metabolic response from neuronal activation. They are both indirect measures of neuronal activation with low temporal resolution on the order of seconds. PET uses a radioactive-labeled tracer, tagged to glucose (blood sugar) that is injected into the subject to detect the parts of the brain that require energy to function (the glucose response). FMRI uses a strong magnetic field and radio waves to look at blood flow in the brain to detect areas of activity that need more oxygen to function (the hemodynamic response). The translation from the hemodynamic response or the glucose response back to the actual synaptic neuronal function is still not fully understood (Goense and Logothetis 2008; Logothetis and Pfeuffer 2004) For a review of connectivity measures of fMRI and PET for neurologic disorders see Rowe (2010).

Functional and effective connectivity techniques are dependent on calculating the communication of active neural signals that are oscillating over short and long periods of time. Techniques such as EEG and MEG, with their excellent temporal resolution, are optimal for calculating connectivity (Sakkalis 2011; Greenblatt et al. 2012). Traditionally to determine connectivity in the EEG or MEG, a frequency analysis was performed to convert the original EEG or MEG data into its frequency content; then, coherence analysis was used to obtain information about the temporal relationships of frequency components at different recording sites (electrode or coil). The results of the coherence analysis were typically displayed in sensor space using a template of the head with lines connecting the electrodes or coils that are coherent with each other. These results can now be displayed in source space due to improve analytical techniques.

MEG and EEG data are usually filtered and have noise artifacts removed prior to advanced analysis. In some cases, it maybe helpful to first decompose a signal in temporal and spatial modes using techniques such as principle component analysis

(PCA) or independent component analysis (ICA). These techniques can be used to extract a particular signal of interest (i.e., an epileptic seizure) or an artifact signal such as cardiac activity to be removed from all the data channels.

3 Network Connectivity Measurement

Network connectivity measurements can be measured in the frequency domain with methods such as coherence and phase synchrony and in the time domain with methods such as correlation and Granger causality.

4 Coherence and Phase Synchrony

First, we look at the frequency domain methods for calculating neuronal networks; these tend to be symmetrical providing no information on directionality.

4.1 Coherence

Coherence is used to determine if different areas of the brain are generating signals that are significantly correlated (coherent) or not significantly correlated (not coherent), using a scale of 0–1. coherence is a measure of the synchronicity of the neuronal patterns of oscillator activity. The coherence analysis is a technique that can be applied to study functional relationships between spatially separated scalp electrodes or coils and to estimate the similarities of waveform components generated by the neurons in the underlying cortical regions (French and Beaumont 1984). Transients and oscillations of brain electric activity are found in MEG and EEG recordings of spontaneous brain activity. Coherence differs from correlation because the assessment of brain synchrony is done for very narrow frequency bands where the band activity is quantified by an amplitude and phase. These transient waveform oscillations can be quantified by first applying a time–frequency decomposition technique such as the fast Fourier transform (FFT), of a contiguous or slightly overlapping sequence of short data segments. This generates a sequence of amplitude/phase components for each narrow frequency bin of the FFT that spans the frequency content of the data. After transformation to a time–frequency representation, the strength of network interactions can be estimated by calculation of coherence, which is a measure of synchrony between signals from different brain regions for each FFT frequency component. This is the most common measure to describe how two or more time series are related. Strictly speaking, coherence is a statistic that is used to determine the relationship between 2 data sets. It is used to determine if the signal content of 2 inputs is the same or different. If the signals measured by 2 electrodes or coils are identical, then they have a coherence value of 1; depending on how dissimilar they are, the coherent value will approach 0. It is commonly used to estimate the spectral densities of two signals and so is equivalent to a correlation coefficient in the frequency domain. Unlike correlation, coherence has a range of 0–1, and since each FFT yields one pair of FFT components, multiple independent segments of data are needed for evaluation (Kelly et al. 1997).

As mentioned earlier, this technique can be applied to the MEG and EEG waveforms in sensor space, or it can be applied to the localized MEG solutions in source space. coherence has been widely used in studying epileptiform activity to determine seizure onset zones. In sensor space, Song et al. showed that EEG coherence can be used to characterize a pattern of strong coherence centered on temporal lobe structures in several patients with epilepsy (Song et al. 2013). In source space, Elisevich et al. showed that MEG coherence source imaging in the brain can provide targets for successful surgical resections in patients with epilepsy (Elisevich et al. 2011). Hinkley et al. used source space MEG to detect decreased and increased connectivity differences between patients with schizophrenia and control subjects that may prove to be important target areas for treatment (Hinkley et al. 2011).

Directionality of network interaction cannot be determined from coherence, and the exact amplitudes of the network interaction are not equal to region-to-region coherence amplitudes. Coherence does provide a global estimate of all important regions of network activity regardless of source amplitudes. Because of the need to minimize bias by increasing the number of data segments in calculations, coherence is not well suited for quantifying rapid temporal changes in synchronized activity. Rather, it is best when used for long time series of data to identify sources of brain network activity that persist for long durations. However, it is desirable that the individual FFT components follow temporal changes in network connectivity. Therefore, the length of segments of data used in the FFT transform should be selected in the same way as recommended for correlation calculations. For MEG data, we have found approximately 0.5 s of data to be near optimal for data filtered 3–50 Hz. When applied to very low-frequency band data, the FFT data segment length needs to be increased. An F-statistic can be used to test statistical significance relative to the hypothesis of true coherence (Kelly et al. 1997).

Coherence is best when used for long time series of data to identify sources of brain activity that are part of the same network. Coherence analysis supplies information on the degree of synchrony of brain activity at different locations for each frequency, independent of power. However, individual time points with large amplitudes are more highly weighted in the FFT transform and subsequently in coherence calculations. This is in contrast to phase synchrony which utilizes instantaneous measurements of only the phase differences between signals. As mentioned earlier, coherence analysis can be applied to the MEG and EEG waveforms in sensor space, or it can be applied to the localized MEG solutions in source space.

4.2 Phase Synchrony

The phase relationship is a way to estimate the synchrony of oscillations in EEG/MEG data. This is the process by which two or more cyclic signals tend to have oscillator activity that are at the same frequency (in phase) or out of synchrony (out of phase) by a relative phase angle. Phase synchrony is used to investigate whether two waveforms with the same narrow frequency band have relatively stable phase differences independent of their amplitude behavior. This is used to determine if the phases are coupled across the brain and to see if they are phase-locked to an external stimuli or event.

One example of phase synchronization of neuronal activity can be seen in the pattern of the oscillatory activity. If the oscillations are all in synchrony and positive at the same time, then the phase is 0°; if they were opposite (one positive and one negative) to each other, they would have a phase angle of 180°, See Fig. 5 (Uhlhaas and Singer 2010). Phase synchrony measures how stable the phase difference (small or large) varies over a short period of time. Phase relationships can be examined by testing the stability of the signals' phase differences across trials (phase-locking) over a single electrode or between pairs of electrodes (Lachaux et al. 1999). This approach can yield estimates of the precision of local and long-range synchrony. Importantly, measures of phase-locking provide estimates of synchrony independent of the amplitude of oscillations. This is in contrast to measures of coherence where phase and amplitude are intertwined (Uhlhaas et al. 2009). Phase synchrony reflects the exact timing of communication between distant neural populations that are related functionally, the exchange of information between global and local neuronal networks, and the sequential temporal activity of neural processes in response to incoming sensory stimuli (Sauseng and Klimesch 2008).

In the extensive study by Gaillard et al., significant long-distance phase synchrony in the beta range was observed after presented words; this increase was the only significant correlate with conscious access to the stimuli (Gaillard et al. 2009). In advance of the stimuli, attentional and working memory functions, in part, were correlated with phase-coupling of prefrontal and posterior brain areas. Gross et al. (2004) found frontal, parietal, and temporal beta coherence was relevant for the processing of stimuli in working memory. In the field of schizophrenia, Uhlhaas and Singer 2010 provided an in-depth review of abnormal neural oscillations and synchrony in this patient group. They review several studies that indicated that patients with schizophrenia have a reduced phase synchrony in the beta and gamma bands. (Uhlhaas and Singer 2010).

Phase synchrony is better used for short-duration events such as in an evoked event. Phase is used to determine how much the two locations (recording sites) are interacting within a very narrow time window (milliseconds). A great analogy for understanding the difference between when to use coherence or phase synchrony is soldiers marching in a parade: Phase synchrony is used to determine how synchronized their feet are marching in unison in a few steps, while coherence is used to see how synchronous their feet were marching in unison over the entire parade

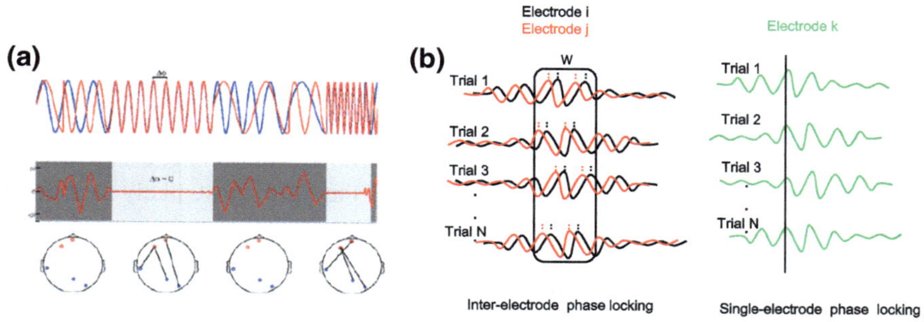

Fig. 5 Measuring neural synchrony in EEG/MEG signals. Measuring phase synchrony in brain signals: The synchrony of oscillations in EEG/MEG data can be estimated by analyzing phase relationships. The *top panel* shows oscillatory brain signals recorded by two different groups of sensors (*red* and *blue*) placed in the positions shown in the *bottom panel*. The *middle panel* shows the difference in oscillatory phase between the red and blue signals. Phase difference values around zero indicate phase synchrony. The *bottom panel* illustrates patterns of synchrony between distant sensor sites at different time points. The *black lines* link synchronous sensors (Uhlhaas and Singer 2010)

route. On a side note, it is possible to determine information flow with a Hilbert transform on the data (to obtain the instantaneous phases). Two measures can be computed from the phases. The first one is the synchronization index, which quantifies the degree of coupling of the phases of two signals [0 (no coupling) to 1 (strongest coupling)]. The second measure is a directionality index which quantifies the direction of coupling between two oscillators [−1 and 1 correspond to unidirectional coupling (away and toward the reference area, respectively), and 0 corresponds to symmetric bidirectional coupling]. The results from the analysis of phase synchronization can be used for the quantification of coupling strength and direction because these measures are more sensitive and robust than coherence and phase difference and independent of amplitude dynamics. This is usually applied to a narrow frequency band, usually less then 5 Hz as opposed to a larger frequency range that desynchronizes rapidly due to many varied frequencies mixed in.

5 Granger Causality and Correlation

Now, we turn to the time domain methods. Long-range connections between different brain structures involve time delays due to the finite conduction velocity of axons. Since most connections in the brain are reciprocal, they form feedback loops that support oscillatory activity. Neuronal activity recorded from multiple cortical areas around the brain can become synchronized and form a large-scale network. The effective connectivity (direction of information) of a brain network can be studied using a Granger causality measure (Brovelli et al. 2004; Gross et al. 2002), while the strength of the connection can be assessed using a correlation measure between two sites.

5.1 Correlation

Correlations is a mathematical technique to measure the similarity of 2 signals on a scale of −1 to +1. Correlation is one of the most commonly used methods to determine the strength of an interaction between two locations or signals. When two areas of the brain are active at the same time, they are most likely talking or communicating with each other. A correlation between one specific region in the brain and the entire brain can be analyzed; this can also be extended to all possible correlations within the brain. Correlations can be determined in several different ways: The Pearson product moment correlation, the Spearman rank order correlation, the Kendall rank order correlation, or by mutual information methods. The Pearson product moment correlation quantifies the linear correlations between two signals or locations, where the Spearman rank order correlation is a nonparametric measure of correlation between two signals based on the rank (i.e., the similarity of the orderings of the data when ranked by each individual quantity) and the Kendall tau (τ) coefficient which is a non parametric test that uses the relative ordering of ranks. The mutual information of two time series is a measure of their mutual dependence where the unit of measure is a bit. These measures have been widely used to quantify correlations between EEG or MEG recordings from healthy participants (Bonita et al. 2014) or patients with neurologic disorders such as traumatic brain injury (Castellanos et al. 2011), Alzheimer's (Stam et al. 2007), epilepsy (Ponten et al. 2007), and with schizophrenia (Rubinov et al. 2009).

One advantage of computing correlation matrices is that these correlations can be further studied using graph theory to evaluate the topological properties of the functional networks or wavelet analysis applied to the temporal signals to compute frequency-dependent correlation matrices. Graph theory is used to create models of the complex functional brain networks that can be further studied (Stam et al. 2007; Stam and Reijneveld 2007; Bassett and Bullmore 2006). Graphs are composed of vertices (corresponding to neurons or brain regions) and edges (corresponding to synapses or pathways, or statistical dependencies between neural elements). Graphs of brain networks can be quantitatively examined for vertex degrees and strengths, degree correlations, clustering coefficients, path lengths (distances), and vertex and edge centrality, among other characteristics. Graph theory provides models of complex networks in the brain and allows one to better understand the relations between network structure and the processes taking place on those networks.

Despite its usefulness for detecting linear statistical dependencies, the correlation analysis has certain limitations. The most important relies on the fact that some networks are not spatially independent and can overlap. In other words, the same cortical region can belong to more than one active network (i.e., rest and memory). Therefore, the activation pattern of that region may turn out to be a sum of several simultaneously active networks, limiting the ability to capture the one functional network using correlation-based approaches (Smith et al. 2012).

5.2 Granger Causality

Granger causality as well as other methods such as directed transfer function (DTF) and partial directed coherence (PDC) are designed to determine causality of network activities (Sakkalis 2011). The basis for these techniques is a multiple linear regression model of future activity at a given site, determined by past activity at sites and times within the network. Granger causality is a statistical method for determining the direction of information flow in a brain network. By looking at the interaction between time series, information can be provided on how one signal may affect another signal. There is no a priori information used. Granger causality assumes that a cause precedes an event. Therefore, a signal can be predicated based on the past information of a second signal. Simply stated X causes Y if X provides information that predicts the future of Y better then any information already known about Y.

Granger causality has been used to study the effective connectivity in patients with schizophrenia using fMRI (Demirci et al. 2009) more so than with EEG or MEG. When used with FMRI data, an independent component analysis (ICA) is initially used to extract the time courses of spatially independent components, and then, a Granger causality test is used to investigate causal relationships between brain activation networks. This study found evidence that distributed networks are organized in a fashion suggestive of hubs of activity within specific circuits that directly or indirectly influence other neural function in normal controls but that this connectivity is abnormal in patients with schizophrenia (Demirci et al. 2009).

Granger causality analysis for studying effective connectivity in neural systems has a few limitations; one is that it relies heavily on sensor space analyses when source space analysis may provide more information, also there is the use of an unrealistically small sets of factors that make analyses vulnerable to spurious correlation, and finally the failure to address the complex and chaotic nature of neural processes.

Both directed transfer function (DTF) (Kaminski and Blinowska 1991) and partial directed coherence (PDC) (Sameshima and Baccala 1999) have been developed based on network model approaches such as Granger causality. DTF extends Granger causality to multichannel MEG and EEG. It has been used to estimate functional connectivity in neurological disorders such as epilepsy (Franaszczuk and Bergey 1998). Another more recent measure called transfer entropy (Schreiber 2000) was designed to detect directed exchange of information between two systems by considering the effects of the state of one element on the state transition probabilities of the other element. Seibenhuhner et al. found lower entropy, suggesting decreased MEG signal content, but increased functional connectivity in patients with schizophrenia compared to control subjects (Siebenhühner et al. 2013).

6 Conclusion

The basis of brain functioning is the neuronal oscillations. In this section, we attempted to review several of the most common methods used to measure the brain's synchronous oscillations which make up the network of brain connectivity. Many of these techniques are currently being used to expand clinical knowledge in many fields with psychiatry being one of the most prominent fields. We have highlighted some of the types of information that can be derived from the varied techniques as well as provided some of the limitations of each technique. In the future, a combined anatomical, functional, and effective connectivity mapping will become the mainstay of the neurosurgeon, neurologist, and psychiatrist for assessing and diagnosing normal and abnormal brain networks. These techniques will not only provide biomarkers of diseases but also help to provide individualized treatment therapies based on pre- and post-treatment connectivity imaging. With the evolution of computers and mathematics, we expect to see more sophisticated and powerful analytical neuroimaging methods developed and applied to the functional neuroimaging data. The primary functional neuroimaging results will continue to be provided from the excellent high temporal resolution of MEG and EEG, and the high spatial resolution of fMRI and PET as well as the anatomical connectivity maps derived from MRI-DTI imaging.

References

Bassett DS, Bullmore E (2006) Small-world brain networks. Neuroscientist 12(6):512–523

Bonita JD, Ambolode LCC II, Rosenberg BM, Cellucci CJ, Watanabe TAA, Rapp PE, Albano AM (2014) Time domain measures of inter-channel EEG correlations: a comparison of linear, nonparametric and nonlinear measures. Cogn Neurodyn 8:1–15

Brovelli A, Ding M, Ledberg A, Chen Y, Nakamura R, Bressler SL (2004) Beta oscillations in a large-scale sensorimotor cortical network: Directional influences revealed by Granger causality. PNAS 101(26):9849–9854

Cabral J, Kringelbach ML, Deco G (2014) Exploring the network dynamics underlying brain activity during rest. Prog Neurobiol 114:102–131

Castellanos NP, Leyva I, Buldu JM, Bajo R, Paul N, Cuesta P, Ordonez VE, Pascua CL, Bocaletti S, Maestu E, del-Pozo F (2011) Principles of recovery from traumatic brain injury: reorganization of functional networks. Neuroimage 55(3):1189–1199

Chana G, Bousman CA, Money TT, Gibbons A, Gillett P, Dean B, Everall IP (2013) Biomarker investigations related to pathophysiological pathways in schizophrenia and psychosis. Front Cell Neurosci 7(95):1–18

Demirci O, Stevense MC, Andreasend NC, Michaela A, Liua J, Whiteh T, Pearlsone JD, Clark VP, Calhouna VD (2009) Investigation of relationships between fMRI brain networks in the spectral domain using ICA and Granger causality reveals distinct differences between schizophrenia patients and healthy controls. Neuroimage 46(2):419–431

Elisevich K, Shukla N, Moran JE, Smith B, Schultz L, Mason K, Barkley GL, Tepley N, Gumenyuk V, Bowyer SM (2011) An assessment of MEG coherence imaging in the study of temporal lobe epilepsy. Epilepsia 52(6):1110–1119

Franaszczuk PJ, Bergey GK (1998) Application of the directed transfer function method to mesial and lateral onset temporal lobe seizures. Brain Topogr 11(1):13–21

French CC, Beaumont JG (1984) A critical review of EEG coherence studies of hemisphere function. Int J Psychophysiol 1(3):241–254

Friston KJ, Frith CD, Liddle PF, Frackowiak RS (1993) Functional connectivity: the principal-component analysis of large (PET) data sets. J Cereb Blood Flow Metab 13:5–14

Gaillard R, Dehaene S, Adam C, Clemenceau S, Hasboun D, Baulac M, Cohen L, Naccache L (2009) Converging intracranial markers of conscious access. PLoS Biol 7:472–492

Goense JB, Logothetis NK (2008) Neurophysiology of the BOLD fMRI signal in awake monkeys. Curr Biol 18(9):631–640

Greenblatt RE, Pflieger ME, Ossadtchi AE (2012) Connectivity measures applied to human brain electrophysiological data. J Neurosci Methods 2007(1):1–16

Gross J, Schmitz F, Schnitzler I, Kessler K, Shapiro K, Hommel B, Schnitzler A (2004) Modulation of long-range neural synchrony reflects temporal limitations of visual attention in humans. Proc Natl Acad Sci USA 101(35):13050–13055

Gross J, Timmermann L, Kujala J, Dirks M, Schmitz F, Salmelin R, Schnitzler A (2002) The neural basis of intermittent motor control in humans. PNAS 99(4):2299–2302

Haenschel C, Linden D (2011) Exploring intermediate phenotypes with eeg: working memory dysfunction in schizophrenia. Behav Brain Res 216:481–495

Hamalainen M, Hari R, Ilmoniemi J, Knuutila J, Lounamaa OV (1993) Magnetoencephalography-theory, instrumentation and applications to noninvasive studies of the working human brain. Rev Mod Phys 65(2):413–497

Hinkley LB, Vinogradov S, Guggisberg AG, Fisher M, Findlay AM, Nagarajan SS (2011) Clinical symptoms and alpha band resting-state functional connectivity imaging in patients with schizophrenia: implications for novel approaches to treatment. Biol Psychiatry 70 (12):1134–1142

Honey CJ, Kötter R, Breakspear M, Sporns O (2007) Network structure of cerebral cortex shapes functional connectivity on multiple time scales. Proc Natl Acad Sci USA 104(24):10240–10245

Horwitz B (2003) The elusive concept of brain connectivity. Neuroimage 19:466–470

Kaminski MJ, Blinowska KJ (1991) A new method of the description of the information flow in the brain structures. Biol Cybern 65(3):203–210

Kelly EF, Lenz JE, Franaszczk PJ, Truong YK (1997) A general statistical framework for frequency-domain analysis of EEG topographic structure. Comput Biomed Res 30:129–164

Kubicki M, Westin CF, McCarley RW, Shenton ME (2005) The application of DTI to investigate white matter abnormalities in schizophrenia. Ann NY Acad Sci 1064:134–148

Lachaux JP, Rodriguez E, Martinerie J, Varela FJ (1999) Measuring phase synchrony in brain signals. Hum Brain Mapp 8:194–208

Le Bihan D, Mangin JF et al (2001) Diffusion tensor imaging: concepts and applications. J Magn Reson Imaging 13(4):534–546

Llinas RR (1988) The Intrinsic electrophysiological properties of mammalian neurons: a new insight into CNS function. Science 242(4886):1654–1664

Llinas RR, Grace AA, Yarom Y (1991) In vitro neurons in mammalian cortical layer 4 exhibit intrinsic oscillatory activity in the 10- to 50-Hz frequency range. Proc Natl Acad Sci USA 88 (3):897–901

Logothetis NK, Pfeuffer J (2004) On the nature of the BOLD fMRI contrast mechanism. Magn Reson Imaging 22(10):1517–1531

Ogawa S, Lee TM, Kay AR, Tank DW (1990) Brain magnetic resonance imaging with contrast dependent on blood oxygenation. Proc Natl Acad Sci USA 87:9868–9872

Ponten SC, Bartolemi F, Stam CJ (2007) Small-world networks and epilepsy: graph theoretical analysis of intracerebrally recorded mesial temporal lobe seizures. Clin Neurophysiol 118:918–927

Rotarska-Jagiela A, Schönmeyer R, Oertel V, Haenschel C, Vogeley K, Linden DE (2008) The corpus callosum in schizophrenia-volume and connectivity changes affect specific regions. Neuroimage 39(4):1522–1532

Rowe JB (2010) Connectivity analysis is essential to understand neurological disorders. Front Syst Neurosci 4(144):1–13

Rubinov M, Knock SA, Stam CJ, Micheloyannis S, Harris AWF, Williams LM, Breakspear M (2009) Small world properties of nonlinear brain activity in schizophrenia. Hum Brain Mapp 30:403–416

Sakkalis V (2011) Review of advanced techniques for the estimation of brain connectivity measured with EEG/MEG. Comput Biol Med 41:1110–1117

Sameshima K, Baccala LA (1999) Using partial directed coherence to describe neuronal ensemble interactions. J Neurosci Meth 94(1):93–103

Sauseng P, Klimesch W (2008) What does phase information of oscillatory brain activity tell us about cognitive processes? Neurosci Biobehav Rev 32(5):1001–1013

Schreiber T (2000) Measuring information transfer. Phys Rev Lett 85:461–464

Siebenhühner F, Weiss SA, Coppola R, Weinberger DR, Bassett DS (2013) Intra- and inter-frequency brain network structure in health and schizophrenia. PLoS ONE 8(8):1–13

Smith SM, Miller KL, Moeller S, Xu J, Auerbach EJ, Woolrich MW, Beckmann CF, Jenkinson M, Andersson J, Glasser MF, Van Essen D, Feinberg D, Yacoub E, Ugurbil K (2012) Temporally-independent functional modes of spontaneous brain activity. PNAS 109(8):3131–3136

Song J, Tucker DM, Gilbert T, Hou J, Mattson C, Luu P, Holmes MD (2013) Methods for examining electrophysiological coherence in epileptic networks. Front Neurol 4:55

Sporns O (2007) Brain connectivity. Scholarpedia, 2(10):4695

Stam CJ, Jones BF, Nolte G, Breakspear M, Scheltens PH (2007) Small-world networks and functional connectivity in Alzheimer's disease. Cereb Cortex 17:92–99

Stam CJ, Reijneveld JC (2007) Graph theoretical analysis of complex networks in the brain. Nonlin Biomed Phys 1(3):1–18

Stephan KE, Friston KJ, Frith C (2009) Dysconnection in schizophrenia: from abnormal synaptic plasticity to failures of self-monitoring. Schizophr Bull 35:509–527

Towle VL, Hunter JD, Edgar JC, Chkhenkeli SA, Castelle MC, Frim DM, Kohrman M, Hecox KE (2007) Frequency domain analysis of human subdural recordings. J Clin Neurophysiol 24 (2):205–213

Uhlhaas PJ, Roux F, Rodriguez E, Rotarska-Jagiela A, Singer W (2009) Neural synchrony and the development of cortical networks. Trends in Cogn Sci 14(2):72–80

Uhlhaas PJ, Singer Wolf (2010) Abnormal neural oscillations and synchrony in schizophrenia. Nat Rev Neurosci 11(2):100–113

Van Essen DC, Smith SM, Barch DM, Behrens TEJ, Yacoub E, Ugurbil K, for the WU-Minn HCP Consortium (2013) The WU-Minn human connectome project: an overview. NeuroImage 80:62–79

Varela F, Lachaux JP, Rodriguez E, Martinerie J (2001) The brainweb: phase synchronization and large-scale integration. Nat Rev Neurosci 2(4):229–239

Wang XJ (2010) Neurophysiological and computational principles of cortical rhythms in cognition. Physiol Rev 90:1195–1268

Wedeen VJ, Wanga RP, Schmahmannb JD, Bennera T, Tsengc WYI, Daia G, Pandyad DN, Hagmanne P, D'Arceuila H, de Crespignya AJ (2008) Diffusion spectrum magnetic resonance imaging (DSI) tractography of crossing fibers. Neuroimage 41(4):1267–1277

White T, Magnotta VA, Bockholt HJ, Williams S, Wallace S, Ehrlich S, Mueller BA, Ho BC, Jung RE, Clark VP, Lauriello J, Bustillo JR, Schulz SC, Gollub RL, Andreasen NC, Calhoun VD, Lim KO (2011) Global white matter abnormalities in schizophrenia: a multisite diffusion tensor imaging study. Schizophr Bull 37(1):222–232

Nonlinear Measures and Dynamics in Psychophysiology of Consciousness

Petr Bob

Abstract According to recent findings nonlinear dynamic processes related to neural chaos and complexity likely play a crucial role in neural synchronization of distributed neural activities that enable information integration and conscious experience. Disturbances in these interactions produce patterns of temporal and spatial disorganization with decreased or increased functional connectivity and complexity that underlie specific changes of perceptual and cognitive states. These perceptual and cognitive changes may be characterized by neural chaos with significantly increased brain sensitivity that may underlie sensitization and kindling, and cognitive hypersensitivity in some mental disorders. Together these findings suggest that processes related to more irregular neural states with higher complexity that may lead to neural chaos, negatively affect information integration and processing in the brain, and may influence disintegrated conscious experience.

Keywords Brain · Chaos · Complexity · Consciousness · Nonlinear dynamics · Self-organization

Contents

P. Bob (✉)
1st Faculty of Medicine, Department of Psychiatry and UHSL,
Center for Neuropsychiatric Research of Traumatic Stress, Charles University, Prague,
Ke Karlovu 11, 128 00 Prague, Czech Republic
e-mail: petrbob@netscape.net

P. Bob
CEITEC Masaryk University, Brno, Czech Republic

Curr Topics Behav Neurosci (2014) 21: 331–343
DOI: 10.1007/7854_2014_321

Published Online: 3 June 2014

1 Introduction

The concept of dynamical chaos was established by French mathematician Henri Poincaré (1854–1912), who studied predictability in a system behavior and found that chaotic pseudo-randomness is caused by high system sensitivity leading to disproportional changes as a response to stimuli that influence behavior of the system (Poincaré 1908/1998; Peterson 1993). As a consequence, the sensitivity significantly decreases ability to predict system's behavior which leads to information loss about later system's development. In his "Science and method" Poincaré (1908/1998, p. 68) wrote: "A very small, unnoticeable cause can determine a visible very large effect; in this case we claim that this effect is a product of random…. However, even if the natural laws were perfectly known, we will never be able to know the initial conditions with some approximation. If this allows us to know the future with the same approximation that is all we want. We will say that the phenomenon is foreseeable, that it is governed by laws; however this is not always the case, it is possible that very small initial differences lead to very large one in the final state…". This basic principle applied to brain dynamics may explain certain specific brain processes where brain sensitivity significantly increases. This type of brain dynamics likely occurs in sensitization and kindling phenomena that might underlie psychological hypersensitivity, which is typical for various mental disorders. This basic principle of chaotic sensitivity is in agreement with data that kindling leads to epileptiform processes characterized by chaotic EEG activity that might explain mental disintegration and increased perceptual sensitivity in several psychiatric disorders (Elbert et al. 1994; Freeman 2000; Bob 2011).

2 Mechanisms of Consciousness and Chaos Theory

Because of sensitivity and unpredictability nonlinear dynamical systems, although they may be also deterministic, exhibit complex and random-like behavior. Experimental research indicates that values of measured properties of many biological systems look random and their determinants are frequently unknown because of high complexity of factors that influence state of living organisms (Elbert et al. 1994; Freeman 2000; Dokoumetzidis et al. 2001; Korn and Faure 2003). Concept of randomness relies on evidence that every complex system has a large number of degrees of freedom which cannot be directly observed and are manifested through the system's fluctuations (Elbert et al. 1994; Dokoumetzidis et al. 2001; Freeman 1991, 2000, 2001). Recent research shows that the chaotic deterministic dynamical systems display the random-like behavior often indistinguishable from truly random processes (Elbert et al. 1994; Dokoumetzidis et al. 2001). On the other hand there is evidence that chaotic dynamics tends to produce a spontaneous order and patterns of organization—self-organization (Elbert et al. 1994; Freeman 2001; Dokoumetzidis et al. 2001; Korn and Faure 2003; Bob 2011).

The self-organization patterns typically are linked to instability states with high sensitivity that may result into new modes of behavior. The sudden phase transitions called bifurcations represent a typical form of system's behavior which is deterministic and is characterized by typical modes of behavior called "attractor". The attractor represents typical non-random behavior of a system which is represented by compressed and limited subset of all possible states in the "state space" [state space = all possible states that a system in principle could have] (Elbert et al. 1994; Bob 2011). On the other hand, a random system has no restrictions on its behavior ("it is random") and there are no limitations for its behavior in the "state space".

In this sense, it is possible to use the term state space for various phenomena describing less complex processes (e.g. position of a particle), or very complex states of the human brain. Deterministic systems are in their behavior strictly limited and resulting behavior such as movement of a very small body (for example in gravitational field) is exactly defined and predictable, which means that in constant conditions (of gravitation) the body falls and then does not move. Behavior of such a system is strictly limited and defined in the time and space (which together defines the state space of the body which is compressed because of limited possibilities and "freedom"). The state space may be visualized by state space diagram in which every possible state of the system corresponds to a unique point in the space and the number of dimensions or parameters of this space represent degree of freedom of the system and every dimension may be represented as an axis. Of course in cases of complex behavior it is difficult or impossible to imagine the state space, similarly like we cannot imagine cube or sphere which has more than three dimensions. In this context, the term multidimensional space represents analogy with usual experience which is used for definition of mathematical terms. For example, the state space (that includes spatial and time dimension) of the mechanical system may be described by all possible values of position and momentum or in thermodynamics by various states of a chemical system which may be described as function of pressure, temperature or composition (Elbert et al. 1994; Dokoumetzidis et al. 2001; Bob 2011).

What is specific for dynamical and chaotic systems is that they have limited behavior and therefore limited space of occurrence in the state space similarly as other deterministic systems but their behavior has limited predictability or it is unpredictable in the space and time. Specific form of behavior of the chaotic system defined by "attractor" includes spatial and time dimensions of all its possible states in the past and future that can be described as a "geometrical object" in the state space. In another words this means that the dynamic and chaotic systems are neither deterministic nor random. Scientific description of complex macrosystems such as living organisms may be defined by various complex "state functions" such as temperature, blood pressure, blood flow or electrical activity such as EEG, ECG and electrodermal activity (EDA), and also other physiological, behavioral or cognitive characteristics (Freeman 1991, 2000; Elbert et al. 1994; Globus and Arpaia 1994; Gottschalk et al. 1995; Huber et al. 1999; Melancon and Joanette 2000; Faure and Korn 2001; Meyer-Lindenberg et al. 2002; Korn and Faure 2003;

Paulus and Braff 2003; Breakspear 2006; Bob 2007, 2011; Bob et al. 2009a, b; Gao et al. 2011).

In comparison to linear relationships where parameters that determine development of a system are proportional, a purpose to use nonlinear methods in neuroscience and psychology is to understand relatively short periods in the behavior of a system which are extremely sensitive to very little changes (the so-called sensitivity to initial conditions). This sensitivity during critical times initiates new trends in the system's evolution which later may emerge as very different macroscopic patterns of neural activity and mental processes (Elbert et al. 1994; Freeman 1983, 1991, 2000; Birbaumer et al. 1995; Kantz and Schreiber 1997; Meyer-Lindenberg et al. 2002; van Putten and Stam 2001; Faure and Korn 2001; Globus and Arpaia 1994; Korn and Faure 2003; Gao et al. 2011).

Several authors proposed that chaotic transitions may emerge in a wide variety of cognitive phenomena and possibly may be linked to specific changes during development of mental disorders such as depression, schizophrenia, dissociation and other mental disorders (Pediaditakis 1992; Schmid 1991; Barton 1994; Gottschalk et al. 1995; Huber et al. 1999; Melancon and Joanette 2000; Korn and Faure 2003; Paulus and Braff 2003; Bob 2007; Bob et al. 2009a, b) and might underlie psychological hypersensitivity to outside stimuli and their pathological processing (Bob 2011). Because epileptic (or epileptiform) activity represents a typical form of chaos in the brain (Freeman 1991, 2000; Elbert et al. 1994; Korn and Faure 2003) it is likely that kindling process as a typical form of progressive sensitization to subconvulsive stimuli may be related to psychopathological symptoms (Kraus 2000; Teicher et al. 2003, 2006) and specifically linked to chaotic processes on neural level (Bob et al. 2009a; Bob 2003, 2007, 2011).

In this context, relationship between chaos and sensitization likely could explain clinical evidence that repeated stressful events also may determine sensitization similar to kindling that leads to an increase in responsiveness to repeated stressors with significantly increased vulnerability to the stress stimuli that may lead to dissociative disorders, schizophrenia, depression and other mental disorders (Post et al. 1995; Post and Weiss 1998; Kraus 2000; Teicher et al. 2003, 2006). This increased sensitivity likely could be related to experience of conflict and mental instability that produces disintegration of conscious experience (Bob 2007, 2011; Bob et al. 2009a).

3 Neural Correlate of Consciousness and Complexity

According to recent growing evidence neural correlate of consciousness is related to processing of distributed information that is represented by integration through levels of neural synchronization among multiple brain regions that is related to large-scale integration, or "binding" (John 2002; Crick and Koch 2003; Bob 2011). Seminal contribution to discussions about the mechanisms of large scale integration reported Crick and Koch (1992), who proposed that the problem of binding cannot be simply resolved as a simple consequence of synchronization among large groups of neurons. As a basis for that opinion they emphasize the binding problem of distributed information represented by different modalities (such as form, motion, color, smell, sound etc.). Processing of information related to a perceived object produces synchronous activities in separate areas of the brain but there is no evidence of that spatial convergency in the brain that would represent neural correlate of consciousness. The hypothetical center for information convergency was termed "Cartesian theatre" (Crick and Koch 1992; Dennett 1991), but recent neuroscience has not located a distinct place in which distributed information in the brain comes together.

Recent findings suggest that a candidate mechanism for the integration or binding of distributed brain activities is the so-called gamma activity (most frequently high frequency oscillations about 40 Hz). Although there is growing evidence that EEG gamma waves enable different neuronal circuits to enter into synchrony with the perceptual information and oscillate together during transient periods of synchronized firing there is no explanation as to what mechanism is behind this synchronization and information convergency integrating various percepts, memories and associations. From this point of view the phenomenon of synchronization and functional integration represents intriguing evidence that information convergency is successfully achieved but what is behind this process of synchronization remains a puzzle.

A solution of the binding problem may reside within the fundamental problem of consciousness in modern neuroscience. The predominant opinion is that consciousness emerges from a dynamical nucleus of persisting reverberation and interactions of neural groups (John 2002; Bob 2011). For example, Tononi and Edelman (2000) emphasized that consciousness is represented by re-entry of neural signals via changes of complexity and entropy in the central nervous system. Libet (1998) suggests that subjective experience represents a field emerging from neural synchronization and coherence, and is not reducible to any physical process (see also John 2002). In accordance with Libet, Squires (1998) maintains that consciousness may be understood as a primitive (irreducible) component of the world including specific qualities of subjective experience (qualia) that cannot be reduced to any other physical quality (see also Duch 2005; John 2002; Bob 2011). According to Freeman (1991, 2000, 2001), mental images of the world emerge as a consequence of creating order from non-linear chaotic activity of large groups of neurons. These nonlinear chaotic processes represent a consequence of high system complexity, when the system involves a large number of complex interlinked and simultaneously active neural assemblies and runs in a desynchronized parallel distributed mode which can lead to chaos and self-organization (Freeman 1991, 2000, 2001; Velazquez et al. 2003; Atmanspacher and Fach 2005).

Similarly, Tononi and Edelman (1998) proposed that consciousness is not a thing, but a process or changing streams on a time scale of fractions of seconds. In agreement with James (1890) they emphasized that a fundamental aspect of consciousness is an integrated state although at the same time there is evidence that distributed neural activity, particularly in the thalamocortical system, is essential for conscious experience (Edelman 1989; Picton and Stuss 1994; Newman 1995). Therefore it is possible to suppose that interactions among neuronal assemblies in distributed brain areas create a unified neural process corresponding to a multimodal conscious scene (Edelman 1989; Tononi et al. 1992; Lumer et al. 1997; Tononi and Edelman 1998). Altogether these studies suggest that a specific sign of effective reentrant interactions are short-term temporal correlations of distributed neural interactions between involved neuronal groups (Tononi and Edelman 1998; Tononi et al. 1998a; Bressler 1995), which may explain why feeble, degraded, or short-lasting stimuli often are not consciously perceived, even though they may produce a behavioral response such as perception without awareness (Marcel 1983; Merikle et al. 2001). In agreement with this current evidence also changes in conscious experience driven by external stimuli, memories, mental images or dreams are related to changes in activities or deactivations of specific widely distributed brain areas (Roland 1993; Frackowiak 1997; Tononi and Edelman 1998). Tononi and Edelman (1998) applied methods of functional clustering and found that a subset of distributed elements within a system gives rise to a single, integrated process in cases when these elements interact significantly more strongly among themselves than with the rest of the system. This interaction means that they form a functional cluster which is possible to measure by mutual information as a level of integration (Tononi et al. 1998b). When the level of integration calculated among all neurons within the subsystem is higher than level of integration that the same neurons of this subsytem have with neurons out of the subsystem then the subsystem presents the functional cluster (Papoulis 1991; Tononi and Edelman 1998). For example, it is possible to compare levels of synchronous firing among cortical regions and between cortex and thalamus (Tononi et al. 1992; Lumer et al. 1997; Tononi and Edelman 1998). Functional clustering also enable to define system complexity as a number of its parts (i.e. number of clusters) that have higher level of integration within the subsystem than is a level of integration that neurons of this subsystem have with neurons out of the subsystem (Tononi and Edelman 1998).

In this context, concept of complexity can be applied to neurophysiological data and enable to evaluate the degree to which neural processes are integrated and/or differentiated (Friston et al. 1995; Sporns et al. 2000, 2002; Sporns 2013, 2014; Balduzzi et al. 2009; Casali et al. 2013). It is also possible to compare the values of neural complexity in different cognitive and arousal states or empirically test the relationships between brain complexity and levels of conscious experience (Tononi and Edelman 1998; Bob et al. 2009a, b; Bob 2011).

Because consciousness is related to high level of functional integration among neurons, it is possible to predict that the complexity could correlate with subjective conscious states and several studies reported relationship between attentional functions and brain EEG complexity. For example, Lutzenberger et al. (1995) reported that increased EEG complexity indicates an increase in simultaneously activated neural assemblies. Further studies also reported that EEG complexity was substantially higher during imagery than during actual sensory stimulation (Birbaumer et al. 1993; Schupp et al. 1994; Molle et al. 1997) or experimental cognitive load (Bizas et al. 1999; Meyer-Lindenberg 1996; Micheloyannis et al. 1998, 2002; Mölle et al. 1995, 1997; Stam et al. 1996; Tomberg 1999). Further studies also indicate that complexity is significantly lower during full alertness than during drowsiness (Matousek et al. 1995) and similarly also clear alertness during a state of meditation has shown to be associated with a decrease in EEG complexity (Aftanas and Golocheikine 2002). Consistently with these data it has been reported that divergent creative thought is associated with higher EEG complexity whereas convergent analytical thought was related to the lower complexity (Molle et al. 1996). In this context also other studies show that neural EEG complexity reflects the attentional mode related to processing of cortical stimuli (Pritchard and Duke 1995; Molle et al. 1995, 1996, 1997; Balduzzi et al. 2009; Casali et al. 2013). Taken together these results suggest that attentional narrowing decreases complexity and causes a reduction in neural competition in connection with an inhibition of neural assemblies irrelevant for task completion during selective attention (Lutzenberger et al. 1992).

Preliminary data and models suggest that EEG complexity in principle could reflect typical attentional changes related to dissociative states and dissociation can be described as a kind of divided or parallel neural process where several information processors within the brain system may have a higher level of independence (Li and Spiegel 1992; Bob 2003, 2011). From this point of view dissociated consciousness is related to increased independence among neural assemblies and higher EEG complexity. But also the opposite is true because these states of increased neural complexity and decreased connectivity are interupted by time periods when dissociated state is released into consciousness that leads to narrowing attention with decreased complexity and increased connectivity and information integration. These transient periods related to actual experience of aversive events or during reliving of a dissociative state lead to a greater allocation of attention that may cause changes in the ordinary integrative functions of consciousness (Guralnik et al. 2000; Vermetten and Bremner 2004; Bob 2007, 2008). Also high anxiety and arousal related to disturbing past experience intensely narrow the attention (Vermetten and Bremner 2004) and higher degree of functional connectivity and activation in certain brain regions was found in clinical forms of dissociation or hypnosis in functional imaging studies (Faymonville et al. 2006; Cojan et al. 2009).

Together these data suggest that dissociation in usual states of consciousness is related to increased complexity and on the contrary extreme levels of attention related to reliving of dissociated experiences or during hypnosis are linked to higher connectivity, lower complexity and increased autonomic and emotional arousal (Bob et al. 2009a, b, 2010a, b; Bob 2011). In this context, it is possible to suppose that complexity could be also related to a number of independent clusters of mental associations that are related to specific emotional states and autonomic arousal (Jung 1907; Weingartner et al. 1977; Stern and Riegel 1970; Bob et al. 2009a, b, 2010a, b). These complex patterns connecting mental and physiological states produce specific patterns of temporal organization or disorganization with increased or decreased functional connectivity that may underlie specific perceptual, emotional and cognitive states and in highly complex states may lead to neural chaos (Sporns et al. 2000, 2002; Sporns 2013, 2014; Balduzzi et al. 2009; Bob 2011; Casali et al. 2013).

4 Nonlinear Dynamics, Chaos and Intentionality

Seminal contributions to this field of research were provided by Walter Freeman (Freeman 1991, 2000, 2001), who was particularly interested in exploring how the brain generates cognitive processing, intentionality and meaning. His main body of research has been focused on EEG study of perceptual processing in rabbits. In his research, Freeman found that activity in the olfactory cortex is chaotic, and proposed that chaos could underlie basic forms of collective neural activity in perceptual processing including ability to access memorized sensory patterns and learning novel sensory information (Freeman 1991, 2000, 2001; Skarda and Freeman 1987). Freeman also proposed that chaos may explain brain ability to respond flexibly to the outside world and to generate novel activity patterns that are subjectively experienced as "novel" ideas, generated by unpredictable attractors that enable complex dynamic behavior of the brain and intentional behavior (Freeman 1991, 2000, 2001; Skarda and Freeman 1987). In this context, complexity and chaos theory enables to understand collective neural activity and brain functions as a global integrative process based on dynamic collections of attractors which form an "attractor landscape" generated in the web of synaptic connections and representing behavioral "intentional" patterns that may be modified by learning (Skarda and Freeman 1987; Freeman 2000).

Within this framework Freeman (1999) proposed that the linear view of the stimulus-response reflex determinism is not appropriate concept for the behavioral dynamics and suggested that it is necessary to study behavioral responses and intentional behavior as consequences of nonlinear chains of various stimuli and responses. Freeman (1999, 2000) also suggested that chaotic and complex self-organization of multilevel interactions between microscopic neurons in assemblies and the macroscopic emergent states is not possible within the concept of "linear causality", and must be replaced by "circular causality" (or reciprocal causality) that enables to reflect extensive relations of mutual dependencies, actions and influences. Although neurophysiological basis of these integrative processes is only partially understood, the concept of circular causality as a formal semantic description of brain dynamics related to chaotic self-organization and multimodal macroscopic patterns of neural activations may help to explain some functions of consciousness and intentional actions.

5 Conclusions

Recent evidence indicates that brain functions related to consciousness and attention require multiregional functional interaction and large scale integration and binding of multiple neural assemblies (Crick and Koch 1992, 2003; Varela et al. 2001; Fries 2005; Bob 2011). According to several findings the interacting neuronal assemblies represent basic functional units in brain information processing that may behave independently with lower level of binding among the units or may be dynamically integrated into large subsets of neurons that behave coherently with synchronous activity (Seth et al. 2006; Edelman 2003; Elbert et al. 1994; Lutzenberger et al. 1995; Molle et al. 1997; Stam 2005; Balduzzi et al. 2009; Casali et al. 2013). This number of complex interlinked and simultaneously active neural states is reflected in nonlinear processes characterized by neural complexity that also may be linked to chaotic phenomena in the brain.

In summary, nonlinear measures of complexity and chaos in the brain likely can explain time and space changes in neural synchronization and coherence. These changes in brain complexity likely represent basis for discrete mental states that through differences between them enable recognition and awareness of the external and internal worlds. According to this concept the image of the world emerges as a consequence of creating order arising from non-linear activities of large groups of neurons. These highly organized nonlinear processes represent a consequence of high system complexity that occurs when the system involves a large number of interlinked and simultaneously active neural assemblies and runs in a desynchronized parallel distributed mode, which can lead to self-organization and chaos. These levels of complexity within the brain likely represent basic code that enables to connect mental and physical space and to define corresponding differences and their recognition in mental and physical space that through neural complexities can represent external space in the mental space.

References

Aftanas LI, Golocheikine SA (2002) Non-linear dynamic complexity of the human EEG during meditation. Neurosci Lett 330:143–146

Atmanspacher H, Fach W (2005) Acategoriality as mental instability. J Mind Behav 26:181–205

Balduzzi D, Tononi G (2009) Qualia: the geometry of integrated information. PLoS Comput Biol 5:e1000462. doi:10.1371/journal.pcbi.1000462

Barton S (1994) Chaos, self-organization, and psychology. Am Psychol 49:5–14

Birbaumer N, Lutzenberger W, Elbert T (1993) Imagery and brain processes. In: Birbaumer N, Ohman A (eds) The structure of emotion. Hogrefe and Huber, Toronto, pp 122–138

Birbaumer N, Flor H, Lutzenberger W, Elbert T (1995) Chaos and order in the human brain. Electroencephalogr Clin Neurophysiol Suppl 44:450–459

Bizas E, Simos PG, Stam CJ, Arvanitis S, Terzakis D, Micheloyannis S (1999) EEG correlates of cerebral engagement in reading tasks. Brain Topogr 12:99–105

Bob P (2003) Dissociation and neuroscience: history and new perspectives. Int J Neurosci 113:903–914
Bob P (2007) Hypnotic abreaction releases chaotic patterns of electrodermal activity during dissociation. Int J Clin Exp Hypn 55:435–456
Bob P (2008) Pain, dissociation and subliminal self-representations. Conscious Cogn 17:355–369
Bob P (2011) Brain, mind and consciousness: advances in neuroscience research. Springer, New York
Bob P, Susta M, Gregusova A, Jasova D (2009a) Dissociation, cognitive conflict and nonlinear patterns of heart rate dynamics in patients with unipolar depression. Prog Neuropsychopharmacol Biol Psychiatry 33:141–145
Bob P, Susta M, Chladek J, Glaslova K, Palus M (2009b) Chaos in schizophrenia associations, reality or metaphor? Int J Psychophysiol 73:179–185
Bob P, Susta M, Glaslova K, Boutros NN (2010a) Dissociative symptoms and interregional EEG crosscorrelations in paranoid schizophrenia. Psychiatry Res 177:37–40
Bob P, Susta M, Gregusova A, Jasova D, Mishara A, Raboch J (2010b) Traumatic stress, dissociation, and limbic irritability in patients with unipolar depression being treated with SSRIs. Psychol Rep 107:685–696
Breakspear M (2006) The nonlinear theory of schizophrenia. Aust NZ J Psychiatry 40:20–35
Bressler SL (1995) Large-scale cortical networks and cognition. Brain Res Rev 20:288–304
Casali AG, Gosseries O, Rosanova M, Boly M, Sarasso S, Casali KR, Casarotto S, Bruno MA, Laureys S, Tononi G, Massimini M (2013) A theoretically based index of consciousness independent of sensory processing and behavior. Sci Transl Med 5:198ra105. doi: 10.1126/scitranslmed.3006294
Cojan Y, Waber L, Schwartz S, Rossier L, Forster A, Vuilleumier P (2009) The brain under self-control: modulation of inhibitory and monitoring cortical networks during hypnotic paralysis. Neuron 62:862–875
Crick F, Koch C (1992) The problem of consciousness. Sci Am 267(3):153–159
Crick F, Koch C (2003) A framework for consciousness. Nat Neurosci 6:119–126
Dennett D (1991) Consciousness explained. Little, Brown, Boston
Dokoumetzidis A, Iliadin A, Macheras P (2001) Nonlinear dynamics and chaos theory: Concepts and applications relevant to pharmacodynamics. Pharmacol Res 18:415–426
Duch W (2005) Brain-inspired conscious computing architecture. J Mind Behav 26:1–21
Edelman G (1989) The remembered present. Basic Books, New York
Edelman GM (2003) Naturalizing consciousness: a theoretical framework. Proc Natl Acad Sci USA 100:5520–5524
Elbert T, Ray WJ, Kowalik ZJ, Skinner JE, Graf KE, Birbaumer N (1994) Chaos and physiology: deterministic chaos in excitable cell assemblies. Physiol Rev 74:1–47
Faure P, Korn H (2001) Is there chaos in the brain? I. Concepts of nonlinear dynamics and methods of investigation. CR Acad Sci III 324:773–793
Faymonville ME, Boly M, Laureys SJ (2006) Functional neuroanatomy of the hypnotic state. J Physiol Paris 99:463–469
Frackowiak RSJ (1997) Human brain function. Academic Press, San Diego
Freeman WJ (1983) The physiological basis of mental images. Biol Psychiatry 18:1007–1025
Freeman WJ (1991) The physiology of perception. Sci Am 264:78–85
Freeman WJ (1999) Consciousness, intentionality, and causality. J Conscious Stud 6:143–172
Freeman WJ (2000) Mesoscopics neurodynamics: from neuron to brain. J Physiol Paris 94:303–322
Freeman WJ (2001) Biocomplexity: adaptive behavior in complex stochastic dynamical systems. Biosystems 59:109–123
Fries P (2005) A mechanism for cognitive dynamics: neuronal communication through neuronal coherence. Trends Cogn Sci 9:474–480
Friston KJ, Tononi G, Sporns O, Edelman GM (1995) Characterising the complexity of neuronal interactions. Hum Brain Mapp 3:302–314

Gao J, Hu J, Tung WW (2011) Complexity measures of brain wave dynamics. Cogn Neurodyn 5:171–182. doi:10.1007/s11571-011-9151-3
Globus GC, Arpaia JP (1994) Psychiatry and the new dynamics. Biol Psychiatry 32:352–364
Gottschalk AM, Bauer MS, Whybrow PC (1995) Evidence of chaotic mood variation in bipolar disorder. Arch Gen Psychiatry 52:947–959
Guralnik O, Schmeidler J, Simeon D (2000) Feeling unreal: cognitive processes in depersonalization. Am J Psychiatry 157:103–109
Huber MT, Braun HA, Krieg JC (1999) Consequences of deterministic and random dynamics for the course of affective disorders. Biol Psychiatry 46:256–262
James W (1890) The principles of psychology. Holt, New York
John ER (2002) The neurophysics of consciousness. Brain Res Rev 39:1–28
Jung CG (1907) On psychophysical relations of the associative experiment. J Abnorm Psychol 1:247–255
Kantz H, Schreiber T (1997) Nonlinear time series analysis. Cambridge University Press, Cambridge
Korn H, Faure P (2003) Is there chaos in the brain? II. Experimental evidence and related models. CR Biol 326:787–840
Kraus JE (2000) Sensitization phenomena in psychiatric illness: lessons from the kindling model. J Neuropsychiatry Clin Neurosci 12:328–343
Li D, Spiegel D (1992) A neural network model of dissociative disorders. Psychiatr Ann 22:144–147
Libet B (1998) Do the models offer testable proposals of brain functions for conscious experience. In: Jasper HH, Descarries L, Costelucci VC, Rossignol S (eds) Advances in neurology: consciousness at the frontiers of neuroscience. Lippincott-Raven, Philadelphia, pp 213–217
Lumer ED, Edelman GM, Tononi G (1997) Neural dynamics in a model of the thalamocortical system. I. Layers, loops and the emergence of fast synchronous rhythms. Cereb Cortex 7:207–227
Lutzenberger W, Birbaumer N, Flor H, Rockstroh B, Elbert T (1992) Dimensional analysis of the human EEG and inteligence. Neurosci Lett 143:10–14
Lutzenberger W, Preissl H, Pulvermuller F (1995) Fractal dimension of electroencephalographic time series and underlying brain processes. Biol Cybern 73:477–482
Marcel AJ (1983) Conscious and unconscious perception: an approach to the relations between phenomenal experience and perceptual processes. Cogn Psychol 15:238–300
Matousek M, Wackermann J, Palus M, Berankova A, Albrecht V, Dvorak I (1995) Global dimensional complexity of the EEG in healthy volunteers. Neuropsychobiology 31:47–52
Melancon G, Joanette Y (2000) Chaos, brain and cognition: toward a nonlinear order? Brain Cogn 42:33–36
Merikle PM, Smilek D, Eastwood JD (2001) Perception without awareness: perspectives from cognitive psychology. Cognition 79:115–134
Meyer-Lindenberg A (1996) The evolution of complexity in human brain development: an EEG study. Electroencephalogr Clin Neurophysiol 99:405–411
Meyer-Lindenberg A, Zeman U, Hajak G, Cohen L, Berman KF (2002) Transitions between dynamical states of differing stability in the human brain. Proc Natl Acad Sci USA 99:10948–10953
Micheloyannis S, Flitzanis N, Papanikolau E, Bourkas M, Terzakis D, Arvanitis S et al (1998) Usefulness of non-linear EEG analysis. Acta Neurol Scand 97:13–19
Micheloyannis S, Papanikolaou E, Bizas E, Stam CJ, Simos PG (2002) Ongoing electroencephalographic signal study of simple arithmetic using linear and non-linear measures. Int J Psychophysiol 44:231–238
Molle M, Marshall L, Pietrowsky R (1995) Dimensional complexity of the EEG indicates a right frontocortical locus of attentional control. J Psychophysiol 9:45–55
Molle M, Marshall L, Pietrowsky R (1996) Enhanced dynamic complexity in the human EEG during creative thinking. Neurosci Lett 208:1–4

Molle M, Albrecht C, Mashall L (1997) Adrenocorticotropin widens the focus of attention in humans. A nonlinear electroencephalographic analysis. Psychosom Med 59:497–502
Newman J (1995) Thalamic contributions to attention and consciousness. Conscious Cogn 4:172–193
Papoulis A (1991) Probability, random variables, and stochastic processes. McGraw-Hill, New York
Paulus MP, Braff DL (2003) Chaos and Schizophrenia: does the method fit the madness? Biol Psychiatry 53:3–11
Pediaditakis N (1992) Deterministic non-linear chaos in brain function and borderline psychopathological phenomena. Med Hypotheses 39:67–72
Peterson I (1993) Newton's clock: Chaos in the solar system. W.H. Freeman, New York
Picton TW, Stuss DT (1994) Neurobiology of conscious experience. Curr Biol 4:256–265
Poincaré H (1908/1998) Science and method. Thomas Nelson and Sons, London
Post RM, Weis SR, Smith MA (1995) Sensitization and kindling. In: Friedman MJ, Charney DS, Deutch AY (eds) Neurobiological and clinical consequences of stress: from normal adaptation to posttraumatic stress disorder. Lipincott-Raven, Philadelphia
Post RM, Weiss RB (1998) Sensitization and kindling phenomena in mood, anxiety, and obsessive compulsive disorders: the role of serotonergic mechanisms in illness progression. Biol Psychiatry 44:193–206
Pritchard WS, Duke D (1995) Measuring, "chaos" in the brain: a tutorial review of EEG dimension estimation. Brain Cogn 27:353–397
Roland PE (1993) Brain activation. Wiley-Liss, New York
Schmid GB (1991) Chaos theory and schizophrenia: elementary aspects. Psychopathology 24:185–198
Schupp HT, Lutzenberger W, Birbaumer N (1994) Neurophysiological differences between perception and imagery. Cogn Brain Res 2:77–86
Seth AK, Izhikevich E, Reeke GN, Edelman GM (2006) Theories and measures of consciousness: an extended framework. Proc Natl Acad Sci USA 103:10799–10804
Skarda CHA, Freeman WJ (1987) How brains make chaos in order to make sense of the world. Behav Brain Sci 10:161–195
Sporns O (2013) Network attributes for segregation and integration in the human brain. Curr Opin Neurobiol 23:162–171. doi:10.1016/j.conb.2012.11.015
Sporns O (2014) Contributions and challenges for network models in cognitive neuroscience. Nat Neurosci. doi:10.1038/nn.3690
Sporns O, Tononi G, Edelman GM (2000) Connectivity and complexity: the relationship between neuroanatomy and brain dynamics. Neural Netw 13:909–922
Sporns O, Tononi G, Edelman GM (2002) Theoretical neuroanatomy and the connectivity of the cerebral cortex. Behav Brain Res 135:69–74
Squires EJ (1998) Why are quantum theorists interested in consciousness. In: Hameroff SR, Kaszriak A, Scott AC (eds) Toward a science of consciousness II: the second Tucson discussions and debates. MIT Press, Cambridge, pp 609–618
Stam CJ (2005) Nonlinear dynamical analysis of EEG and MEG: review of an emerging field. Clin Neurophysiol 116:2266–2301
Stam CJ, van Woerkom TC, Pritchard WS (1996) Use of non-linear EEG measures to characterize EEG changes during mental activity. Electroencephalogr Clin Neurophysiol 99:214–224
Stern EJ, Riegel KF (1970) Comparisons of the restricted association of chronic schizophrenic and normal control subjects. J Abnorm Psychol 75:164–171
Teicher MH, Andersen SL, Polcari A, Anderson CM, Navalta CP, Kim DM (2003) The neurobiological consequences of early stress and childhood maltreatment. Neurosci Biobehav Rev 27:33–44
Teicher MH, Tomoda A, Andersen SE (2006) Neurobiological consequences of early stress and childhood maltreatment: are results from human and animal studies comparable? Ann NY Acad Sci 1071:313–323

Tomberg C (1999) Focal enhancement of chaotic strange attractor dimension in the left semantic (Wernicke) human cortex during reading without concomitant change in vigilance level. Neurosci Lett 263:177–180

Tononi G, Edelman GM (1998) Consciousness and complexity. Science 282:1846–1851

Tononi G, Edelman GM (2000) Schizophrenia and the mechanisms of conscious integration. Brain Res Rev 31:391–400

Tononi G, Sporns O, Edelman GM (1992) Reentry and the problem of integrating multiple cortical areas: simulation of dynamic integration in the visual system. Cereb Cortex 2:310–335

Tononi G, Edelman GM, Sporns O (1998a) Complexity and coherency: integrating information in the brain. Trends Cogn Sci 2:474–484

Tononi G, McIntosh AR, Russell DP, Edelman GM (1998b) Functional clustering: identifying strongly interactive brain regions in neuroimaging data. Neuroimage 7:133–149

Van Putten MJAM, Stam CJ (2001) Is the EEG really "chaotic" in hypsarrhythmia. IEEE Eng Med Biol Mag 20:72–79

Varela FJ, Lachaux JP, Rodriguez E, Martinerie J (2001) The brainweb: phase synchronization and largescale integration. Nat Rev Neurosci 2:229–239

Velazquez JLP, Cortez MA, Snead OC III, Wennberg R (2003) Dynamical regimes underlying epileptiform events: role of instabilities and bifurcations in brain activity. Physica D 186:205–220

Vermetten E, Bremner JD (2004) Functional brain imaging and the induction of traumatic recall: a crosscorrelational review between neuroimaging and hypnosis. Int J Clin Exp Hypn 52:218–312

Weingartner H, Miller H, Murphy D (1977) Mood-state-dependent retrieval of verbal associations. J Abnorm Psychol 86:276–284

Tomberg C (1999) Focal enhancement of chaotic strange attractor dimension in the left semantic (Wernicke) human cortex during reading without concomitant change in vigilance level. Neurosci Lett 263:177–180
Tononi G, Edelman GM (1998) Consciousness and complexity. Science 282:1846–1851
Tononi G, Edelman GM (2000) Schizophrenia and the mechanisms of conscious integration. Brain Res Rev 31:391–400
Tononi G, Sporns O, Edelman GM (1992) Reentry and the problem of integrating multiple cortical areas: simulation of dynamic integration in the visual system. Cereb Cortex 2:310–335
Tononi G, Edelman GM, Sporns O (1998) Complexity and coherency: integrating information in the brain. Trends Cogn Sci 2:474–484
Tononi G, McIntosh AR, Russell DP, Edelman GM (1998) Functional clustering: identifying strongly interactive brain regions in neuroimaging data. Neuroimage 7:133–149
Van Putten MJAM, Stam CJ (2001) Is the EEG really "chaotic" in hypsarrhythmia? IEEE Eng Med Biol Mag 20:72–79
Varela FJ, Lachaux JP, Rodriguez E, Martinerie J (2001) The brainweb: phase synchronization and large-scale integration. Nat Rev Neurosci 2:229–239
Velazquez JLP, Cortez MA, Snead OC III, Wennberg R (2003) Dynamical regimes underlying epileptiform events: role of instabilities and bifurcations in brain activity. Physica D 186:205–220
Vermetten E, Bremner JD (2004) Functional brain imaging and the induction of traumatic recall: a cross-correlational review between neuroimaging and hypnosis. Int J Clin Exp Hypn 52:280–312
Weingartner H, Miller H, Murphy DL (1977) Mood-state-dependent retrieval of verbal associations. J Abnorm Psychol 86:276–284

Personalized Medicine in ADHD and Depression: Use of Pharmaco-EEG

Martijn Arns and Sebastian Olbrich

Abstract This chapter summarises recent developments on personalised medicine in psychiatry with a focus on ADHD and depression and their associated biomarkers and phenotypes. Several neurophysiological subtypes in ADHD and depression and their relation to treatment outcome are reviewed. The first important subgroup consists of the 'impaired vigilance' subgroup with often-reported excess frontal theta or alpha activity. This EEG subtype explains ADHD symptoms well based on the EEG Vigilance model, and these ADHD patients responds well to stimulant medication. In depression this subtype might be unresponsive to antidepressant treatments, and some studies suggest these depressive patients might respond better to stimulant medication. Further research should investigate whether sleep problems underlie this impaired vigilance subgroup, thereby perhaps providing a route to more specific treatments for this subgroup. Finally, a slow individual alpha peak frequency is an endophenotype associated with treatment resistance in ADHD and depression. Future studies should incorporate this endophenotype in clinical trials to investigate further the efficacy of new treatments in this substantial subgroup of patients.

Keywords ADHD · Depression · QEEG · Personalised medicine · Phenotype · Biomarker

M. Arns (✉)
Department of Experimental Psychology, Utrecht University, Utrecht, The Netherlands
e-mail: martijn@brainclinics.com

M. Arns
Research Institute Brainclinics, Nijmegen, The Netherlands

S. Olbrich
Department for Psychiatry and Psychotherapy, Universtiy Leipzig, Leipzig, Germany

Curr Topics Behav Neurosci (2014) 21: 345–370

DOI: 10.1007/7854_2014_295

Published Online: 11 March 2014

Contents

1 Introduction

The landscape in psychiatry recently underwent a dramatic change. Large-scale studies investigating the effects of conventional treatments for ADHD and depression in clinical practise have demonstrated, at the group-level, limited efficacy of antidepressant medication and cognitive behavioural therapy in depression (STAR*D: Rush et al. 2006), an overestimation of the effects of cognitive behavioural therapy for depression as a result of publication bias (Cuijpers et al. 2010) and limited long-term effects of stimulant medication, multicomponent behaviour therapy and multimodal treatment in ADHD (NIMH-MTA trial: Molina et al. 2009), although latent class analysis reported a subgroup consisting of children who demonstrated sustained effects of treatment at 2 years follow-up (Swanson et al. 2007). Furthermore, several large pharmaceutical companies announced that they would '*...pull the plug on drug discovery in some areas of neuroscience...*' (Miller 2010). This can be considered a worrying development, since there is still much to improve in treatments for psychiatric disorders. The conclusions about limitations in efficacy and long-term effects are all based on the interpretation of group-averaged data, but also demonstrate that there is a percentage of patients responding to antidepressants (Rush et al. 2006) and there is a subgroup of patients demonstrating long-term effects (Swanson et al. 2007). Therefore, a move beyond data regarding the average effectiveness of treatments to identify the best treatment for any individual (Simon and Perlis 2010) or personalised medicine is crucial.

The fact that only subgroups of patients respond to treatment raises important questions about the underlying assumptions of neurobiological homogeneity within psychiatric disorders, and is rather suggestive of neurobiological heterogeneity. Therefore, a move beyond data regarding the average effectiveness of treatments, to identify the best treatment for a given individual (Simon and Perlis 2010) or personalised medicine is highly relevant. In personalised medicine it is the goal to prescribe the right treatment, for the right person at the right time as opposed to the current 'trial-and-error' approach, by using biomarkers of endophenotypes.

From the point of view that biomarkers should be cost-effective, easy applicable and implemented within the routine diagnostic procedure, the quantitative EEG (QEEG) seems to be appropriate. Still the question is whether it should be considered a diagnostic or prognostic technique? Although several EEG-biomarkers have shown robust discriminative power regarding neuropsychiatric conditions (for depression also see: Olbrich and Arns 2013) it seems not within reach that biomarkers will replace the clinical diagnosis (Savitz et al. 2013). As another illustration, consider any psychiatric disorder as defined according to the DSM-IV or DSM-V (DSM). Besides a list of behavioural symptoms, there is always the final criterion that the complaints result in 'impairments in daily life'. Specifically, this criterion makes it almost impossible to devise any neurobiological test to replace diagnosis based on the DSM, since for one person the same level of impulsivity and inattention is considered a blessing (i.e. artist or CEO), whereas for another person the same levels of impulsivity and inattention is considered a curse, and hence results in a diagnosis only for the latter subject.

Given the recent development of personalised medicine (in line with the NIMH Strategic Plan on Research Domain Criteria or RdoC, and termed Precision Medicine) and the above limitations of current psychiatric diagnosis and treatments, in this chapter we will focus on the prognostic use of QEEG in psychiatry.

This prognostic use of EEG or QEEG has a long history. For example, Satterfield et al. (1971, 1973) were the first to investigate the potential use of EEG in predicting treatment outcome to stimulant medication (main results outlined further on). In 1957 both Fink, Kahn and Oaks (Fink and Kahn 1957) and Roth et al. (1957) investigated EEG predictors to ECT in depression. Fink recently summarised these findings eloquently as: '*Slowing of EEG rhythms was necessary for clinical improvement in ECT*' (Fink 2010).

2 Personalised Medicine: Biomarkers and Endophenotypes

Personalised medicine aims to provide the right treatment to the right person at the right time as opposed to the current 'trial-and-error' approach. Genotypic and phenotypic information (or 'biomarkers') lie at the basis of this approach. However, 2011 marked the 10th year anniversary of the completion of the Human

Genome project, which has sparked numerous large-scale Genome Wide Association studies (GWA) and other genotyping studies in psychiatric disorders, only accounting for a few percent of the genetic variance (Lander 2011). This suggests that a strictly genetic approach to personalised medicine for psychiatry will not be as promising as initially expected. The notion of personalised medicine suggests *heterogeneity* within a given DSM-IV disorder, rather than *homogeneity*, at least from a brain-function-based perspective. Therefore, a variety of 'endophenotypes' or 'biomarkers' are expected within a single DSM-IV disorder to require a different treatment.

The National Institutes of Health declared a biomarker as 'A characteristic that is objectively measured and evaluated as an indicator of normal biologic processes, pathogenic processes or pharmacologic responses to a therapeutic intervention' (De Gruttola et al. 2001). However, the idea behind an endophenotype is that it is the intermediate step between genotype and behaviour and thus is more closely related to genotype than behaviour alone. Therefore, endophenotypes can be investigated to yield more information on the underlying genotype. Given the interest in the last couple of years for genetic linkage studies, this term has become more topical again. In parallel, there have also been many studies using the term biological marker, trait, biomarker etc. Here it is important that, in line with Gottesman and Gould (2003), an 'endophenotype' refers to a marker when also certain heritability indicators are fulfilled, whereas a 'biomarker' simply refers to differences between patient groups, which do not necessarily have a hereditary basis.

Older studies attempting to aid the prescription process with more objective knowledge have studied biological (e.g. neurotransmitter metabolites), psychometric (personality questionnaires), neuropsychological (cognitive function) and psychophysiological (EEG, ERP) techniques (Joyce and Paykel 1989). Biological techniques (such as neurotransmitter metabolites) have to date shown little promise as reliable predictors of treatment response and are not yet recommended for routine clinical practise (Joyce and Paykel 1989; Bruder et al. 1999). Similarly, the clinical utility of 'behavioural phenotypes' remains poor and, at this moment, none of these predictors have clinical use in predicting treatment outcome to various anti-depressive treatments (Simon and Perlis 2010; Cuijpers et al. 2012; Bagby et al. 2002).

However, there has been renewed interest in the use of other measures such as pharmacogenomics (Frieling and Tadić 2013) and pharmacometabolomics (Hefner et al. 2013), which are speculated to show promise in the use of personalised medicine. However to date pharmacogenomics have not shown promising results in predicting treatment outcome in psychiatric disorders (Johnson and Gonzalez 2012; Ji et al. 2011; Menke 2013) and pharmacometabolomics is considered potentially promising at most at this moment, with few reports on its role in personalised medicine (Johnson and Gonzalez 2012; Quinones and Kaddurah-Daouk 2009).

Recent studies suggest that more direct measures of brain function, such as psychophysiology and neuropsychology, may be more reliable in predicting treatment response in depression (Olbrich and Arns 2013). The underlying idea behind this concept is that for example neurophysiological data from EEG capture ongoing neuronal activity at the timescale it takes place, outpassing any other modality such as neuroimaging techniques like fMRI or PET. Further, the EEG is not a surrogate marker of neuronal activity (such as the blood desoxygenation level dependent signal in fMRI or the glucose utilisation in PET) but gives insight into the actual cortical activity. Therefore, the EEG can help to define stable endophenotypes incorporating both the effects of nature and nurture. This potentially makes the EEG an ideal candidate biomarker, which has the potential to predict treatment outcome.

3 EEG as an Endophenotype?

Many studies have investigated the heritability of the EEG in twin studies and family studies (Vogel 1970), and found that many aspects of the EEG are heritable. In a meta-analysis van Beijsterveldt and van Baal (2002) demonstrated high heritability for measures such as the alpha peak frequency (81 %), alpha EEG power (79 %), P300 amplitude (60 %) and P300 latency (51 %), all suggesting that EEG and ERP parameters fulfil the definition of an endophenotype. Below two examples of EEG Phenotypes are discussed in more detail.

3.1 Low-Voltage (Alpha) EEG (LVA) and Alpha Power

LVA is the most well-described EEG phenotype to date and was first described by Adrian and Matthews (1934). The latter author exhibited an EEG in which alpha rhythm '*…may not appear at all at the beginning of an examination, and seldom persists for long without intermission…*'. The LVA EEG has been known to be heritable (autosomal dominant) and the heritability of alpha power is estimated at 79–93 % (Smit et al. 2005, 2010; Anokhin et al. 1992). Low-voltage EEG is a well-described endophenotype in anxiety and alcoholism (Enoch et al. 2003; Ehlers et al. 1999; Bierut et al. 2002). Alpha power and LVA have been successfully associated with a few chromosome loci (Ehlers et al. 1999; Enoch et al. 2008) but also with single genes: a serotonin receptor gene (HTR3B) (Ducci et al. 2009), corticotrophin-releasing binding hormone CRH-BP (Enoch et al. 2008), a gamma-amino butyric acid (GABA)-B receptor gene (Winterer et al. 2003) and with the BDNF Val66Met polymorphism (Gatt et al. 2008; Zoon et al. 2013).

3.2 Alpha Peak Frequency (APF)

The APF has been shown to be the most reproducible and heritable EEG characteristic (van Beijsterveldt and van Baal 2002; Smit et al. 2005; Posthuma et al. 2001) and has been associated with the COMT gene, with the Val/Val genotype marked by a 1.4 Hz slower APF as compared to the Met/Met group (Bodenmann et al. 2009) which could not be replicated in two large independent samples in our lab (Veth et al. submitted), casting doubt on this specific linkage and requiring further studies to unravel the genetic underpinnings of this measure.

In summary, the EEG has a long history in identifying biomarkers or endophenotypes aiding the prediction of treatment outcome and the EEG can be considered a stable, reproducible measure of brain activity with high heritability.

4 ADHD

Considerable research has been carried out for investigating the neurophysiology of ADHD. The first report describing EEG findings in 'behavior problem children' stems from 1938 (Jasper et al. 1938) when the authors described a distinct EEG pattern: '*…There were occasionally two or three waves also in the central or frontal regions at frequencies below what is considered the normal alpha range, that is, at frequencies of 5–6/s…*' (Jasper et al. 1938, p. 644), which we now know to be frontal theta, although the term theta was not introduced until 1944 by Walter and Dovey (1944). In this group of 'behavior problem children' they described a 'Class 1' as 'hyperactive, impulsive and highly variable' which closely resembles the current diagnosis of ADHD. The most predominant features in this group were the occurrence of slow waves above one or more regions and an 'abnormal EEG' in 83 % of the cases. Within 'Class 1' they also reported a subgroup which they termed as 'sub-alpha rhythm' with slow frontal regular activity which occurred in a similar way as the posterior alpha ('*…In other cases a 5–6/s rhythm would predominate in the anterior head regions simultaneous with an 8–10/s rhythm from the posterior regions…*'), thus already hinting at the heterogeneity of EEG findings that has continued to date and will be explained further below. Satterfield and colleagues (1971, 1973) were the first to investigate the potential use of EEG in predicting treatment outcome to stimulant medication. They found that children with excess slow wave activity and large amplitude evoked potentials were more likely to respond to stimulant medication (Satterfield et al. 1971) or, more generally, that abnormal EEG findings could be considered as predictor for positive treatment outcome (Satterfield et al. 1973). Below, the literature on ADHD will be reviewed in more detail focusing on some main subtypes for which at least replication studies have been published.

4.1 'Excess Theta' and 'Theta/Beta Ratio': Impaired Vigilance Regulation

The most consistent findings reported in the literature on ADHD since the introduction of quantitative EEG are those of increased absolute power in theta and increased theta/beta ratio (TBR). The clearest demonstration of the 'diagnostic utility' of this measure is from Monastra et al. (1999), who showed in a multi-centre study of 482 subjects that using a single electrode location (Cz) they could classify with an accuracy of 88 % children with ADHD based on the TBR. Note that most of these studies focused on the EEG as a diagnostic tool for ADHD, which is not automatically compatible with the notion of using the EEG for predictive purposes (as part of personalised medicine) as these two aims have conflicting implications, where the diagnostic use of EEG assumes homogeneity among patients with ADHD, while the predictive approach assumes heterogeneity ('A predictive biomarker is a baseline characteristic that categorises patients by their likelihood for response to a particular treatment' (Savitz et al. 2013)).

Three meta-analyses have investigated the diagnostic value of theta power and the TBR in ADHD compared to healthy controls. Boutros and colleagues (2005) concluded that increased theta power in ADHD is a sufficiently robust finding to warrant further development as a diagnostic test for ADHD, with data suggesting that relative theta power is an even stronger discriminator than absolute theta power. In 2006, Snyder and Hall conducted a meta-analysis specifically investigating the TBR, theta and beta and concluded that an elevated TBR is '*…a commonly observed trait in ADHD relative to controls… by statistical extrapolation, the effect size of 3.08 predicts a sensitivity and specificity of 94 %…*' (Snyder and Hall 2006, p. 453)). However, there is a problem with this extrapolation from an effect sizes (ES) to a sensitivity and specificity measure [see: (Arns et al. 2013a, b) for details] and hence these extrapolated values from Snyder and Hall (2006) should not be considered accurate. A recent meta-analysis incorporating more recent studies refines these findings further and shows a clear 'time effect' of studies, where earlier studies demonstrated the largest ES and more recent studies found the lowest ES between ADHD and non-ADHD groups (Arns et al. 2013a). This chronological effect in the findings was mostly related to the TBR being increased in the non-ADHD control groups which was interpreted by the authors as possibly being related to a decreasing sleep duration observed for non-ADHD children over time (Arns et al. 2013a, b; Iglowstein et al. 2003; Dollman et al. 2007) also found in a meta-analysis covering the last 100 years (Matricciani et al. 2011). Reduced sleep duration can result in prolonged sleep restriction, which results in increased fatigue and increased theta [see Arns and Kenemans (2012) for a review]. However, it was concluded that a substantial subgroup of ADHD patients (estimated between 26–38 %) are characterised by an increased TBR, even in recent studies (Arns et al. 2013a, b). Excess theta and elevated TBR are also favourable predictors for treatment outcome to stimulant medication (Arns et al. 2008; Clarke et al. 2002; Suffin and Emory 1995) and neurofeedback (Arns et al. 2012a; Monastra et al. 2002), thereby demonstrating the predictive value of this measure.

Conceptually, the EEG subtype with excess theta and/or enhanced TBR in ADHD are consistent with the EEG Vigilance model originally developed by Bente (1964) and further developed by Hegerl et al. (2012), which also overlaps with what is sometimes referred to as 'underarousal' and also with the EEG cluster described as 'cortical hypoarousal' (Clarke et al. 2011).

The EEG Vigilance framework can be regarded as an extension of the sleep stage model with a focus on eyes-closed resting period with transitions from relaxed wakefulness through drowsiness to sleep onset, which is seen in stage N2. The EEG allows classifying different functional brain stages at a time scale of, e.g. 1-s epochs, which reflect decreasing levels of vigilance from W to A1, A2, A3, B1, B2 and B3. W stage reflects a desynchronized low amplitude EEG which occurs, e.g. during arithmetic. The three A stages reflect stages where alpha activity is dominant posteriorly (A1), equally distributed (A2), followed by alpha anteriorisation (A3), whereas B stages are reflective of the lowest vigilance stages, which are characterised by an alpha drop-out or low-voltage EEG with slow horizontal eye movements (B1) followed by increased frontal theta and delta activity (B2/3). These vigilance stages are followed by sleep onset with the occurrence of K-complexes and sleep spindles, which mark the transition to stage C in the vigilance model, or classically to stage N2 sleep.

The sequence of EEG vigilance stages that can be assessed in an individual reflect the ability of relaxing or falling asleep. Due to its high temporal resolution of 1-s epochs it is sensitive to short drops of vigilance in contrast to traditional sleep medicine measures. Using a clustering method, three types of EEG vigilance regulation have been defined in a group of healthy subjects: a stable type, a slowly declining type and an unstable type (Olbrich et al. 2012). A stable or rigid EEG Vigilance regulation means that an individual remains in higher vigilance stages for an extended time and does not exhibit lower vigilance stages. This would be seen as rigid parietal/occipital alpha (stage A1), which is often seen in depression (Olbrich et al. 2012; Ulrich et al. 1990; Hegerl et al. 2012). On the other hand, unstable EEG Vigilance regulation suggests that an individual very quickly drops to lower EEG Vigilance stages, displaying the characteristic drowsiness EEG patterns such as frontal theta (stage B2/3), and they switch more often between EEG Vigilance stages. This labile or unstable pattern is often seen in ADHD (Sander et al. 2010). The often-reported 'excess theta' in ADHD mentioned above should thus be viewed as a predominance of the low B2/3 vigilance stages.

A summary of this model is depicted in Fig. 1. An unstable vigilance regulation explains the cognitive deficits that characterise ADHD and Attention Deficit Disorder (ADD), such as impaired sustained attention. Vigilance stabilisation behaviour explains the hyperactivity aspect of ADHD as an attempt to upregulate vigilance.

To summarise, in the majority of ADHD patients an EEG pattern is observed illustrative of a reduced and unstable vigilance regulation (i.e. the same EEG signature a healthy, but fatigued person would possibly demonstrate at the end of the day). The interpretation of increased theta activity as patterns of decreased tonic arousal suggests that the hyperactive behaviour of ADHD patients can be seen as a counter mechanism to auto-stabilisation via externalising behaviour that increases vigilance by riskful and sensation-seeking behaviour. Further, a decreased vigilance in a subgroup of patients with ADHD explains the positive effects of stimulant medication: vigilance is shifted to a high and stable level without the need for externalising behaviour. Interestingly, a similar pattern of

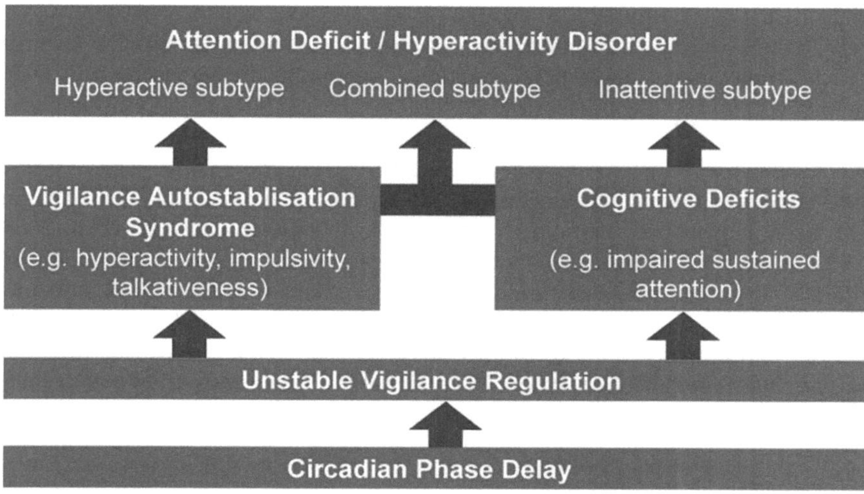

Fig. 1 Overview of the relation between an unstable vigilance regulation and behavioural symptoms of ADHD. Circadian phase delay can be considered one cause for the unstable vigilance regulation, but more in general sleep disorders are known to result in unstable vigilance regulation

reduced EEG vigilance can be found in manic patients, which sometimes also show patterns of reduced vigilance (Small et al. 1997) along with sensation-seeking behaviour. Again, this subtype with reduced vigilance seems responsive to stimulant medication (Schoenknecht et al. 2010).

Recent reviews are increasingly focusing on the role of sleep problems as the underlying aetiology of ADHD, in at least a subgroup of patients (Arns and Kenemans 2012; Miano et al. 2012). A majority of ADHD patients can be characterised by sleep onset insomnia, caused by a delayed circadian phase (van der Heijden et al. 2005; Van Veen et al. 2010). Although this cannot be considered a full-blown sleep disorder, chronic sleep onset insomnia can result in chronic sleep restriction which is known to result in impaired vigilance, attention and cognition (Van Dongen et al. 2003; Axelsson et al. 2008; Belenky et al. 2003). This is further evidenced by a recent meta-analysis incorporating data from 35,936 healthy children, reporting that sleep duration is positively correlated with school performance, executive function, and negatively correlated with internalising and externalising behaviour problems (Astill et al. 2012). Furthermore, it is known that symptoms associated with ADHD can be induced in healthy children by sleep restriction (Fallone et al. 2001; Golan et al. 2004), which also resulted in increased theta EEG power after a week of sleep restriction (effect size=0.53; Beebe et al. (2010)). These studies demonstrate that sustained sleep restriction results in impaired vigilance regulation (excess theta) as well as impaired attention, suggesting an overlap between ADHD symptoms and sleep disruptions. Chronobiological treatments normalising this delayed circadian phase, e.g. early morning bright light (Rybak et al. 2006) and sustained melatonin treatment (Hoebert et al. 2009) have been shown to normalise this sleep onset insomnia and also result in clinical improvement on ADHD symptoms. Therefore, this subgroup of ADHD patients with excess theta and elevated TBR is considered a group with impaired vigilance regulation caused by a delayed circadian phase (also see Fig. 1 and Arns and Kenemans (2012) for a review and Arns et al. (2013a, b)). Thereby it is understandable that vigilance stabilising treatments such as stimulant medication have been shown to be particularly effective in this subgroup (Arns et al. 2008; Clarke et al. 2002; Suffin and Emory 1995), whereas chronobiological treatments with sustained treatment (resulting in long-term normalisation) as well as neurofeedback treatment are also expected to be efficacious (Arns and Kenemans 2012).

Thus, conceptually, this excess theta subgroup can be interpreted as a subgroup with impaired vigilance regulation, likely caused by sleep restriction and/or other factors systematically influencing sleep duration.

4.2 The 'Slow Individual Alpha Peak Frequency' Subgroup

As pointed out above from the old Jasper et al. (1938) study in behavioural problem children, a cluster was identified which most closely resembles what we would now refer to as ADHD. In this 'Class 1' cluster they also reported an additional subgroup, which they termed a 'sub-alpha rhythm' with slow frontal regular activity, which occurred in a similar way as the posterior alpha ('*…In other cases a 5–6/s rhythm would predominate in the anterior head regions simultaneous with an 8–10/s rhythm from the posterior regions…*'). Nowadays, we would consider this a slowed Alpha Peak Frequency or slowed APF. Interestingly since the introduction of quantitative EEG in the 1960s, almost no studies have reported on the APF in ADHD whereas older studies have consistently reported on this measure (Arns 2012). Since it has been shown that ADHD children with a slow APF do not respond well to stimulant medication (Arns et al. 2008), whereas ADHD children with excess theta do (Clarke et al. 2002; Suffin and Emory 1995), it is crucial to dissociate these two different EEG subtypes, which tend to overlap in the EEG frequency domain. As pointed out by Arns et al. (2008) and further demonstrated in Lansbergen et al. (2011), the often-reported increased TBR in ADHD actually combines both the excess frontal theta group (interpreted as the 'impaired vigilance regulation subgroup') as well as a slow APF subgroup, due to the alpha frequency slowing to such a degree that it overlaps with the theta frequency band (4–8 Hz). Therefore, in addition to the limited validity of TBR presented above, this is a further reason why the TBR is probably not a specific measure since it incorporates different subtypes of ADHD. From a personalised medicine perspective this is not optimal, since these subtypes respond differentially to medication and are hypothesised to have a different underlying pathophysiology.

Several studies have now demonstrated that a slow APF is associated with non-response to several treatments such as stimulant medication in ADHD (Arns et al. 2008), rTMS in depression (Arns et al. 2010; Arns 2012), antidepressant medication (Ulrich et al. 1984), comorbid depressive symptoms in ADHD after neurofeedback (Arns et al. 2012a) and antipsychotic medication (Itil et al. 1975). Since alpha peak frequency is associated with a heritability of 81 % (van Beijsterveldt and van Baal 2002), this suggests that a slow APF might be considered a non-specific predictor or even endophenotype for non-response to treatments across a range of disorders. This subgroup comprises a substantial proportion of patients (28 % in ADHD: Arns et al. (2008), 17 % in depression: Arns et al. (2012b)) for whom currently no known treatment exists [see: Arns (2012) for a review].

4.3 Paroxysmal EEG Abnormalities and Epileptiform Discharges

Older studies preceding the era of quantitative EEG have mainly employed visual inspection of the EEG such as identification of epileptiform or paroxysmal activity, estimating the incidences of paroxysmal patterns in ADHD (or former diagnostic classes of ADHD) to be from between 12–15 % (Satterfield et al. 1973; Capute et al. 1968; Hemmer et al. 2001) to approximately 30 % (Hughes et al. 2000), which is high compared to 1–2 % in normal populations (Goodwin 1947; Richter et al. 1971). Note that these individuals did not present with convulsions and thus did not have a clinical diagnosis of epilepsy, but simply exhibited a paroxysmal EEG without a history of seizures. In autism, a prevalence of 46–86 % for paroxysmal EEG activity or epileptiform EEG abnormalities has been reported (Parmeggiani et al. 2010; Yasuhara 2010), hence the findings in the old research on 'abnormal' EEG might have been partly confounded by a subgroup with autism, since autism was not included as a diagnostic entity in the DSM until 1980 when the DSM-III was released.

The exact implications of this paroxysmal and epileptiform EEG activity in subjects without a history of clinical signs of seizures are not very well understood and it is good clinical practise not to treat these subjects with anticonvulsive medication ('*Treat the patient, not the EEG*'). In a very large study among healthy jet fighter pilots, Lennox-Buchthal et al. (1960) classified 6.4 % as 'marked and paroxysmally abnormal'. Moreover, they found that pilots with such EEGs were three times more likely to be involved in a plane crash due to pilot error, indicating that even though these people are not 'epileptic' their brains are 'not normal' and hence the presence of paroxysmal EEG continues to be an exclusion criterion for becoming a pilot to this day. It is interesting to note that several studies found that ADHD patients (Itil and Rizzo 1967; Davids et al. 2006; Silva et al. 1996) and patients with autism (Yasuhara 2010) do respond to anticonvulsant medication. The reported effect size for carbamazepine in the treatment of ADHD was 1.01, which is quite similar to the effect size for stimulant medication (Wood et al. 2007). Furthermore, some studies have demonstrated that interictal and/or subclinical spike activity has detrimental effects on neuropsychological, neurobehavioural, neurodevelopmental, learning and/or autonomic functions and some of these children with subclinical spike patterns do respond to anticonvulsant medication both with a reduction of spikes measured in the EEG and with improvements on memory and attention (Mintz et al. 2009). Like in other psychiatric disorders such as panic disorders (Adamaszek et al. 2011) these findings suggest the existence of a subgroup with paroxysmal EEG, who might better respond to anticonvulsant medication; however further research is required to substantiate this.

4.4 Excess Beta Subgroup

There is clear evidence for a subgroup of ADHD patients that are characterised by excess beta or beta-spindles, and make up 13–20 % (Chabot and Serfontein 1996; Clarke et al. 2001a). Several studies demonstrated that these patients do respond to stimulant medication (Clarke et al. 2003; Chabot et al. 1999; Hermens et al. 2005). Relatively little is known about this excess beta group and about the occurrence of beta-spindles. The latter are generally observed as a grapho-element that indicate sleep onset (AASM Manual) and can also be found in patients with mania (Small et al. 1997). Further, they occur as medication effect due to vigilance decreasing agents like benzodiazepines (Blume 2006) or barbiturates (Schwartz et al. 1971). Furthermore, Clarke et al. (2001b) reported this ADHD subgroup was more prone to moody behaviour and temper tantrums and Barry et al. (2009) reported that the ERP's of this subgroup differed substantially from ADHD children without excess beta, suggesting a different dysfunctional network explaining their complaints. Interestingly the ERP's of the excess beta subgroup appear more normal than those of the ADHD subgroup without excess beta.

Originally Gibbs & Gibbs in 1950 (see: Niedermeyer and Lopes da Silva 1993) distinguished two types of predominantly fast EEG, a moderate increased beta, which they termed 'F1' and a marked increased beta, which they termed 'F2'. Records of the F1 type were initially considered as 'abnormal' until the 1940s, whereas since that time Gibbs & Gibbs only considered the F2 type as 'abnormal'. However, currently electroencephalographers have shown a more lenient philosophy towards the interpretation of fast tracings (Niedermeyer and Lopes da Silva 1993, p. 161). At this moment, the only EEG pattern in the beta range considered abnormal is the 'paroxysmal fast activity' or 'beta band seizure pattern', which most often occurs during non-REM sleep, but also during waking (Stern and Engel 2004). This pattern is quite rare (4 in 3,000) and is most often seen in Lennox–Gastaut syndrome (Halasz et al. 2004). Vogel (1970) also described an EEG pattern of 'occipital slow beta waves' or also termed 'quick alpha variants 16–19/s' which responds in the same way as alpha to eyes opening and also has a similar topographic distribution. This pattern was only found in 0.6 % of a large population of healthy air force applicants and given its very low prevalence and occipital dominance, this subtype is unlikely the explanation of the 'excess beta' or 'beta spindling' subtype observed in ADHD. Therefore, the ADHD subgroup with excess beta or beta spindling (assuming the paroxysmal fast activity has been excluded) can neurologically be considered a 'normal variant'. However, neurophysiologically this can be considered a separate subgroup of ADHD, which does respond to stimulant medication (Chabot et al. 1999; Hermens et al. 2005). Probably, the 'beta-spindle' group also represents a subgroup with impaired vigilance '(see above), as beta-spindles are common signs for sleep onset. More research is required to investigate the exact underlying neurophysiology of this subtype and if other treatments could more specifically target this excess beta or beta spindling.

5 Depression

Lemere published the first description of EEG findings related to depression in 1936 (Lemere 1936). After inspecting the EEG of healthy people and several psychiatric patients he concluded: '*…The ability to produce "good" alpha waves seems to be a neurophysiological characteristic which is related in some way to the affective capacity of the individual…*'. This increased alpha power is to date still considered a hallmark of depression (e.g. see Itil (1983)) and recent studies suggest this endophenotype to be the mediator between the BDNF Val66Met polymorphism and trait depression (Gatt et al. 2008; Zoon et al. 2013).

One of the first attempts at using the EEG as a prognostic tool in depression stems from 1957. Roth et al. (1957) investigated barbiturate-induced EEG changes (delta increase) and found this predicted to some degree the long-term outcome (3–6 months) of ECT in depression. Many subsequent studies have demonstrated that greater 'induced' delta EEG power predicts favourable outcome to ECT (Ictal EEG power (Nobler et al. 2000); ECT-induced delta (Fink and Kahn 1957; Fink 2010; Volavka et al. 1972) and barbiturate-induced delta (Roth et al. 1957)). Or, as Max Fink concluded in a recent review, '*slowing of EEG rhythms was necessary for clinical improvement in ECT*' (Fink 2010).

5.1 Metabolic Activity in the Anterior Cingulate (ACC) and Other Structures

In 1997 Mayberg et al. [see: Mayberg et al. (1997)] reported that pre-treatment increased resting glucose metabolism of the rostral anterior cingulate (BA 24a/b) and predicted favourable treatment response to antidepressants. Two earlier studies already demonstrated a similar finding for the relation between increased ACC metabolism and response to sleep deprivation (Ebert et al. 1994; Wu et al. 1992), which was also confirmed in later studies (Smith et al. 1999; Wu et al. 1999). Since then this has sparked a huge research interest into the link between the ACC and treatment response in depression, and to date this is the most well-investigated finding in treatment prediction in depression. In order to integrate all these findings recently a meta-analysis was performed that included 23 studies (Pizzagalli 2011). Nineteen studies reported that responders to antidepressant treatments demonstrated increased ACC activity pre-treatment whereas the remaining four studies found the opposite. The overall effect size (ES) was a large effect size (ES=0.918). The relationship between increased ACC activity and favourable antidepressant response was found consistently across treatments (SSRI, TCA, ketamine, rTMS and sleep deprivation) and imaging modalities, and did not depend upon medication status at baseline (Pizzagalli 2011). No clear relationship between activity in the anterior cingulate and specific neurotransmitter systems has been reported (Mulert et al. 2007) and treatment-resistant depressive patients have also been shown to respond to deep brain stimulation of ACC areas (see: Hamani et al. (2011) for a review) suggesting that ACC activity reflects a reliable biomarker for antidepressant treatment response in general.

Most studies have used PET, SPECT and fMRI for assessing activity in the ACC. However with LORETA (low resolution brain electromagnetic tomography) as an algorithm for computation of intracortical EEG source estimates it is also possible to assess activity in the ACC using scalp-EEG time series (Pascual-Marqui et al. 1994). Increased theta in the ACC assessed with LORETA has been shown to reflect increased metabolism in the ACC (Pizzagalli et al. 2003). Furthermore, several studies have used this technique to probe ACC activity successfully [reviewed in Pizzagalli (2011)].

5.2 EEG Markers in Depression

In QEEG research, various pre-treatment differences in EEG measures have been reported to be associated with improved antidepressant treatment outcomes. The following summarises findings that have been replicated in at least one study and relate to baseline measures predicting treatment outcome. It should be the goal to identify biomarkers that not only yield valid and effective prediction of treatment response but also can be linked to the underlying pathomechanisms of depression. Only a marker that can be integrated into the prevailing view of pathogenesis or even widens the scope of our understanding will be trusted in the field of clinical routine diagnostic. Therefore EEG research on prediction biomarkers has to bridge the gap between the mere analyses of electrophysiological time series on the one side and psychopathology, behaviour and clinical picture on the other side.

Decreased theta has consistently been reported to be related to a favourable treatment outcome to different antidepressant treatments (Arns et al. 2012b;

Iosifescu et al. 2009; Olbrich and Arns 2013) (with the exception of (Cook et al. 1999)) as well as lower delta power (Knott et al. 2000). Given that most LORETA studies found an association between increased theta in the ACC and treatment response, these findings appear contradictory. However, Knott et al. (2000) and Arns et al. (2012b) analysed the EEG activity across all sites and Iosifescu et al. (2009) only looked at Fp1, Fpz and Fp2. Given that frontal-midline theta has been localised to the medial pre-frontal cortex and anterior cingulate (Ishii et al. 1999; Asada et al. 1999), one would thus expect that only frontal-midline sites would reflect the increased theta, which was indeed reported by Spronk and colleagues who found increased theta at Fz to be associated with favourable treatment outcome (Spronk et al. 2011). Hence these findings have to be interpreted in that increased generalised slow EEG power is a predictor for non-response, whereas increased ACC theta or frontal-midline theta is a positive predictor for response. These reflect different types of theta activity: ACC theta, also referred to as phasic theta, reflective of frontal-midline theta related to information processing versus tonic theta, reflective of widespread frontal theta and related to drowsiness or unstable vigilance regulation (for a review of the different roles of tonic and phasic theta refer to [Klimesch 1999]).

Hegerl et al. (2012) and Olbrich et al. (2012) demonstrated a clear difference in EEG vigilance regulation in patients with depression compared to matched controls. Depressed patients exhibited a hyperstable vigilance regulation expressed by increased A1 stages (parietal alpha) and decreased B2/3 and C stages (frontal theta) which is consistent with a study by Ulrich and Fürstenberg (1999) and other studies demonstrating increased parietal alpha (Itil 1983; Pollock and Schneider 1990), as first observed by Lemere (1936). Vogel (1970) described a pattern of 'Monotonous High Alpha Waves', with a simple autosomal dominance of inheritance. The description of this EEG pattern found by Vogel ('Kontinuität') is very similar to the 'hyperrigid' or 'hyperstable' EEG vigilance found by Hegerl and Hensch (2012) and hence suggests this indeed reflects a 'trait' like EEG vigilance regulation.

Furthermore, increased pre-treatment alpha has been associated with improved treatment outcome to antidepressant medication (Ulrich et al. 1984; Bruder et al. 2001; Tenke et al. 2011) and most antidepressants also result in a decrease of alpha activity [see: Itil (1983) for an overview]. Therefore, the subgroup of non-responders characterised by frontal theta might be interpreted as a subgroup characterised by a *decreased* EEG vigilance regulation (Hegerl and Hensch 2012; Olbrich and Arns 2013), as opposed to the typically reported *increased* or *hyperstable* vigilance regulation ('hyperstable' parietal alpha). Given that patients with a decreased EEG vigilance regulation respond better to stimulant medication (manic depression: Hegerl et al. 2010; Bschor et al. 2001; Schoenknecht et al. 2010; ADHD: Arns et al. 2008; Sander et al. 2010), it is tempting to speculate whether this subgroup of non-responders might respond better to stimulant medication or other vigilance stabilising treatments. Although a recent Cochrane review did report significant improvements of depressive and fatigue symptoms for short-term stimulant medication as add on therapy in depressed patients (Candy et al. 2008), the clinical significance remained unclear and there were very few controlled studies which could be included, thus limiting the generality of this finding (Candy et al. 2008). However, stimulant medication has been applied successfully in a subgroup of depression with excess theta by Suffin and Emory (1995), which was replicated in a prospective randomised controlled trial (Debattista et al. 2010). Therefore, along the same lines as discussed above in relation to sleep problems as the core pathophysiology of ADHD, future research should focus on investigating EEG vigilance regulation and the existence of sleep problems in this subgroup of non-responders in order to develop an appropriate treatment for these patients, who are found to be non-responders to gold-standard antidepressant treatments.

In summary, responders to antidepressant treatments such as antidepressants and rTMS are generally characterised by increased parieto-occipital alpha (or a 'hyperstable' vigilance regulation) and increased theta in the rostral anterior cingulate (Pizzagalli 2011) reflected as frontal-midline theta. A subgroup of non-responders to antidepressant treatments are characterised by generalised increased frontal theta reflective of decreased EEG vigilance regulation. It is hypothesised that this latter group might be better responders to vigilance stabilising treatments such as psychostimulants or chronobiological treatments such as melatonin or early morning bright light.

5.3 Alpha Peak Frequency in Depression

In one of the earliest studies investigating EEG predictors of treatment response in depression, Ulrich et al. (1984) found that non-responders to a tricyclic antidepressant (TCA), specifically amitryptiline, and pirlindole (a tetracyclic compound) demonstrated slower APF (8 Hz) as compared to responders (9.5 Hz). Furthermore, they also found that after 4 weeks of treatment only responders demonstrated an increase of 0.5 Hz in their APF, whereas the non-responders did not. More recently, it has also been shown that depressed patients with a pretreatment slow APF also respond less well to rTMS (Arns et al. 2010, 2012b). Furthermore, as discussed above, a slow APF could represent a generic biomarker for non-response.

5.4 Treatment Emergent or Pharmacodynamic Biomarkers in Depression

The measures discussed above all involved baseline measures, which were investigated for their capability of predicting treatment outcome. However, another well-investigated line of research relates to 'treatment emergent biomarkers' or 'pharmacodynamic' biomarkers (Savitz et al. 2013) which measure the EEG at baseline and subsequently after treatment for several days, with the changes used to predict treatment outcome. This approach has been mainly applied to antidepressants as, given that this class of drugs generally takes 4–6 weeks to demonstrate its clinical effects, knowing whether a drug is likely to prove efficacious within several days has clinical relevance. Two of these methods will be discussed in more detail in the following, namely EEG cordance and the Antidepressant Treatment Response.

5.5 EEG Cordance

The EEG cordance method was initially developed by Leuchter and colleagues to provide a measure, which had face-validity for the detection of cortical deafferentation (Leuchter et al. 1994a, b). They observed that the EEG over a white-matter lesion often exhibited decreased absolute theta power, but increased relative theta power, which they termed 'discordant'. Therefore, the EEG cordance method combines both absolute and relative EEG power. Negative values of this measure (discordance)—specifically in theta or beta—reflect low perfusion or metabolism, whereas positive values (concordance)—specifically in alpha—reflect high perfusion or metabolism (Leuchter et al. 1994a, b). This has been confirmed by comparing cordance EEG with simultaneous measuring perfusion employing PET scans (Leuchter et al. 1999).

In a first study, it was found that depressive patients characterised by a 'discordant' brain state at baseline could be characterised as non-responders (Cook et al. 1999). Subjects were classified into 'discordant' if >30 % of all electrodes exhibited discordance or if fewer electrodes that are highly deviant. Furthermore, central (Cz, FC1, FC2) theta cordance was related to treatment outcome after ECT (Stubbeman et al. 2004). More recent studies have focused on EEG cordance in the theta frequency band at pre-frontal electrodes (Fp1, Fp2, Fpz) and have found that theta cordance *change (decrease)* across 48 h to 2 weeks of treatment predicted longer-term treatment outcome (Cook et al. 2002, 2005). In an independent replication study, Bares et al. (2007, 2008, 2010) also found that responders were characterised by a decrease in pre-frontal (Fp1, Fp2, Fz) theta cordance after 1 week. Furthermore, Cook et al. (2005) demonstrated that a medication wash-out period for assessing the quantitative EEG is not critical in reliably using EEG cordance. This further suggests that change in frontal theta cordance is a reflection of the early beneficial effects of the treatment and is hence not dependent upon treatment type since the same cordance effects have been observed with SSRI, SNRI, TCA, rTMS and ECT. Across studies of depressive patients treated with various antidepressant medications, decreases in pre-frontal theta cordance 1 week after start of medication have consistently predicted response, with overall accuracy ranging from 72 to 88 % (Iosifescu et al. 2009).

A pre-frontal theta cordance *increase* was found in placebo-responders (Leuchter et al. 2002). A more recent study from this group refined this further by examining right-medial frontal sites and found that theta cordance after 1 week was only decreased in the medication responders but not in the placebo-responders (Cook et al. 2009), hence demonstrating specificity of this measure to treatment outcome and not to placebo response.

As a limitation of this measure it should be noted that the mentioned mixture of absolute and relative EEG power values for calculation of the cordance measure lowers the possibility for interpretation of the underlying neuronal activities (Kuo and Tsai 2010).

5.6 Antidepressant Treatment Response

The ATR measure was also developed by Leuchter and colleagues (2009a, b) and is commercialised by Aspect Medical Systems. The first results of this measure were published in 2009 by Iosifescu et al. (2009), demonstrating that the ATR measure was able to predict treatment outcome to an SSRI or Velafaxine with an accuracy of 70 % (82 % sensitivity; 54 % specificity). Recently, the results of a large clinical trial (BRITE-MD) investigating the ATR were published (Leuchter et al. 2009a, b). This measure is based on EEG recorded from Fpz (FT7 and FT8) and is the non-linear weighted combination of (1) combined relative alpha and theta (3–12 Hz/2–20 Hz) at baseline and (2) the difference between absolute alpha1 power (8.5–12 Hz) at baseline and absolute alpha2 power (9–11.5 Hz) after 1 week of treatment (Leuchter et al. 2009a, b). It was demonstrated that a high ATR value predicted response to an SSRI with 74 % overall accuracy (58 % sensitivity, 91 % specificity). Interestingly, in another study, they reported that patients with a low ATR responded better to the atypical antidepressant bupropion (Leuchter et al. 2009a, b) thereby demonstrating that this measure identified two subgroups of depressive patients with subsequent implications for two types of antidepressants.

The disadvantage of this method is that patients already need to be prescribed the medication before any prediction can be made and this method could not be used on 15 % of the patients due to ECG artefacts (Leuchter et al. 2009a, b), hence also reflecting a ‘treatment emergent biomarker’.

6 Conclusion

Much research has been conducted in ADHD and depression to investigate the potential of predicting treatment outcome using EEG as a marker, and the results are promising. The next step would be to integrate these different metrics further, make advantage of the different information they provide about the underlying neuronal activity and investigate the similarities and differences in order to further our knowledge, so that EEG- and ERP-based data can be used in practise to predict treatment outcome. Finally, some examples have been presented where the identification of EEG-based subgroups sheds more light on the underlying pathology of the disease state, and can thus be used to develop more effective treatments for the different subgroups.

Acknowledgment This work has been adapted and updated from the PhD thesis: 'Personalized medicine in ADHD and Depression: A quest for EEG treatment predictors' by Martijn Arns.

References

Adamaszek M, Olbrich S, Gallinat J (2011) The diagnostic value of clinical EEG in detecting abnormal synchronicity in panic disorder. Clin EEG Neurosci 42(3):166–174

Adrian ED, Matthews BHC (1934) The Berger rhythm: potential changes from the occipital lobes in man. Brain: J Neurol 57(4):355

Anokhin A, Steinlein O, Fischer C, Mao Y, Vogt P, Schalt E, Vogel F (1992) A genetic study of the human low-voltage electroencephalogram. Hum Genet 90(1–2):99–112

Arns M (2012) EEG-based personalized medicine in ADHD: Individual alpha peak frequency as an endophenotype associated with nonresponse. J Neurotherapy 16(2):123–141

Arns M, Kenemans JL (2012) Neurofeedback in ADHD and insomnia: vigilance stabilization through sleep spindles and circadian networks. Neurosci Biobehav Rev. doi:10.1016/j.neubiorev.2012.10.006

Arns M, Conners CK, Kraemer HC (2013a) A decade of EEG theta/beta ratio research in ADHD: a meta-analysis. J Attention Disord 17(5):374–383

Arns M, Drinkenburg WHIM, Kenemans JL (2012a) The effects of QEEG-informed neurofeedback in ADHD: an open-label pilot study. Appl Psychophysiol Biofeedback 37(3):171–180

Arns M, Drinkenburg WHIM, Fitzgerald PB, Kenemans JL (2012b) Neurophysiological predictors of non-response to rTMS in depression. Brain Stimulation 5:569–576

Arns M, Gunkelman J, Breteler M, Spronk D (2008) EEG phenotypes predict treatment outcome to stimulants in children with ADHD. J Integr Neurosci 7(3):421–438

Arns M, Spronk D, Fitzgerald PB (2010) Potential differential effects of 9 and 10 Hz rTMS in the treatment of depression. Brain Stimulation 3(2):124–126

Arns M, van der Heijden KB, Arnold LE, Kenemans JL (2013b) Geographic variation in the prevalence of attention-deficit/hyperactivity disorder: the sunny perspective. Biol Psychiatry. doi:10.1016/j.biopsych.2013.02.010

Asada H, Fukuda Y, Tsunoda S, Yamaguchi M, Tonoike M (1999) Frontal midline theta rhythms reflect alternative activation of prefrontal cortex and anterior cingulate cortex in humans. Neurosci Lett 274(1):29–32

Astill RG, Van der Heijden KB, Van Ijzendoorn MH, Van Someren EJ (2012) Sleep, cognition, and behavioral problems in school-age children: a century of research meta-analyzed. Psychol Bull 138:155–159

Axelsson J, Kecklund G, Åkerstedt T, Donofrio P, Lekander M, Ingre M (2008) Sleepiness and performance in response to repeated sleep restriction and subsequent recovery during semi-laboratory conditions. Chronobiol Int 25(2–3):297–308

Bagby RM, Ryder AG, Cristi C (2002) Psychosocial and clinical predictors of response to pharmacotherapy for depression. J Psychiatry Neurosci: JPN 27(4):250–257

Bares M, Brunovsky M, Kopecek M, Novak T, Stopkova P, Kozeny J, Sos P, Krajca V, Höschl C (2008) Early reduction in prefrontal theta QEEG cordance value predicts response to

venlafaxine treatment in patients with resistant depressive disorder. European Psychiatry: J Assoc Eur Psychiatrists 23(5):350–355
Bares M, Brunovsky M, Kopecek M, Stopkova P, Novak T, Kozeny J, Höschl C (2007) Changes in QEEG prefrontal cordance as a predictor of response to antidepressants in patients with treatment resistant depressive disorder: a pilot study. J Psychiatr Res 41(3–4):319–325
Bares M, Brunovsky M, Novak T, Kopecek M, Stopkova P, Sos P, Krajca V, Höschl C (2010) The change of prefrontal QEEG theta cordance as a predictor of response to bupropion treatment in patients who had failed to respond to previous antidepressant treatments. Eur Neuropsychopharmacol 20(7):459–466
Barry RJ, Clarke AR, McCarthy R, Selikowitz M, Brown CR (2009) Event-related potentials in children with attention-deficit/hyperactivity disorder and excess beta activity in the EEG. Acta Neuropsychologica 7(4):249
Beebe DW, Rose D, Amin R (2010) Attention, learning, and arousal of experimentally sleep-restricted adolescents in a simulated classroom. J Adolesc Health 47(5):523–525
Belenky G, Wesensten NJ, Thorne DR, Thomas ML, Sing HC, Redmond DP, Russo MB, Balkin TJ (2003) Patterns of performance degradation and restoration during sleep restriction and subsequent recovery: a sleep dose-response study. J Sleep Res 12(1):1–12
Bente D (1964) Die Insuffizienz des Vigilitätstonus eine klinische und elektroencephalographische Studie zum Aufbau narkoleptischer und neurasthenischer Syndrome, Habilitationsschrift
Bierut LJ, Saccone NL, Rice JP, Goate A, Foroud T, Edenberg H, Almasy L, Conneally PM, Crowe R, Hesselbrock V, Li TK, Nurnberger J, Porjesz B, Schuckit MA, Tischfield J, Begleiter H, Reich T (2002) Defining alcohol-related phenotypes in humans, the collaborative study on the genetics of alcohol. Alcohol Res Health 26(3):208–213
Blume WT (2006) Drug effects on EEG. J Clin Neurophysiol 23(4):306
Bodenmann S, Rusterholz T, Dürr R, Stoll C, Bachmann V, Geissler E, Jaggi-Schwarz K, Landolt HP (2009) The functional Val158Met polymorphism of COMT predicts interindividual differences in brain alpha oscillations in young men. J Neurosci 29(35):10855–10862
Boutros N, Fraenkel L, Feingold A (2005) A four-step approach for developing diagnostic tests in psychiatry: EEG in ADHD as a test case. J Neuropsychiatry Clin Neurosci 17(4):455–464
Bruder GE, Stewart JW, Tenke CE, McGrath PJ, Leite P, Bhattacharya N, Quitkin FM (2001) Electroencephalographic and perceptual asymmetry differences between responders and nonresponders to an SSRI antidepressant. Biol Psychiatry 49(5):416–425
Bruder GE, Tenke CE, Stewart JW, McGrath PJ, Quitkin FM (1999) Predictors of therapeutic response to treatments for depression: a review of electrophysiologic and dichotic listening studies. CNS Spectr 4(8):30–36
Bschor T, Müller-Oerlinghausen B, Ulrich G (2001) Decreased level of EEG-vigilance in acute mania as a possible predictor for a rapid effect of methylphenidate: a case study. Clinical EEG (Electroencephalography) 32(1):36–39
Candy M, Jones L, Williams R, Tookman A, King M (2008) Psychostimulants for depression. *Cochrane Database* Syst Rev (Online) 16(2): p. CD006722
Capute AJ, Niedermeyer EFL, Richardson F (1968) The electroencephalogram in children with minimal cerebral dysfunction. Pediatrics 41(6):1104
Chabot RJ, Serfontein G (1996) Quantitative electroencephalographic profiles of children with attention deficit disorder. Biol Psychiatry 40(10):951–963
Chabot RJ, Orgill AA, Crawford G, Harris MJ, Serfontein G (1999) Behavioral and electrophysiologic predictors of treatment response to stimulants in children with attention disorders. J Child Neurol 14(6):343–351
Clarke AR, Barry RJ, Dupuy FE, Heckel LD, McCarthy R, Selikowitz M, Johnstone SJ (2011) Behavioural differences between EEG-defined subgroups of children with attention-deficit/hyperactivity disorder. Clin Neurophysiol 122(7):1333–1341
Clarke AR, Barry RJ, McCarthy R, Selikowitz M (1998) EEG analysis in attention-deficit/hyperactivity disorder: a comparative study of two subtypes. Psychiatry Res 81(1):19–29
Clarke AR, Barry RJ, McCarthy R, Selikowitz M (2001a) EEG-defined subtypes of children with attention-deficit/hyperactivity disorder. Clin Neurophysiol 112(11):2098–2105

Clarke AR, Barry RJ, McCarthy R, Selikowitz M (2001b) Excess beta activity in children with attention-deficit/hyperactivity disorder: an atypical electrophysiological group. Psychiatry Res 103(2–3):205–218

Clarke AR, Barry RJ, McCarthy R, Selikowitz M, Croft RJ (2002) EEG differences between good and poor responders to methylphenidate in boys with the inattentive type of attention-deficit/hyperactivity disorder. Clin Neurophysiol 113(8):1191–1198

Clarke AR, Barry RJ, McCarthy R, Selikowitz M, Clarke DC, Croft RJ (2003) Effects of stimulant medications on children with attention-deficit/hyperactivity disorder and excessive beta activity in their EEG. Clin Neurophysiol 114(9):1729–1737

Cook IA, Hunter AM, Abrams M, Siegman B, Leuchter AF (2009) Midline and right frontal brain function as a physiologic biomarker of remission in major depression. Psychiatry Res: Neuroimaging 174:152–157

Cook IA, Leuchter AF, Morgan M, Witte E, Stubbeman WF, Abrams M, Rosenberg S, Uijtdehaage SH (2002) Early changes in prefrontal activity characterize clinical responders to antidepressants. Neuropsychopharmacology 27(1):120–131

Cook IA, Leuchter AF, Morgan ML, Stubbeman W, Siegman B, Abrams M (2005) Changes in prefrontal activity characterize clinical response in SSRI nonresponders: a pilot study. J Psychiatr Res 39(5):461–466

Cook IA, Leuchter AF, Witte E, Abrams M, Uijtdehaage SH, Stubbeman W, Rosenberg-Thompson S, Anderson-Hanley C, Dunkin JJ (1999) Neurophysiologic predictors of treatment response to fluoxetine in major depression. Psychiatry Res 85(3):263–273

Cuijpers P, Reynolds CF, Donker T, Li J Andersson G, Beekman A (2012) Personalized treatment of adult depression: medication, psychotherapy or both? a systematic review. Depression and anxiety 29(10):855–864

Cuijpers P, Smit F, Bohlmeijer E, Hollon SD, Andersson G (2010) Efficacy of cognitive-behavioural therapy and other psychological treatments for adult depression: meta-analytic study of publication bias. Br J Psychiatry 196:173–178

Davids E, Kis B, Specka M, Gastpar M (2006) A pilot clinical trial of oxcarbazepine in adults with attention-deficit hyperactivity disorder. Prog Neuropsychopharmacol Biol Psychiatry 30(6):1033–1038

Debattista C, Kinrys G, Hoffman D, Goldstein C, Zajecka J, Kocsis J, Teicher M, Potkin S, Preda A, Multani G, Brandt L, Schiller M, Iosifescu D, Fava M (2010) The use of referenced-EEG (rEEG) in assisting medication selection for the treatment of depression. J Psychiatr Res 45(1):64–75

Dollman J, Ridley K, Olds T, Lowe E (2007) Trends in the duration of school-day sleep among 10–15 year-old South Australians between 1985 and 2004. Acta Paediatr 96(7):1011–1014

Ducci F, Enoch MA, Yuan Q, Shen PH, White KV, Hodgkinson C, Albaugh B, Virkkunen M, Goldman D (2009) HTR3B is associated with alcoholism with antisocial behavior and alpha EEG power–an intermediate phenotype for alcoholism and co-morbid behaviors. Alcohol 43(1):73–84

Ebert D, Feistel H, Barocka A, Kaschka W (1994) Increased limbic blood flow and total sleep deprivation in major depression with melancholia. Psychiatry Res 55(2):101–109

Ehlers CL, Garcia-Andrade C, Wall TL, Cloutier D, Phillips E (1999) Electroencephalographic responses to alcohol challenge in native American mission Indians. Biol Psychiatry 45(6):776–787

Enoch MA, Schuckit MA, Johnson BA, Goldman D (2003) Genetics of alcoholism using intermediate phenotypes. Alcohol Clin Exp Res 27(2):169–176

Enoch MA, Shen PH, Ducci F, Yuan Q, Liu J, White KV, Albaugh B, Hodgkinson CA, Goldman D (2008) Common genetic origins for EEG, alcoholism and anxiety: the role of CRH-BP. PLoS ONE 3(10):e3620

Fallone G, Acebo C, Arnedt JT, Seifer R, Carskadon MA (2001) Effects of acute sleep restriction on behavior, sustained attention, and response inhibition in children. Percept Mot Skills 93(1):213–229

Fink M (2010) Remembering the lost neuroscience of pharmaco-EEG. Acta Psychiatr Scand 121(3):161–173

Fink M, Kahn RL (1957) Relation of electroencephalographic delta activity to behavioral response in electroshock; quantitative serial studies. AMA Arch Neurol Psychiatry 78(5):516–525

Frieling H, Tadić A, (2013) Value of genetic and epigenetic testing as biomarkers of response to antidepressant treatment. Int Rev Psychiatry (Abingdon, England) 25(5):572–578

Gatt JM, Kuan SA, Dobson-Stone C, Paul RH, Joffe RT, Kemp AH, Gordon E, Schofield PR, Williams LM (2008) Association between BDNF Val66Met polymorphism and trait depression is mediated via resting EEG alpha band activity. Biol Psychol 79(2):275–284

Golan N, Shahar E, Ravid S, Pillar G (2004) Sleep disorders and daytime sleepiness in children with attention-deficit/hyperactive disorder. Sleep 27(2):261–266

Goodwin JE (1947) The significance of alpha variants in the EEG, and their relationship to an epileptiform syndrome. Am J Psychiatry 104(6):369–379

Gottesman II, Gould TD (2003) The endophenotype concept in psychiatry: etymology and strategic intentions. Am J Psychiatry 160(4):636–645

De Gruttola VG, Clax P, DeMets DL, Downing GJ, Ellenberg SS, Friedman L, Gail MH, Prentice R, Wittes J, Zeger SL (2001) Considerations in the evaluation of surrogate endpoints in clinical trials, summary of a national institutes of health workshop. Control Clin Trials 22(5):485–502

Halasz P, Janszky J, Barcs G, Szcs A (2004) Generalised paroxysmal fast activity (GPFA) is not always a sign of malignant epileptic encephalopathy. Seizure 13(4):270–276

Hamani C, Mayberg H, Stone S, Laxton A, Haber S, Lozano AM (2011) The subcallosal cingulate gyrus in the context of major depression. Biol Psychiatry 69(4):301–308

Hefner G, Laib AK, Sigurdsson H, Hohner M, Hiemke C (2013) The value of drug and metabolite concentration in blood as a biomarker of psychopharmacological therapy. Int Rev Psychiatry 25(5):494–508

Hegerl U, Hensch T (2012) The vigilance regulation model of affective disorders and ADHD. Neurosci Biobehav Rev. doi:10.1016/j.neubiorev.2012.10.008

Hegerl U, Himmerich H, Engmann B, Hensch T (2010) Mania and attention-deficit/hyperactivity disorder: common symptomatology, common pathophysiology and common treatment? Curr Opin Psychiatry 23(1):1–7

Hegerl U, Wilk K, Olbrich S, Schoenknecht P, Sander C (2012) Hyperstable regulation of vigilance in patients with major depressive disorder. World J Biol Psychiatry 13(6):436–446

Hemmer SA, Pasternak JF, Zecker SG, Trommer BL (2001) Stimulant therapy and seizure risk in children with ADHD. Pediatr Neurol 24(2):99–102

Hermens DF, Cooper NJ, Kohn M, Clarke S, Gordon E (2005) Predicting stimulant medication response in ADHD: evidence from an integrated profile of neuropsychological, psychophysiological and clinical factors. J Integr Neurosci 4(1):107–121

Hoebert M, van der Heijden KB, van Geijlswijk IM, Smits MG (2009) Long-term follow-up of melatonin treatment in children with ADHD and chronic sleep onset insomnia. J Pineal Res 47(1):1–7

Hughes JR, DeLeo AJ, Melyn MA (2000) The electroencephalogram in attention deficit-hyperactivity disorder: emphasis on epileptiform discharges. Epilepsy Behav: E&B 1(4): 271–277

Iglowstein I, Jenni G, Molinari L, Largo H (2003) Sleep duration from infancy to adolescence: reference values and generational trends. Pediatrics 111(2):302–307

Iosifescu DV, Greenwald S, Devlin P, Mischoulon D, Denninger JW, Alpert JE, Fava M (2009) Frontal EEG predictors of treatment outcome in major depressive disorder. Eur Neuropsychopharmacol 19(11):772–777

Ishii R, Shinosaki K, Ukai S, Inouye T, Ishihara T, Yoshimine T, Hirabuki N, Asada H, Kihara T, Robinson SE, Takeda M (1999) Medial prefrontal cortex generates frontal midline theta rhythm. NeuroReport 10(4):675–679

Itil TM (1983) The discovery of antidepressant drugs by computer-analyzed human cerebral bio-electrical potentials (CEEG). Prog Neurobiol 20(3–4):185–249

Itil TM, Rizzo AE (1967) Behavior and quantitative EEG correlations during treatment of behavior-disturbed adolescents. Electroencephalogr Clin Neurophysiol 23(1):81

Itil TM, Marasa J, Saletu B, Davis S, Mucciardi AN (1975) Computerized EEG: predictor of outcome in schizophrenia. J Nerv Ment Dis 160(3):118–203

Jasper HH, Solomon P, Bradley C (1938) Electroencephalographic analyses of behavior problem children. Am J Psychiatry 95(3):641

Ji Y, Hebbring S, Zhu H, Jenkins GD, Biernacka J, Snyder K, Drews M, Fiehn O, Zeng Z, Schaid D, Mrazek DA, Kaddurah-Daouk R, Weinshilboum RM (2011) Glycine and a glycine dehydrogenase (GLDC) SNP as citalopram/escitalopram response biomarkers in depression: pharmacometabolomics-informed pharmacogenomics. Clin Pharmacol Ther 89(1):97–104

Johnson CH, Gonzalez FJ (2012) Challenges and opportunities of metabolomics. J Cell Physiol 227(8):2975–2981

Joyce PR, Paykel ES (1989) Predictors of drug response in depression. Arch Gen Psychiatry 46(1):89–99

Knott V, Mahoney C, Kennedy S, Evans K (2000) Pre-treatment EEG and it's relationship to depression severity and paroxetine treatment outcome. Pharmacopsychiatry 33(6):201–205

Kuo CC, Tsai JF (2010) Cordance or antidepressant treatment response (ATR) index?. Psychiatry Res 180(1):60 (author reply 61–2)

Lander ES (2011) Initial impact of the sequencing of the human genome. Nature 470(7333): 187–197

Lansbergen M, Arns M, van Dongen-Boomsma M, Spronk D, Buitelaar JK (2011) The increase in theta/beta ratio on resting state EEG in boys with attention-deficit/hyperactivity disorder is mediated by slow alpha peak frequency. Prog Neuropsychopharmacol Biol Psychiatry 35:47–52

Lemere F (1936) The significance of individual differences in the Berger rhythm. Brain: J Neurol 59:366–75

Lennox-Buchtal M, Buchtal F, Rosenfalck P (1960) Correlation of electroencephalographic findings with crash rate of military jet pilots. Epilepsia 1:366–372

Leuchter AF, Cook IA, Gilmer WS, Marangell LB, Burgoyne KS, Howland RH, Trivedi MH, Zisook S, Jain R, Fava M, Iosifescu D, Greenwald S (2009a) Effectiveness of a quantitative electroencephalographic biomarker for predicting differential response or remission with escitalopram and bupropion in major depressive disorder. Psychiatry Res 169(2):132–138

Leuchter AF, Cook IA, Lufkin RB, Dunkin J, Newton TF, Cummings JL, Mackey JK, Walter DO (1994a) Cordance: a new method for assessment of cerebral perfusion and metabolism using quantitative electroencephalography. NeuroImage 1(3):208–219

Leuchter AF, Cook IA, Marangell LB, Gilmer WS, Burgoyne KS, Howland RH, Trivedi MH, Zisook S, Jain R, McCracken JT, Fava M, Iosifescu D, Greenwald S (2009b) Comparative effectiveness of biomarkers and clinical indicators for predicting outcomes of SSRI treatment in major depressive disorder: results of the BRITE-MD study. Psychiatry Res 169(2):124–131

Leuchter AF, Cook IA, Mena I, Dunkin JJ, Cummings JL, Newton TF, Migneco O, Lufkin RB, Walter DO, Lachenbruch PA (1994b) Assessment of cerebral perfusion using quantitative EEG cordance. Psychiatry Res 55(3):141–152

Leuchter AF, Cook IA, Witte EA, Morgan M, Abrams M (2002) Changes in brain function of depressed subjects during treatment with placebo. Am J Psychiatry 159(1):122–129

Leuchter AF, Uijtdehaage SH, Cook IA, O'Hara R, Mandelkern M (1999) Relationship between brain electrical activity and cortical perfusion in normal subjects. Psychiatry Res 90(2):125–140

Matricciani L, Olds T, Williams M (2011) A review of evidence for the claim that children are sleeping less than in the past. Sleep 34(5):651–659

Mayberg HS, Brannan SK, Mahurin RK, Jerabek PA, Brickman JS, Tekell JL, Silva JA, McGinnis S, Glass TG, Martin CC, Fox PT (1997) Cingulate function in depression: a potential predictor of treatment response. NeuroReport 8(4):1057–1061

Menke A (2013) Gene expression: biomarker of antidepressant therapy? Int Rev Psychiatry 25(5):579–591

Miano S, Parisi P, Villa MP (2012) The sleep phenotypes of attention deficit hyperactivity disorder: the role of arousal during sleep and implications for treatment. Med Hypotheses 79(2):147–53

Miller G (2010) Is pharma running out of brainy ideas? Science 329(5991):502

Mintz M, Legoff D, Scornaienchi J, Brown M, Levin-Allen S, Mintz P, Smith C (2009) The underrecognized epilepsy spectrum: the effects of levetiracetam on neuropsychological functioning in relation to subclinical spike production. J Child Neurol 24(7):807–815

Molina BS, Hinshaw SP, Swanson JM, Arnold LE, Vitiello B, Jensen PS, Epstein JN, Hoza B, Hechtman L, Abikoff HB, Elliott GR, Greenhill LL, Newcorn JH, Wells KC, Wigal T, Gibbons RD, Hur K, Houck PR, MTA Cooperative Group (2009) The MTA at 8 years: prospective follow-up of children treated for combined-type ADHD in a multisite study. J Am Acad Child Adolesc Psychiatry 48(5):484–500

Monastra VJ, Lubar JF, Linden M, VanDeusen P, Green G, Wing W, Phillips A, Fenger TN (1999) Assessing attention deficit hyperactivity disorder via quantitative electroencephalography: an initial validation study. Neuropsychology 13(3):424–433

Monastra VJ, Monastra DM, George S (2002) The effects of stimulant therapy, EEG biofeedback, and parenting style on the primary symptoms of attention-deficit/hyperactivity disorder. Appl Psychophysiology Biofeedback 27(4):231–249

Mulert C, Juckel G, Brunnmeier M, Karch S, Leicht G, Mergl R, Möller HJ, Hegerl U, Pogarell O (2007) Prediction of treatment response in major depression: integration of concepts. J Affect Disord 98(3):215–225

Niedermeyer E, Lopes da Silva FH (1993) Electroencephalography : basic principles, clinical applications, and related fields. Williams & Williams, Baltimore

Nobler MS, Luber B, Moeller JR, Katzman GP, Prudic J, Devanand DP, Dichter GS, Sackeim HA (2000) Quantitative EEG during seizures induced by electroconvulsive therapy: relations to treatment modality and clinical features. I. Global analyses. The journal of ECT 16(3):211

Olbrich S, Arns SM (2013) EEG biomarkers in major depressive disorder: discriminative power and prediction of treatment response. Int Rev Psychiatry 25(5):604–618

Olbrich S, Sander C, Minkwitz J, Chittka T, Mergl R, Hegerl U, Himmerich H (2012) EEG vigilance regulation patterns and their discriminative power to separate patients with major depression from healthy controls. Neuropsychobiology 65(4):188–194

Parmeggiani A, Barcia G, Posar A, Raimondi E, Santucci M, Scaduto MC (2010) Epilepsy and EEG paroxysmal abnormalities in autism spectrum disorders. Brain Dev 32(9):783–789

Pascual-Marqui RD, Michel CM, Lehmann D (1994) Low resolution electromagnetic tomography: a new method for localizing electrical activity in the brain. Int J Psychophysiol: Official J Int Organ Psychophysiol 18(1):49–65

Pizzagalli DA (2011) Frontocingulate dysfunction in depression: toward biomarkers of treatment response. Neuropsychopharmacology 36:183–206

Pizzagalli DA, Oakes TR, Davidson RJ (2003) Coupling of theta activity and glucose metabolism in the human rostral anterior cingulate cortex: an EEG/PET study of normal and depressed subjects. Psychophysiology 40(6):939–949

Pollock VE, Schneider LS (1990) Quantitative, waking EEG research on depression. Biol Psychiatry 27(7):757–780

Posthuma D, Neale MC, Boomsma DI, de Geus EJ (2001) Are smarter brains running faster? Heritability of alpha peak frequency, IQ, and their interrelation, Behavior genetics 31(6):567–579

Quinones MP, Kaddurah-Daouk R (2009) Metabolomics tools for identifying biomarkers for neuropsychiatric diseases. Neurobiol Dis 35(2):165–176

Richter PL, Zimmerman EA, Raichle ME, Liske E (1971) Electroencephalograms of 2,947 United States air force academy cadets (1965–1969). Aerosp Med 42(9):1011–1014

Roth M, Kay DW, Shaw J, Green J (1957) Prognosis and pentothal induced electroencephalographic changes in electro-convulsive treatment; an approach to the problem of regulation of convulsive therapy. Electroencephalogr Clin Neurophysiol 9(2):225–237

Rush AJ, Trivedi MH, Wisniewski SR, Nierenberg AA, Stewart JW, Warden D, Niederehe G, Thase ME, Lavori PW, Lebowitz BD, McGrath PJ, Rosenbaum JF, Sackeim HA, Kupfer DJ, Luther J, Fava M (2006) Acute and longer-term outcomes in depressed outpatients requiring one or several treatment steps: a STAR*D report. Am J Psychiatry 163(11):1905–1917

Rybak YE, McNeely HE, Mackenzie BE, Jain UR, Levitan RD (2006) An open trial of light therapy in adult attention-deficit/hyperactivity disorder. J Clin Psychiatry 67(10):1527–1535

Sander C, Arns M, Olbrich S, Hegerl U (2010) EEG-vigilance and response to stimulants in paediatric patients with attention deficit/hyperactivity disorder. Clin Neurophysiol 121:1511–1518

Satterfield JH, Cantwell DP, Saul RE, Lesser LI, Podosin RL (1973) Response to stimulant drug treatment in hyperactive children: prediction from EEG and neurological findings. J Autism Child Schizophr 3(1):36–48

Satterfield JH, Lesser LI, Podosin RL (1971) Evoked cortical potentials in hyperkinetic children. Calif Med 115(3):48

Savitz JB, Rauch SL, Drevets WC (2013) Clinical application of brain imaging for the diagnosis of mood disorders: the current state of play. Mol Psychiatry 18(5):528–539

Schoenknecht P, Olbrich S, Sander C, Spindler P, Hegerl U (2010) Treatment of acute mania with modafinil monotherapy. Bio Psychiatry 67:e55–e57

Schwartz J, Feldstein S, Fink M, Shapiro DM, Itil TM (1971) Evidence for a characteristic EEG frequency response to thiopental. Electroencephalogr Clin Neurophysiol 31(2):149–153

Silva RR, Munoz DM, Alpert M (1996) Carbamazepine use in children and adolescents with features of attention-deficit hyperactivity disorder: a meta-analysis. J Am Acad Child Adolesc Psychiatry 35(3):352–358

Simon GE, Perlis RH (2010) Personalized medicine for depression: can we match patients with treatments? Am J Psychiatry 167(12):1445–1455

Small JG, Milstein V, Medlock CE (1997) Clinical EEG findings in mania. Clinical EEG (Electroencephalography) 28(4):229–235

Smit DJ, Boersma M, van Beijsterveldt CE, Posthuma D, Boomsma DI, Stam CJ, de Geus EJ (2010) Endophenotypes in a dynamically connected brain. Behav Genet 40(2):167–177

Smit DJ, Posthuma D, Boomsma DI, Geus EJ (2005) Heritability of background EEG across the power spectrum. Psychophysiology 42(6):691–697

Smith GS, Reynolds CF, Pollock B, Derbyshire S, Nofzinger E, Dew MA, Houck PR, Milko D, Meltzer CC, Kupfer DJ (1999) Cerebral glucose metabolic response to combined total sleep deprivation and antidepressant treatment in geriatric depression. Am J Psychiatry 156(5):683–689

Snyder SM, Hall JR (2006) A meta-analysis of quantitative EEG power associated with attention-deficit hyperactivity disorder. J Clin Neurophysiol 23(5):440–455

Spronk D, Arns M, Barnett KJ, Cooper NJ, Gordon E (2011) An investigation of EEG, genetic and cognitive markers of treatment response to antidepressant medication in patients with major depressive disorder: a pilot study. J Affect Disord 128:41–48

Stern JM, Engel J (2004) Atlas of EEG patterns. Lippincott Williams & Wilkins, Baltimore

Stubbeman WF, Leuchter AF, Cook IA, Shurman BD, Morgan M, Gunay I, Gonzalez S (2004) Pretreatment neurophysiologic function and ECT response in depression. J ECT 20(3):142–144

Suffin SC, Emory WH (1995) Neurometric subgroups in attentional and affective disorders and their association with pharmacotherapeutic outcome. Clinical EEG (Electroencephalography) 26(2):76–83

Swanson JM, Hinshaw SP, Arnold LE, Gibbons RD, Marcus S, Hur K, Jensen PS, Vitiello B, Abikoff HB, Greenhill LL, Hechtman L, Pelham WE, Wells KC, Conners CK, March JS, Elliott GR, Epstein JN, Hoagwood K, Hoza B, Molina BS, Newcorn JH, Severe JB, Wigal T (2007) Secondary evaluations of MTA 36-month outcomes: propensity score and growth mixture model analyses. J Am Acad Child Adolesc Psychiatry 46(8):1003–1014

Tenke CE, Kayser J, Manna CG, Fekri S, Kroppmann CJ, Schaller JD, Alschuler DM, Stewart JW, McGrath PJ, Bruder GE (2011) Current source density measures of electroencephalographic alpha predict antidepressant treatment response. Biol Psychiatry 70(4):388–394
Ulrich G, Fürstenberg U (1999) Quantitative assessment of dynamic electroencephalogram (EEG) organization as a tool for subtyping depressive syndromes. Eur Psychiatry 14(4):217–229
Ulrich G, Herrmann WM, Hegerl U, Müller-Oerlinghausen B (1990) Effect of lithium on the dynamics of electroencephalographic vigilance in healthy subjects. J Affect Disord 20(1):19–25
Ulrich G, Renfordt E, Zeller G, Frick K (1984) Interrelation between changes in the EEG and psychopathology under pharmacotherapy for endogenous depression. Contrib Predictor Question, Pharmacopsychiatry 17(6):178–183
van Beijsterveldt CE, van Baal GC (2002) Twin and family studies of the human electroencephalogram: a review and a meta-analysis. Biol Psychol 61(1–2):111–138
van der Heijden KB, Blok MJ, Spee K, Archer SN, Smits MG, Curfs LM, Gunning WB (2005) No evidence to support an association of PER3 clock gene polymorphism with ADHD-related idiopathic chronic sleep onset insomnia. Biol Rhythm Res 36(5):381–388
Van Dongen HP, Maislin G, Mullington JM, Dinges DF (2003) The cumulative cost of additional wakefulness: dose-response effects on neurobehavioral functions and sleep physiology from chronic sleep restriction and total sleep deprivation. Sleep 26(2):117–126
Van Veen MM, Kooij JJ, Boonstra AM, Gordijn MC, Van Someren EJ (2010) Delayed circadian rhythm in adults with attention-deficit/hyperactivity disorder and chronic sleep-onset insomnia. Biol Psychiatry 67(11):1091–1096
Vogel F (1970) The genetic basis of the normal human electroencephalogram (EEG). Hum Genet 10(2):91–114
Volavka J, Feldstein S, Abrams R, Dornbush R, Fink M (1972) EEG and clinical change after bilateral and unilateral electroconvulsive therapy. Electroencephalogr Clin Neurophysiol 32(6):631–639
Walter W, Dovey VJ (1944) Electro-encephalography in cases of sub-cortical tumour. J Neurol Neurosurg Psychiatry 7(3–4):57
Winterer G, Mahlberg R, Smolka MN, Samochowiec J, Ziller M, Rommelspacher HP, Herrmann WM, Schmidt LG, Sander T (2003) Association analysis of exonic variants of the GABA B-receptor gene and alpha electroencephalogram voltage in normal subjects and alcohol-dependent patients. Behav Genet 33(1):7–15
Wood JG, Crager JL, Delap CM, Heiskell KD (2007) Beyond methylphenidate: nonstimulant medications for youth with ADHD. J Attention Dis, 11(3):341–350
Wu J, Buchsbaum MS, Gillin JC, Tang C, Cadwell S, Wiegand M, Najafi A, Klein E, Hazen K, Bunney WE, Fallon JH, Keator D (1999) Prediction of antidepressant effects of sleep deprivation by metabolic rates in the ventral anterior cingulate and medial prefrontal cortex. Am J Psychiatry 156(8):1149–1158
Wu JC, Gillin JC, Buchsbaum MS, Hershey T, Johnson JC, Bunney WE (1992) Effect of sleep deprivation on brain metabolism of depressed patients. Am J Psychiatry 149(4):538–543
Yasuhara A (2010) Correlation between EEG abnormalities and symptoms of autism spectrum disorder (ASD). Brain Dev 32(10):791–798
Zoon HF, Veth CP, Arns M, Drinkenburg WH, Talloen W, Peeters PJ, Kenemans JL (2013) EEG alpha power as an intermediate measure between brain-derived neurotrophic factor Val66Met and depression severity in patients with major depressive disorder. J Clin Neurophysiol 30(3):261–267

Psychophysiology-Informed (Multimodal) Imaging

Nikolaj Bak and Bob Oranje

Abstract Electroencephalography (EEG) and magnetic resonance imaging are two popular methodologies for brain research. While EEG has a high temporal resolution, yet a low spatial resolution, MRI has the complete opposite, a high spatial resolution, yet a low temporal resolution. Obviously therefore, researchers have been searching for ways combining the two methodologies, for more than two decades. However, there are many issues that have to be solved before the methodologies can be successfully and, more importantly reliably, combined. Here, we give an overview of these issues, and present strategies that have been used over the past two decades to overcome them. We start with a general description of EEG and (f)MRI methodology, then present the difficulties involved in combining both methodologies, and lastly present and discuss the most popular strategies that have been used over the past two decades to solve these problems. We conclude that in spite of the many issues, the two methodologies can be combined successfully, provided that the correct procedures are followed.

Keywords EEG · (f)MRI · P50 suppression · Schizophrenia

N. Bak · B. Oranje
Center for Clinical Intervention and Neuropsychiatric Schizophrenia Research (CINS), Copenhagen University Hospital, Psychiatric Center Glostrup, Glostrup, Denmark

N. Bak · B. Oranje
Center for Neuropsychiatric Schizophrenia Research (CNSR), Copenhagen University Hospital, Psychiatric Center Glostrup, Glostrup, Denmark

B. Oranje
Department of Neurology, Faculty of Health Sciences, Psychiatry, and Sensory Sciences, University of Copenhagen, Copenhagen, Denmark

B. Oranje (✉)
Department of Psychiatry, Brain Center Rudolf Magnus, University Medical Center Utrecht, Heidelberglaan 100, 3584 CX Utrecht, The Netherlands
e-mail: B.Oranje@cnsr.dk

Curr Topics Behav Neurosci (2014) 21: 371–385
DOI: 10.1007/7854_2014_323

Published Online: 21 June 2014

Contents

1 Introduction to EEG

Electroencephalography (EEG) is a methodology to assess electrical activity of the brain. It was first discovered by Hans Berger (Berger 1929; See also: Jung and Berger 1979). EEG makes use of the fact that neurons produce electricity (pre- and post-synaptic potentials) which spread throughout the brain until it reaches the scalp. This electrical activity can be measured by for instance placing electrodes on the scalp, by placing them on the brain tissue itself or, even more invasive, by inserting them in the brain (Niedermeyer and Lopes da Silva 2005). Obviously, the two last techniques are mostly used in animal research, although sometimes it is used in humans as well, e.g. when a patient needs brain surgery. From the scalp, it is not possible to assess the electrical activity of one single neuron. For this, the electrical current is simply too small. EEG recordings from the scalp generally consist of populations of neurons "firing" at the same time, in synchrony. Before the electrical activity arrives at the scalp, it usually has travelled through a number of brain tissues, including cerebrospinal fluids, dura mater, bone, skin, which all affect the signal: where cortical discharges can reach amplitudes up to 1.5 mV, typical activity recorded by scalp electrodes lies between 10 and 100 μV, but more generally do not exceed 50 μV (Niedermeyer and Lopes da Silva 2005). This means that the signals that reach the scalp need to be amplified, typically 10^5–10^6 times, to enable assessment. In clinical settings, EEG is most frequently investigated without further processing. Unprocessed EEG is for instance used to determine whether an individual suffers from epileptic seizures, or is used in sleep research (e.g. Keshavan et al. 1990). Unprocessed EEG can also be decomposed in frequency bands (e.g. alpha, beta, gamma bands) by means of Fourier transformation (Kooi et al. 1978; see also Fig. 1), from which certain psychological activities or mental states are commonly inferred (Barry et al. 2003a; Kiloh et al. 1972; Niedermeyer and Lopes da Silva 2005).

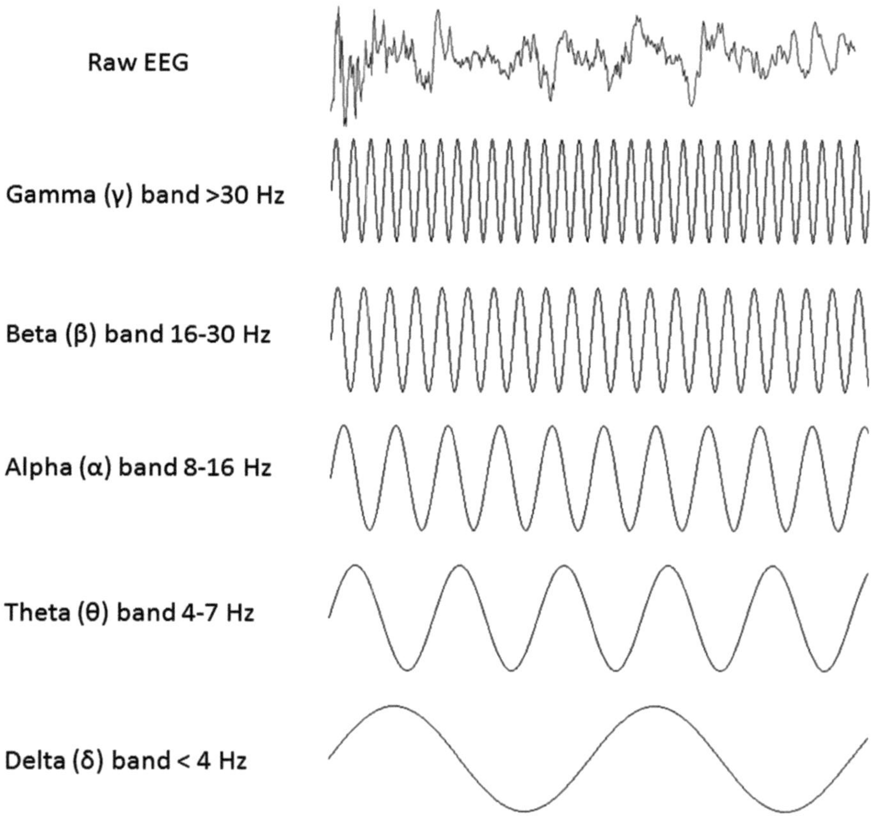

Fig. 1 Raw EEG decomposed with Fourier transformation, displaying the main frequency bands

However, EEG can also be processed: If identical sensory stimuli are presented a number of times, then so-called event-related brain potentials (ERPs, sometimes also referred to as evoked potentials, EPs) can be derived from them, by averaging the elicited EEG signals (e.g. Pfefferbaum et al. 1995). By averaging, one accomplishes that all randomly generated "noise" in the brain is cancelled out, because in contrast to the presented stimulus, the noise is not time locked to the EEG signal, thereby ensuring that the activity left is solely generated by the presentation of the sensory stimulus. Obviously, the more the trials are repeated, the more it will assure that the ERP is indeed generated by the stimulus in question, and the less it will be affected by randomly generated brain activity (e.g. Fabiani et al. 2000; Pfefferbaum et al. 1995). ERPs are for instance clinically used to determine the extent of auditory loss with so-called brain auditory evoked potentials (BAEPs) (Markand 1994) and also frequently used in research, such as research on schizophrenia or other psychiatric disorders (Aggernaes et al. 2010; Barry et al. 2003b; Coull 1998; Madsen et al. 2013; Näätänen 1990; Oranje et al. 2013; Sumich et al. 2006). An auditory

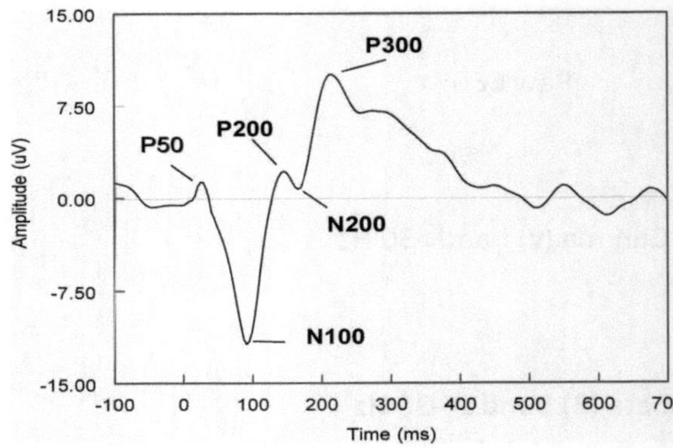

Fig. 2 Example of an ERP, following an auditory stimulus

stimulus can, also depending on the paradigm, among others elicit a P50, N100, P200, N200 and even later components such as a P300 ERPs, (see Fig. 2). About the nomenclature: the "P" and "N" stand for positivity, respectively negativity, the number stand for time in ms. Hence, P50 stands for a positive deflection in the EEG, 50 ms following stimulus presentation; N100 stands for a negativity appearing 100 ms post-stimulus, etc.

2 Introduction to (f)MRI

2.1 MRI

Magnetic Resonance Imaging (MRI) is presently favored over other imaging techniques for investigation of brain structure and function. This is primarily because MRI is a non-invasive methodology, uses no radioactive radiation, has relatively short acquisition times, and obtains high spatial resolution. A further advantage of MRI methodology is that both brain structure and function can be assessed in one single session. Inside an MRI scanner, there is a strong homogenous static magnetic field (B_0) usually ranging from 1.5 to 3 Tesla (T), although rather recently 7 T scanners have also become available for research. Nuclei with gyromagnetic properties (such as protons) will have a tendency to align the axis of their intrinsic spin along the B_0 field. The axis of the spins will precess, or "spin around", the direction of B_0 (z-direction) with a frequency, the "Larmor frequency", that is specific for each type of nuclei and dependent on the surrounding field strength. Imaging with magnetic resonance typically uses hydrogen atoms because the abundance of hydrogen in biological systems ensures a high signal. The alignment of the spins results in a net magnetization of the subject while he or she remains inside the field. With a radio frequency (RF) pulse in resonance with the Larmor frequency of the nuclei of interest (as mentioned before, usually

Hydrogen), the net magnetization can be "pushed" out of equilibrium (excitation). The net magnetization will then precess while it returns to equilibrium (relaxation). This results in emission of radio waves with the Larmor frequency but only as long as there exists a magnetization component perpendicular to the B_0 field (along the xy-plane). These radio waves can be received as a signal. The relaxation is determined by two time constants T1 (spin-lattice relaxation time) and T2 (spin–spin relaxation time). Both T1 and T2 are dependent on the environment that the nuclei are part of and, therefore, specific for each type of tissue. T1 describes how the magnetization returns to equilibrium along the longitudinal axis (*z*-axis). T2 describes the loss of magnetization in the xy-plane due to dephasing of the spins as a result of molecular interactions. Inhomogenities in the field and susceptibility losses, e.g. due to the presence of deoxyhemoglobin or air/tissue boundaries, cause a faster loss of signal than the pure T2; the combined time constant is called T2*. The timing and strength of the excitation pulses, as well as the delay between excitation and signal reception, determines whether a sequence is more sensitive for T1 or T2.

When a gradient in the magnetic field is applied using an additional electromagnet, the Larmor frequency is also affected. This can be used for imaging. By applying a magnetic gradient superimposed on B_0 during the excitation, only the 2D-plane with Larmor frequency corresponding to the RF pulse is excited (slice selection). Applying a gradient along the excited plane determines the precession of the protons in a controlled way. This process is repeated with different gradients, varying the direction and gradient strength (each variation of direction and strength corresponding to a point or line in "k-space") in a predetermined sequence saving the signal each time. The selected 2D plane can then be imaged by a mathematical technique called Fourier transformation. For further information on the basic principles of MRI, the reader is referred to textbooks on the subject (e.g. Jezzard et al. 2008).

2.2 *fMRI*

In the early days of its use in brain research, MRI was predominantly used to study brain structure only. Nowadays, a number of equally ingenuous as advanced techniques have been developed, resulting in a variety of additional uses of MRI, such as determining structural brain networks with diffusion tensor imaging (DTI), functional brain networks by using regression approaches, or even biochemistry of the brain by means of spectroscopy. However, a discussion on these approaches is beyond the scope of this chapter. Besides these techniques, MRI can also be used to map changes in brain hemodynamics. This is termed functional magnetic resonance imaging or fMRI. The basis for performing fMRI is the Blood Oxygen Level Dependent (BOLD) signal, first used by Ogawa et al. (1990). The technique is based on the fact that oxyhemoglobin is diamagnetic while deoxyhemoglobin is paramagnetic. Paramagnetic molecules disrupt the homogeneity of the magnetic field and decrease T2*. If blood flow increases due to increased cellular activity, and a comparable increase in oxygen consumption is not present, oxygenation increases. This measurable increase in signal is termed the BOLD contrast. Increased blood flow followed by neuronal activity in the brain is referred to as the hemodynamic response. The BOLD contrast is the sum of changes in the cerebral blood flow (CBF), cerebral blood volume (CBV) and oxygen metabolic rate that occurs in response to neural activity. It is still debated what kind of neural activity specifically determines these changes, although increasing evidence suggest that BOLD is more a measure of local field potential (LFP) changes rather than spikes in activity (Logothetis et al. 2001), i.e. the hemodynamic response follows the level of neuronal input to a given area more than the actual firing of the neurons (output) in the area. When measuring fMRI, the same sequence is repeated consecutively throughout the scan. Each of the repeated sequences images the whole, or a part of, the brain (termed a volume). It is then possible to detect changes in the BOLD response following different types of stimuli (conditions). These conditions, or responses, can be compared in a contrast, by subtracting one condition from another or by comparing a condition to baseline. The faster the repetition time (TR), the higher the temporal resolution becomes. This is why the so-called "echo planar imaging" (EPI) sequence is frequently used in fMRI, because this sequence can sample the entire two-dimensional k-space following a single RF excitation pulse. For more detailed information on BOLD imaging, please see reviews (e.g. Logothetis and Wandell 2004; Raichle and Mintun 2006) or textbooks on the subject (e.g. Jezzard et al. 2008).

3 EEG and fMRI Combined

3.1 General

Although there have been many decades of research devoted to the behavioral and psychological processes behind ERPs, research on the generators of these ERPs in the brain is lagging behind. Initially, attempts were made to locate these generators by EEG source localization, using software such as BESA® (MEGIS Software GmbH, Gräfelfing, Germany; e.g. Scherg et al. 2002; Scherg and Picton 1991). Later, software was developed that could integrate EEG with MRI, such as Curry® (Compumedics, Neuroscan). Although source localization with EEG has high temporal resolution, its spatial resolution is rather poor: this resolution is approximately half of the average electrode distance, i.e. approximately 1 cm when 64 electrodes are used (Scherg 1990). In contrast to EEG methodology, fMRI has much higher spatial resolution, yet much lower temporal resolution, because the hemodynamic response is relatively slow—it peaks at approximately 6 s following a stimulus. In other words, the combination of EEG and fMRI methodology complements each other, i.e. potentially it could result in data with both high temporal resolution due to the EEG part, and high spatial resolution due to the fMRI part. However, combined use of EEG and MRI methodology has many issues and although being performed for approximately a decade or two, it is still developing (Im et al. 2006). In this chapter, we shall predominantly focus on through electrophysiology informed MRI imaging techniques and the many issues that need to be dealt with when combining the two methodologies.

3.2 Issues

While a typical EEG environment is devoid of any other stimulation than the stimuli presented in the paradigms, the typical MRI environment is rather noisy. This noise is likely to interfere with the paradigms used in EEG settings, especially if they happen to make use of auditory stimuli. In addition, an MRI assessment frequently triggers anxiety, much more than is the case in a typical EEG assessment: the procedures are intimidating, the scanner space is claustrophobic and as mentioned before, rather noisy. This anxiety most likely impacts on the results. Another issue is that responses to stimuli in close temporal proximity of each other, such as in typical EEG paradigms, cannot be separated with fMRI, because the hemodynamic response is much slower than these interstimulus intervals. Moreover, some ERPs are rather small in amplitude, such as that of the P50 ERP (on average between 4 and 5 μV, e.g. Oranje et al. 2006) and hence its generators in the brain may be fewer, or at least much less active, than those of larger waveforms, such as the N100 or the P300. The generators of these larger waveforms may therefore dominate the results. A further issue is that metal can affect MRI images, i.e. the EEG electrodes or their wires can affect the quality of the MRI data. Finally, an MRI scanner produces huge electrical artefacts in EEG data when both techniques are used concurrently (see below for more detail). Over the years, several methods and strategies have evolved to overcome the above-mentioned issues. Broadly speaking, these strategies can be categorized into two approaches.

3.3 Associative Approach (Separate EEG and FMRI Assessments)

In this approach, EEG and (f)MRI are assessed separately from each other. Roughly, there are two strategies that make use of this approach. The first, based solely on associations, takes the least effort in preparation and data processing: It simply asks for testing the same individuals in an EEG setting and in an (f)MRI setting, after which the results of both techniques are analyzed for possible associations. This strategy has for instance been used to study associations between structural brain correlates and several electrophysiological parameters, such as sensorimotor gating (assessed with a pre-pulse inhibition of the startle reflex paradigm or PPI) (e.g. Hammer et al., 2012; Kumari et al. 2005, 2008), sensory gating (assessed with a P50 suppression paradigm) (e.g. Arciniegas et al. 2001), mismatch negativity (e.g. Rasser et al. 2011) or P300 amplitude (e.g. Egan et al. 1994; Fusar-Poli et al. 2011; McCarley et al. 2002); an interesting variant of this approach is when so-called low-resolution electromagnetic tomography (LORETA) is coupled with MRI, such as for instance was used in the study of Pae et al. (2003) to locate generators behind the P300 amplitude.

This is currently the most widely used approach for pragmatic reasons; neither the compatibility between EEG and MRI equipment nor that of the paradigms used are relevant when using this approach. However, obvious disadvantages of this strategy are that it is still difficult to find the actual generators behind a specific ERP amplitude in the brain, because it will remain uncertain whether the MRI derived brain activations that correlate with a specific ERP of interest are not related to other (earlier or later appearing) ERPs, or to other processes behind the paradigm altogether.

An alternative strategy that deals with some of these issues makes use of functional instead of structural MRI, and in which identical, or at least very similar, paradigms are used in both the EEG and fMRI setting. The difficulty with this approach is that a typical EEG setting is quite different from a typical MRI setting (see Sect. 3.2). Where EEG is usually recorded in sound isolated and electrically shielded rooms, an MRI setting is quite the opposite: recording MRI images makes a lot of noise. For paradigms using visual stimulation, this may seem less of a problem. However, in the EEG setting, subjects are usually sitting upright, while subjects lie down in an MRI scanner, which means that visual stimuli are usually presented by mirrors or goggles. Paradigms using auditory stimulation call for more drastic procedures. In one of the strategies, scanner noise is as much avoided as possible by making use of so-called interleaved scanning: in this method, the MRI recordings are stopped a period just before and after a trial of a paradigm. During this "silent" period, the auditory stimuli of the trial can be presented. The MRI scanning then starts again before the expected peak of the hemodynamic response, which enables identification of the activated brain areas. This method was for instance used by Tregellas et al. (2007). Another strategy ingeniously made use of the scanner gradient coils to create auditory stimuli with a duration of 50 ms to locate sources of P50 suppression in the brain (Mathiak et al. 2011). Other researchers tried to reduce the noise of the scanner to such levels that it did not interfere with the auditory stimulation of the paradigm. This is rather difficult, due to the intensity of the scanner noise, which usually exceeds that of the auditory stimuli of the paradigms by manifold: typical intensities of auditory stimuli in EEG paradigms (PPI paradigms not included) are at most 90 dBa, whereas typical MRI scanner sequences may reach sound intensities up to 114 dBa or even beyond (Counter et al. 1997). This means that at most MRI noise can be reduced to such levels that the stimuli of the paradigms can be heard, yet the residual scanner noise could still very well have an impact on the results. There are several examples of studies that made use of this approach, with the investigators frequently taking drastic additional steps to minimize the effects of residual scanner noise (e.g. Hazlett et al. 2001, 2008; Mayer et al. 2009).

Alternatively, stimuli of a different modality can be used altogether, e.g. by using visual or tactile stimuli instead of auditory stimuli. Frequently, these alternative paradigms are first validated in EEG sessions to verify that they yield similar results as their auditory versions. This strategy, together with often more advanced procedures, has been used in several studies (Bak et al. 2011, 2013; Kumari et al. 2003). Although this strategy is more accurate in relating EEG or EMG (electromyography) signals to fMRI derived sources in the brain than the first mentioned strategy, it still has several limitations. Importantly, it is not certain that the paradigms in the EEG and fMRI setting produce similar results, e.g. does P50 suppression or PPI of the startle reflex actually occur in both settings, and if so will it occur in similar levels. The results may after all still be influenced by non-paradigm related processes, such as anxiety due to MRI assessment or scanner noise. Another issue is that a certain ERP cannot be directly coupled to a specific activity in the brain due to the low temporal resolution of (f)MRI: An auditory stimulus gives not only rise to P50 amplitudes, but also to N100, P200, N200, etc. amplitudes; it is impossible with this approach to directly determine which ERPs are coupled with which activated areas in the brain as indicated by the fMRI scan; this can only be done by approximation, e.g. by applying associative statistical analyses. The next category of approaches go one step further, they assess EEG concurrently with MRI, effectively solving most of the above-mentioned issues.

3.4 Concurrent Approach (Simultaneous EEG and FMRI Assessment)

In this approach, the psychophysiological measurements are assessed concurrently with the imaging technique, i.e. EEG is assessed inside the MRI scanner. The most important advantage of assessing EEG and MRI concurrently is that one does not need to assume that the same paradigm triggers the same processes in the brain in the (f)MRI session as in the EEG session, as would be the case in the associative approach, but one can directly deduce from the data whether this is the case or not (see: Herrmann and Debener 2008; Huster et al. 2012; Ritter and Villringer 2006, for introductory reviews of concurrent EEG-fMRI). However, concurrent assessment is certainly not easy, given the many issues mentioned above: it not only sets limitations for the paradigms and even the EEG equipment that can be used in the MRI environment, but it also means more complicated processing of the data. The MRI scanner will produce serious artefacts in the EEG data that will need to be removed, and vice versa, the electrodes and cables necessary for EEG assessment can produce artefacts in the MRI images. Nevertheless, with the right equipment and appropriate processing of the data, EEG and EMG can actually be assessed concurrently with fMRI with only minor effects on the quality of the data.

Starting with the concept that the quality of the MRI data could be affected by the wires and electrodes of the EEG cap, there are studies showing that this impact can be minimized provided suitable equipment is used (Krakow et al. 2000; Mullinger et al. 2008). The impact of the MRI scanner on EEG signals, however, requires more attention. Back in 1824, William Sturgeon was the first to show that a coil of wire that is moved inside a magnetic field produces an electric current. Indeed, the shifting magnetic gradients that are necessary for fMRI produce artefacts in the EEG data due to the induced electrical current in the wires from the electrodes. Fortunately, these magnetic gradients are applied in a controlled and specific pattern, determined by the fMRI sequence. Due to this repetitiveness, the artefacts can be removed by averaging the EEG signal from the volumes. Typically, this is done by using a sliding average, and subtracting this average from the original data (Allen et al. 2000). However, movement of a subject also creates artefacts. Besides disturbing the MRI images, these movements will induce electrical currents in the wires of the electrodes, due to the same principles as mentioned above: moving conductive materials in the ever present steady state, magnetic field of an MRI scanner produces currents. Usually, these movement artefacts can be minimized by securing the wires from the electrodes and by preventing head movement as much as possible, e.g. by fixation of a subject's head in the head coil together with strict instruction of the subject. Nevertheless, even the small movements due to the heartbeat will produce EEG artefacts: movement of blood (a conductive fluid) through vessels creates pulse-related (cardioballistic) artefacts. Again, due to their repetitiveness also these artefacts can be relatively easy removed (Allen et al. 1998). By properly applying these procedures, all scanner artefacts can be quite efficiently removed (see Fig. 3): even information on ERP level is preserved, showing virtually no differences between EEG assessed in an EEG setting, or assessed in an MRI scanner (Bregadze and Lavric 2006). As an example of this strategy, we will describe the approach we recently used in our laboratory, where we made an attempt to locate the sources of P50 suppression with concurrent EEG and fMRI methodology. We started by adapting a typical auditory P50 suppression paradigm to an fMRI friendly version, among others by replacing the auditory stimuli with somatosensory stimuli. This somatosensory paradigm was first validated for showing actual P50 suppression in a pilot study, in which healthy subjects were assessed in a normal EEG setting. Following this validation study, subjects were assessed with the fMRI friendly paradigm first in a normal EEG setting (assessing only EEG) and later in an MRI setting, where both EEG and fMRI was assessed concurrently. Following the above described removal of MRI artefacts (see Fig. 3), we first statistically compared the EEG results of the two settings with each other.

After validating that these data sets did not significantly differ from each other, we first located by proxy, the generators of P50 suppression in the EEG data with BESA software. In the fMRI data, a contrast depicting differences between a trial type with suppression and one without was calculated. Following this, we imported the through EEG-derived sources in the fMRI recordings, and searched for significant clusters in the contrast that overlapped with these sources. This enabled us to

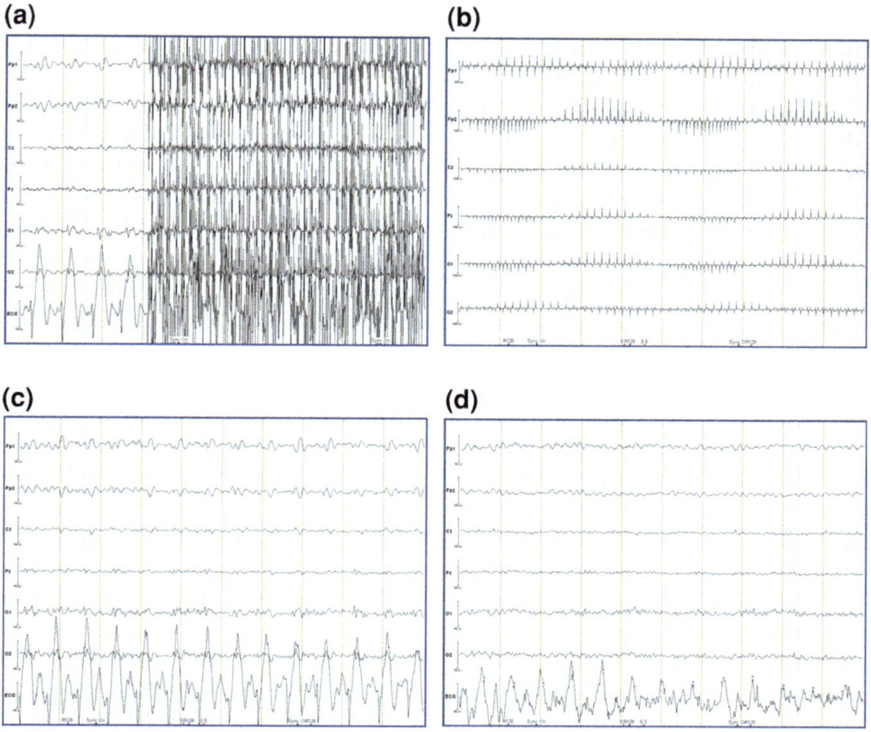

Fig. 3 Removal of MRI artefacts from EEG data during pre-processing. For simplicity, only six EEG channels are shown: FP1, FP2, Cz, Pz, O1, and O2. The channel below the EEG channels (in **a**, **c** and **d**) represents recording of ECG. **a** Representation of the beginning of MRI scanning in the EEG recording, showing the huge artefacts that this causes (*note* the scale represents 500 μV). **b** Representation of the EEG signal during the MRI scan (*note* the scale represents 5,000 μV), revealing the repetitiveness of the gradient artefacts. **c** The same data as **b**, with the gradient artefacts removed (same scale as in **a**); note that the contribution of the heartbeat and pulse can still be observed in the six channels. **d** In this last step, also the pulse artefacts are removed (same scale as **a** and **c**); the data is now ready for the usual further processing of EEG data

determine which of the many brain activations that followed the presentation of a trial were actually time locked to the P50 ERP. As such, we could successfully locate the generators of P50 suppression, first in healthy subjects (Bak et al. 2011) and later in the aberrant P50 suppression of patients with schizophrenia (Bak et al. 2013). An interesting alternative to this approach is to combine the above-described interleaved approach with concurrent EEG and fMRI assessment. The advantage of this is that the EEG artefacts due to the shifting magnetic gradients can be avoided. However, the shape of the hemodynamic response cannot be detected with this method, and the above-described cardioballistic artefacts still need to be removed. This procedure has for instance successfully been used to locate sources of PPI (Kumari et al. 2003) and P300 amplitude (Mulert et al. 2004).

Summarized, these studies show that it is possible to deal with the many issues that are involved in concurrent scanning of MRI and EEG, but it also shows that the procedures are quite laborious and challenging.

Attempts to extract more and more information out of the concurrent EEG and fMRI measurements have moved the field into yet another interesting area, directed towards incorporating single EEG trials in the statistical analysis of the data sets, rather than the averaged ERPs. This method requires additional expertise in the more advanced mathematical and computational models involved. A growing number of reports within this field have emerged over the last decade. A popular method is using specific components based on amplitudes of the waveforms from a specific electrode and entering these as explanatory variables in the general linear model (GLM) of the MRI analyses (e.g. Benar et al. 2007; Debener et al. 2005; Fuglo et al. 2012; Mobascher et al. 2012). This is presently only feasible for larger, robust components in the EEG signals, such as the N100 or P300 amplitudes, because they can be detected in single trials. A variant on this method makes use of an algorithm to identify components in the EEG signal corresponding to parts of the ERP. This can be done with independent component analysis (ICA) as suggested by Debener et al. (2006) but in principle any algorithm to extract relevant components or factors from the EEG signal could be used. Components representing artifacts can then be discarded, while all the relevant components can be used as predictors in the GLM analysis of the fMRI data or integrated with results from a similar approach based on performing parallel ICA, i.e. on both fMRI and EEG data and hemodynamic deconvolution (Eichele et al. 2009). The advantage of this strategy is that more complicated tasks can be analyzed in depth or specific parts of an ERP can be modelled. However, only part of the brain activity can be used in the fMRI analysis; therefore, components should be chosen to address the specific research question.

4 Conclusion

Obviously, the associative approach is by far the less laborious of the two approaches. An additional advantage of this approach is that the resulting EEG data can be compared to those of previous EEG studies. The assumption that the ERP is correlated with the average hemodynamic response is reasonable, even though EEG and fMRI measures represent different aspects of neuronal activity. However, the associative approach does not make optimal use of the strengths of the EEG and (f)MRI methodology, i.e. combining the high temporal resolution of EEG with the high spatial resolution of MRI, such as can be achieved by concurrent assessment of EEG and fMRI. Although this last (concurrent) approach is definitely preferable, it is also rather complicated and laborious. Nevertheless, when properly executed, the results are equally spectacular as reliable.

Acknowledgments The authors have no conflicts of interest.

References

Aggernaes B, Glenthoj BY, Ebdrup BH, Rasmussen H, Lublin H, Oranje B (2010) Sensorimotor gating and habituation in antipsychotic-naïve, first-episode schizophrenia patients before and after six months treatment with quetiapine. Int J Neuropsychopharmacol 13(10):1383–1395

Allen PJ, Josephs O, Turner R (2000) A method for removing imaging artifact from continuous EEG recorded during functional MRI. Neuroimage 12(2):230–239

Allen PJ, Polizzi G, Krakow K, Fish DR, Lemieux L (1998) Identification of EEG events in the MR scanner: the problem of pulse artifact and a method for its subtraction. Neuroimage 8(3):229–239

Arciniegas DB, Topkoff JL, Rojas DC, Sheeder J, Teale P, Young DA, Sandberg E, Reite ML, Adler LE (2001) Reduced hippocampal volume in association with p50 nonsuppression following traumatic brain injury. J Neuropsychiatry Clin Neurosci 13(2):213–221

Bak N, Glenthoj BY, Rostrup E, Larsson H, Oranje B (2011) Source localization of sensory gating: a combined EEG and fMRI study in healthy volunteers. Neuroimage 54(4):2711–2718

Bak N, Rostrup E, Larsson HB, Glenthoj BY, Oranje B (2013) Concurrent functional magnetic resonance imaging and electroencephalography assessment of sensory gating in schizophrenia. Hum Brain Mapp

Barry RJ, Clarke AR, Johnstone SJ (2003a) A review of electrophysiology in attention-deficit/hyperactivity disorder: I Qualitative and quantitative electroencephalography. Clin Neurophysiol 114(2):171–183

Barry RJ, Johnstone SJ, Clarke AR (2003b) A review of electrophysiology in attention-deficit/hyperactivity disorder: II. Event-related potentials. Clin Neurophysiol 114(2):184–198

Benar CG, Schon D, Grimault S, Nazarian B, Burle B, Roth M, Badier JM, Marquis P, Liegeois-Chauvel C, Anton JL (2007) Single-trial analysis of oddball event-related potentials in simultaneous EEG-fMRI. Hum Brain Mapp 28(7):602–613

Berger H (1929) Uber das Elektrenkephalogramm des Menschen. Archiv für Psychiatrie und Nervenkrankheiten 87(1):527–570

Bregadze N, Lavric A (2006) ERP differences with vs. without concurrent fMRI. Int J Psychophysiol 62(1):54–59

Coull JT (1998) Neural correlates of attention and arousal: insights from electrophysiology, functional neuroimaging and psychopharmacology. Prog Neurobiol 55(4):343–361

Counter SA, Olofsson A, Grahn HF, Borg E (1997) MRI acoustic noise: sound pressure and frequency analysis. J Magn Reson Imaging 7(3):606–611

Debener S, Ullsperger M, Siegel M, Engel AK (2006) Single-trial EEG-fMRI reveals the dynamics of cognitive function. Trends Cogn Sci 10(12):558–563

Debener S, Ullsperger M, Siegel M, Fiehler K, von Cramon DY, Engel AK (2005) Trial-by-trial coupling of concurrent electroencephalogram and functional magnetic resonance imaging identifies the dynamics of performance monitoring. J Neurosci 25(50):11730–11737

Egan MF, Duncan CC, Suddath RL, Kirch DG, Mirsky AF, Wyatt RJ (1994) Event-related potential abnormalities correlate with structural brain alterations and clinical features in patients with chronic schizophrenia. Schizophr Res 11(3):259–271

Eichele T, Calhoun VD, Debener S (2009) Mining EEG-fMRI using independent component analysis. Int J Psychophysiol 73(1):53–61

Fabiani M, Gratton G, Coles MG (2000) Event-related brain potentials. In: Cacioppo JT (ed) Cambridge University Press, New York, pp 53–84

Fuglo D, Pedersen H, Rostrup E, Hansen AE, Larsson HB (2012) Correlation between single-trial visual evoked potentials and the blood oxygenation level dependent response in simultaneously recorded electroencephalography-functional magnetic resonance imaging. Magn Reson Med 68(1):252–260

Fusar-Poli P, Crossley N, Woolley J, Carletti F, Perez-Iglesias R, Broome M, Johns L, Tabraham P, Bramon E, McGuire P (2011) White matter alterations related to P300 abnormalities in individuals at high risk for psychosis: an MRI-EEG study. J Psychiatry Neurosci 36(4):239–248

Hammer TB, Oranje B, Skimminge A, Aggernaes B, Ebdrup BH, Glenthoj BY, Baare W (2012). Structural brain correlates of sensorimotor gating in antipsychotic-naïve men with first-episode schizophrenia. J Psychiatry Neurosci 37(4):8

Hazlett EA, Buchsbaum MS, Tang CY, Fleischman MB, Wei TC, Byne W, Haznedar MM (2001) Thalamic activation during an attention-to-prepulse startle modification paradigm: a functional MRI study. Biol Psychiatry 50(4):281–291

Hazlett EA, Buchsbaum MS, Zhang J, Newmark RE, Glanton CF, Zelmanova Y, Haznedar MM, Chu KW, Nenadic I, Kemether EM, Tang CY, New AS, Siever LJ (2008) Frontal-striatal-thalamic mediodorsal nucleus dysfunction in schizophrenia-spectrum patients during sensorimotor gating. Neuroimage 42(3):1164–1177

Herrmann CS, Debener S (2008) Simultaneous recording of EEG and BOLD responses: a historical perspective. Int J Psychophysiol 67(3):161–168

Huster RJ, Debener S, Eichele T, Herrmann CS (2012) Methods for simultaneous EEG-fMRI: an introductory review. J Neurosci 32(18):6053–6060

Im CH, Liu Z, Zhang N, Chen W, He B (2006) Functional cortical source imaging from simultaneously recorded ERP and fMRI. J Neurosci Methods 157(1):118–123

Jezzard P, Matthews PM, Smith SM (2008) Functional MRI. In: Jezzard P, Matthews PM, Smith SM (eds) Oxford University Press, New York

Jung R, Berger W (1979) Fiftieth anniversary of Hans Berger's publication of the electroencephalogram. His first records in 1924–1931 (author's transl). Arch Psychiatr Nervenkr 227(4):279–300

Keshavan MS, Reynolds CF, Kupfer DJ (1990) Electroencephalographic sleep in schizophrenia: a critical review. Compr Psychiatry 31(1):34–47

Kiloh LG, McComas AJ, Osselton JW (1972) Clinical electroencephalography. In: Kiloh LG, McComas AJ, Osselton JW (eds) vol 3. Butterworth & co, London

Kooi KA, Tucker RP, Marshall RE (1978) Basic electrical principles-genesis and regulation of electrical activity of the brain. Fundam Electroencephal 2(1–4):3–48

Krakow K, Allen PJ, Symms MR, Lemieux L, Josephs O, Fish DR (2000) EEG recording during fMRI experiments: image quality. Hum Brain Mapp 10(1):10–15

Kumari V, Antonova E, Zachariah E, Galea A, Aasen I, Ettinger U, Mitterschiffthaler MT, Sharma T (2005) Structural brain correlates of prepulse inhibition of the acoustic startle response in healthy humans. Neuroimage 26(4):1052–1058

Kumari V, Fannon D, Geyer MA, Premkumar P, Antonova E, Simmons A, Kuipers E (2008) Cortical grey matter volume and sensorimotor gating in schizophrenia. Cortex 44(9): 1206–1214

Kumari V, Gray JA, Geyer MA, Ffytche D, Soni W, Mitterschiffthaler MT, Vythelingum GN, Simmons A, Williams SC, Sharma T (2003) Neural correlates of tactile prepulse inhibition: a functional MRI study in normal and schizophrenic subjects. Psychiatry Res 122(2):99–113

Logothetis NK, Pauls J, Augath M, Trinath T, Oeltermann A (2001) Neurophysiological investigation of the basis of the fMRI signal. Nature 412(6843):150–157

Logothetis NK, Wandell BA (2004) Interpreting the BOLD signal. Annu Rev Physiol 66:735–769

Madsen GF, Bilenberg N, Cantio C, Oranje B (2013) Increased prepulse inhibition and sensitization of the startle reflex in autistic children. Autism Res (Accepted Epub)

Markand ON (1994) Brainstem auditory evoked potentials. J Clin Neurophysiol 11(3):319–342

Mathiak K, Ackermann H, Rapp A, Mathiak KA, Shergill S, Riecker A, Kircher TT (2011) Neuromagnetic oscillations and hemodynamic correlates of P50 suppression in schizophrenia. Psychiatry Res 194(1):95–104

Mayer AR, Hanlon FM, Franco AR, Teshiba TM, Thoma RJ, Clark VP, Canive JM (2009) The neural networks underlying auditory sensory gating. Neuroimage 44(1):182–189

McCarley RW, Salisbury DF, Hirayasu Y, Yurgelun-Todd DA, Tohen M, Zarate C, Kikinis R, Jolesz FA, Shenton ME (2002) Association between smaller left posterior superior temporal gyrus volume on magnetic resonance imaging and smaller left temporal P300 amplitude in first-episode schizophrenia. Arch Gen Psychiatry 59(4):321–331

Mobascher A, Warbrick T, Brinkmeyer J, Musso F, Stoecker T, Jon SN, Winterer G (2012) Nicotine effects on anterior cingulate cortex in schizophrenia and healthy smokers as revealed by EEG-informed fMRI. Psychiatry Res 204(2–3):168–177

Mulert C, Jager L, Schmitt R, Bussfeld P, Pogarell O, Moller HJ, Juckel G, Hegerl U (2004) Integration of fMRI and simultaneous EEG: towards a comprehensive understanding of localization and time-course of brain activity in target detection. Neuroimage 22(1):83–94

Mullinger K, Debener S, Coxon R, Bowtell R (2008) Effects of simultaneous EEG recording on MRI data quality at 1.5, 3 and 7 tesla. Int J Psychophysiol 67(3):178–188

Näätänen R (1990) The role of attention in auditory information processing as revealed by event-related potentials and other brain measures of cognitive function. Behav and Brain Sci 13(2): 201–288

Niedermeyer E, Lopes da Silva F (2005) Electroencephalography, basic principles, clinical applications, and related fields. In: Niedermeyer E, Lopes da Silva F (eds) vol 5. Lippincott WIlliams & Wilkins, Philadelphia

Ogawa S, Lee TM, Kay AR, Tank DW (1990) Brain magnetic resonance imaging with contrast dependent on blood oxygenation. Proc Natl Acad Sci USA 87(24):9868–9872

Oranje B, Aggernaes B, Rasmussen H, Ebdrup BH, Glenthoj BY (2013) P50 suppression and its neural generators in antipsychotic-naïve, first-episode schizophrenia before and after 6 months of quetiapine treatment. Schizophr Bull 39(2):472–480

Oranje B, Geyer MA, Kenemans JL, Verbaten MN (2006) Prepulse inhibition and P50 suppression: commonalities and dissociations. Psychiatry Res 143(2–3):147–158

Pae JS, Kwon JS, Youn T, Park HJ, Kim MS, Lee B, Park KS (2003) LORETA imaging of P300 in schizophrenia with individual MRI and 128-channel EEG. Neuroimage 20(3):1552–1560

Pfefferbaum A, Roth WT, Ford JM (1995) Event-related potentials in the study of psychiatric disorders. Arch Gen Psychiatry 52(7):559–563

Raichle ME, Mintun MA (2006) Brain work and brain imaging. Annu Rev Neurosci 29:449–476

Rasser PE, Schall U, Todd J, Michie PT, Ward PB, Johnston P, Helmbold K, Case V, Soyland A, Tooney PA, Thompson PM (2011) Gray matter deficits, mismatch negativity, and outcomes in schizophrenia. Schizophr Bull 37(1):131–140

Ritter P, Villringer A (2006) Simultaneous EEG-fMRI. Neurosci Biobehav Rev 30(6):823–838

Scherg M (1990) Fundamentals of dipole source potential analysis. In: Grandori, F, Hoke M, Romani GL (eds) auditory evoked magnetic fields and electric potentials. Karger, Advanced Audiology 6, Basel, pp 40–69

Scherg M, Ille N, Bornfleth H, Berg P (2002) Advanced tools for digital EEG review: virtual source montages, whole-head mapping, correlation, and phase analysis. J Clin Neurophysiol 19(2):91–112

Scherg M, Picton TW (1991) Separation and identification of event-related potential components by brain electric source analysis. Electroencephalogr Clin Neurophysiol Suppl 42:24–37

Sumich AL, Kumari V, Heasman BC, Gordon E, Brammer M (2006) Abnormal asymmetry of N200 and P300 event-related potentials in subclinical depression. J Affect Disord 92(2–3):171–183

Tregellas JR, Davalos DB, Rojas DC, Waldo MC, Gibson L, Wylie K, Du YP, Freedman R (2007) Increased hemodynamic response in the hippocampus, thalamus and prefrontal cortex during abnormal sensory gating in schizophrenia. Schizophr Res 92(1–3):262–272

Index

Curr Topics Behav Neurosci (2014) 18: 387–390
DOI: 10.1007/978-3-319-12769-9

MIX
Papier aus verantwortungsvollen Quellen
Paper from responsible sources
FSC® C105338

If you have any concerns about our products,
you can contact us on
ProductSafety@springernature.com

In case Publisher is established outside the EU,
the EU authorized representative is:
Springer Nature Customer Service Center GmbH
Europaplatz 3, 69115 Heidelberg, Germany

Printed by Libri Plureos GmbH
in Hamburg, Germany